PROCESS TECHNOLOGY AND FLOWSHEETS
VOLUME II

CHEMICAL ENGINEERING BOOKS

Sources and Production Economics of Chemical Products
Calculator Programs for Chemical Engineers
Calculator Programs for Multicomponent Distillation
Controlling Corrosion in Process Equipment
Effective Communication for Engineers
Fluid Movers: Pumps, Compressors, Fans and Blowers
Industrial Air Pollution Engineering
Industrial Wastewater and Solid Waste Engineering
Modern Cost Engineering: Methods & Data
Practical Process Instrumentation & Control
Process Energy Conservation
Process Heat Exchange
Process Piping Systems
Process Technology and Flowsheets: Volume I
Safe and Efficient Plant Operation and Maintenance
Selecting Materials for Process Equipment
Separation Techniques 1: Liquid-Liquid Systems
Separation Techniques 2: Gas/Liquid/Solid Systems
Skills Vital to Successful Managers
Solids Handling
Supplementary Readings in Engineering Design
Synfuels Engineering
You and Your Job

BOOKS PUBLISHED BY CHEMICAL ENGINEERING

Fluid Mixing Technology: James Y. Oldshue
Industrial Heat Exchangers: G. Walker
Physical Properties: Carl L. Yaws
Pneumatic Conveying of Bulk Materials: Milton N. Kraus

PROCESS TECHNOLOGY AND FLOWSHEETS VOLUME II

Edited by

Richard Greene

and

the Staff of Chemical Engineering

McGraw-Hill Publications Co., New York, N.Y.

Library of Congress Cataloging in Publication Data
(Revised for vol. 2)
Main entry under title:

Process technology and flowsheets.

   Vol. 1 contains articles which appeared in Chemical
engineering over the last five years.
   Vol. 2: By Richard Greene and the staff of Chemical
engineering.
   Includes index.
   1. Chemical processes.   I. Cavaseno, Vincent.
II. Greene, Richard, 1944–   .   III. Chemical
engineering.
TP155.7.P76   1979   660.2   79-12117
   ISBN 0-07-606578-2 (v. 1)
   ISBN 0-07-606869-2 (v. 2)
   ISBN 0-07-024388-3 (case)

Introduction

This book contains information on significant processes that have appeared in the pages of CHEMICAL ENGINEERING over the last four years. It is a continuation of *Process Technology and Flowsheets,* which was published in 1979. Volume II contains all new material. Together, these two books present the results of nine years of new techniques that have been developed to serve the chemical process industries.

We have attemped to follow the same format in dividing subjects into sections as was done in Volume I. As before, a section is devoted to winners of CHEMICAL ENGINEERING's biennial Kirkpatrick Chemical Engineering Award. This award is given to developers of those processes judged by a panel of prominent engineering educators to be the most significant additions to the body of chemical engineering technology.

Along with the first volume, this book details the search chemical engineers have been pursuing to meet environmental regulations, reduce energy costs, and cope with changing feedstock and product requirements. This is information that is at once both interesting and useful. *Process Technology and Flowsheets: Volume II* should help you in doing business today, and provide guidance for the future.

Contents

Section I
Coal Processing and Conversion

Coal technology reigns at AIChE gathering

Two processes for solvent de-ashing of liquefied coal highlighted a continuing symposium on coal conversion, which attracted the biggest audience at the meeting. Other highly-rated sessions dwelt on automotive plastics, career planning, and feedstocks.

☐ Under sunny Miami Beach skies, chemical engineers attending AIChE's 71st annual meeting (Nov. 12-16) had their pick of interesting topics to choose from. Sessions on metrication (see previous article), automotive plastics, feedstocks outlook, career planning and education, and regulatory matters drew solid audiences. But a four-day session on the conversion of coal to synthetic fuel or feedstock was by far the most popular event.

DE-ASHING LURES VISITORS—Back-to-back papers dealing with competing concepts for solvent de-ashing of liquefied coal were presented. C-E Lummus (Bloomfield, N.J.) discussed its findings on "antisolvent" de-ashing.

The company described a test unit installed last fall at the Ft. Lewis, Wash., pilot plant designed around the solvent-refined coal (SRC) process and operated by Pittsburg and Midway Coal Mining Co. under a U.S. Dept. of Energy (DOE) grant. Startup of Lummus' unit was to have begun in November, the same time as the paper was delivered. (A much larger Lummus unit is currently under construction at the 600-ton/d H-Coal pilot plant being built at Catlettsburg, Ky., scheduled to start up in 1980.)

The Lummus process employs an "antisolvent" that causes micron-sized ash particles to agglomerate, leaving behind a product containing less than 0.1% (by weight) ash (Fig. 1). (Such separation of solids from liquids is a major problem in the development of coal liquefaction routes.) Antisolvent is recycled, except for a small portion of purge material. The exact nature of this material is proprietary, as is the design of the de-ashing settler.

Kerr-McGee Corp. (Oklahoma City, Okla.) described its "critical solvent" de-ashing route (Fig. 2), which also achieves a claimed ash reduction to 0.1% by weight in the liquid product. In this process, the critical solvent works in two settling stages of different temperature levels. The first stage, at the lower temperature, results in a heavy phase of underflow that is stripped to recover entrained de-ashing solvent and yield a free-flowing ash concentrate. Light phase from the first settler is heated to decrease the critical solvent's density; coal values are rejected in the second stage.

The dissolving power of the solvent changes roughly in direct proportion to its density, according to the Kerr-McGee paper. The firm has tested the route on a bench scale for over five years, and in a pilot-plant for one year. This spring, the pilot plant was moved to another SRC test facility, at Wilsonville, Ala., and operated there this past summer. SRC recovery rates in the 76-81% range were reported for

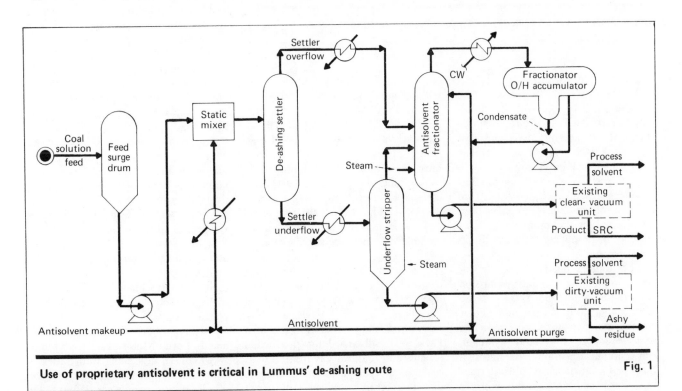

Use of proprietary antisolvent is critical in Lummus' de-ashing route

Fig. 1

Originally published January 1, 1979

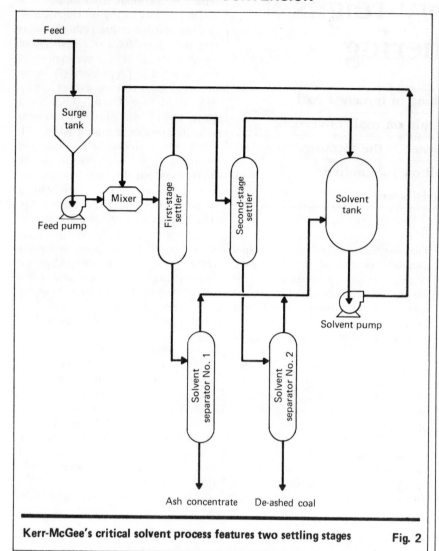

Kerr-McGee's critical solvent process features two settling stages Fig. 2

tests completed just this past October.

Another paper dealt not with the de-ashing of SRC, but with the lique-faction step itself. A group from the chemical engineering department of Auburn University, led by Diwakar Garg, discussed its examination of a "short residence" SRC route that produces a solid-fuel product. (Two versions of SRC have already been investigated in the pilot plant, SRC I and SRC II, producing solid and liquid fuels, respectively.)

The authors observed that a proposed tightening of sulfur-emission regulations could doom SRC I, because of its higher pollution potential. Their remedy is a two-stage process: In the first, residence time and hydrogen consumption are minimized "through the use of inexpensive mineral catalysts that have been treated to improve their selectivity for desulfurization over hydrogenation." The second stage employs hydrotreating with a cobalt-molybdenum catalyst.

OTHER ROUTES—A number of other papers detailed processes under development. A discussion of the H-Coal route, under construction at Catlettsburg, Ky., along with Lummus' de-ashing unit, was presented by Hydrocarbon Research, Inc., the process' developer. The study features a broad macroeconomic cost justification based on some assumed, legislated economic incentives.

Mobil Oil described its methanol-to-gasoline process piloted at its research and development facility at Paulsboro, N.J., while Exxon Research and Engineering Co. (Florham Park, N.J.) discussed the production of substitute natural gas from Illinois coal via catalytic gasification—a route that it recently revealed will get a 1-ton/d tryout under a contract with DOE (*Chem. Eng.,* Nov. 20, p. 82).

In addition, the Institute of Gas Technology (Chicago, Ill.) described the "Coal Conversion Systems Technical Data Book" now in preparation for DOE, and illustrated how the book (though still incomplete) could be used for process design.

CAREERS TOP METRICATION—Although AIChE's announcement about metrication (see previous article) created relatively little excitement, the session on career planning, which emphasized the importance of planning to both individuals and corporations, drew a sizable group. (Many of the approximately 300 ChE students attending the meeting undoubtedly sat through that session.)

Outgoing AIChE president William H. Corcoran (James Y. Oldshue takes over the helm from Corcoran this month, beginning his one-year term) opened the session with a paper advising engineers to set career goals marked by the decades of their working lives. During the first decade, he suggested, engineers should establish their fundamental career paths, as well as their new family lives. Then, during later decades, participation in professional societies and assistance to newcomers to the field should become more important. Though such plans can only be roughly sketched, Corcoran says, something of this sort is needed to achieve basic goals in life.

Arnold A. Bondi, of Shell Development Co. (Houston), remarked that risk-taking can pay off in the long run. Risk-taking, he explained, could involve learning about new processes and other matters at the expense of knowing about things of current commercial importance. Over the course of a lifetime, risk-takers generally earn higher salaries, Bondi has observed, on the basis of his own statistical studies.

Others emphasized the company viewpoint, stressing the need for free flow in information between employee and employer regarding career paths and performance evaluations.

AUTOMOTIVE PLASTICS—Today, the average automobile contains 185 lb of plastic materials, most of which goes for trim and decorative applications, noted Ford Motor Co.'s Seymour Newman. In describing future uses—ones that will raise the amount of plastics in the average car to 350 lb by 1985—Newman mentioned a number of large structural components, including doors, roof and floor panels, fire-walls, and even some chassis parts, such as wheels, springs, suspension-control arms and radiator supports. He emphasized, however, that high-volume low-cost production techniques must be developed before these goals can materialize. He also stressed the need for rapid online quality control for the monitoring of materials consistency and the nature and location of materials defects.

John A. Svera, of General Motors, assured attendees that plastics, in addition to saving weight, will ease manufacturing difficulties, particularly by increasing opportunities to integrate several components into larger parts. He sees especially big opportunities for fiberglass-reinforced plastics.

Nevertheless, substituting plastics for metals presents some problems, says Svera. For example, certain properties of plastics—e.g., fatigue strength, resistance to impact—do not measure up to those of currently-used metals. Also, the surface quality of sheet-molded plastic materials needs to be improved. And, Svera noted, the tricky assembly techniques required with plastics might result in slower production rates, at least until auto manufacturers become more familiar with the materials.

Svera pointed out that current development work to solve these problems is underway. Graphite-fiber reinforcement, for instance, is being used to improve resistance to fatigue.

FEEDSTOCKS—Du Pont's Kenneth N. McKelvey reported on a study written by an AIChE ad hoc taskforce for the U.S. Office of Technology Assessment, discussing the outlook for chemical feedstocks. Basically, it reiterated a now familiar line that there will be a gradual shift away from natural gas and distillate fuel oil, toward coal, and ultimately, biomass. McKelvey emphasized that synthesis gas and ethylene, which are the predominant feedstocks of today, will remain as such, though the raw materials for generating them will change. Production of synthesis gas from coal will result in the construction of large, coal-based synthesis-gas complexes (*Chem. Eng.*, Nov. 6, pp. 73-75).— **John C. Davis; Vincent Cavaseno; Richard W. Greene.**

Synthetic gas and chemicals from coal: economic appraisals

Conventional economic analyses cast
doubt on the commercial potential of
producing synthetic gas from coal.
However, making chemicals via
coal gasification appears more hopeful.

Joseph P. Leonard, Chem Systems Inc.

☐ Early economic evaluations of coal-based synthetic fuels and chemicals were made before and during the worst period of plant construction hyperinflation in the history of the U.S. Equipment and material costs were escalating rapidly.

The impact on plant construction

For a long period, fabricators were refusing to offer firm price quotations. Final costs were geared to date of delivery rather than date of purchase.

During the worst period, it was not uncommon for prices to increase from 1 to 2% per month between date of purchase order and date of delivery. Material shortages also plagued the construction industry. In many instances, certainty of supply became more important than price.

This situation radically changed the engineering construction business for refinery, chemical, petrochemical and fledgling coal-gasification plants. Engineering contractors were no longer willing to bid on a lump-sum basis. Even on a cost-plus basis, clients were told to include large contingencies in their budget figures because of continually escalating material costs.

Against this backdrop, it was not surprising that capital cost estimates filed with the Federal Power Commission (FPC) by the first gas companies to consider commercial coal-gasification projects soon became economically outdated. Similarly, technical brochures put out by promoters of new and existing coal-based technologies for producing synthetic fuels or chemicals understated plant investment and, therefore, product costs.

The experience of the first natural-gas company to

announce plans for a commercial-scale coal-gasification plant is dramatic but not unique. The cost of El Paso Natural Gas Co.'s projected Four Corners Plant was estimated in early 1973 by the National Petroleum Council to be $209 million. Later that year, however, FPC revised the figure to $437 million; in mid-1974, the estimate reached $740 million; and in early 1975, El Paso executives indicated a cost of $1 billion.

Of course, not all of the cost escalation has been due to inflation. Obviously, as a project advances from the planning stage to commercial reality, its scope broadens and its true cost begins to emerge. At any rate, early studies dealing with coal-based synthetic fuels and chemicals proposed attractive but illusory economics based on early-1970s investment figures.

In this article, economic appraisals of the coal-based technologies closest to commercial reality will be based on the startup of plants in the early 1980s, not ten years earlier. The most likely candidates for commercialization are: (1) giant SNG plants, and (2) large-volume synthesis-gas chemicals, specifically ammonia and methanol.

Costs of SNG from giant plants

Estimating the future cost of a giant SNG-from-coal plant must necessarily be speculative. No commercial plant of this kind is currently in operation, so cost data cannot be based on actual plant construction and operation. Further, much of the cost data that are available are inconsistent regarding how much one of these plants would cost today, much less in 1980 and beyond. Probably the most reliable data at hand come from three SNG-from-coal projects that have progressed considerably past the planning stage—those of El Paso Natural Gas Co., Wesco (Western Gasification Co.) and American Natural Gas Service Co.*

All three projects are based on the Lurgi process, with lignite the feedstock. Because it would take about four years to build one of these commercial plants, any large-scale plant coming onstream in the early 1980s would be based on Lurgi technology. So all the economic data presented are based on that process.

Table I itemizes the capital investment by plant sections for 1980 and 1985 startups of a commercial coal-based SNG plant. The total capital required would be approxi-

---

*The chances of project completion for the El Paso and Wesco projects appear quite slim at this time. Great Plains Gasification Associates (a recently formed consortium that has expanded participation in American Natural Resources' high-Btu coal-gasification project to five major U.S. natural-gas systems) has asked the Federal Energy Regulatory Commission for a certificate to build a 137.5-million-ft³/d high-Btu coal-gasification plant, using the Lurgi process, in Mercer County, North Dakota. Under the present schedule, the plant should start commercial operation by December 1983.

This article is based on a paper presented at a Delaware Valley AIChE Symposium on chemical feedstock alternatives, Drexel University, Mar. 14, 1978. Although the cost figures given would not necessarily be those that the author would present today, they would certainly be similar, and probably within the range of engineering accuracy in comparison to earlier estimates. Certainly, the conclusions reached remain valid today.

Originally published March 26, 1979

Estimated capital investment for SNG from coal by Lurgi process* Table I

	Plant investment million $†	
	1980 startup	1985 startup
Coal preparation	89	125
Oxygen plant	134	188
Coal gasification	148	207
Shift conversion	36	51
Gas purification	181	253
Methanation	52	73
Dehydration and compression	19	27
Total onsites	**659**	**924**
Utilities and offsites	341	479
Contractor's overhead and profit	Included above	
Engineering and design costs	Included above	
Subtotal	**1,000**	**1,403**
Project contingency, 15%	150	210
Total plant investment	1,150	1,613
Initial charge of catalyst and chemicals	6	8
Royalties	25	34
Interest during construction	242	339
Startup	31	43
Working capital	27	38
Total capital requirement	**1,481**	**2,075**

*Based on information from El Paso, Wesco and American Natural Gas, including original FPC filings and correspondence updating the filings.

†Based on a 4-year construction period of a 250-million-Btu/d plant with the following construction schedule: 1st year—5%, 2nd year—20%, 3rd year—50%, 4th year—25%.

mately $1.58 billion in 1980, and $2.18 billion in 1985. The design capacity would be 250 billion Btu/d (950 Btu/std ft³ gas), with an onstream factor of 90%. The total capital represents the entire cost of building the plant and getting it ready for startup. It includes all process and general facilities, and utilities.

The plant would require only water and coal. A coal mine and any additions to the gas pipeline system are not considered part of the investment. However, because the plant is assumed to be located near a coal mine in the Western U.S. (in all probability, in a very dry area), the cost of a water pipeline is included.

These cost figures are typical yet generalized capital-cost estimates. There will obviously be differences between these figures and those of actual projects.

Estimates of operating costs for producing SNG from coal via gasification in 1980 and 1985 are summarized in Table II.

The cost of SNG from coal can be calculated a number of ways, depending on the method of financing. Table II summarizes the economics of coal-based SNG calculated via: utility financing, first-year cost; utility financing, 20-year average cost; and private investor financing, 12% discounted-cash-flow-rate-of-return (DCFRR).*

Also shown is the cost of SNG from coal if the U.S. government were to put up half the total capital requirement with no return or interest taken on that portion of the investment. These government-subsidized gas costs were calculated using the 20-year average cost via utility financing. The gas cost has been calculated for two different coal prices, to permit sensitivity analyses on the

*For an explanation of utility financing, see Robert Skanser's "Coal Gasification: Commercial Concepts, Gas Cost Guidelines," C. F. Braun & Co., Alhambra, CA 91802, Jan. 1976, prepared for the U.S. Energy Research & Development Admin. and American Gas Assn., under Contract No. E(49-18)-1235. Copies are available free from ERDA, AGA and C. F. Braun.

Estimated annual net operating costs for SNG from coal by the Lurgi process Table II

	1980 startup		1985 startup	
	Million $	$/million Btu	Million $	$/million Btu
Coal (7.466 million tons/yr)*	59.7	0.72	76.2	0.92
Catalyst and chemicals	5.0	0.06	6.4	0.08
Raw water (2 billion gal/yr)	1.1	0.01	1.4	0.02
Labor:				
Process operating	5.4	0.07	7.1	0.09
Maintenance	18.2	0.22	25.5	0.31
Supervision	4.7	0.06	6.5	0.08
Administration and general overhead	17.0	0.21	23.5	0.28
Supplies:				
Operating	1.6	0.02	2.1	0.03
Maintenance	12.1	0.15	17.0	0.21
Local taxes and insurance	31.1	0.38	43.6	0.53
Total gross operating costs	**155.9**	**1.90**	**209.3**	**2.55**
Byproduct credits:				
Sulfur (48,700 long tons/yr)	1.7	0.02	2.0	0.03
Ammonia (73,700 short tons/yr)	14.0	0.17	17.5	0.21
Total byproduct credits	15.7	0.19	19.5	0.24
Total net operating cost	**140.2**	**1.71**	**189.8**	**2.31**

*Based on an overall thermal efficiency of 65.0% and a coal heating value of 17 million Btu/ton, using western coal at $0.47/million Btu in 1980 and $0.60/million Btu in 1985.

Economic summary of SNG via coal-gasification, 250-billion-Btu/d (82.5-trillion-Btu/yr) plant*   Table III

	Investment, million $		Annual net operating cost, million $/yr	Gas cost			
				Utility financing‡			Investor financing, 12% DCFRR
				$/million Btu		20 yr avg., 50% govt. subsidy	
	Total plant†	Total capital**		1st yr	20-yr avg.		
1980 startup							
Coal at $0.47/million Btu	1,150	1,481	140.2	5.03	3.86	2.78	5.36
Coal at $0.94 million Btu	1,150	1,481	199.9	5.75	4.58	3.50	6.08
1985 startup							
Coal at $0.60/million Btu	1,613	2,075	189.8	6.96	5.32	3.81	7.43
Coal at $1.20/million Btu	1,613	2,075	266.0	7.88	6.25	4.74	8.36

*Based on Lurgi coal gasification and ERDA/AGA guidelines.

†Includes all offsites, water facilities, coal handling and contingency, but not cost of coal mine or additions to gas pipeline system.

**Includes interest during construction, working capital, catalyst, royalties and capitalized portion of startup costs.

‡Financing scheme: debt-equity ratio of 75/25, interest on debt of 9% and after-tax return on equity of 15%. (For explanation of utility financing, see Robert Skanser's "Coal Gasification: Commercial Concepts, Gas Cost Guidelines," C.F. Braun & Co., Alhambra, CA 91802, Jan. 1976. Prepared for U.S. Energy Research & Development Admin. and American Gas Assn., under Contract No. E (49-18)-1235. Copies are available free from ERDA, AGA and C.F. Braun.)

effect of coal prices on gas costs for both the 1980 and 1985 startups (Fig. 1 and 2).

Energy legislation passed last year by Congress does not, until Jan. 1, 1985, remove price controls from newly discovered interstate and intrastate natural gas or from hard-to-get gas from wells more than 5,000 ft. deep (if drilled after Feb. 19, 1977).

Upon enactment of the legislation, the ceiling price of new gas was allowed to jump to $2.05/million Btu (for gas having a heating value of 1,020 Btu/ft³). Through

Apr. 20, 1981, the ceiling will climb annually at the rate of inflation plus an additional increment of 3.7%. Thereafter, it will increase at the inflation rate plus 4.2%, until decontrol occurs.

Depending on inflation, the formula is thus expected to boost gas prices 9% to 12% a year. An inflation rate of 6% will hike the price by 1985 to $3.66/million Btu; at 8%, the price will be $4.18/million Btu. For a plant starting up in 1985, the lowest selling price for SNG (without government subsidy) would be about $5.32/million Btu

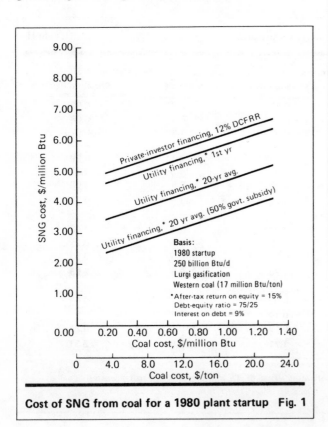

Cost of SNG from coal for a 1980 plant startup   Fig. 1

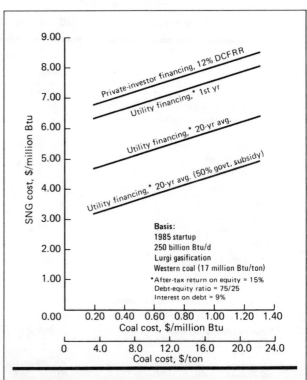

Cost of SNG from coal for a 1985 plant startup   Fig. 2

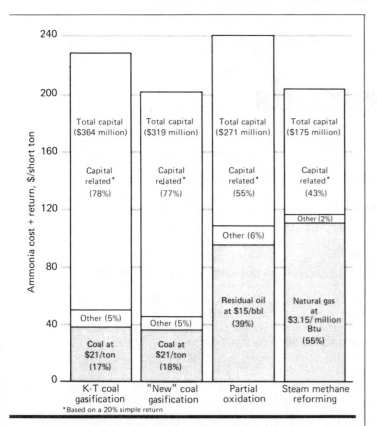

Ammonia costs in 1980 for 2,000-short-ton/d plants   **Fig. 3**

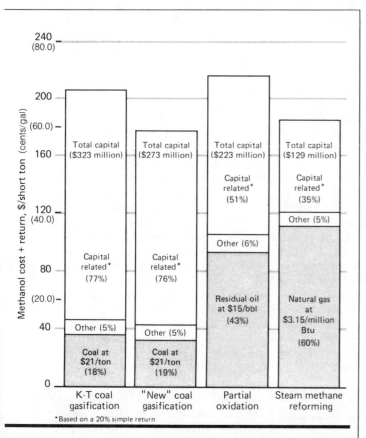

Methanol costs in 1980 for 2,000-short-ton/d plants   **Fig. 4**

(Table III), which clearly would not be competitive with natural gas.

This unfavorable economic analysis does not preclude the possibility that some gas companies may go ahead and complete one of these projects if the FPC were to approve gas prices such that the higher-priced gas could be "rolled in" with wellhead natural gas. This would result in a higher market clearing-price, but one nowhere near the level for gas solely from coal gasification. If, however, the U.S. government were to share 50% of the investment with a gas company, the economics of SNG from coal would begin to compare favorably with those of wellhead natural gas.

Large-volume chemicals from coal

Ammonia and methanol from coal certainly seem to be likely candidates for commercialization in early 1980s. Although natural gas has been the traditional feedstock for ammonia and methanol in the U.S., 25 plants have been built in different parts of the world to convert coal to ammonia or methanol, with capacities ranging up to 1,000 tons/d. The largest is a plant built for African Explosives and Chemicals in South Africa, based on Koppers-Totzek (K-T) gasifiers, which produces 1,000 tons/d of ammonia and 60 tons/d of methanol.

Estimating the cost of coal-based ammonia and methanol units is a little easier than for coal-based SNG plants. One reason is that such plants have actually been built (although not in the U.S.). Other reasons are that typical coal-based ammonia and methanol projects will be smaller in scale than SNG ones, and that the ammonia (or methanol) synthesis step will be essentially identical to that in conventional natural-gas-based plants.

Capital- and operating-costs estimates for the production of ammonia and methanol (chemical-grade methanol, not methyl fuel) from coal in 1980 are presented in Fig. 3 and 4, along with conventional natural-gas-based economics and fuel-oil-based, partial-oxidation economics. The figures indicate that ammonia or methanol based upon "second generation" gasifier technologies (pressurized Koppers-Totzek, Texaco, slagging Lurgi, or equivalent) could compete with steam methane-reforming in 1980. The much higher capital cost of coal-based plants is largely compensated for by lower raw-materials cost.

For comparison, similar-sized (2,000-short-ton/d) units are considered. If a more realistic size of 1,000 tons/d had been assumed for the steam methane-reforming unit, the economics of coal-based ammonia and methanol would look even better.

Jay Matley, *Editor*

The author

Joseph P. Leonard is a senior member of Chem Systems' Process Technology Group (747 Third Ave., New York, NY 10017), involved in economic/commercial planning studies and process technical, economic evaluations. He previously worked for Pullman-Kellogg, Bechtel Corp. and Gordian Associates. His degrees include an M.S. in chemical engineering from Manhattan College and an M.B.A. in finance from Fairleigh-Dickinson University.

Hygas at the crossroads

It's either the very best, or one of the least reliable processes for high-Btu substitute natural gas from coal—depending on who's talking. Caught in the middle is the U.S. Dept. of Energy, which must soon decide on continued research and development funding for Hygas.

☐ This is a story about a disputed technology—the Hygas coal-gasification process. It involves: the organization that has devoted more than 15 years developing the technology to its current pilot-plant level; an engineering-design firm that has performed a critical technical audit; and the U.S. Dept. of Energy, which is caught in the middle of the debate, along with an assortment of engineering firms and consultants.

The auditor, Scientific Design Co. (SD), a unit of Halcon International, Inc. (New York City), has drafted a report for DOE that states that Hygas is not yet ready for demonstration scaleup. According to SD engineers who worked on the project, the study cites in particular a poor record of steady-state operations at the pilot plant in Chicago, not just because plant equipment is underdesigned but because of a number of inherent process flaws—particularly, uncontrolled clinkering of char, uneconomic pretreatment of caking-type coals, and generally unstable conditions in fluidized beds.

Hygas' developer, the Institute of Gas Technology (Chicago), is vigorously rebutting SD, saying that the firm expects too much of the pilot plant—i.e., performance typical of a fully integrated demonstration unit rather than of an investigative pilot plant. The Institute (IGT) points to a wide variety of data, all correlated in a computerized kinetic model, that indicate scaleup feasibility.

At stake is not just the technical integrity of the principals, who have impugned each other's ability, but

millions of taxpayer dollars—money that's to be spent on an expensive demonstration plant to try one or more "second generation" technologies for making high-Btu substitute natural gas (SNG) from coal. The debate represents energy-development policy in microcosm, and demonstrates DOE's problems in promoting technology to commercialization.

HYGAS' PRECARIOUS POSITION—One of the most scrutinized of all the new coal-gasification technologies, Hygas has advanced the furthest while still the child of government funding. Other techniques—particularly "slagging" Lurgi and COED/Cogas—while getting some public support have received a good deal of private-industry funding. No doubt they and other processes pose many technical problems. But less is known about these, because many of the process data are proprietary.

The Hygas debate carries a certain urgency today because DOE is under great pressure to choose the technologies it wants to fund at the demonstration* level. Furthermore, the agency must decide soon on whether to spend $12-14 million in interim research and development to keep the Hygas pilot-plant investigation going another year. That would keep the process warm for a possible demonstration trial, something that may prove likely if Congress extends authorization for such trials to more than one plant.

* The purpose of a demonstration plant, according to DOE's definition, is "to demonstrate and validate economic, environmental and productive capacity of a near commercial-size plant by integrating and operating a single, modular unit using commercial-sized components."—DOE/ET-0087, March 1979, p. 450.

A major go/no-go decision regarding construction of the first SNG demonstrator based on "second generation" technology will probably come this year. The final contenders in this competition are Conoco Coal Development Co. (Stamford, Conn.), leader of a consortium of pipeline companies supporting the slagging Lurgi process, and the Illinois Coal Gasification Group (Chicago), the umbrella for several Midwest utilities that are pushing the COED/Cogas process. DOE had intended to choose by late last year from site-specific conceptual designs done for both, but has delayed.

Aggressively pursuing a possible second demonstration grant is IGT and its Hygas process. The method lost out to the other two in the semi-final round, at least partly because its lead company, Ken-Tex—a subsidiary of Texas Gas Transmission Corp. in partnership with the State of Kentucky—offered less than the 1:1 "up front" construction money put up by the other two groups. IGT supporters contend that DOE (through its predecessor, the Energy Research and Development Administration) chose budgetary savings over technical superiority. Additionally, IGT claims that on a fully commercial basis its process can make high-Btu gas more cheaply than the others.

Though Congress has not authorized funds for a second demonstrator, IGT forces are nevertheless lobbying heavily for one, arguing that Hygas is a logical choice on its own, or at the very least in competition with the loser of the first plant's contest. And if indeed only one plant were to be built with government funding, Hygas proponents would still want to be considered, even though they are officially out of the running. Their arguments have been loud enough to cause DOE to continue Hygas research and development, and to contract with UOP Inc. (Des Plaines, Ill.) to perform a non-site-specific, "generic" conceptual de-

Originally published July 2, 1979

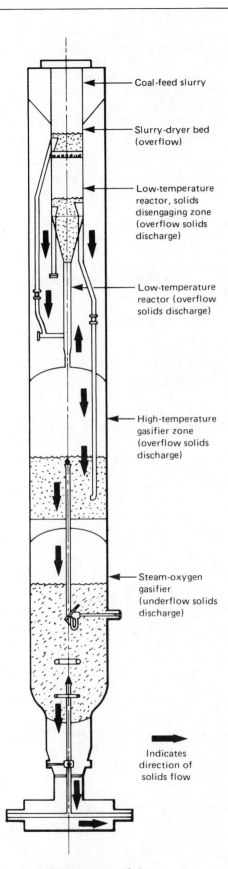

Coal-feed slurry

Slurry-dryer bed (overflow)

Low-temperature reactor, solids disengaging zone (overflow solids discharge)

Low-temperature reactor (overflow solids discharge)

High-temperature gasifier zone (overflow solids discharge)

Steam-oxygen gasifier (underflow solids discharge)

➡ Indicates direction of solids flow

Simplified cutaway of the main Hygas reactor section

sign of a Hygas plant running on eastern coal, keeping Hygas hopes alive.

Further, the outlook for demonstration funds is, if anything, brighter this year. Congress is considering authorization of a second solvent-refined coal demonstrator (a liquefaction project), disregarding the Carter Administration's policy to build just one. This could provide a precedent for a second high-Btu SNG demonstrator.

In early June, Energy Secretary James R. Schlesinger told a Washington conference of economists that the government may have to subsidize development of synthetic oil and gas to rush development. In addition, a study done recently for DOE urges a multi-purpose coal-gasification test facility, with a common coal-preparation and feeding system, that would serve a number of developing technologies at the site, including possibly Hygas.

So while IGT sweats out interim R&D support, and the future of some 160 persons employed at the Chicago pilot plant hangs in the balance, IGT still feels solidly in the running for the demonstration program.

ENTER THE AUDITOR—A technical dispute has arisen amid all the administrative wrangling. From August 1977 until just recently, Scientific Design performed a technical audit for DOE, monitoring pilot-plant test data with personnel stationed at the Chicago site. SD carried out its task completely at the behest of DOE, accepting little of the advice and comment of IGT personnel.

In criticizing Hygas, SD is waging a fairly lonely crusade. Its engineers say that scaleup on the basis of existing data poses too many technical risks. Some consultants and members of the academic community have expressed similar misgivings, but the company doing the conceptual design, UOP, says that it considers existing data good enough for its work. And an engineer for Pullman Kellogg, which has been called in by DOE for another technical audit (begun last fall and to run through February 1980) thinks that IGT's optimistic position is "not quite as far out of line" as SD contends.

Information from these companies, as well as from C. F. Braun & Co., which assessed the economics of vari-

How Hygas works

Of the three basic coal-gasification methods—fixed-bed, entrained-bed and fluidized-bed—Hygas falls into the last category. Its chief advantage is an ability to produce a high concentration of methane right in a high-pressure gasifier (about 1,000 psi), thereby cutting down on the size of methanation equipment needed downstream. It does this by setting up three separate zones of gasification to optimize various intermediate reactions.

The heart of the process is the multibed gasifier, which houses three fluidized reaction beds and one fluidized dryer—all in a single vertical array. Slurried, pulverized coal (which may be pretreated) is sprayed onto the dryer located at the top. The fluidizing medium in this dryer is product gas rising from the low-temperature pipe reactor located just below. From the dryer, this stream goes on to particulate scrubbing and methanation.

Overflow from the dryer falls to the bottom of the low-temperature leg, and is then swept in an entrained manner to a disengaging section above, from which it overflows to a downcomer serving the high-temperature reactor. Residence time in the low-temperature reactor is a matter of seconds, but here about a third of the methane is produced, much of it by means of a hydrogasification reaction between carbon and hydrogen. Temperature is 1,200 to 1,400°F.

In the high-temperature reactor, operating at temperatures approaching 1,750°F, another third of the methane make is produced, along with hydrogen and carbon oxides. The char produced here overflows to the steam-oxygen gasifier below, and then is discharged as ash. Temperatures here approach 1,850°F.

ous coal-gasification routes in a report issued in 1976, will all be considered by DOE this summer, at which time the agency must decide whether to continue research and development support.

Such interim funding is crucial to IGT. The Institute has already lost the support of the Gas Research Institute (Chicago)—a funding body that derives its finances from a small tariff on gas deliveries of member companies. GRI prefers to support more-fundamental research, and believes that Hygas is ready for commercial demonstration. The organization, and its predecessor in R&D funding, the American Gas Assn., have until now provided IGT with one third of the roughly $1 million/mo needed to keep the Chicago pilot plant running.

CLINKERING—SD's main concern is the steam-oxygen gasifier (SOG), the hottest reactor in the vertical multistage Hygas array. To avoid sintering or "clinkering" of char in the SOG, SD says, the steam/O_2 ratio must be held to at least 6.5—an uneconomically high level. SD scoffs at IGT's claim that 5.5 is sufficient (a level that translates into savings of 30¢/million Btu of gas), pointing out that the 5.5 ratio hasn't been tested at the design pressure of about 1,000 psi. Without the extra steam, the temperature in this stage would get too hot, leading to clinkering and a forced shutdown.

IGT's president, Bernard S. Lee—who along with Paul B. Tarman and W. G. Bair (vice-president and assistant vice-president for process research, respectively) rebutted SD in an interview with *CE*—counters that although the Chicago pilot plant is incapable of simultaneously achieving the desired mass velocity and space velocity in the SOG, because of an unsuitable length/diameter ratio in the reactor, the optimum conditions have been proven by means of a computerized kinetic model. Lee adds that current non-optimum simulations are sufficient to confirm the model.

The real reason for injecting large amounts of steam into the SOG, IGT claims, is not so much to cool the reaction zone as to achieve a fluid velocity that is better insurance against clinkering. The proof of this contention lies with the kinetic model, a proprietary model developed with IGT's own funds, and one that SD has not been allowed to examine in detail. Not having been convinced of the

model's repeatability, SD sticks to its position, and complains that the model has never been demonstrated with conditions determined prior to test runs, only with data obtained upon hindsight.

COAL PREPARATION—At the front end of the process, SD is particularly critical of two initial processing stages, coal pretreatment and slurry drying.

Pretreatment of pulverized coal is necessary with agglomerating coals typical of the eastern U.S. By heating the coal to about 750°F, a hard shell develops around particles, serving to prevent caking in the gasifier. The trouble is, SD says, that pretreatment releases too much volatile material, thereby reducing the yield of synthetic gas. The company doubts that any high-Btu gasification process that depends on pretreatment can prove economical. IGT counters with the claim that evolved volatiles can be burned for plant energy needs (not done in the pilot plant), augmenting the energy generated in the exothermic gasification reaction anyway. Current estimates of Hygas SNG costs include a pretreatment penalty.* But SD says that volatiles come off so dilute as to be useless as fuel gas.

Before going to the gasifier, feed coal—pretreated or not—is slurried with oil (ideally, recycle oil). Slurrying helps get around expensive lock hoppers that would otherwise be needed to get coal into the high-pressure reactor. But once up to pressure, the oil must be volatilized and recycled. For this step, IGT and SD—both highly experienced with fluid-bed design—have differing philosophies. The current design has the slurry sprayed onto the top of the fluid-bed dryer (top zone in the vertical array). SD believes that this results in too much fines overflow and urges placing the inlet within the bed. IGT says it makes no difference. Even though fines carryover into gas cleaning apparatus has been a major operational problem at Chicago, the Institute believes that more suitable crushing of feed, as well as an improvement in the velocity in the dryer, plus better screening will solve the problem

WHAT'S STEADY STATE?—One of the hottest areas of debate is over what constitutes steady-state operation. The

* The 1976 Braun report estimated $3.69/million Btu as the cost of gas from a 250-million-Btu/d plant feeding on eastern coal. IGT says only 18¢ of this is the pretreatment penalty.

point is crucial to Hygas, since scaleup depends on a convincing demonstration of steady state in the pilot plant, in accordance with a set of performance criteria set up by C. F. Braun in 1976 that cover all candidate demonstration routes. Furthermore, the government wants a smooth-running "turnkey" type plant for demonstration.

In all but a few of the nearly 80 test runs made at the Hygas pilot plant startup in 1971, operations haven't come close to seven days of steady state, which in SD's view is the minimum needed to assure that material balances are true and that the plant won't strangle itself with clinkers. While many of the runs recorded by IGT have lasted seven days or more, seldom has actual steady state been more than a few hours, SD charges.

In response, IGT states that only three to four residence times, or a total of around three hours, are necessary to prove steady operations. Anything more would give only a false sense of security, IGT president Lee points out, again focusing on the Institute's kinetic model as the answer to any clinkering problem. Furthermore, the bulk of Hygas test runs have been made to gather specific data, not to demonstrate steady state.

Until SD drafted its report, only lignite (no pretreatment necessary) had been processed under steady-state conditions that demonstrated "technical feasibility" under the Braun-developed criteria. However, IGT reports that in late spring it succeeded in an extended test with run-of-mine Illinois No. 6 coal, a caking coal. Lowell Miller, head of gasification research and development at DOE, says that from what he's heard of the test results, this could indeed result in a finding of technical (though not necessarily economic) feasibility with caking coal, but he's waiting to see the data before making any firm judgment. He especially wants to know if material balances close to ±5%.

OTHER CRITICISMS—While disagreements over clinkering, pretreatment and steady state form the core of SD's criticism, a lot of lesser problems are also at issue. Many of these are due to the insufficient design of the pilot plant, which cannot meet all performance goals at one time; others are due to an alleged inadequacy in the process' concept.

The feed system is incapable of

delivering a steady flow of coal, SD says, and the result is an unsteady operation for which no material balance is reliable. Furthermore, oxygen injection at the steam-oxygen gasifier is not directly controlled by variations in coal feedrates. The result is often an injection rate higher than desired, and the inevitable clinkering. IGT has worked to correct some of these deficiencies, particularly by installing a large amount of storage capacity for pulverized coal, and claims that the deficiencies in the pilot-plant design can easily be corrected in the scaled-up plant.

SD thinks that the multiple fluid-bed concept of the gasifier is overly complicated and unstable. IGT counters that there is really only one level control in the entire vertical array—at the SOG outlet—and that the levels of the intermediate beds are automatically controlled at fixed overflow points. Any possibility of a "blow-through" of gas from a lower bed to a higher one is termed impossible.

And SD claims that Hygas cannot achieve the desired methanation rate of 65-70% in the gasifier, getting no more than 60%. IGT says that despite actual operating percentages, the results agree well with its model.

OUTSIDER'S VIEW—To Ravi Nadkarni, of Arthur D. Little, Inc. (Cambridge, Mass.) and a one-time associate of Bernard Lee at ADL, the Hygas process is "very elegant in terms of concepts, but I've always felt it is a little too complicated." The route's strong point is its solids feed system—much more reliable than lock hoppers, for instance—but solids exit is another matter. Nadkarni is particularly concerned that the current SOG design is contrived (the reactor unit was, after all, originally designed for a different purpose), and that there are other more-established methods of handling char other than a single-stage fluidized bed. "I don't know why you would chose IGT's method other than you're emotionally committed to a fluid bed," Nadkarni says.

On the question of steady state, he seconded SD's concern that operations haven't been smooth enough to close material balances. If so, as SD contends, scaleup is too scary. Material balances are very important at this stage of development, Nadkarni cautions.

John C. Davis

LETTERS

In defense of Hygas

Sir: I have read with great interest your editorial on synthetic fuels (see *Chem. Eng.*, July 2, 1979, p.39) and I agree with your views. However, I take exception to the article entitled "Hygas at the crossroads," which discusses the differences in technical opinion concerning the Hygas technology.

Obviously, there always will be differences concerning any process during its development. I would like to point out that the Hygas technology and its pilot plant operation have been thoroughly analyzed by at least four other major engineering firms—C.F. Braun, Procon, Bechtel, and Pullman Kellogg—over the last 5 years. All of these organizations have provided positive reviews and have expressed confidence in the technology and its prospects for future commercialization.

Hygas is the most efficient process today for converting coal to high-Btu gas and has the lowest cost of gas production. It can handle all ranks of coal, including highly caking eastern bituminous coal. Indeed, as noted in the article, we recently completed the first successful testing of run-of-mine bituminous coal in our pilot plant. This is the first time that run-of-mine bituminous coal from the U.S. has been successfully gasified by any process.

In reference to steady-state, I would like to point out as an example a recently completed test that entailed 7½ days of steady production of gas of a constant quality. We did not encounter "uncontrolled clinkering of char" or "unstable conditions in fluidized beds," as implied in the article.

In regard to the use of a kinetic model, we, like other process developers, find it a valuable tool to correlate the data we have amassed in laboratory, pilot plant, or other developmental operations. Based on this experience, we are confident that the model can accurately predict behavior in larger-scale demonstration plants and have used it for this purpose. The model does not, nor do we claim it to, predict sintering or other mechanical aspects of the system. These aspects have been successfully demonstrated in extended pilot plant operations.

Finally, I am puzzled at the choice of Mr. Ravi Nadkarni, at the close of the article, as an unbiased third party, qualified to assess the efficacy of the Hygas technology. Although I have met Mr. Nadkarni at several technical meetings, he joined Arthur D. Little, Inc., after I left that organization; thus he was never my one-time associate at ADL, as the article stated. To my knowledge, Mr. Nadkarni has never analyzed operation of the Hygas facility on a first-hand basis, nor am I acquainted with how thoroughly he has reviewed and analyzed the immense Hygas data base.

BERNARD S. LEE
President
Institute of Gas Technology
Chicago, Ill.

Originally published August 27, 1979

Entrained-bed coal gasifiers handle double throughput

New design comprising four, instead of two, feed injectors makes higher capacity possible. And a higher-pressure gasifier is being tested in a demonstration plant.

Hermann Staege, Krupp-Koppers GmbH

□ The Koppers-Totzek process, which accounts for 89% of the world's total coal-based ammonia capacity of 4,500 tons/d, has been modified to double the capacity of its gasifier. Developed by Krupp-Koppers GmbH (Essen, West Germany), the modified entrained-bed system is about to see its first commercial service at a pair of fertilizer plants in India.

Krupp-Koppers has also joined with Shell Oil Co. to develop a higher-pressure gasification process. Koppers-Totzek operates at atmospheric pressure; the Shell-Koppers process will operate at 30 bar and promises to have six to twelve times the capacity of the older system.

MORE VOLUME—To get the larger capacity, the standard two-headed Koppers-Totzek gasifier has been replaced with a four-headed design that can produce a maximum of 50,000 Nm³/h of raw synthesis gas and has an operating range of 40 to 120% of design. Two-headed units have a maximum capacity of 25,000 Nm³/h of syn gas and a design range of 60-120%

The new four-headed type has other advantages. For a specific volume of gas, investment costs are lower for four-headed units than for two-headed gasifiers. Also the new type requires only 60% of the space needed for a two-headed unit, and it has lower specific heat losses.

A pair of four-headed gasifiers will supply 1.3 billion m³/d of synthesis gas for a 600-ton/d ammonia plant in San Jeronimo, Brazil, by 1983. Brazil's state-owned company, Petro-

bras, will operate the $200-million facility.

The new design was first employed by the Fertilizer Corporation of India Ltd. Here, the largest gasifiers in the world are providing feedstock for two identical 900-ton/d ammonia plants—one located in Ramaundam, started up in September 1978, and one in Talcher commissioned in June 1978. Although these plants were ordered in 1969 and 1970, respectively, the project moved slowly until the 1973 oil crisis and the subsequent increases in the price of oil.

COAL DUST—In the Koppers-Totzek process, raw coal is pulverized to particles smaller than 0.1 mm. For bituminous coal, the allowable amount of oversize particles can be approximately 10%, and for lignite about 15-20%. This dust is then dried with flue gas produced by burning a portion of the coal.

The dried dust is separated from the flue gas in cyclones. Fines are collected in an electrostatic precipitator. The dried coal dust is then stored in bunkers ahead of the gasifiers.

The gasifiers consist of conical reactor sections with common gas and slag outlets. In the two-headed design, two reactor sections are opposite each

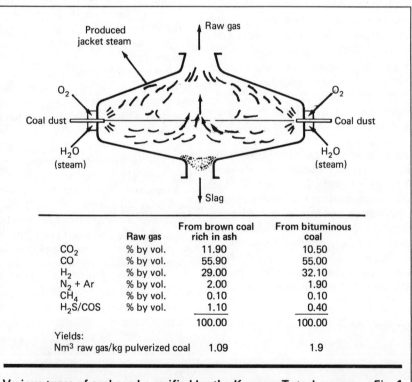

	Raw gas	From brown coal rich in ash	From bituminous coal
CO_2	% by vol.	11.90	10.50
CO	% by vol.	55.90	55.00
H_2	% by vol.	29.00	32.10
N_2 + Ar	% by vol.	2.00	1.90
CH_4	% by vol.	0.10	0.10
H_2S/COS	% by vol.	1.10	0.40
		100.00	100.00
Yields: Nm³ raw gas/kg pulverized coal		1.09	1.9

Various types of coal can be gasified by the Koppers-Totzek process   Fig. 1

Originally published September 10, 1979

other; with four-headed units, four sections are arranged in a cross.

Oxygen (98% pure) and steam are mixed with the coal dust and blown into the gasifier, which operates in an entrained mode at a pressure of 1.05 bar.

Gasification takes place in less than one second in the following rapid sequence: oxygen, steam and coal are preheated; coal pyrolyzes; then partial oxidation takes place. These steps do not occur consecutively but rather in parallel, since the process is continuous, with feed constantly entering the gasifier (see Fig. 1).

During the oxidation phase, at about 2,000°C, the volatile constituents formed during pyrolysis react immediately to form carbon dioxide, carbon monoxide, steam and hydrogen. Methane formation at this temperature is virtually impossible.

The coke that remains as a solid after pyrolysis reacts not only with the feed oxygen but also with CO_2 and steam produced during the pyrolysis reactions. The solids that do not react at this point are taken off through the bottom of the reactor as slag.

The raw gas produced consists of CO, CO_2, H_2, H_2S, COS and H_2O. The product contains some nitrogen that is released from the organic compounds in the coal. This raw gas leaves the gasifier at 1,400 to 1,600°C and passes through a waste-heat boiler to generate saturated steam at pressures up to 100 bar. The gas then is washed and cooled by direct contact with water. This wash removes most of the fly ash. Electrostatic precipitators then reduce the residual dust content of the syn gas to 0.3 to 0.5 mg/Nm^3.

YIELDS AND COSTS—For the amount of coal equivalent to a net caloric value of 10,000 kj, the process consumes 0.19-0.23 Nm^3 of oxygen and 0.045 kg of steam, to produce 0.67 to 0.72 Nm^3 of gas and 0.72-0.79 kg waste-heat steam. Yield of actual synthesis-gas components—CO and H_2, which may constitute 87% of the raw gas (see table)—would be 0.57 to 0.62 Nm^3.

Various solid fuels have been gasified by this process, including peat, anthracite, petroleum coke, and coals with ash contents of up to 50%.

The amount of carbon actually gasified by the system can be expressed by the ratio of carbon in the product gas to carbon in the fuel charged. This ratio would be approximately 82% for anthracite and something like 98% for lignite.

After sulfur removal, the product gas can be used in a number of ways (see Fig. 2), such as for a chemical feedstock or as a medium-Btu fuel gas.

The coal required (free of water and ash) for the production of 1,000 kg of ammonia would be about 1.2 to 1.4 tons. Synthesis-gas requirements for various chemical products are given in the table (see Fig. 2).

Based on construction in Western Europe, the costs for a plant using four-headed gasifiers would be approximately $45 million for enough gas to manufacture 900 tons/d ammonia; for a 600-ton/d ammonia facility, a gasification plant would cost about $33 million.

HIGHER PRESSURE—The gas produced by this route must be compressed, if it is to be used for chemical processes or fuel. Part of the energy need for compression could be saved if the gasification process could be operated at a higher pressure. Also, with a high-pressure process, it would be possible to increase capacity and reduce the size of piping and other equipment.

Krupp-Koppers has joined with Shell Oil to finance a $50-million demonstration plant based on the Shell-Koppers high-pressure gasification process. A 150-ton/d plant was started up in November 1978 on the site of Deutsche Shell's Hamburg-Harburg oil refinery. A prototype with a reactor that can process 1,000 tons/d coal is scheduled to follow around 1985.

Based on this new process, it is hoped that gasification plants can be built with a reactor able to process 4,000 to 5,000 tons/d of coal. This would be equivalent to 300,000 Nm^3/h—approximately six times the capacity of the modified Koppers-Totzek process and twelve times the capacity of the older two-headed design.

Also, calculations have shown that raw gas produced with the Shell-Koppers process can be 10% cheaper than gas produced by the Koppers-Totzek route.

Reginald Berry, Editor

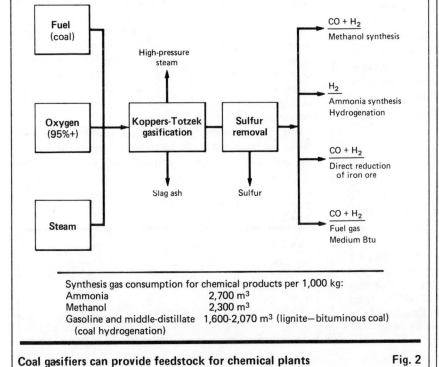

Synthesis gas consumption for chemical products per 1,000 kg:

Ammonia	2,700 m³
Methanol	2,300 m³
Gasoline and middle-distillate (coal hydrogenation)	1,600-2,070 m³ (lignite—bituminous coal)

Coal gasifiers can provide feedstock for chemical plants   Fig. 2

The Author

Hermann Staege holds a degree in mechanical engineering and chemical process engineering from the Technical College in Essen. And for more than 25 years he has been involved with coal gasification and synthesis gas production at Krupp-Koppers GmbH, Essen, West Germany.

Methanol from coal: cost projections to 1990

Future costs of fuel-grade methanol are estimated, taking into account such factors as plant size and efficiency, capital investment schedule, cost of coal, and rate of inflation.

R. I. Kermode, *University of Kentucky,*
A. F. Nicholson, *Kentucky Institute for Mining and Minerals Research* and **J. E. Jones, Jr.,** *Kentucky Dept. of Energy*

☐ At least four studies evaluating the economics of converting coal to methanol were completed between 1976-1978 [1,2,3,5]. Comparison of three of these, each for producing fuel-grade (95%) methanol, yields estimated product prices ranging from 18.8 to 71.0¢/gal [1,2,3].

The well-developed state of this technology ensures that most of the differences would be reconciled by putting the studies on the same basis; that is, when differences in assumed efficiencies, coal characteristics, plant life, year of construction, interest rate, etc., are accounted for, the cost of methanol should be essentially the same.

In an earlier study, these adjustments were màde by converting all costs (coal, operating and capital) to a 1977 base [4]. It was next assumed that all of the capital was invested instantaneously and that full plant capacity was achieved immediately. Application of the same economic model and parameters to these "instant plant" numbers produces methanol values that are quite consistent.

In this article, the adjusted best values from each of the three studies—Badger [1], Du Pont [2] and Exxon* [3]—for operating costs, plant efficiency, coal characteristics, etc., are used to project the cost of methanol up to 1990. The scaleup for different plant sizes is based on the tons of coal feed per day, and the capital costs reported in the Badger and Du Pont studies.

A description of this procedure is available [4].

Scaling up capital costs

As a rule, plant capital investment can be scaled up by means of the equation: Cost of Plant B = Cost of Plant A (Size of Plant B/Size of Plant A)^x. The exponent x ranges from 0.6 to 0.7.

Comparisons of the scaleup of capital costs for the three plants were made, in each case using the total coal feed in tons per day and the 1977 "instant plant" total investments. Table I lists the scaleup exponent for the three plants [1,2,3]. Two plants closest in size (Exxon and Du

*Chem-Systems Inc. did the economic calculations, which are included in the appendix of the Exxon report [3].

Originally published February 25, 1980

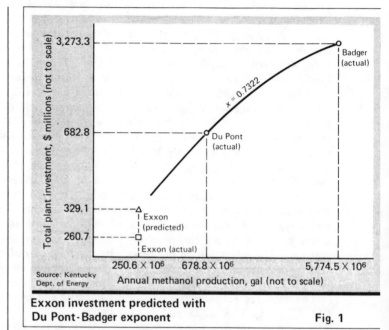

Exxon investment predicted with Du Pont - Badger exponent

Fig. 1

Pont) show virtually no economies of scale, as x is almost 1.0. Comparison of the Du Pont and Badger plants, scaleup exponent $x = 0.73$, indicates much consistency in total plant investment despite the large difference in size.

Because of the depth of the Badger and Du Pont studies, the scaleup exponent of $x = 0.73$ was selected as the basis for estimating total plant investment over a broad range of plant size [4]. Using the Du Pont and Badger capital investment and the scaleup exponent of 0.73, the scaleup equation becomes: Cost ($ millions) = 0.9121 (tons of coal/stream day)$^{0.73}$.

With this equation, the total capital investment for any size of methanol plant can be estimated, given the tons of coal feed per day. For example, a plant using 30,000

Economy-of-scale exponents based on total coal feed		Table I
Scaleup of plant	**Known plant***	**Rounded exponent x**
Du Pont	Badger	0.73
Exxon	Badger	0.81
Exxon	Du Pont	0.97

Source: Kentucky Dept. of Energy
*Plant A

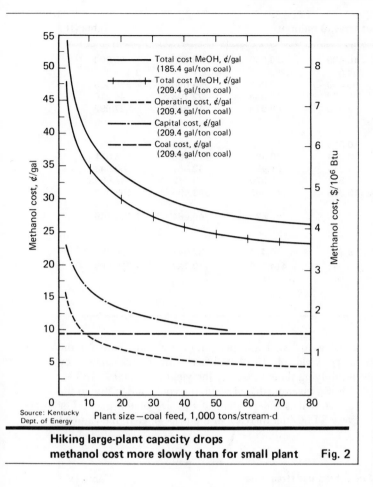

**Hiking large-plant capacity drops
methanol cost more slowly than for small plant** Fig. 2

Basis for scaling up 1977 "instant plant methanol costs	Table II
Coal	$20/ton, Du Pont composition
Plant efficiency	Two cases: Badger—57.9% (highest) and Exxon—51.2% (lowest)
Utilities	Badger 1977 "instant plant" cost of $40.9 x 10⁶ scaled proportionally by the ratio of tons of coal per day
Catalyst and chemicals	Badger 1977 "instant plant" (including lime) cost of $25.8 x 10⁶ scaled proportionally by the ratio of tons of coal per day
Operating and supervisory labor	Badger 1977 "instant plant" cost of $18.0 x 10⁶ scaled by the ratio of the tons of coal per day raised to the 0.25 power
Maintenance	Du Pont 4.0% of total capital investment
General services	Du Pont 1.0% of total capital investment
Taxes and insurance	Du Pont 1.75% of total capital investment
Sulfur credit	Du Pont 0.0416 tons of sulfur per ton coal at $60/ton
Capital charges	20-year plant life, 65/35 debt-equity ratio, 12% return on investment, 9% cost of money
Methanol yield	57.9% efficiency = 209.7 gal/ton of coal; 51.2% efficiency = 185.4 gal/ton of coal

Source: Kentucky Dept of Energy

tons/stream-d of coal will cost (in 1977 dollars): 0.9121 $(30,000)^{0.73}$ = $1,692 million.

Fig. 1 shows how this equation can be used to predict the total capital necessary to build a methanol plant of the size in the Exxon study.

Scaleup of operating costs

Operating costs fall into three categories:

1. Those approximately proportional to plant size, such as for coal and catalyst.

2. Those that are some percentage of the plant investment, such as for maintenance, taxes and insurance, and general services.

3. Those that show economies of scale different from capital costs, such as for operating labor and supervision, for which normally $01.5 \leq x \leq 0.25$.

Following these guidelines, and having the 1977 "instant plant" operating costs for each of the studies, the operating costs for any two plants can be estimated if the third plant serves as a base case.

This comparison is hindered by differences in the items that each study included in the various categories [4]. However, differences are reconciled by combining items in each report and then comparing the totals. For example, the Badger study lists a separate charge of $8.4 x 10⁶ for lime. This is combined with the reported $17.4 x 10⁶ for catalysts and chemicals, to yield a new total of $25.8 x 10⁶. Because this falls into Category 1, direct scaling by the ratio of tons/stream-d of coal consumed should produce costs that are comparable.

Comparing Du Pont with Badger, this ratio is 8,702/74,000. The estimated Du Pont cost for catalyst and chemicals becomes $3.034 x 10⁶, versus the reported 1977 Du Pont cost of $2.961 x 10⁶. This combination produces consistent results between the two studies for this category.

Table II lists the import cost categories, as well as the basis for scaling for changes in plant size. The use of this table is aided by two examples:

1. Estimated Du Pont utility cost—Because the Du Pont plant uses 8,702 tons/stream-d of coal, it is 8,702/74,000 the size of the Badger plant. Table III shows that the Du Pont utility cost should be directly proportional to this ratio, or: ($40.9 x 10⁶) (8,702/74,000) = $4.809 x 10⁶ per year.

2. Estimated Du Pont cost for operating and supervisory labor—This same ratio should be raised to the 0.25 power, or: ($18.6 x 10⁶) $(8,702/74,000)^{0.25}$ = $10,892 x 10⁶ per year.

Estimating costs for different plant sizes

The procedure for estimating the coal, operating and capital costs for producing methanol from coal has been presented. Because all of the costs are in the form of equations and ratios, regularly spaced plant sizes can be chosen. A convenient starting point is a plant using 2,500 tons/stream-d of coal. This size is successively doubled until an 80,000-ton plant size is reached.

Table III presents total annual costs, which range from $83.051 x 10⁶ (2,500 tons/stream-d) to $1,289.8 x 10⁶

Projected 1977 "instant plant" cost summary for various plant sizes, $ millions						Table III
Tons coal/stream-d	2,500	5,000	10,000	20,000	40,000	80,000
Total capital investment	275.8	457.4	758.6	1,258.3	2,087.1	3,461.8
Raw materials, coal*	16.500	33.000	66.000	132.000	264.000	528.000
Utilities	1.382	2.763	5.527	11.054	22.108	44.216
Catalyst and chemicals	0.872	1.743	3.486	6.973	13.946	27.892
Direct labor and overhead:						
Operations and supervision	7.717	9.177	10.913	12.978	15.434	18.354
Maintenance	10.983	18.240	30.284	50.281	75.088	138.611
Taxes and insurance	4.806	7.980	13.249	21.998	32.851	60.643
General services	2.746	4.560	7.511	12.570	18.772	34.653
Interest on capital	40.104	66.510	110.308	182.969	303.485	503.380
Total manufacturing cost	85.110	143.973	247.278	430.823	745.684	1,355.749
Credits:						
Sulfur	2.059	4.118	8.237	16.474	32.947	65.894
Total net cost	83.051	139.855	239.041	414.349	712.737	1,289.855

*At $20/ton

Source: Kentucky Dept. of Energy

(80,000 tons/stream-d), all costs being expressed in 1977 dollars. The highest and lowest efficiencies—209.7 and 185.4 gal of methanol per ton of coal, respectively—are used with the total net annual cost in Table III to calculate the cost of methanol. Table V and Fig. 2 show these costs, in ¢/gal and $/million Btu, for the yield of 209.7 gal of methanol per ton of coal. The costs for the lower yield of 185.4 gal/ton can be calculated by multiplying these values by the yield ratio 209.7/185.4.

For a given methanol yield, the contribution due to the cost of coal is independent of the size of the plant. For every $4.00/ton increase in the price of coal, the value of

Future annual costs of methanol production based on tons of coal feed per stream-day							Table IV
	Tons coal/stream-d	2,500	5,000	10,000	20,000	40,000	80,000
1977	Cost, $ x 10⁶/yr						
	Coal	16.500	33.000	66.000	132.000	264.000	528.000
	Operating	26.447	40.345	62.733	99.380	145.252	258.475
	Capital	40.104	66.510	110.308	182.969	303.485	503.380
	Total	83.051	139.855	239.041	414.349	712.737	1,289.855
	¢/gal*	47.99	40.38	34.53	29.92	25.75	23.30
	$/10⁶ Btu	7.63	6.42	5.49	4.76	4.09	3.70
1980	Cost, $ x 10⁶/yr						
	Coal	19.100	38.200	76.402	152.803	305.606	611.213
	Operating	27.672	49.423	76.848	121.741	177.934	316.632
	Capital	51.936	86.132	142.852	236.950	393.022	652.551
	Total	98.708	173.755	296.102	511.494	876.562	1,580.396
	¢/gal*	57.03	50.14	42.75	36.93	31.66	28.55
	$/10⁶ Btu	9.06	7.97	6.79	5.87	5.03	4.54
1985	Cost, $ x 10⁶/yr						
	Coal	24.377	48.754	97.508	195.017	390.034	780.067
	Operating	45.441	69.321	107.787	170.755	249.572	444.112
	Capital	79.910	132.525	219.795	364.577	604.712	1,003.015
	Total	149.728	250.600	425.090	730.349	1,244.318	2,227.194
	¢/gal*	86.52	72.35	61.40	52.74	44.96	40.23
	$/10⁶ Btu	13.74	11.50	9.76	8.38	7.14	6.40
1990	Cost, $ x 10⁶/yr						
	Coal	31.112	62.225	124.449	248.899	497.798	995.597
	Operating	63.732	97.223	151.174	239.486	350.028	622.873
	Capital	122.951	203.906	338.182	560.946	930.424	1,543.262
	Total	217.795	363.354	613.805	1,049.331	1,778.250	3,161.732
	¢/gal*	125.85	104.91	88.66	75.77	64.24	57.11
	$/10⁶ Btu	20.00	16.67	14.09	12.04	10.21	9.07

*Based on 209.7 gal methanol/ton of coal

Source: Kentucky Dept. of Energy

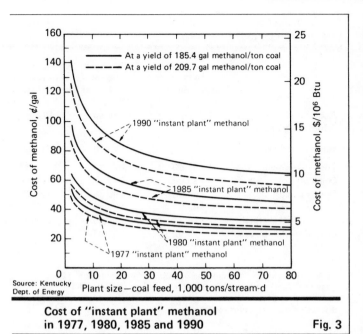

Cost of "instant plant" methanol
in 1977, 1980, 1985 and 1990   Fig. 3

"Instant plant" methanol values for various plant sizes				Table V	
Plant size, tons/stream-d	Coal cost, ¢/gal	Capital, ¢/gal	Operating cost, ¢/gal	Total cost, ¢/gal	Total cost, $/10⁶ Btu
2,500	9.54	23.17	15.28	47.99	$7.63
5,000	9.54	19.22	11.62	40.38	$6.42
10,000	9.54	15.93	9.06	34.53	$5.49
20,000	9.54	13.21	7.17	29.92	$4.76
40,000	9.54	10.96	5.25	25.75	$4.09
80,000	9.54	9.09	4.67	23.30	$3.70

Source: Kentucky Dept. of Energy

methanol will rise 2.16¢/gal when the yield of methanol is 185.4 gal/ton of coal, and 1.91¢/gal when the yield is 209.7 gal/ton. These numbers correspond to increases of 34.34 and 30.36¢/million Btu for the 185.4 and 209.7 yields, respectively. These numbers or any fractions of them can be added to the methanol values given in Table IV and Fig. 2.

Fig. 2 shows that, for smaller-capacity plants, the selling price of methanol decreases more sharply with increased coal throughput. Increasing the plant size from 2,500 to 5,000 tons/stream-d of coal decreases the price of methanol 15.8%, whereas increasing it from 40,000 to 80,000 tons/stream-d lowers the price only 9.5%.

If the overall plant efficiency predicted by the Badger study can be achieved, the lower total-cost curve in Fig. 2 applies. This shows that the cost of methanol drops by 24.69¢/gal when plant size is boosted from 2,500 to 80,000 tons. Of this total possible cost reduction, 73%, 83% and 90% is achieved by a plant of 20,000, 30,000 and 40,000 tons, respectively.

It appears that a 20,000-ton plant would be the best size to build. It would have the advantages of: (1) requiring a coal feedrate comparable to that of existing large power plants; (2) achieving 75% of the potential economies of scale; (3) establishing the overall plant thermal efficiency; (4) demonstrating coal conversion to a clean fuel on a large scale; (5) producing 1977 "instant plant" methanol at 29.9 to 33.9¢/gal (or $4.76 to $5.38/million Btu), depending on the yield.

Future "instant plant" methanol costs

In order to evaluate the future cost of methanol, the costs in Table III have been scaled up for inflation to the years 1980, 1985 and 1990. The inflation rate between now and any future date is, of course, difficult to predict. It is unlikely that coal price, operating cost and fixed plant investment will increase at the same rate. For these cost estimates, it has been assumed that the price of coal will go up at 5%/year, operating costs will rise at 7%/year,

and fixed plant investment will increase at 9%/year.

Table IV and Fig. 3 show the three major cost items for each of the three future years mentioned, as well as for 1977. These numbers have been translated into a cost of methanol in ¢/gal and $/million Btu (shown in parentheses), for a yield of 209.7 gal/ton.

The assumed rates of inflation of 5.0%, 7.0%, and 9.0% for coal, operating cost and plant investment, respectively, average out to a 7.3% annual rate between 1977 and 1990. It is hoped that this average rate of inflation will prove adequate, although the 1978 and 1979 rates have exceeded it.

Effect of six-year investment schedule

Case comparisons to this point have been based on the "instant plant" concept of all capital being invested instantaneously and 100% production being reached immediately. This method of costing methanol has advantages when several cases are compared. However, the procedure obviously understates investment costs, because interest during the construction period is ignored, as is the rise in construction costs due to inflation.

To account for costs during construction and startup, the following need to be specified: (1) the length of the construction period; (2) the fraction of plant investment spent in each year; (3) the annual rate of inflation for plant investment; and (4) the plant-capacity attainment schedule.

Table VI shows a hypothetical 6-yr investment schedule for a 20,000-ton/stream-d plant, beginning in 1977. The last column shows the effect of a 9% inflation rate on the capital investment. The dollars invested have increased from $1,258 x 10⁶ to $1,642 x 10⁶, or approximately 30%.

The Kentucky Center for Energy Research economic model was used to estimate the capital charges based on the two investment schedules in Table VI, using the same economic parameters. Full plant capacity was achieved in the first year of operation, 1983, and the 20-yr plant life did not include any construction time.

Table VII has three columns, the first of which is the previously described 1977 "instant plant" cost for a 20,000-ton coal plant. This yields 1,382 x 10⁶ gal of methanol/yr at a cost of 29.92¢/gal.

The second column shows the effect of charging interest on debt, and the required return on equity, during

Investment schedule for a 20,000-ton plant			Table VI
Year	Percent invested	Annual increment, $ million (No inflation)	Annual increment, $ million (9% inflation)
1977	2	25.17	25.17
1978	8	100.66	109.72
1979	20	251.66	299.00
1980	35	440.40	570.34
1981	25	314.58	444.05
1982	10	125.83	193.60
Total	**100**	**1,258.30**	**1,641.88**

Source: Kentucky Dept. of Energy

Effect on cost of methanol of 6-yr construction period, $ millions			Table VII
	1977 "instant plant"	6-yr construction (no inflation)	6-yr construction (with inflation)*
Plant investment	1,258.30	1,258.30	1,641.88
Added investment due to 6-yr construction period	–	719.00	917.00
Total investment	1,257.00	1,977.30	2,558.88
Coal cost	132.00	132.00	168.47
Operating cost	99.38	99.38	139.38
Capital charges	182.97	287.51	372.06
Total annual charges	414.35	518.89	679.91
Cost of methanol, ¢/gal	29.92	37.47	49.09

*Coal cost inflated to beginning of 1983 (plant operation begins) at 5%/yr, operating cost at 7%/yr, and capital costs at 9%/yr.

Source: Kentucky Dept. of Energy

construction. If the cost of coal and operating costs are kept constant, the added investment raises the price of methanol to 37.47¢/gal, an increase of 7.55¢/gal.

(The assumptions for Tables VI and VII are: 20,000 ton/d conversion plant, 20-yr life, no net salvage value, 35%/65% equity/debt financing, 12% return on equity, 9% return on debt, $1,257 x 10^6 plant cost (1977 "instant plant" investment), $20/ton coal cost, $132 x 10^6 coal-cost/yr (330 d), $99.380 x 10^6 operating cost after sulfur credit, 1,382 x 10^6 gal/yr of methanol, and capital expenditures occur over 6 yr.)

Including financial charges during the construction period will increase the cost 25.2%. This rise is a function of the interest rate, return required on equity, credit given for unspent capital, length of time required for construction, the investment schedule, etc. Although the actual cost increase for a specific case would depend upon many factors, 25% could be used as a typical figure in making cost estimates.

The last column shows the effect of inflation rates of 9% for capital, 7% for operating costs and 5% for coal between 1978 and 1983. It has been assumed that the 1977 costs reflect that year's costs and that the inflation rates have to be taken into account for only five of the six years. Interest charges during construction increased 29.4%, reflecting the almost 30% higher investment. The effect of all three inflation factors is to raise the price an additional 11.62¢ to 49.09¢/gal, or almost 64% over the base-case cost.

A report by the Electric Power Research Institute (published about the time the study on which this article is based was being made) analyzes four cases for producing methanol from coal—using a (1) Foster Wheeler, (2) BGC-Lurgi-slagging, (3) Koppers-Totzek and (4) Texaco gasifier [5]. In two cases, only methanol is produced; in the other two, fuel gas also is a product. Comparisons of total capital investments reported by the Institute (suitably adjusted) with investments that would be predicted using the generalized "instant plant" equation that was given earlier show very good agreement: +0.2%, +3.8%, −12.2% and +8.5%, respectively.

Acknowledgement

Financial support for this research project was provided by the State of Kentucky through its Div. of Technology Assessment, Dept. of Energy.

References

1. "Conceptual Design of a Coal to Methanol Commercial Plant," Badger Plants, Inc., FE-2416-24 Feb. 1978.
2. McGeorge, A., "Economic Feasibility Study, Fuel Grade Methanol from Coal," DuPont Co., TID-27606 1976.
3. Cornell, H. G., Heinzelmann, F. J., and Nicholson, E. W. S., "Production Economics for Hydrogen, Ammonia, and Methanol During the 1980-2000 Period," Exxon Research and Engineering Co., Apr. 1977.
4. Kermode, R. I., Nicholson, A. F., Holmes, D. F., and Jones, J. E., Jr., "The Potential for Methanol from Coal: Kentucky's Perspective on Costs and Markets," Div. of Technology Assessment, Kentucky Center for Energy Research, Lexington, Ky., Mar. 1979.
5. "Screening Evaluation: Synthetic Liquid Fuels Manufacture by Ralph M. Parsons Co.," Electric Power Research Inst., AF-523 Project 715, Final Report, Aug. 1977.

The authors

R. I. Kermode   A. F. Nicholson   J. E. Jones

Richard I. Kermode is a professor in the Dept. of Chemical Engineering University of Kentucky, Lexington, KY 40506. Previously, he had taught at Carnegie-Mellon University. His research interests include process control, systems modeling, process economics, as well as energy. He holds a B.S. degree from Case - Western Reserve University, and M.S. and Ph.D. degrees from Northwestern University, all in chemical engineering.

Arthur F. Nicholson is an energy consultant with the Kentucky Dept. of Energy. During the past five years, he has been engaged in energy research and has written several reports on energy and energy costs. He holds a B.A. from Wichita State University and an M.B.A from Harvard Business School, and is enroute to a Ph.D. in business administration at New York University. He is also enrolled in a Doctor of Business Administration program at the University of Kentucky.

James E. Jones, Jr., is Director of the Div. of Technology Assessment with the Kentucky Dept. of Energy, responsible for energy technical/economic studies and planning for coal conversion projects. While with the Institute of Mining and Minerals Research, University of Kentucky, he was principal investigator on several coal-conversion research projects that emphasized environmental aspects. He received his Ph.D. in chemical engineering from the University

Guide to coal-cleaning methods

A list of available routes, prepared by CHEMICAL ENGINEERING, gives details on stage of development, sulfur-removal capabilities, and preliminary cost information.

☐ There's much to be said these days about techniques that clean coal before it is burned in industrial and utility boilers, or in gasification or liquefaction plants. As the table* shows, many firms are announcing or piloting new cleaning processes to remove such impurities as sulfur and ash, and are building demonstration units to develop models for conventional methods. (The table, compiled from *CE* interviews and other sources, contains the most noteworthy technology available on a worldwide basis.)

Many of these techniques have some distance to go before reaching the commercial stage, as there are doubts about economics and sulfur-removal abilities. Testing will undoubtedly answer many of these questions, although sulfur-control experts already have some insights. For example, according to *CE* sources, chemical cleaning routes may not be able to compete with flue-gas desulfurization (FGD) in the U.S., where regulations demand the removal of up to 90% of all the sulfur in coal.

However, as shown in a recent study prepared for the U.S. Environmental Protection Agency by Pedco Environmental Inc. (Cincinnati), coal cleaning combined with FGD may be cheaper than FGD alone.

Another plus for coal cleaning is its potential for reducing the consumption of hydrogen used in gasification and liquefaction plants to remove sulfur (as H_2S) during processing, and for standardizing the raw material (coal) fed to such facilities.

PARTIAL ROUNDUP—Pros and cons aside, the pace of research work on coal cleaning appears to be accelerating, as the following partial list of activities suggests:

- In March, an Australian firm, Energy Recycling Corp. (Sydney) plans to complete a 960,000-metric-ton/yr unit 100 miles north of Sydney to test its new cleaning route (*Chem. Eng.*, Oct. 22, 1979, p. 80) based on linear accelerator technology.
- Last August, at the Second Chemical Congress of the North American Continent, held in Las Vegas, two new microbial desulfurization routes were disclosed: one devised by William Finnerty of the University of Georgia in Athens (*Chem. Eng.*, Sept. 8, 1980, p. 19) and the other by researchers at New Mexico Institute of Mining and Technology (Socorro).
- Last April, the Ohio Dept. of Energy (Columbus) signed a contract with Advanced Energy Dynamics, Inc. (Natick, Mass.) to develop its new desulfurization technique that electrically separates sulfur components from the rest of the coal.
- Last March, researchers at the University of Toledo announced a new chemical cleaning route based on liquid SO_2, at ACS's 179th annual meeting held in Houston (*Chem. Eng.*, Apr. 7, 1980, p. 35).
- And last January, TRW Inc. (Redondo Beach, Calif.) announced that it would begin conducting the first pilot tests of three chemical oxydesulfurization routes developed by Ames Laboratory, Ledgemont Laboratories and the Dept. of Energy's Pittsburgh Energy Technology Center (*Chem. Eng.*, Jan. 14, 1980, p. 51).

There is also activity to refine existing technology. This June, the Electric Power Research Institute (Palo Alto, Calif.) will complete, at a cost of over $10 million, its Homer City, Pa., coal cleaning test facility, which will have the capacity to treat 20-100 tons/h of coal (depending on the amount of fines present). This facility, which will employ conventional cleaning techniques, will be used to test the feasibility of cleaning different types of coals under varying conditions prior to building large-scale cleaning units for utilities.

VARIOUS TECHNIQUES—To remove sulfur, three basic approaches are being applied:

Physical. Conventional physical coal cleaning is essentially just washing with water to separate ash and pyrites that are heavier than the rest of the coal. Some of the newer routes are, however, somewhat exotic and do not use water but are based on fluid dynamics and behavior of particles in electric or magnetic fields. Yet, the physical routes (even though some of the newer ones boast high removal efficiencies) do not remove organic sulfur, which is found in the organic matrix of the coal.

Chemical. These systems try to remove organic and inorganic sulfur by reaction or extraction, while maintaining the essential character of the coal. Temperatures and pressures are not extreme, and ordinary unit operations are used.

Biological. Various strains of bacteria have been identified that metabolize sulfur-containing compounds, oxidizing them into soluble sulfates. Some strains attack pyrites, and others attack organic components. "An advantage of this technology is that you don't use a reactant that you must get rid of," says Patrick R. Dugan, Dean of Microbiology, Ohio State University. And these processes offer low cost and operate at room temperature. But the routes are slow—the organisms can take eight days or more to remove all the sulfur—and the method is in the earliest stages of development.

Reginald I. Berry

*Table appears on pp. 22–23.

Originally published January 26, 1981

Coal cleaning methods have various degrees of success in removing both organic and inorganic sulfur

Developer	Technique/status	Sulfur removal (i—inorganic; o—organic)	Estimated cost, $/ton (t—product cost* p—processing cost)
Chemical			
Ames Laboratory (Iowa State University, Ames)	*Oxydesulfurization.* Coal is oxidized with O_2 at 200 psi in a $150^\circ C$ solution of Na_2CO_3. Soluble sulfates are removed. Bench.	70—90%(i); up to 40%(o)	N.A.
Atlantic Richfield Co. (Philadelphia, Pa.)	*Oxydesulfurization.* Coal is oxidized in an aqueous solution with a complexing agent. Sulfur compounds are converted to soluble sulfates. Pilot.	95—100%(i); 25—37%(o)	46.—58.(t)
Battelle Columbus Laboratories (Columbus, Ohio)	*Hydrothermal.* Mineral and organic sulfur compounds are leached out by sodium and calcium hydroxides at high temperature and pressure. Mini plant (1/3 ton/d).	80—90%(i); 10—30%(o)	55.90(t)
General Electric Co. (Valley Forge, Pa.)	*Microwaves.* Sodium hydroxide is combined with coal. This mix is irradiated with microwaves. Soluble sodium sulfide and polysulfides are formed.	95%(i); up to 60%(o). Sulfur removal depends on the number of irradiation cycles.	41.80(t)
Hazen Research, Inc. (Golden, Colo.)	*Magnex.* Ground coal is contacted with iron carbonyl vapor at $150^\circ C$ and atmospheric pressure. The carbonyl reacts with pyrites to form a magnetic iron sulfide that is separated magnetically. Now being developed by Nedlog Technology Group (Arvada, Colo.).	More than 80%(i); does not remove organic sulfur.	40.70(t)
Jet Propulsion Laboratory (Pasadena, Calif.)	*Chlorinolysis.* Chlorine is reacted with pulverized coal suspended in methyl form. Soluble sulfates are formed. Bench.	70—90(i); 30—60(o)	46.00(t)
KVB, Inc. (Tustin, Calif.)	*Oxidation.* Dry pulverized coal is treated with hot NO_2 gas. Sulfur compounds are oxidized. Some sulfur is removed as SO_2, some as sulfates. Bench.	95—100%(i); 30—70%(o)	47.50(t)
Ledgemont Laboratories (Lexington, Mass.)	*Oxygen Leaching.* Ground coal is leached in an acidic solution or an ammonia solution. Work has been postponed.	90—100%(i)	46.9(t)
Pittsburgh Energy Technology Center (Pa.) of the Department of Energy	*Oxydesulfurization.* A coal/water slurry is combined with compressed air at 150-200°C. Sulfur is removed as soluble compounds. Laboratory scale.	80—90%(i); 20—50%(o)	51.60(t)
Pentanyl Technologies, Inc. (Arvada, Colo.)	*Hydrodesulfurization.* Iron carbonyl catalyst is introduced to a coal slurry at 60-70°C. Hydrogen-bearing species created react with sulfur-containing compounds to form solubles. Laboratory.	Organic sulfur is removed but process must be preceded by a physical cleaning to remove pyrites. Up to 98% total sulfur has been removed.	10—15(p)
Syracuse Research Corp. (Syracuse, N.Y.)	Crushed coal is exposed to ammonia gas, then ground again. A physical coal cleaning follows. Pilot.	Up to 80(i)	37.00(t)

Developer	Technique/status	Sulfur removal (i-inorganic; o-organic)	Estimated cost, $/ton (t—product cost* p—processing cost)
TRW, Inc. (Redondo Beach, Calif.)	*Meyers.* A hot solution of ferric sulfate is used to extract pyritic sulfur, which is oxidized to ferrous sulfate and sulfuric acid. Dormant.	60–90%(i)	43.40(t)
	Gravimelt. Powdered potassium hydroxide and sodium hydroxide are mixed with dried coal and heated to 375°C. The salts react with sulfur-bearing compounds to form sulfides, which are removed. Bench.	90–100%(i); 40–70%(o)	20 (p) (total cost $2.80/ MMBtu)
University of Toledo (Ohio)	*Liquid SO_2.* Coal is contacted under pressure at 150°C with liquid SO_2, which attacks co-valently bound organic S compounds, allowing them to be extracted.	50–60% total sulfur beginning with 3-4% S, coal residual will have as little as 0.8% S.	N.A.

Physical

Developer	Technique/status	Sulfur removal (i-inorganic; o-organic)	Estimated cost, $/ton (t—product cost* p—processing cost)
Advance Energy Dynamics, Inc. (Natick, Mass.)	*Electronic.* Ground coal via a carrier gas enters an ionizer, which charges the particles and keeps them separated before they enter an electrostatic separator where pyrites and ash are removed. Bench scale.	Total sulfur removal is 33–68%.	Cost of desulfurization is reported to be one-half to one-quarter that of (FDG.)
Auburn University (Auburn, Ala.)	*High-gradient Magnetic.* Pulverized coal passes into a fluidized bed that is subjected to a high-gradient magnetic field. Paramagnetic pyrites are separated from the rest of the coal. Pilot stage.	Up to 94%(i). Total desulfurization of 55-70% of Eastern coals has been attained.	6–15(p)
Energy Recycling Corp. (Sydney, Australia)	*Linear Acceleration.* Hot high-velocity gas (5,000 ft/s) pulverizes coal. Linear accelerator technology is used to remove ash by centrifugal force. A chemical process believed to involve fluorine can follow. A 960,000-metric-ton/yr facility is under construction.	Sulfur content of processed coal averages between 0.05 and 0.5%.	Lower Colorado River Authority (Austin) was promised in 1978 a delivered product cost of $1.48/ MMBtu.)

Microbial

Developer	Technique/status	Sulfur removal (i-inorganic; o-organic)	Estimated cost, $/ton (t—product cost* p—processing cost)
Jet Propulsion Laboratory, Ohio State University (Columbus), New Mexico Institute of Mining and Technology (Socorro) and Union Carbide Corp. (New York.)	Microorganisms oxidize pyrites in a reaction that takes several days. Developers working independently are at lab stage.	Up to 90 (i)	5-14 (p)
University of Georgia (Athens)	Organisms such as Acinetobacter and a member of the family Micrococcaceae will oxidize organic sulfur to soluble sulfates. Laboratory.	Can convert 95% of dibenzothiophenes to solubles.	N.A.

*Product cost includes cost of coal based on 1977 dollars; processing costs are in 1979 dollars and are costs associated with the process but not including the initial cost of the coal.

Sources: *CE* interviews; Dept. of Energy, papers presented at the 73rd annual meeting of AIChE by A. Attar (North Carolina State University), T.G. Squires et al. (Ames Laboratory), A.B. Tipton (Occidental Research Corp.), Warzinski et al. (DOE's Pittsburgh Energy Technology Center).

Coal combustor retrofits to gas- and oil-fired boilers

This fluidized-bed combustion system controls

SO₂ and NOₓ emissions without the need for major

pollution-abatement equipment. It is reported

to require less space than other systems

and have a payback period of two to three years.

Leonard Kaplan, Assistant Editor

☐ Wormser Engineering, Inc. (Middleton, Mass.) has developed a pollution-free, automatic coal burner that is designed to operate in conjunction with either new or existing gas or oil boilers. According to Gordon Baty, the company's president, the system is the first of its kind available in the U.S. for boilers with capacities from 10,000-100,000 lb/h of steam.

The company recently contracted to provide a 70,000-lb/h (250 psig) steam capacity system for Iowa Beef Processing Co.'s (Dakota City, Neb.) Amarillo, Tex. packing plant. Scheduled for installation during mid-1982, the system not only will provide steam for processing and refrigeration but will cogenerate 1.25 MW of electrical power for use at the plant.

A second system, scheduled for startup in late 1982, will be retrofitted to an existing boiler at the University of Lowell (Lowell, Mass.) to generate steam for heating and cooling. The two installations will operate without research subsidies, both being purchased on a commercial-performance guarantee basis.

The heart of the system, the Wormser Grate, is an atmospheric, fluidized-bed coal combustor (FBC) of second-generation design, which overcomes the drawbacks of earlier FBC systems, according to the company. Some of these problems: poor operating reliability, inefficient coal combus-

tion, incomplete desulfurization, large equipment size, and the systems' inability to be retrofitted to existing boilers. In addition, older fluidized-bed units could not modulate (turndown) their output from full load by much more than a factor of 3:1—severely limiting their ability to follow fluctuating demands.

PROVEN OPERATION—A test unit, with a capacity of 3,000 lb/h (15 psig) steam, has been operating since 1979. Retrofitted to an oil-fired boiler, located at the USM Corp. (Beverly, Mass.), it produces steam for space heating. To date, the unit has logged several thousand hours, the longest continuous test being 1,000 h, completed in June 1980. The pilot-plant work, sponsored by the U.S. Dept. of Energy, initially tested the operating reliability and desulfurizing and combustion efficiencies of the system. The reported results showed an overall operating availability of 94%; desulfurizing efficiencies of 88-97% (depending on the limestone used); and combustion efficiencies exceeding 97%.

Most commercial and industrial manufacturing operations rely on steam boilers to produce steam to heat processes, generate electricity, and heat and cool facilities. Today, the dominant fuels used are gas and oil; where coal is used, there are three types of combustion units:

■ *Stokers*—In the oldest way to

burn coal, these units use lumps of coal that have been spread onto a perforated plate, or grate, through which combustion air passes. They are preferred for small to medium boilers (100,000 lb/h).

■ *Pulverized-coal burners*—Introduced in the 1920's, these units work much like a domestic oil-burner. Powdered coal is blown into a combustion chamber, where it is burned in suspension. This is the preferred method of burning coal in large boilers (those above 100,000 lb/h).

■ *Fluidized-bed combustors*—Invented in the 1960s, they use sand-sized materials to form a bed. The flow of combustion air causes these particles, which are supported by a grate, to bubble, giving the bed the appearance of a boiling fluid.

FLUIDIZED-BED ADVANTAGES—Conventional coal combustors control emissions of flyash (smoke) and various gases, including SO₂ and NOₓ, by means of wet and dry scrubbers, baghouses and electrostatic precipitators. According to this company, these systems can account for up to 40% of the cost of building and operating conventional coal-combustion systems.

FBCs can produce stack gases as clean as those from using natural gas, and without the need for wet scrubbers. The boiling action of the bed causes coal to burn at a much lower temperature than is possible in a stoker—typically 1,500-1,600°F versus 3,000°F. The lower temperature allows the FBC to operate below the ash fusion point of coal (the temperature at which the ash in the coal melts), avoiding the formation of "clinkers" (chunks of fused ash) and making automatic operation of the bed possible. In addition, this temperature is below the point at which most NOₓ pollutants are formed.

Originally published February 22, 1982

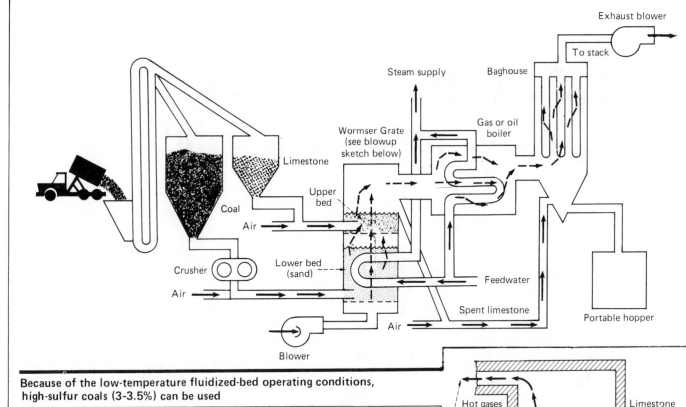

Because of the low-temperature fluidized-bed operating conditions, high-sulfur coals (3-3.5%) can be used

The low temperature also enables the burner to remove SO_2 without the use of a stack-gas scrubber, by employing ground limestone as the bed material. The limestone reacts with the SO_2 that is emitted as the coal burns, forming calcium sulfate (gypsum). This reaction takes place only at these relatively low temperatures.

SYSTEM OPERATION—Flowsheet shows the basic layout of the Wormser Grate process. The fluidized-bed boiler produces 60% of the steam output, while the remaining 40% is produced by the existing or new oil or gas boiler. While the firm will provide information on entire systems, from storage silos to baghouses, it provides only the Wormser Grate combustor and the coal feed system.

Coal passes through a coal processing train, in which it is crushed (to less than 1/4 in.) and dried, and then pneumatically conveyed to the lower bed of the Wormser Grate. This lower bed contains sand at a temperature of about 1,700°F, where the coal is burned. The resulting gases flow through the upper bed, containing limestone, where they are desulfurized. Combustion temperature is controlled by extracting heat to make steam in boiler tubes immersed in the

sand bed. The remaining useful heat is removed from the hot gases (approximately 1,500°F) in the existing boiler. When the gases leave this boiler, they pass to a baghouse where most of the particulates are removed.

OUTPUT CONTROL—Unlike conventional fluidized beds, the Wormser Grate uses two separate, shallow beds (on the order of 12-18 in.), one for combustion and the other for desulfurization (see drawing of Wormser Grate). According to the company, shallow beds are much less active than a single deep-bed design (3-5 ft), where large amounts of bed material and unburnt coal are thrown upward out of the turbulent bed, making necessary a large vertical height to allow this material to fall back into the bed.

To control the steam output of the combustor, a sand storage bed enables the unit to vary the height of the combustion bed, changing the number of boiler tubes covered. If the tubes are left completely covered and a reduction of output is attempted, by reducing the coal feedrate, the steam tubes—being so much cooler than the bed—will continue to extract heat, and the bed temperature will fall below the ignition point. To eliminate this problem, this unit automatically, by means of

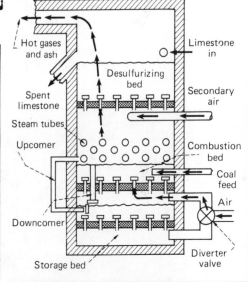

Wormser Grate: output capacity is controlled by combustion-bed height

diverted combustion air, regulates the amount of hot sand passing from the storage compartment to the combustion bed (when steam demand goes up). Sand is removed from the combustion bed through a downcomer when steam demand goes down.

Not shown in drawing, but necessary for the equipment startup, is a gas- or oil-fired preheater, which heats the combustion bed to the coal ignition temperature (about 1,000°F). Startup takes about 1½ h, said to be less than half that of conventional FBCs.

Methane from coal aided by use of potassium catalyst

This process combines two reaction steps—gasification and methanation—in one fluidized-bed reactor to produce synthetic natural gas at a greater overall efficiency than do conventional processes, says the company.

Leonard J. Kaplan, Assistant Editor

☐ Exxon Research and Engineering Co. (Florham Park, N.J.) is currently designing a large pilot plant that will use its Catalytic Coal Gasification (CCG) process to produce methane from coal. Data collected from a one-ton/d (of coal) process-development

1-ton/d unit has operated since 1979

Originally published March 22, 1982

unit, located at Baytown, Tex., and in operation since 1979, will be used for scaleup. Construction of the new 100-ton/d facility, expected to cost over $500 million, is scheduled to start sometime in 1983. The facility will be located in the Netherlands at Rotterdam's Europoort (*Chem. Eng.*, June 16, 1980, p. 68), adjacent to an existing Esso chemical plant, which will supply some of the utilities and support operations.

Exxon Research expects the plant to be completed in 1985-1986, and then operated for about three and a half years to commercialize the technology. Allowing time for design and construction, and assuming there is adequate economic incentive, the company believes the first commercial plant (3,000-4,000 tons/d of coal) might be onstream in the mid-1990s. The firm chose the Netherlands because it felt the outlook for substitute natural gas looks better in Europe than in the U.S., and because the Netherlands has an established gas-distribution system and coal import facilities.

The CCG technology (see Fig. 1) employs a potassium catalyst (K_2CO_3) to selectively produce methane by reacting coal and steam at one temperature and in a fluidized-bed reaction vessel. The overall simplified reaction using carbon follows:

$$2C + 2H_2O \rightarrow CH_4 + CO_2$$

When the equation is written for coal,

which contains hydrogen, it becomes:

$$1.2H_2O + 1.7CH_{0.9}O_{0.1} \rightarrow CH_4 + 0.7CO_2$$

CONVENTIONAL PROCESSES—Thermal-gasification approaches require that coal be reacted with steam at a high temperature (925°C) to produce synthesis gas ($CO + H_2$), which then undergoes additional reactions at a lower temperature (425°C), producing CH_4 and CO_2. These reactions include a "water gas shift" (required because the proportions of CO and H_2 produced in the gasification are different from what is needed in the next step), and a methanation, as shown below with their heats of reaction (ΔH_r):

Gasification
$$C + H_2O \rightarrow H_2 + CO$$
$$\Delta H_r = 64 \text{ kcal/mol}$$
Shift
$$CO + H_2O \rightarrow CO_2 + H_2$$
$$\Delta H_r = -8 \text{ kcal/mol}$$
Methanation
$$3H_2 + CO \rightarrow CH_4 + H_2O$$
$$\Delta H_r = -54 \text{ kcal/mol}$$

These reactions are not the only ones taking place during the gasification and methanation steps, but they do represent the main results.

The thermal gasification methods (such as the Winkler, Koppers-Totzek and Lurgi processes) use oxygen to burn some of the coal to liberate thermal energy needed for the gasification (about 31 kcal/mol).

This method results in significant inefficiencies, says the company, because the temperature of the second stage is lower than that of the first; therefore, the heat liberated by the methanation cannot be used to provide the high-temperature heat required for the gasification.

Fig. 2 compares the basic conventional processes with Exxon's CCG. If a

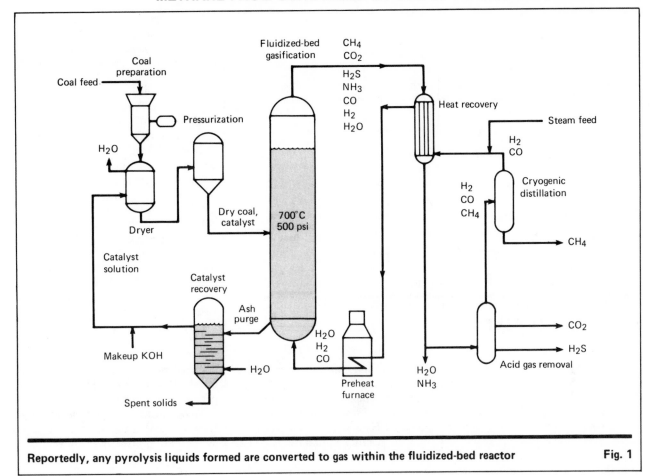

Reportedly, any pyrolysis liquids formed are converted to gas within the fluidized-bed reactor — Fig. 1

unit of CH_4 produced contains 1,000 kcal of energy, the thermal gasification requires approximately 300 kcal (obtained from oxidation of a portion of the coal). In addition, 280 kcal will be given off in the exothermic methanation. For the same 1,000 kcal of CH_4 produced, the CCG process requires only 20 kcal, says the firm.

CATALYST PERMITS SELECTIVITY — The use of alkali-metal salts of weak acids (e.g., K_2CO_3 and Na_2CO_3) for catalyzing the steam gasification of coal has been known for years. In 1921, Taylor and Neville[*] found that increasing the amount of K_2CO_3 on charcoal increased the quantity of synthesis gas produced when the mixture was reacted with steam at a constant temperature. In the early 1970s, Exxon researchers confirmed that commercially acceptable gasification rates could be obtained at a lower temperature (700°C) with a K_2CO_3 catalyst (with concentrations of 10-20% by weight) than without a catalyst (925°C).

[*]H. S. Taylor and H. A. Neville, *J. Am. Chem. Soc.*, **43**, 2055 (1921).

A more important result of their work was the discovery that the potassium catalyst on char (coal that has been partially depleted of carbon and hydrogen) would catalyze the methanation reaction at 700°C, the temperature at which the catalyzed gasification takes place (see Fig. 3). (If the catalysis could be carried out at 400°C, there would be negligible H_2 and CO at equilibrium, and no recycle would be necessary. Further research is needed on this for the next generation of gasifiers.)

At 700°C, studies found that if steam is brought to equilibrium with carbon, at an elevated pressure of 3.5 MPa (500 psig), five compounds— H_2O, CH_4, CO_2, H_2 and CO—will be present in significant quantities, produced as a result of the chemical reactions shown earlier. When the heats of reaction for the three reactions are added together, the result is a small energy input of 2 kcal/mol. According to the company, this means that in addition to yielding the required product, CH_4, essentially all the heat of combustion in the carbon appears as

the heat of combustion of the CH_4, and it is not necessary to supply reaction heat to the gasifier (except for enough heat to raise the temperature of the recycle gases before they enter the gasifier).

But the selectivity of the overall reaction cannot be achieved by chemistry alone at this temperature; recycle and separation steps are required. The H_2 and CO are separated and recycled to the gasifier so that there is no net formation of endothermic products (H_2 and CO).

PROCESS OPERATION — The entire process consists of four main operations; coal preparation, fluidized-bed gasification, catalyst makeup and recovery, and product separation and recovery.

The coal, crushed to a size suitable for fluidization (approximately 1-2 mm) is mixed with an aqueous solution of KOH and K_2CO_3 to yield approximately 8% (by weight) of catalyst on the coal. Dried and pressurized, the coal is fed through a series of lock hoppers to the gasifier, which operates at 700°C and 500 psig (35 atm).

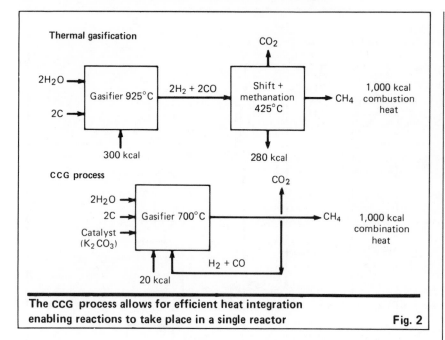

The CCG process allows for efficient heat integration enabling reactions to take place in a single reactor   Fig. 2

It is in the gasifier that the coal—fluidized by gases at a velocity of about 30 cm/s—reacts with steam to produce CH_4 and CO_2 as net products. The raw-gas product also contains unreacted H_2 and CO plus steam as major constituents. Sulfur and nitrogen in the coal appear as minor quantities of H_2S and NH_3. In addition, the exiting gas contains fines of unreacted coal and ash that are separated by a cyclone and returned to the bed.

A combination of heat recovery (the 1,300°F gases are cooled to about 500°F) and wet scrubbing removes the

H_2O, NH_3 and entrained particulates. A conventional solvent process for acid scrubbing removes the H_2S and CO_2. Separation of the methane from the unreacted H_2 and CO takes place in a cryogenic distillation unit, with the CO and H_2 being recycled back to the gasifier after being mixed with steam and preheated to the reaction temperature in a gas-fired furnace.

CATALYST RECOVERY—Residual solids (ash and unreacted coal) must be continuously removed from the gasifier and the catalyst recovered. This is accomplished by means of a staged,

countercurrent water-wash process. Because some of the catalyst reacts with clay minerals in the coal ash to form insoluble catalytically-inactive potassium aluminosilicates, makeup catalyst is added as KOH to the recovery solution. This method reportedly recovers 65-70% of the potassium, primarly as K_2CO_3.

During March and April 1981, a 23-day run was made on the process development unit, in which operations were said to be generally stable. The gasifier operated at 500 psig and 690°C, with the following results: steam conversion was 35%; carbon conversion ran 85-90%; the average methane content of the product gas reached 20-25%, said to be somewhat lower than the targeted value of 25%. Even so, the run was considered a major success.

The average results, on the 1-ton/d unit, indicated that 1.2 lb of steam is required per lb of coal, and approximately 18,000 ft³ of CH_4 is produced per ton of coal.

ADVANTAGES AND SOME PROBLEMS— Exxon feels that the major advantages of the CCG process over conventional routes for producing methane from coal are:

■ Using a catalyst permits efficient heat integration and minimizes heat input requirements, as well as promotes a one-step gasification and methanation.

■ Oxygen, which costs money and can cause coal-ash slagging, is not required.

■ No pyrolysis liquids are produced.

■ The process operates at moderate temperatures.

■ Scaling up fluidized-bed reactors to very large sizes is relatively easy, so that comparatively few units are required for a commercial facility.

In scaling up more than 3,000 times from the 1-ton/d unit to a commercial system, there are a number of design uncertainties. For instance, there is concern that if a gas-scrubbing system is designed for a maximum expected loading of entrained solids from the gasifier, it may have so much overcapacity that normal rates will not give meaningful performance data. Also, in the catalyst recovery system, spent solids must be separated from the wash water, but a filter sized for the worst solids expected will not provide good information when the loading is low.

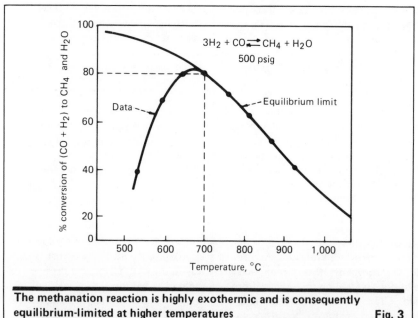

The methanation reaction is highly exothermic and is consequently equilibrium-limited at higher temperatures   Fig. 3

Better fluid-bed units ready to make debuts

Three kinds of designs promise better efficiencies, more flexibility and less corrosion. Commercial-size projects are underway in the U.S. and Europe.

A second generation of coal-fired, fluidized-bed combustors (FBCs) are poised for a shot at the lucrative gas- and oil-boiler market. Several units in the U.S. and Europe (see table) have either made their commercial bows or will do so in coming months. Other designs are either in the feasibility-study or pilot-project phases (the U.S. Dept. of Energy has asked industry to submit proposals for advanced FBC concepts).

The new models, which are mainly of the circulating or multistage variety, are said to provide better combustion efficiency, pollution control and turndown capability* than earlier FBCs. In addition, say some manufacturers, the

*The ability to function at less than full boiler rating. E.g., a boiler with a 3:1 turndown ratio can operate at a minimum capacity of 33% of full rating.

new designs take up less space and are able to solve such known FBC pitfalls as steam-tube erosion and corrosion, and nonuniform heat distribution within the bed. Use of advanced FBCs often allows for simpler coal preparation, note the systems' proponents.

However, despite the high praise, the jury is still out on the new FBCs. Although some large-scale trials have been successful, as yet no commercial advanced-design unit that uses limestone as a desulfurization agent has been proved in operation. And boiler experts have yet to become true believers. For one thing, they note, utilities have had no experience in the operation of the huge cyclones associated with the use of circulating FBCs. As for multistage units, there is considerable potential for fouling.

A STEP UP—The latest advances in FBC technology stem from the need to correct the acknowledged defects of earlier models—steam-tube problems, turndown ratios rarely exceeding 3:1, and average combustion efficiencies of 87-89%.

The novel units improve FBC performance by:

■ Putting the steam tubes outside the firing chamber. This separates the heat-transfer and combustor portions of the boiler, reducing erosion and corrosion.

■ Establishing separate hot-solids (coal, limestone) and hot-gases exchange systems (Fig. 1). This enables the operator to turn down the hot solids and achieve turndown ratios of 6:1 and higher.

■ Recycling solids through the combustion chamber. This allows longer residence times, improving combustion efficiency and pollution control. Since the coal is passed through the system until it burns, there is no need to grind it very finely.

PIONEER WORK—The initial studies of the circulating-bed concept have been done by two European firms: Lurgi Chemie und Hüttentechnik GmbH (Frankfurt, West Germany) and Ahlstrom Co. (Helsinki, Finland). Both have used the advanced models in applications other than coal burning—Lurgi to calcine alumina, and Ahlstrom to burn biomass. However, they

---

A number of advanced-design fluidized-bed combustors are being developed

Designer	Location	Type	Status	Rating (million Btu/h)	Turndown ratio	Combustion efficiency	SO$_2$ removal
Lurgi	Lünen, West Germany	Circulating	Startup slated for May 1982	280	3:1	99%	85%
Pyropower	Bakersfield, Calif.	Circulating	Started up in December 1981	50	3:1	98-99%	>90%
Struthers Wells	Uvalde, Tex.	Circulating	Started up last month	50	4:1	N.A.	—
Conoco/Stone & Webster	Lake Charles, La.	Circulating	Awaiting final approval	60	4:1	>94%	—
Wormser Engineering	Dakota City, Neb.	Multistage	Startup slated for late 1982	70	30:1	88-97%	—

Originally published April 5, 1982

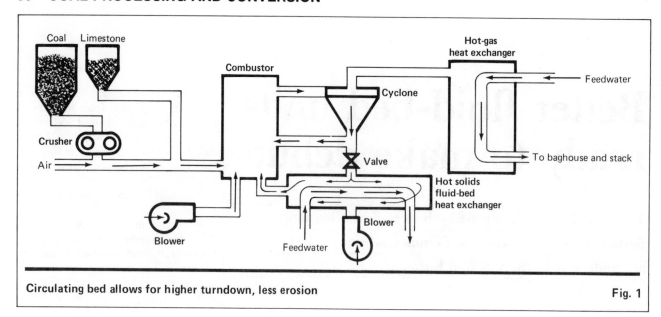

Circulating bed allows for higher turndown, less erosion   Fig. 1

have been working with coal/limestone feed for the past several years.

Ahlstrom, for example, has used that feed in a 200,000-lb/h steam cogeneration unit at Kauttua, Finland, that usually is fueled by peat or wood. The tests have been termed successful, but few details are available.

Lurgi has performed more extensive trials at two pilot plants located in Frankfurt. And by mid-summer, the West German concern expects to start up a demonstration/commercial unit for Vereinigte Aluminium Werke at Lünen. The facility, which will be jointly financed by VAW and the West German government, will direct most of its output into making process steam.

Last year, Lurgi, along with Combustion Engineering Inc. (Windsor, Conn.) and the Tennessee Valley Authority, designed a 200-MW electric utility circulating-FBC, with an eye toward eventually building a demonstration facility. However, due to reservations about the efficiency of the cyclones, whose size would have reached more than 30 ft in dia., TVA has vetoed further development.

"Cyclones that big just haven't been built," comments a TVA engineer. "We weren't sure we could get enough collection efficiency."

U.S. INSTALLATIONS—Through a San Diego, Calif., firm owned jointly by Ahlstrom and General Atomic Co. (San Diego), one of the first U.S. commercial applications of coal-fired circulating FBCs is going ahead. Pyropower Inc. is installing a 50-million-

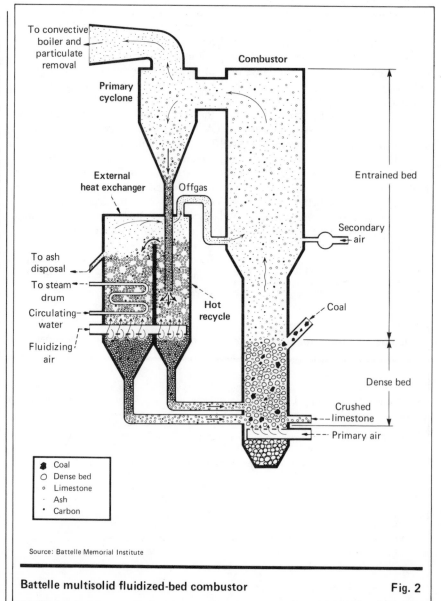

Source: Battelle Memorial Institute

Battelle multisolid fluidized-bed combustor   Fig. 2

Btu/h unit, based on the Ahlstrom design, to make steam for an enhanced-oil-recovery project near Bakersfield, Calif., run by Gulf Oil Exploration and Production Co. U.S.A.—a subsidiary of Gulf Oil Corp. (Pittsburgh, Pa.).

At Uvalde, Tex., another circulating-FBC to be used for enhanced oil recovery was started up last month for Conoco Inc. (Stamford, Conn.) by Struthers Wells Corp. (Warren, Pa.). Since last November, Struthers Wells has operated a 5-million-Btu/h model at Winfield, Kan., for the same use.

This design, created by Battelle Memorial Institute (Columbus, Ohio), has what is called a dense bed (see Fig. 2). According to John Fanaritos, a project engineer at Struthers Wells, this feature enhances the combustion of fines, which in conventional FBCs are often blown right out the stack, and in other circulating beds must sometimes be recycled numerous times for complete burning.

In this system, fines, after going through one cycle, are blocked by the dense bed's barrier of compact limestone and coal, which promotes long residence times. Fanaritos also says that the system burns chunks of up to minus 2-in. mesh, which is larger than what other units burn.

Another circulating-FBC project still has not received the final go-ahead. At presstime, Conoco officials have not decided whether to continue or abandon plans for a 60-billion-Btu/h demonstration model at the firm's Lake Charles, La., chemical complex.

The proposed unit, which has been delayed because of the Conoco merger with Du Pont, would be a scaleup of a pilot system, developed jointly with Stone and Webster Engineering Co. (Houston), and demonstrated last year (*Chementator*, Nov. 30, 1981, p. 17).

On the threshold of marketing a circulating system is a Swedish firm, Studsvik Energiteknik AB (Stockholm). After completing a four-year pilot project, the company is gearing up to offer FBCs of 25 MW and larger to municipalities, for district heating, and to industry. A joint production and marketing agreement is expected to be sealed with boiler maker Generator Industri AB (Göteborg), within the next few months.

ANOTHER APPROACH—Along with circulating beds, the most developed advanced-design FBC is the multistage

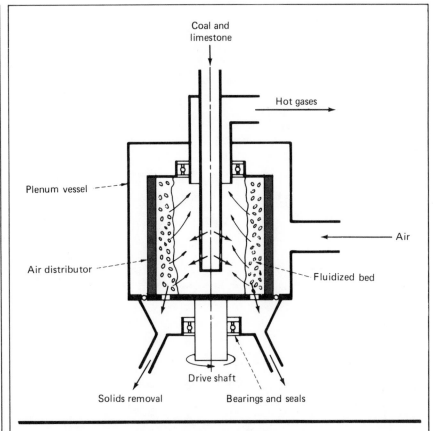

Rotating-bed design is still far from commercialization   Fig. 3

model, which is closer to conventional FBCs than is the circulating type. While some multistage units have multiple beds, others simply have several air inlets, allowing for staged combustion in one bed.

Only one system, from Wormser Engineering Inc. (Middleton, Mass.), is currently on the market (see *Chem. Eng.*, Feb. 22, pp. 22-25), but several other groups are exploring the concept. The University of Natal (South Africa) and Howard University (Washington, D.C.) have each come up with design proposals, and the former has built and run an experimental unit.

Some boiler experts maintain that the multistage design has poor combustion efficiencies, caused by brief residence times in the shallow beds (1-1.5 ft deep), and that there is fouling or erosion and corrosion of steam tubes and air distributor plates (of the upper stages). However, Gordon Baty, president of Wormser Engineering, claims that this is not true.

Wormser beds may be shallow, he concedes, but 98-99% carbon burnout is assured, nonetheless, because the

unburned coal must pass through an upper bed packed with incandescent limestone. As for deterioration of steam tubes, Baty answers that his company's multistage test units have run for over 1,000 h without encountering such difficulties.

A BED THAT TURNS—Further from commercialization than the circulating or multistage design is the rotating-FBC. Development has gone on at United Technologies Corp. (Hartford, Conn.), Lehigh University (Bethlehem, Pa.) and the U. of Aston, U.K.

In these units, inward drag created by air inflow is balanced by the centrifugal motion of the bed material (see Fig. 3), thus potentially permitting larger airflow and better coal combustion per unit of volume than is possible with a system working against the force of gravity, say the Lehigh researchers.

Rotating beds are a pie-in-the-sky idea right now, points out Floyd W. Crouse, of DOE's Energy Technology Center at Morgantown, W.Va. "It'll be a long time until this gets out of the research phase."

Eric P. Johnson

Section II
Inorganic Chemicals

Ammonia Manufacture
Hydrogen Production
Other Chemicals

Tightening the loop in ammonia manufacture

☐ As the cost of natural-gas and light-hydrocarbon feedstocks has soared since the Arab oil embargo of five years ago, so has the incentive for developing more-energy-efficient routes to ammonia. Although process developers are working on many fronts to achieve such savings, the most recent advances have come primarily in one area—around the synthesis loop, the tail end of the chemical sequence to ammonia.

In the past, when feedstock was cheap and plentiful, concern over hydrogen wasn't as strong as it is today, with hydrogen being "downgraded" to fuel status because part of the synthesis-gas stream must be purged to remove unwanted inert compounds (see box for NH_3 process details). With natural-gas prices running upward of $2.5 to $3/million Btu, it is apparent that hydrogen is too valuable to burn.

For some time, ammonia producers have been retrofitting cryogenic and pressure-swing-adsorption units to recover hydrogen from the purge stream. But only recently have H_2-recovery units—cryogenic and others—been designed into ammonia processes available from licensors. In the news:

■ Pullman Kellogg (Houston) has performed commercial tests of purge-gas-system improvements that it has developed. In December it confirmed this for the first time.

■ In November, at the Second Latin American Petrochemical Congress in Cancun, Mexico, C-E Lummus (Bloomfield, N.J.) announced development of a new high-efficiency plant design that uses cryogenics to recover normally wasted hydrogen from synthesis purge gas.

■ In October, at an Institution of Chemical Engineers meeting in London, Humphreys & Glasgow Ltd. (London) took the wraps off a new route that features molecular sieving to recover purified hydrogen from shift gas, and essentially eliminates conventional synthesis-gas purging.

■ And in 1977, Petrocarbon Developments, Inc. (Houston) started using the first cryogenic hydrogen-recovery unit specifically designed to recirculate H_2 to the synthesis loop.

PARING ENERGY, UPPING YIELDS—Pullman Kellogg, as do most process developers, remains tight-lipped about the specifics of its improved technologies, preferring to discuss general efforts. Says the company's vice-president of process operations, James Finneran: "The plants we are building and bidding today include . . . advances that result in improved yields, tremendous energy savings, or both."

One of the numerous approaches Kellogg has taken to produce ammonia more efficiently is improvement of the synthesis loop. According to Finneran, "We have recently placed a cryogenic purge-gas recovery system in service. This recovers hydrogen in the purge stream (normally about 60% of it is hydrogen), as hydrogen is just too valuable to be used as fuel. The cryogenic system adds about 100 tons of capacity to the plant.

"Also, catalytic purge-gas conversion units enable us to make additional ammonia from hydrogen that might otherwise be used as fuel. Generally, this can add about 6% to a plant's capacity."

Finneran further notes that most process developers are working on various approaches to upgrade ammonia production. Regarding Kellogg, he says: "I think it is important to point out that under our latest advanced designs—now I'm talking about developments that are being presented to clients—we have ways of making extra ammonia, or we can orient the process to save energy and keep the same ammonia production. In the plant already mentioned, we are saving the client three quarters of a million Btu per ton of product." Putting this in perspective, many "typical" steam-methane reforming processes require approximately 31 million Btu/ton of product.

ENERGY SLASHER—Also relying on cryogenics to cut energy consumption is C-E Lummus, which says its Ammonia Design-I process uses just 28 million Btu/ton of product, and its Ammonia Design-II just 26 million Btu/ton (some 16% less than the "typical" 31 million). Both designs, Lummus points out, use process units that have been commercially proven—but the "II" design incorporates approaches that deviate from the conventional to a greater degree.

Key to the two designs is a cryogenic unit that recovers hydrogen from purge gas, allowing approximately 99.5% of the H_2 to be converted to product, compared with the 92 to 95% that is normally utilized. This allows syngas purity criteria to be relaxed, since hydrogen loss from the synthesis reaction is minimal.

Other process changes combine with the cryogenic unit to help cut energy requirements. For instance, a 3:1 steam-to-carbon ratio (vs. 4.5:1 conventionally) reduces low-level waste heat; a more efficient carbon dioxide removal system makes the ratio change possible.

Overall, the cryogenic step adds no capital investment compared with conventional plants, because reduced syngas generation and steam-system equipment costs offset the cost of the cryogenic unit and other equipment, according to the firm. Lummus estimates that a 1,500-ton/d plant would run about $70 million, based on mid-1978 prices. Annual manufacturing-cost savings at such a plant would be about $4 million with the "I" design and approximately $7.5 million with the "II" design, Lummus estimates (see table).

MORE CRYOGENICS—Another cryogenic H_2-recovery method has been commercialized by Petrocarbon Developments. The technique has been used for debottlenecking since mid-1977 at the Lima, Ohio, 1,500-ton/d ammonia plant of Vistron Corp. (Cleveland). Now, two American Cyanamid plants (with a combined capacity of 1,750 tons/d) at Fortier, La., are set to use the cryogenic technique.

The unit at Lima has increased ammonia production by approxi-

Originally published January 29, 1979

Cryogenic hydrogen-recovery unit is said to pare manufacturing costs for 1,500-ton/d NH₃ plants

Variable costs	Unit	$/unit	Conventional design		Ammonia Design – I*		Ammonia Design-II†	
			Unit/ton	$/ton	Unit/ton	$/ton	Unit/ton	$/ton
Feedstock	million Btu	3.0	19.8	59.4	19.38	58.14	19.38	58.14
Fuel	million Btu	3.0	11.2	33.6	8.62	25.86	6.63	19.89
Electricity	KWh	0.02	40	0.80	40	0.80	40	0.80
Boiler feedwater	thousand gal	1.0	0.42	0.42	0.20	0.20	0.19	0.19
Cooling water (circulating)	thousand gal	0.01	73.0	0.73	62.8	0.63	55.2	0.55
Catalysts and chemicals	—	—	—	0.92	—	0.92	—	0.92
Total variable costs				95.87		86.55		80.49
Assumed associated costs				39.20		39.20		39.20
Total manufacturing cost				135.07		125.75		119.69

*Includes cryogenic H₂-recovery, lower steam/carbon ratio, improved CO₂ removal, and preheating of reformer combustion air.
†Includes all of above, plus further improvements in CO₂ removal, and radial-flow reactor.
Source: "High efficiency ammonia plant design," by S. Talbert et al., C-E Lummus (Nov. 1978).

Traditional ammonia production — a brief refresher

Steam reforming, with feedstock natural gas or LPG or naphtha, is the most widely accepted process for ammonia manufacture. The route includes four major chemical steps:

Reforming $CH_4 + H_2O \rightarrow CO + 3H_2$
Shift $CO + H_2O \rightarrow CO_2 + H_2$
Methanation $CO + 3H_2 \rightarrow H_2O + CH_4$
Synthesis $3H_2 + N_2 \rightarrow 2NH_3$

In operation, feedstock is desulfurized and mixed with 1,500-psi steam in a steam/carbon molar ratio of about 4.5:1. The mixture is then fed at high temperature to Ni-containing tubes in a primary reformer. Reaction products pass at 1,400°F to 1,500°F to a secondary reformer, into which air is introduced to provide oxygen (and nitrogen for the later synthesis reaction). The oxygen exothermically reacts with hydrogen and natural gas to help drive the reforming reaction at about 1,800°F.

The two-step water-gas shift reaction follows. First, a high-temperature shift reactor provides a CO concentration of about 2 to 3%. Then, because shift equilibrium is favored at lower temperatures, shift effluent is passed through a fixed bed of copper and zinc oxide catalyst in a low-temperature shift reactor. Exit gas is further cooled and CO₂ subsequently removed.

Next, the process gas is preheated and enters the methanator, where unreacted CO (and the remaining CO₂) reacts exothermically with hydrogen to form methane and water vapor. Upon cooling, water is removed and the steam is sent to the suction side of the synthesis-gas compressor. At this point, the syngas stream is approximately a 3:1 mixture of hydrogen to nitrogen, with the remainder about 0.9% methane and 0.3% argon.

The compressors step up the gas to about 3,000 psi — the high pressure that is needed to favor synthesis equilibrium. Once pressurized and mixed with recycle gas, the stream enters the synthesis converter, where ammonia is catalytically formed at 750° to 950°F. Product is recovered as liquid via refrigeration, and the unreacted syngas is recycled.

In the synthesis step, however, some of the gas stream must be purged to prevent buildup of argon and methane. Purging causes a significant loss of hydrogen that could be used for additional ammonia manufacture. It's this loss that process designers are seeking to minimize.

---

mately 75 tons/d, according to Petrocarbon, with the installed cost of the recovery plant and ancillaries at less than $2 million. Energy requirements of the incremental production are just 25 million Btu/ton of production, the developer claims. (For process details, see *Chem. Eng.*, Oct. 10, 1977, pp. 90-92.)

HYDROGEN ADSORPTION THE KEY— For its new process—as yet untried commercially—Humphreys & Glasgow claims up to a 25% reduction in capital costs, plus much better process economics due to feedstock savings.

This time, the key to the development is not cryogenics, but a pressure-swing-adsorption step based on Union Carbide Corp.'s Polybed process for recovering hydrogen. This is a new departure for PSA: by placing the molecular-sieve PSA unit between the high-temperature shift reactor and the syngas compressor, H&G has eliminated such well-known trouble elements as the secondary reformer, low-temperature carbon monoxide shift, auxiliary boiler, and carbon dioxide absorption.

The initial processing steps are conventional—natural gas is heated and desulfurized, and then mixed with steam in a 1:3 ratio and reformed in a tubular reformer. After cooling in a waste-heat boiler, the gas is fed to a high-temperature shift reactor to reduce CO content to below 4%. Further cooling follows.

Then the new technique comes into play. Cooled gas enters the pressure-swing-adsorption stage, which separates it into a pure hydrogen stream and a purge stream. The purge flow (which contains CO, CO₂ and some methane) is recycled to the primary reformer for supplementary fuel; the hydrogen stream is sent to the ammonia synthesis loop.

According to the developer, the design requires a smaller synthesis loop because the syngas feed is highly pure hydrogen. And, except for a very small inerts-purge that is sent to the primary reformer, the energy-wasting purge-gas system of conventional plants is eliminated.

Most of the elements that comprise the system are well proven commercially; the only novelty to ammonia making is the PSA portion, and even this has been demonstrated on the steam reformer of a 33,000-metric-ton/yr hydrogen plant in Germany, H&G notes. A typical 1,000-mt/d ammonia plant would require two 10-bed PSA units, each slightly smaller than those operating in Germany.

The H&G system has two main drawbacks. First, additional CO₂-recovery equipment is needed to make the process mesh with urea production. And second, since the process eliminates the secondary reformer, air is not introduced into the system. So another nitrogen source is required for the synthesis step. The needed 660 std m³ of nitrogen per metric ton of ammonia could be supplied either by unwanted nitrogen from a nearby air-liquefaction plant, or by a plant built specifically to manufacture N₂.

Overall, capital costs would run about 75 to 85% of the cost of a conventional plant, depending on whether a nitrogen plant were needed, according to the British firm. Natural-gas consumption would range from 30.7 gigajoules/mt of ammonia (26.4 million Btu/ton) to 32.8 gigajoules/mt (28.2 million Btu/ton), with the higher figure resulting if nitrogen production were powered by steam generated by the ammonia process.

Larry J. Ricci

Argon recovery: Aid to ammonia economics

The recovery of liquid argon from synthesis-loop purge gas may provide a significant income to the ammonia producer— particularly if there is a steel mill nearby to provide a constant market for this product.

Reginald I. Berry, Assistant Editor

☐ To reduce costs and increase production in ammonia plants, designers have turned to the recovery of hydrogen from synthesis-loop purge gas (see pp. 35–36). But it may be just as economically attractive, if not more so, to recover argon from this stream.

In the ammonia process, synthesis gas is recycled about the ammonia converter. Some of this gas must be purged to control a buildup of argon and methane. In the past, this purge, containing about 85% N_2 and H_2, has been consumed as a fuel. Now, adsorption or cryogenic processes are being used to recover hydrogen.

ARGON INCOME—But there can be economic incentives to recover argon as well. Argon has a commercial value of from $250 to $380/ton, and about 14 tons/d can be recovered from the purge of a 1,000-ton/d ammonia plant.

Also, significant quantities of this element are used in the steel industry during the melting, casting and annealing of special alloys. Based on the sale of argon, a relationship has been formed between ammonia and steel industries in Europe, but so far this has not happened in the U.S.

Lotepro Corp. (New York), a wholly owned subsidiary of West Germany's Linde AG, offers a cryogenic process that recovers pure argon, along with a synthesis gas (with the right stoichiometric proportions for ammonia production), and a stream containing 98% methane.

This route will be used to recover 35 lb-moles/h of liquid argon from the purge gas of a 1,000-ton/d ammonia plant that is under construction in Baroda, India, for the Gujarat State Fertilizer Co.

Experience with the technology, developed by Linde, has been gained through operations in West Germany for BASF AG and in The Netherlands for DSM N.V., and the route has been running since 1974 in Wratza, Bulgaria, for Technokomplekt, where 29 lb-moles/h of argon are produced.

COLD SEPARATION—In this cryogenic process, fractionation separates the components of the synthesis-loop purge gas. Argon and methane are liquefied and recovered in trayed distillation columns, and synthesis gas is recovered as overheads.

The major portion of the refrigeration required for this technology is supplied by a high-pressure nitrogen cycle with an expander. Additional refrigeration is attained by isenthalpic expansion via the Joule-Thomson effect (at constant enthalpy, a decrease in pressure will cause a drop in temperature).

Purge gas leaving the ammonia process is expanded to about 1,000 psi. At this pressure, the synthesis gas that will be recovered can be fed to one of the intermediate stages of the ammonia process' synthesis-gas compressor. This expansion also permits the use of brazed-aluminum heat exchangers that could not be used at a higher pressure.

Entering the argon recovery process, the purge gas passes through a molecular-sieve adsorber (see flowsheet). Here, in order to prevent plugging and freezing in the cold box, trace amounts of water and ammonia are removed. For this, two adsorbers are rotated: while one is in operation, the other is regenerated with methane that has been recovered.

The dried purge gas then enters the cold box, passing through a plate-fin exchanger with a counterflow design. In this aluminum heat exchanger, synthesis gas and methane leaving the process cool the purge to below $-200°F$.

Synthesis gas and methane also are reclaimed as part of argon recovery route

Investment	$4.6 million
Utilities	
Power	1,570 kW
Cooling water	160 gpm
Steam	680 lb/h
Makeup water	16 lb/h
Products	
Synthesis gas	817 lb-moles/h
Argon	29
Methane	64
Ammonia (3% in water)	666 lb-moles/h

Basis: 1,000-ton/d ammonia plant.
Purge-gas feed of 930 lb-moles/h.
1979 costs.

Originally published July 16, 1979

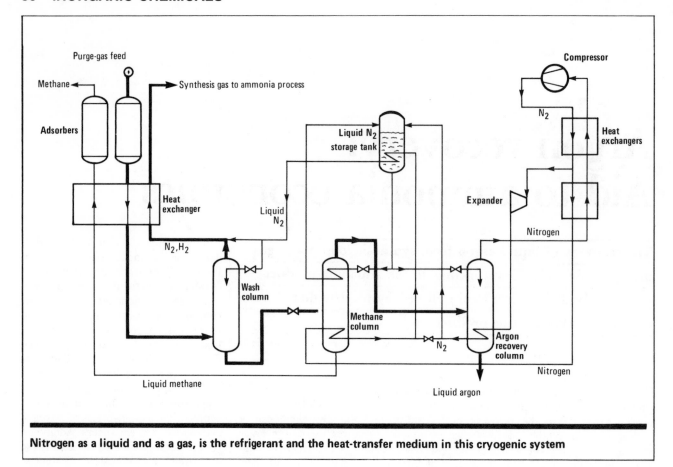

Nitrogen as a liquid and as a gas, is the refrigerant and the heat-transfer medium in this cryogenic system

After cooling, the gas enters a wash column, where liquid N_2 scrubs out argon and methane. The overheads—synthesis-gas components, N_2 and H_2—are recycled back to the ammonia process (after nitrogen from the refrigeration cycle is added to correct the composition).

The bottoms from the nitrogen wash are transferred to a fractionating column, where this liquid—consisting of Ar, CH_4 and N_2—is flashed from 1,000 to less than 100 psi in order to obtain cooling via the Joule-Thomson effect.

With distillation, liquid methane—which is later vaporized while cooling the purge-gas feed—is removed from the bottom of the column at a temperature below $-260°F$. A gaseous nitrogen-argon mixture is taken off as the overheads, and this mixture is transferred to another, colder, column for the separation of argon.

At approximately $-300°F$, pure argon is removed as a liquid from the bottom of the argon recovery column. The overhead of this unit is basically nitrogen, which is used in a refrigeration system.

HEATING AND COOLING—In a series of plate-fin, crossflow heat exchangers, heat is transferred between nitrogen leaving fractionation and nitrogen that has just been compressed. (The lower-pressure stream is warmed as the higher-pressure gas is cooled.)

In the compressor, N_2 is raised to a pressure above that of the purge-gas feed. After cooling with the lower-pressure stream, some of the N_2 is passed through a turbine expander.

Expanded nitrogen (below critical pressure and at approximately $-298°F$) condenses at the same time that it provides heat in the reboiler of the argon recovery column. The N_2 that bypasses the expander provides heat in the methane column by passing through its reboiler.

These two streams are combined, flashed and transferred to a liquid-nitrogen storage tank. This tank provides liquid for cooling in the condenser of the methane column and reflux for the argon recovery column. (In the tank, there is a coil to cool N_2—from the methane column reboiler—that will be added to the recovered synthesis gas.)

HIGH PURITY—All the streams recovered by this route have some economic value. The argon is 99.9995% pure, with less than a total of 4 ppm of N_2, H_2 and CH_4. The methane stream has a value as a fuel or a feedstock and contains about 98% CH_4, 1.25% Ar and 0.8% N_2. In addition, for a 1,000-ton/d ammonia plant (see table) the 817 lb-moles/h of synthesis gas that is reclaimed translates into an additional 90 tons/d of ammonia production.

In the table, estimated costs for the argon recovery system are presented, based on 1979 prices. It is important to note that a cryogenic system to recover hydrogen alone might have an investment cost a third of the figure of $4.6 million given for the argon system. It is the recovery and particularly the sale of the element that is the key to the economics of the process.

Although argon has general metallurgical applications and is employed as a filler gas for incandescent bulbs, transportation costs and an uncertain market may be significant economic factors. However if a large consumer of this element (e.g., a steel plant) were to be located nearby, the conditions for argon recovery might be ideal.

Improved reactor design for ammonia synthesis

By fully utilizing modern catalysts, the energy and cost of producing ammonia can be reduced. However, operating conditions and equipment must change.

Friedrich Förster, Uhde GmbH

☐ In the early 1960s, Uhde GmbH (Dortmund, West Germany) in cooperation with Chemie Linz AG, formerly Österreichische Stickstoffwerke AG Linz, developed a new ammonia reactor with indirect cooling between its three catalyst beds. Here, hot reaction gas leaving one catalyst bed is cooled to the inlet temperature of the next bed by unreacted synthesis gas.

This reactor, which has become known as the ÖSW ammonia converter, was first commissioned in 1965 in an ammonia plant in Ireland. Since then, seven reactors of this type, with capacities of up to 1,000 tons/d of ammonia have been in commercial operation.

These units were designed for a temperature of 420°C at the inlet to the first catalyst bed, and a temperature of 500°C at the outlet of the third. The ammonia concentration attained under these conditions is 16.7% when using a natural-gas-based synthesis gas with 3.4% NH_3 (Fig. 1).

Based on this proven design, development work was carried out with the aim of reducing energy and capital cost requirements by increasing the ammonia formation rate. Uhde has developed and now offers the ICR-25-type modified 3-bed reactor that can attain a concentration of 25% NH_3 at its outlet.

ECONOMIC EDGE—This increased ammonia formation rate with the new design allows a reduction in the required recycle gas rate. With the smaller recycle, there is a 45% reduction in the energy required for the recycle gas and refrigeration compressors compared with the older ÖSW design. Also, due to the lower flowrate, capital investment is 11% less for the synthesis loop and refrigeration unit: heat exchange surfaces and pipe cross-sections will be smaller.

This can be explained as follows: In the manufacture of ammonia, synthesis gas—a 3:1 mixture of hydrogen and nitrogen—is mixed with unreacted recycle gas and passed through a reactor. The ammonia formed (see flowsheet) is condensed from the stream leaving the reactor, and the unreacted gases are recycled.

If the reaction conversion of nitrogen and hydrogen is increased, there will be a higher concentration of ammonia in the gas leaving the reactor. As a result, for any specific ammonia production, the total feed to the reactor can be reduced. And since there will be less unreacted gas leaving the vessel, it is the recycle portion of that total feed that can be reduced.

OLD AND NEW—The improved performance offered by the ICR-25 reactor was accomplished by modifications in operating conditions. In theory, the ammonia formation rate that can be attained in each catalyst bed can be increased by reducing the temperature at the inlet to the individual beds.

However, there is a tradeoff. The lowest temperatures that can be used in practical operation depend on the characteristics of the available catalysts: Lowering the working temperatures will reduce catalyst activity and actually decrease formation rates. To

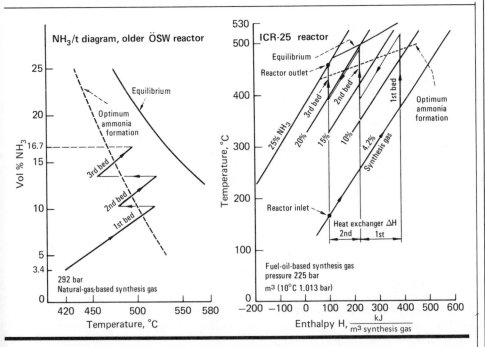

Equilibrium curves indicate advantage of improved reactor design   Fig. 1

Originally published September 8, 1980

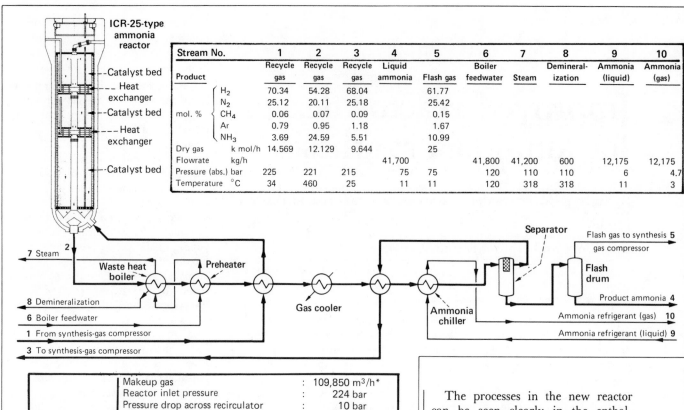

Stream No.		1	2	3	4	5	6	7	8	9	10
Product		Recycle gas	Recycle gas	Recycle gas	Liquid ammonia	Flash gas	Boiler feedwater	Steam	Demineral- ization	Ammonia (liquid)	Ammonia (gas)
mol. %	H_2	70.34	54.28	68.04		61.77					
	N_2	25.12	20.11	25.18		25.42					
	CH_4	0.06	0.07	0.09		0.15					
	Ar	0.79	0.95	1.18		1.67					
	NH_3	3.69	24.59	5.51		10.99					
Dry gas	k mol/h	14.569	12.129	9.644		25					
Flowrate	kg/h				41,700		41,800	41,200	600	12,175	12,175
Pressure (abs.)	bar	225	221	215	75	75	120	110	110	6	4.7
Temperature	°C	34	460	25	11	11	120	318	318	11	3

7 Steam
8 Demineralization
6 Boiler feedwater
1 From synthesis-gas compressor
3 To synthesis-gas compressor

Separator
Flash gas to synthesis 5
gas compressor
Flash drum
Product ammonia 4
Ammonia refrigerant (gas) 10
Ammonia refrigerant (liquid) 9

Ammonia chiller

Gas cooler

Waste heat boiler

Preheater

ICR-25-type ammonia reactor
- Catalyst bed
- Heat exchanger
- Catalyst bed
- Heat exchanger
- Catalyst bed

data	Makeup gas	: 109,850 m³/h*
	Reactor inlet pressure	: 224 bar
	Pressure drop across recirculator	: 10 bar
	Shaft power for recirculation ($\eta = 0.65$)	: 691 kW
	Shaft power for refrigeration	: 870 kW
	Steam production (115 bar)	: 41,200 kg/h
*m³ (0°; 1.013 bar)	Cooling-water consumption (25°C; $\Delta t = 10°C$)	: 1,800 m³/h

A fuel-oil-based 1,000-metric-ton/d ammonia plant using the new reactor design   Fig. 2

counteract this, the catalyst quantity can be increased slightly, since the gas at the inlet to each catalyst bed is not at equilibrium. (This will not significantly affect pressure drop through the reactor, because with increased formation, the total flow will be lower.)

Modern catalysts permit the temperature at the inlet to the first catalyst bed to be reduced to 380°C (compared with the 420°C of the old design). By reducing inlet temperatures to the second and third beds accordingly (see Fig. 1), NH_3 concentration can be increased.

The older ÖSW design was based on the fact that it is easier to optimize reactor conditions by dividing the catalyst into a greater number of beds. However, the number of actual beds was restricted for mechanical reasons. The catalyst inventory was divided into three sections (adiabatic reactions taking place in each) with heat exchangers between them.

In addition to the heat exchangers

between the beds, the older design has a third heat exchanger, located after the last bed. In this exchanger, the reaction gas is first cooled from 500°C before the gas is fed to a waste-heat boiler. Reduction of the temperature to 450°C or less was necessary to protect the boiler, in view of the experience with materials of construction for H_2/NH_3 gas mixtures at high pressures and high temperatures.

But as a result of the modified operating conditions for the ICR-25 reactor, the reaction gas leaves the third catalyst bed at a temperature of only approximately 460°C; in this case, the third heat exchanger can be omitted.

In principle, the major difference between the ICR-25 and the ÖSW is the enlargement of the individual catalyst beds and the absence of the last heat exchanger. Essentially, there is no change in the mechanical design, so that the reliability of the new unit is as good as that of the old, and testing of the new reactor is unnecessary.

The processes in the new reactor can be seen clearly in the enthalpy/temperature diagram. Fig. 1 shows such a diagram for synthesis gas on the basis of heavy fuel oil. The diagram contains a set of curves, $H = f(t)$, for the syn gas and the reaction gas produced as a result; and it also contains the equilibrium curve and the optimum curve for maximum reaction velocity.

The entire course of the reaction in the catalyst beds—including the two intermediate cooling steps—appears on the diagram as a plot that fluctuates around the optimum curve; the adiabatic, exothermal reactions are represented by a vertical line running upward at $H = $ constant. Also, the proximity of the temperature of the reaction gas to the equilibrium temperature can be seen as the difference between the temperature along the line $a_1 = $ constant.

Reginald Berry, Editor

Acknowledgment
The author wishes to acknowledge the kind assistance of BASF (Ludwigshafen) in computing the catalyst volumes required.

The Author
Friedrich Förster is a senior process engineer with Uhde GmbH, D-4600 Dortmund, a subsidiary of Hoechst AG, where he is in charge of R&D projects. For the past eleven years, he has been involved in the design and development of gasification, gas purification and gas conversion processes, and synthesis processes—e.g., for the production of ammonia and methanol. He holds an engineering degree (Dipl.-Ing.) from the University of Karlsruhe.

Grander process changes eyed by ammonia makers

Process designers, who have already made major cuts in ammonia-plant energy usage, continue to come up with new ways to make natural-gas-fed facilities more efficient.

☐ The ammonia industry is entering a period when even bigger process changes will be undertaken to achieve further 10-15% reductions in energy consumption. This is the assessment of György Honti, deputy general manager of the Hungarian Chemical Industry. Honti was the lead speaker at British Sulphur Corp.'s Fourth International Conference on Fertilizer Technology, held Jan. 19-21 in London. (For some time, of course, ammonia makers have been paring energy requirements—see *Chem. Eng.*, Jan. 29, 1979, pp. 54-56.)

Most opportunities for saving energy in the near term, Honti said, will occur in plants fed by natural gas (80% of world capacity), although partial-oxidation routes are promising greater feedstock flexibility (pp. 137–139). According to Honti, ammonia produced from coal still demands about 50% more energy than from gas.

Honti noted that in a natural-gas-based plant, energy consumption is typically 7.2-8.5 million kcal/metric ton of ammonia, and "any effective action to reduce energy consumption in the short term must be achieved principally through improvements in the methane steam-reforming process." Some 55% of the total work loss occurs in the primary reformer.

NEW FLOWSHEETS—To cut such loss, process designers are investigating a number of options, including basic flowsheet changes.

For instance, efforts now in progress at Montedison S.p.A. to revamp the ammonia flowsheet were detailed by Ludovico Mariani, process supervisor for inorganic chemicals for the firm's engineering subsidiary, Tecnimont. Montedison and Tecnimont have taken a multipronged approach that Mariani says will reduce energy requirements by about 20% over existing designs—to around 6.7 million kcal/m.t. of ammonia.

"Montedison has worked out a flowsheet," Mariani reported, "in which the fuel requirement of the primary reformer is reduced by using high-level waste heat in [an added] tertiary reformer. The lower output of steam is compensated for by using a lower-pressure synthesis loop with absorption refrigeration, which substantially cuts energy requirements."

Mariani also noted that Montedison will soon start up two full-scale 300-m.t./d urea trains at San Giuseppe di Cairo, based on the new IDR (Istituto Donegani di Ricerche) process. The process is said to achieve, through isobaric recycling, high conversion rates and energy savings of up to 40% over conventional urea processes.

Another revised flowsheet, known as LEAD (Low Energy Ammonia Design), now being offered was discussed by Humphreys & Glasgow Ltd. (London). Its process group head, Frank C. Brown, estimates that LEAD could reduce energy input to around 7.0 million kcal/m.t. The H&G refinements were arrived at, says Brown, by stipulating that the design not use "any technological novelty likely to require plant-scale testing."

Among other things, the H&G design calls for high preheating of the reactants, and a close temperature approach at the reformer exit. Hydrogen usage is raised to 99% efficiency by purifying the purge gas cryogenically and recovering hydrogen for recycle. The process also uses a new converter (stemming from ICI technology) that enables high conversions at lower pressure.

H&G is also applying for patents on a physical solvent process for absorbing CO_2 following the carbon monoxide shift reaction. Such physical solvent techniques are said to enable substantial energy savings over conventional absorption units. Keith J. Stokes, of James Chemical Engineering Co. (Greenwich, Conn.), estimated that savings of up to 4 megajoules/kg of ammonia are possible.

NEW LOOK FOR REACTORS—Tinkering with synthesis and shift-reactor designs is another energy-saving measure being pursued. A novel synthesis-reactor design, known as ACAR (Ammonia Casale Axial Radial), was described by Ettore Comandini, technical director of Ammonia Casale (Lugano, Switzerland).

The radial design of the fixed-bed catalytic reactor is said to offer less pressure drop to gas flow than more-conventional axial designs. Ammonia Casale claims to have eliminated the tendency of the gas flow to bypass the upper zone of catalyst in these reactors by using an improved distribution system. Patents have been applied for, and commercial-size units have been installed in three 1,000-m.t./d plants.

Haldor Topsoe (Lyngby, Denmark) has also modified its reactor systems. The company's manager for general catalyst research, Ib Dybkjaer, said that by introducing improved catalysts and more-efficient process design, the firm has boosted energy efficiency from 50% in 1973 to 63% in 1980."

COLD RUSH—Other opportunities for energy conservation are being realized through recovery of hydrogen and inert gases from the purge-gas stream, typically using add-on cryogenic or membrane-absorption units (for instance, Monsanto's PRISM gas separators).

Among the firms offering such units is Petrocarbon Developments Ltd. (Manchester, U.K.). Greg J. Ashton, a senior process engineer for Petrocarbon, estimated that at current prices for raw materials and ammonia, a cryogenic unit "pays back in about one year when the recycle hydrogen is used to make more ammonia." About 55 additional short tons of ammonia per day can be made in a 1,150-ton/d plant, he noted.

Other credits can be realized by recovering inert gases—principally argon (*Chem. Eng.*, July 16, 1979, p. 62). Ashton noted that Petrocarbon last year acquired rights to a cryogenic argon-recovery technique developed by East Germany's Chemieanlagenbau (Dresden). At current prices, he said, the payback for a unit in the U.S. can be as short as two years.

James H. Mannon

Originally published February 23, 1981

Hollow fibers recover hydrogen

Gas-stream compositions can be adjusted, or pure gases recovered, via a new unit operation that uses partial pressure as the driving force for mass transfer.

D. L. MacLean and T. E. Graham, Monsanto Co.

☐ A new method for continuously separating hydrogen from gas streams has been developed, based on semipermeable membranes. Devices designed by Monsanto Co. (St. Louis), called Prism separators, can attain hydrogen purities of up to 99% (see *Chem. Eng.*, Dec. 3, 1979, p. 43).

Large-scale industrial applications of semipermeable membranes are quite recent. In 1971, *CE*'s Kirkpatrick award (see *Chem. Eng.*, Nov. 29, 1971, p. 54) was given for the development of hollow-fiber reverse-osmosis separators for water desalting. Now, Monsanto has put into commercial operation gas-separation systems that employ hollow fibers made of polysulfone covered with a proprietary coating.

These fibers have a high permeability rate, exhibit a long operating life with little change in performance, resist impurities such as H_2S, H_2O and NH_3 (see table), and can function at pressures of up to 2,150 psi.

The separation system, which employs differential partial pressure as its driving force, is versatile: It can be used to adjust composition (e.g., a H_2/CH_4 ratio) as well as recover a gas. The process can operate at pressures as low as 150 psi and temperatures from 0 to 40°C, and it can operate even if the feedstream pressure fluctuates rapidly.

There are other advantages: Pressure drop through the separator for the nonpermeating components of the feed is less than 10 psi; the system has no moving parts; and where sufficient pressure is available, the only utilities required are for instrumentation.

Monsanto has used Prism separators in three of its facilities since 1977: In its Texas City, Tex., petrochemical complex, where the ratio of H_2 to CO is adjusted for methanol and acetic acid production; in its Pensacola, Fla., facility (since 1978), where 99% H_2 is recovered from a hydrogenation plant purge steam; and in Luling, La., since 1979, where hydrogen is recovered from an ammonia synthesis-purge loop (see photo).

THROUGH THE TUBE—Basically, the separators, which resemble shell-and-tube heat exchangers, contain a bundle of the hollow fibers that are capped on one end and have an open epoxy tube-sheet on the other (see figure). (The vessels operating in Monsanto's industrial plants are 4 in. or 8 in. dia. by 10 ft long; 4-in.-dia. x 20-ft separators have been successfully field-tested.)

For each vessel, there may be as many as 10,000-100,000 fibers in a bundle. And the fibers, which are 800 μm dia. or less, have a hollow center up to 400 μm in dia.

The separators are oriented vertically, with the tube sheet down. The feed gas enters the shell side. Hydrogen permeates from the higher-pressure shell side through the walls of the fibers and passes downward through the lower-pressure hollow center.

Composition of the product stream can be controlled by adjusting bore (tube) pressure and/or the feedgas flowrate. The basic equation for flow of gas, i, through membranes is:

$$Q = (P/l)_i A (P_s Y_{si}) - P_B X_{Bi})$$

where $(P/l)_i$ is a permeability rate coefficient of component i for transport across the membrane; Q_i is the molar or standard volume flowrate; A is the membrane area; $P_s Y_{si}$ is the shell partial pressure of i; $P_B X_{Bi}$ is the bore partial pressure of i.

Semipermeable membranes recover 90% pure H₂ for an ammonia plant

Originally published February 25, 1980

42

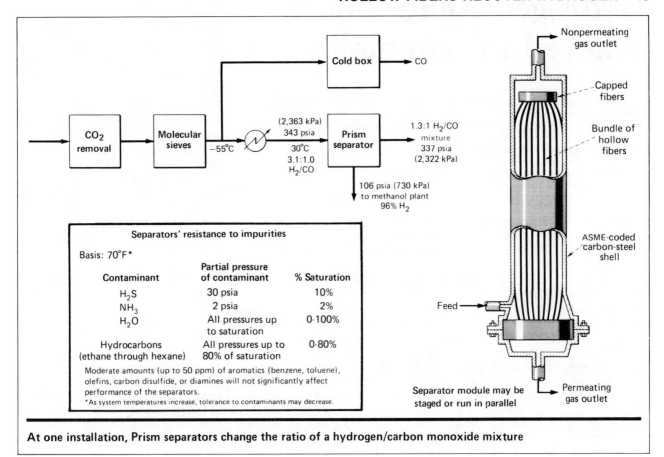

Separators' resistance to impurities

Basis: 70°F*

Contaminant	Partial pressure of contaminant	% Saturation
H_2S	30 psia	10%
NH_3	2 psia	2%
H_2O	All pressures up to saturation	0-100%
Hydrocarbons (ethane through hexane)	All pressures up to 80% of saturation	0-80%

Moderate amounts (up to 50 ppm) of aromatics (benzene, toluene), olefins, carbon disulfide, or diamines will not significantly affect performance of the separators.

*As system temperatures increase, tolerance to contaminants may decrease.

Separator module may be staged or run in parallel

At one installation, Prism separators change the ratio of a hydrogen/carbon monoxide mixture

The separation of one component from another is determined by the rate at which the two substances will permeate through the membrane. (The ratio of two permeability-rate coefficients is defined as the separation factor $a^i_j = (P/l_i / P/l_j)$. If one substance permeates much faster than another, the separation can be made. Hydrogen (and helium) will permeate through the polysulfone fibers much faster than methane, carbon monoxide, oxygen, nitrogen, and aliphatic hydrocarbons.

Since Prism separator modules can be staged or run in series or in parallel, the device can be used to attain high purities or to adjust the composition of synthesis gas or reactor feeds.

ONE APPLICATION — Monsanto's Texas City plant had a need for additional pure carbon monoxide feedstock for acetic acid production. Although ample CO was available in the form of a 3.1:1.0, hydrogen/carbon monoxide mixture, pure CO supply was bottlenecked at the existing cold box (Fig. 2), and a significant part of this carbon monoxide was used to produce an approximate 1.3:1 H_2/CO gas for oxo alcohol manufacturing.

Here, Prism separators were used to produce the oxo-alcohol feedstock directly from the 3.1:1.0 H_2/CO ratio stream, freeing the cold box to produce only the pure CO needed for use in making acetic acid.

Using a 337-psia (2,322 kPa) shell pressure and 106-psia (730 kPa) hollow-fiber bore pressure, the Prism separators changed the H_2/CO ratio from about 3.1:1 to about 1.3:1, while recovering 93% of the CO. Ratios lower than 1.3:1—even lower than 1:1—could have been achieved by the proper choice of bore pressure or separator area.

Hydrogen purity of the permeate has been 96%. And over the three-year period that the separators have been in operation, there has been less than a 10% reduction in the permeability rate coefficient and separation factor.

EASILY ADDED — Because of the retrofit nature of most applications, Prism separators are supplied as shop-constructed skid-mounted units that can be installed on a concrete mounting pad. This reduces installation costs and space requirements.

And unless it is necessary to pretreat the feed gas to add compression or to

remove excess amounts of contaminants, operating costs for the separator would be minimal, restricted to instrumentation requirements and a part of one operator's time.

Total investment cost is very dependent on the application. Feed flowrate and pressure and the amounts of contaminant present, also the required hydrogen purity and recovery, will determine a cost that may vary from $0.5 million to $5 million.

This technology is competitive with other hydrogen-recovery processes (e.g., those based on cryogenic or molecular-sieve technology). And the Prism separator installation can easily be expanded by adding new modules. However, at hydrogen concentrations of less than 30%, hydrogen recovery may not be practical.

Reginald Berry, Editor

The Authors

Tommy Graham is manager, engineering and manufacturing, Separations Business Group of Monsanto Co. (Research Triangle Park, N.C.). He received B.S. and M.S. degrees in chemical engineering from Oklahoma State University.

Donald L. MacLean is engineering manager of the Separations Business Group of Monsanto Co. He has a B.S., an M.S., and a Ph.D. in chemical engineering from the University of Tennessee.

Hydrogen routes' future is keyed to economics

While thermochemical methods remain a distant possibility, electrolysis is emerging as a contender for small-volume markets. Partial oxidation and coal gasification also are seen as important future sources of hydrogen.

☐ After looking at the potential of some esoteric hydrogen-production methods in the 1970s—e.g., thermochemical, electrolytic, hybrid (*Chem. Eng.*, Mar. 14, 1977, p. 86)—researchers now are more preoccupied with determining the economics and possible markets for all available hydrogen routes. The current consensus: Although natural-gas reforming remains the most attractive technology, electrolysis will find specific uses in small-volume applications. Gasification of coal and partial oxidation of residual oil also will become viable

alternatives in some cases. As for thermochemical and hybrid routes, much development work remains to be done.

Reforming, of course, requires the lowest investment and yields the cheapest hydrogen (see Table I), while the investments for partial oxidation and coal gasification are approximately three and five times greater, respectively. Electrolysis depends on the cost of electricity, which can run to more than four times the cost of natural gas.

Yet users of hydrogen as an inter-

mediate for ammonia and methanol are increasingly turning to non-natural-gas sources, perhaps partly spurred by fears about the price and availability of natural gas. Choice of raw material and process is becoming more important, in the light of optimistic projections for growth in hydrogen demand. Indeed, according to the National Research Council (Washington, D.C.), industrial requirements for hydrogen "are expected to double or triple by the year 2000," from a current base of about 3 trillion ft³/yr.

The availability of cheap residual oil (*Chem. Eng.*, Oct. 8, 1979, p. 57) has sparked a number of partial-oxidation projects in the U.S. and elsewhere. Meanwhile, coal-to-ammonia technology, while not flourishing yet in the U.S., is making its mark in such countries as India, South Africa and Turkey. As for coal to methanol, a number of U.S. projects are in the cards (*Chem. Eng.*, Apr. 7, p. 43).

COST IS PARAMOUNT—In recent years, economic evaluations of hydrogen routes have been issued by such entities as Exxon, Pullman Kellogg, Booz, Allen and Hamilton, and the National Aeronautics and Space Ad-

Comparison of hydrogen production alternatives — Table I

	Steam reforming	Coal gasification	Electrolysis	Thermochemical decomposition
Appropriate overall thermal efficiency	70%	60-65%	32% (SPE) 21-25% (KOH)	Upper limit: 55%
State of the art	Well known technology	Mature technology available	Proven, reliable technology	Research stages, not a proven technology
Environmental effects	Depletion of natural gas and other light hydrocarbons	Impacts of coal mining; air pollution likely to be less than via electricity from coal	Pollution problems with electricity generation	High efficiency means less resource use. Possible release of harmful chemicals
Advantages	Currently the cheapest method in the U.S.	Cheapest and most secure near-term alternative to methane reforming; abundant coal reserves in U.S.	Small plant size, can use non-fossil sources	Can use non-fossil sources
Disadvantages	Scant longterm potential due to limitations on methane supply	Ultimate limitation is exhaustion of coal resources, requires large plant size	High cost, low net energy efficiency	Materials problems in reactant containment; large complex plant expected
Estimated hydrogen production costs*, $/10⁶ Btu product (1980 $)	6.80	9.50	19.65	

*Figures are based on an 880-million-std-ft³/d plant on the Gulf Coast; 100% equity financing; and a 10% discounted cash flow return. Electricity is priced at 3¢/kWh, coal at $0.96/10⁶ Btu, and natural gas at $3.15/10⁶ Btu.

Source: National Research Council

Originally published July 14, 1980

U.S. industrial-hydrogen and hydrogen-feedstock needs   Table II

	1975	1980	1985	1990	2000
Hydrogen required, quads/yr					
For oil refining	0.15	0.21	0.28	0.38	0.56
For ammonia	0.35	0.46	0.55	0.66	0.90
For methanol	0.06	0.08	0.11	0.14	0.23
For all small uses	0.02	0.03	0.04	0.05	0.08
Total	0.58	0.78	0.98	1.23	1.77
Natural-gas feed required, k bbl/d crude equivalent					
Hydrogen for oil refining	106	147	197	267	394
Ammonia	268	357	427	511	700
Methanol	63	87	118	155	251
Hydrogen for all small uses	14	21	28	35	56
Total	451	612	770	968	1,401

Source: Exxon Research and Engineering Co.

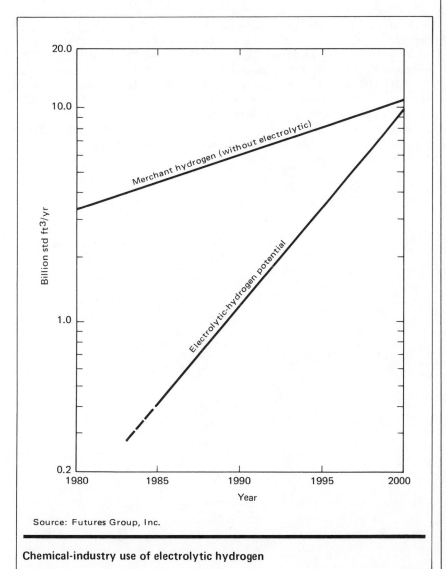

Chemical-industry use of electrolytic hydrogen

Source: Futures Group, Inc.

ministration's Jet Propulsion Lab, the Institute of Gas Technology, and the U.K.'s National Coal Board. Experts agree that specific electrolytic processes based on photovoltaics and ocean thermal energy conversion have the potential for producing electricity very cheaply. But for electrolysis in general, "the most important need is for a reduction in energy investment," says the National Research Council.

Outside the U.S.—in such countries as Norway, Iceland and India—cheap hydroelectric power is being used to operate large hydrogen facilities. (Brown, Boveri & Co., Baden, Switzerland, is currently building an electrolysis plant at Aswan, Egypt, that will produce 36,000 m^3/h of hydrogen for use in fertilizers.) Elsewhere, water-cracking technology is the subject of various R&D programs, including one begun last year by the European Commission. It will run four years and cost more than $9 million. Japan also plans to spend money on several such projects.

So far, the U.S. has had no hydro-electric-powered units in operation. But at presstime, the U.S. Dept. of Energy (DOE) was getting ready to approve funding for a $150,000 feasibility study that will take place at Potsdam, N.Y. Ultimately, the agency plans to spend $3-$5 million on a project to supply hydrogen for the local (within a 50-mi radius of Potsdam) merchant market.

DOE also is backing General Electric Co.'s Solid Polymer Electrolyte (SPE) technology, which is billed as an advanced, high-efficiency electrolysis process boasting an efficiency of 90%, vs. 60-70% for conventional electrolysis. Last April, GE started up a facility in Wilmington, Mass., capable of producing 2,140 std ft^3/h. According to DOE officials, this is the minimum size for a plant that would supply commercial users.

CUTTING ENERGY NEEDS—Researchers at the University of Connecticut are using coal to cut energy requirements for cracking water. The coal supplies some of the energy spent in electrolyzing a coal/water slurry; the process, which takes place at room temperature, uses only half of the energy needed for conventional electrolysis, says researcher Robert W. Coughlin. In addition, the technique can be employed in advanced electrolyzers—e.g., the SPE. Pure hydrogen is generated at the cathode, and carbon oxides at the anode. Sulfur and ash, present in coal, do not contaminate the gaseous product.

A route that uses biomass materials—corn wastes, sugar-cane bagasse, Douglas-fir wood pulp—was described at the 2nd International Conference on Alternative Energy Sources, held last December in Miami, Fla. As explained by A. J. Darnell, of Rockwell International's Energy Systems Group (Canoga Park, Calif.), biomass and water are reacted with bromine at 250°C. Resultant HBr is electrolyzed into hydrogen and bromine; the latter is recycled.

The process can produce about 50,000-60,000 std ft³ of gas from one ton of biomass (on a dry, ash-free basis). Overall thermal efficiency is 67% (including electricity generation), vs. 20-30% for other electrolysis methods. Although these figures are based on bench-scale tests, "it looks like the process can compete with hydrogen-from-coal via water-gas shift reaction," says Darnell.

SMALL NICHE—Some observers believe that electrolysis may be most successful in making hydrogen for the merchant market. A study on the potential of electrolytic hydrogen, completed last year by the Futures Group, Inc. (Glastonbury, Conn.) for the Electric Power Research Institute (Palo Alto, Calif.), found that when hydrogen demand is under 100 million std ft³/yr, capital costs are lower for electrolyzers in comparison with steam reformers. The lower investment offsets the difference between the price of electricity and that of natural gas.

Based on a 33-million-std ft³/yr facility, total cost of hydrogen from an electrolyzer was estimated at $5.51/1,000 std ft³, vs. $8.54 for hydrogen from steam reforming.

The Futures Group report sees a market for hydrogen in pharmaceuticals, chemical synthesis, fats and oils hydrogenation, and in metals, electronics and the float-glass industry. Potential demand for electrolytic hydrogen is seen as rising from less than 1 billion std ft³/yr now to 16 billion by the year 2000.

The chemical industry, which the study says may be the first to adopt electrolysis, will account for 9.7 billion std ft³/yr of the total, pharmaceuticals will consume 0.24 billion, and fats and oils 0.93 billion. All projections are based on the commercialization of an advanced high-efficiency process such as GE's SPE.

IN REVERSE—Another process, which is claimed to be competitive with electrolytic hydrogen, may soon enter the pilot-plant stage. It is based on a reversal of normal synthesis—namely, methanol is steam-reformed to generate hydrogen. Jointly owned by Belgium's Catalysts and Chemicals Europe and France's Société Chimique de la Grand Paroisse, the technique is said to obtain gas product at a cost less than that of bottled merchant hydrogen.

A water/methanol feed is vaporized and reformed in a tubular reactor at temperatures of 280-300°C, vs. 900°C for normal reforming. After removal of condensables, there is hydrogen purification by pressure-swing adsorption. The process is said to require no process heat after startup; startup heat itself can be generated by burning methanol.

The developers believe that the route would be most economical for plants in the 250-m³/h range. Cost of 100 Nm³ of gas product (99.99% pure at 25 bar and 35°C) would be $25.86, based on 1980 dollars. The Belgian and French firms plan to spend $700,000 to build a 100-m³/h pilot plant.

Reginald I. Berry

Metal hydrides selectively remove H_2 from gas streams

Tested in a pilot facility, a new system shows promise for commercially recovering hydrogen from off-gases at purities of 99 mol%. The system uses the ability of the hydride crystals to absorb and desorb H_2 under different operating conditions.

Leonard J. Kaplan, Assistant Editor*

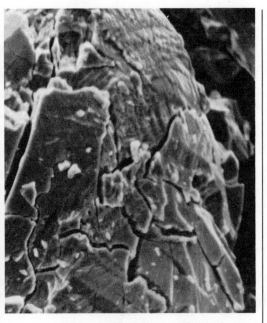

Pilot system (below) uses pellets made from metal hydride crystals (above) to remove H_2 from ammonia purge stream

☐ Air Products and Chemicals, Inc. (Allentown, Pa.) and Ergenics Div., MPD Technology Corp. (Wyckoff, N.J.) have jointly developed a cyclic process for separating H_2 from industrial gas streams (e.g., ammonia process purge-gas) that uses reversible metal hydrides. According to Joseph G. Santangelo, director of new business development at Air Products, a patent application on the process has been applied for.

In part an outgrowth of work done by Air Products on developing a method for storing H_2 for motor vehicle fuel under a U.S. Dept. of Energy contract, the process employs beds of metal hydride pellets that use the heat of reaction from H_2 absorption to provide the energy required for the desorption regeneration.

The immediate goal of Air Products is to take recent pilot-plant results and commercialize the system for use on so-called clean stream applications—gas streams that do not contain contaminants that would poison the metal hydride absorption pellets. Each application will require the development of hydride pellets that will handle gas

* Technical data and process information were taken from a paper entitled "Hydrogen Separation From Mixed Gas Streams Using Reversible Metal Hydrides" by J. J. Sheridan, III, F. G. Eisenberg, E. J. Greskovich of Air Products and Chemicals, Inc. (Allentown, Pa.); G. D. Sandrock of Inco Research and Development Center, Inc. (Suffern, N.Y.); and E. L. Huston, Ergenics Div., MPD Technology Corp. (Wyckoff, N.J.).

streams having specific chemical contaminants.

PILOT UNIT—In operation since June 1981 at Air Products' New Orleans ammonia plant, a pilot unit has reportedly operated satisfactorily, removing H_2 (99 mol% purity) from a 5,000-ft³/d slip stream of the ammonia purge gas. The feed contained 60% H_2, 1-3% ammonia, with a remaining composition of methane, nitrogen and argon. No pretreatment to remove the ammonia was found necessary.

The pilot unit, made up of three pellet beds containing lanthanum nickel ($LaNi_5$), was tested to see how the system's performance and cycle stability would be influenced by parameters such as feedrate (50-150 kg-moles/h m²); pressure (1-4 MPa); percent H_2 recovery (75-95%); cycle time (5-25-min sequence); feed gas and column temperatures; metal hydride type; ammonia concentration in the feed. The fully automated system is said to have operated with no measurable decline in performance.

A NEED FOR H_2 SEPARATION—In a presentation at the American Chemical Soc. meeting held in Las Vegas in April (*Chem. Eng.*, May 17, 1982, p. 51), Santangelo said that hydrogen recovery is becoming of greater interest to many petroleum refineries, and ammonia and petrochemical producers, because of the escalating cost of natural gas, which is used to make H_2 by steam reforming, and the increasing demand for H_2 in petroleum refineries. The greater demand for H_2 in refineries, he went on to say, was in part due to the increased processing of sour crudes in place of sweet crudes, and consumer demand patterns requiring greater yields of light products from each barrel of crude.

Some methods for recovering H_2 from gas streams:

■ *Cryogenic separation*—This classic route for H_2 recovery uses cryogenic temperatures to produce phase changes. The most common applications of this technology include H_2/CO separation and H_2 recovery from purge gas streams from ammonia or toluene hydrodealkylation plants.

■ *Molecular-sieve pressure swing adsorption*—The major applications for this process, developed by Union Carbide, are the purification of (a) H_2 produced in steam reformers, and (b) the H_2 byproduct from naphtha catalytic reformers.

■ *Hollow-fiber separators*—Developed by Monsanto (PRISM separators), this relatively new technology uses hollow-fiber membranes to selectively recover H_2, and has gained acceptance with ammonia producers.

Santangelo believes that each of the above technologies offers both advantages and disadvantages for particular applications. He feels that the new hydride technology will eventually be attractive because (1) no pretreatment of the gas is necessary for ammonia applications, (2) it can work on streams with low H_2 concentrations (possibly as low as 15%, and (3) the estimated operating cost will be reasonable, approximately $0.30-1.50/1,000 std ft^3 of H_2.

HYDRIDE TECHNOLOGY—The new process uses the reversible and selective capability of metal hydrides, shown in the following equation:

$$M + x/2H_2 \rightleftharpoons MH_x + Heat$$

where M represents metal hydrides such as $LaNi_5$, $FeTi$ and Mg_2Cu.

Pellets are formed from the metal

Representative H_2 mass balance of three-column system

	Feed	Byproduct offgas	H_2 product
Temperature (°C)	20.0	50.0	23.0
Composition (mol %)			
H_2	60.0	13.2	99.0
Others	40.0	86.8	1.0
Flowrate (lb-moles/h)	0.55	.25	.30

hydride and a proprietary inert metal (ballast) powder. The components are mixed, compressed and sintered in an inert atmosphere, then crushed to produce irregular-shaped pellets of narrow size range (during use, the pellets exhibit acceptable particle size stability). The ballast material acts as a sink, which stores the heat of absorption. Subsequently, this stored heat is used for the desorption step when the pellets' operating parameters are changed.

Originally, the hydrogen capacity of these pellets was reported to be 76% of theoretical (about 1 g of H_2/100 g of metal hydride), but recent new processing procedures that reduce the oxygen content of the powders before the sintering step now yield capacities in excess of 85%.

In simple terms, the hydride process operates by passing a feed gas containing H_2 through a column packed with the absorbent pellets. But, according to the developers, successful operation of this system depends on the conditions under which the absorption and desorption take place. One alternative would be to use a thermal-swing operating cycle, in which the desorption of the hydrogen is accomplished by increasing the temperature of the pellets (by the addition of heat). The developers of the system rejected this alternative because (1) a significant amount of energy would be needed to desorb the hydrogen (about 5-15

kcal/mol of H_2) from the pellets, and (2) the rates of absorption and desorption would be controlled by the heat-transfer characteristics of the absorber system, leading to a need for what was considered to be complex and expensive hardware.

ADIABATIC PROCESS—To overcome these problems, an adiabatic, pressure-swing process was developed, in which the heat of reaction from the hydrogen absorption step is retained in the composite absorbent to (1) effect desorption of the hydride, and (2) achieve what are said to be high rates of adsorption and desorption.

During the absorption step, the hydride absorbent undergoes an adiabatic temperature rise as a consequence of the bed retaining the heat of adsorption. The H_2 desorption from the pellets is accomplished by (1) stopping the flow of feed to the column, and (2) reducing the column pressure to a prescribed value. The energy retained in the pellets supplies the heat necessary for the desorption.

While Air Products will not reveal the particular operating conditions for its pilot plant, it did provide the H_2 mass balance for a test run through the facility (see table).

No economic figures are yet available for a full commercial system, since, according to Santangelo, accurate evaluations cannot be made from the data obtained from the relatively small pilot unit.

Progress on the hydrogen front

Making hydrogen on a large scale by the solar splitting of water is still considered futuristic by most energy experts (*Chem. Eng.*, July 26, p. 17), but several new developments may well hasten the maturity of the technology. In one case, the developer of a closed-loop solar energy conversion system has begun a cooperative program with the U.S. Dept. of Energy to build prototype units.

On a more basic level, two researchers, working independently, say they've made water-splitting electrodes with new highs in performance and efficiency. But claims by these scientists that their achievements represent breakthroughs are disputed by many in the field.

The subject of the DOE agreement is a combination photovoltaic/electrochemical technology first announced by Texas Instruments, Inc. (Dallas) in early 1979. Since then, TI says, it has plowed in "millions of dollars" to demonstrate the feasibility of the system, develop processes and components, and obtain adequate patent coverage.

The goal of the $18-million four-year TI/DOE effort is to come up with workable commercial-scale devices and systems. TI, which will contribute $4 million toward the program, says that "several years" and a host of engineering and economic studies will be needed before any product reaches the marketplace. Economic projections show that the system "could compete effectively as a principal source of energy for residences and small commercial establishments," says E. L. ("Pete") Johnson, director of TI's solar energy project.

CLOSED CYCLE—The TI system has three basic components. The first consists of silicon solar cell, immersed in a halogen acid, which converts sunlight to an electric current. This current electrolyzes the halogen acid, typically hydrobromic, releasing hydrogen and bromine. The second component is a device for storing the hydrogen and bromine. (The hydrogen is stored in the form of metal hydrides and the bromine is held in solution.) Meanwhile, the solar-heated electrolyte is passed through a heat exchanger, where useful energy is extracted from it. The third

compenent is a fuel cell in which the hydrogen and bromine are reacted to generate an electric current. The HBr thus produced is returned, along with the cooled electrolyte, to the solar cell chamber, where it begins the cycle anew.

While the TI concept is self-contained, several recently disclosed electrodes have potential as components in a wide range of systems. One of the newly announced water-splitting electrode systems is made of ordinary iron oxide. (Previously, rather expensive materials such as platinum or gallium arsenide have been used.) Its developer, Gabor Somorjai, chemistry professor at the University of California (Berkeley), told last September's American Chemical Society meeting that an assembly of the two iron oxide electrodes can catalytically photodissociate water into hydrogen and oxygen without any application of external electrical potential (as is sometimes required with photolytic cells).

Somorjai's electrodes consist of two disks of about 1 cm^2 each. One is doped with magnesium to create a "p-type" semiconducting device; the other is doped with silicon, resulting in an "n-type" counterpart. When connected to an external circuit, immersed in water containing a bit of sodium sulfate, and exposed to sunlight, the device electrolyzes the water. The yield from the two disks, Somorjai told the ACS session, is about 10^{15} molecules of hydrogen per minute. (This is considered a high yield by most standards.) Moreover, he added, the system produces the gas continuously for eight hours without slowing down. And the catalytic effect of the electrodes can be prolonged indefinitely, Somorjai reported, by bubbling air through the electrolysis mixture.

Meanwhile, investigators at Texas A & M University last month unveiled another relatively efficient photolytic electrode system. The researchers, working under the direction of chemistry professor John O'M. Bockris, say their system is based on silicon materials. Although silicon in water rapidly becomes coated with a nonconducting oxide, the problem can be partially overcome by doping the anode with phosphorus and the cathode with

boron. Bockris says his group has improved efficiencies of the silicon electrodes further by doping the anode with an undisclosed material and coating the cathode with platinum.

According to Bockris, when these two treated electrodes are hooked up into a circuit, immersed in dilute sulfuric acid, and exposed to visible light, they electrolyze the water with an efficiency of 10-13%—a remarkably high level for photolysis cells. The only catch is that an external electrical potential must be supplied to the system. Nonetheless, Bockris asserts that his technology could be used to produce hydrogen at a cost equivalent to gasoline that now sells for $1/gallon.

CONTRASTING VIEWS—Somorjai and Bockris are excited about the commercial potential of their electrode systems—an excitement not shared by some other researchers in the same field. While Somorjai concedes that more work must be done to study the effects of doping levels, pH, and electrode geometries on hydrogen yield, he adds enthusiastically: "We have shown that we can continuously produce energy-rich molecules from simple, plentiful and cheap materials, using only the power of sunlight." And Bockris labels his achievement a "breakthrough" that could lead to widespread use of hydrogen as a fuel.

Since the two electrode announcements were made, however, some specialists have expressed skepticism, questioning, for example, whether the Berkeley system is truly catalytic, or whether a cell with even the 12% sunlight-to-hydrogen conversion efficiency of the Texas A & M system could be incorporated into an economical package.

One specialist in photolytic decomposition of water, Adam Heller of Bell Laboratories (Murray Hill, N.J.), cautions that units such as the recently disclosed electrodes, while promising in the long run, may not be developed into viable commercial energy producers until well into the 21st century. And Massachusetts Institute of Technology hydrogen investigator Mark B. Wrighton says the two electrode announcements are impressive, but they are still "very far away" from practical application.

Originally published November 29, 1982

Catalytic purification of diaphragm-cell caustic

A relatively simple system removes sodium chlorate from caustic produced by diaphragm cells. Since this process generates some NaCl, the treated caustic is best suited to applications where sodium chloride concentration is not critical.

Lawrence C. Mitchell and Michael M. Modan, General Electric Co.

☐ One of the undesirable byproducts of diaphragm chlor-alkali cells is sodium chlorate. Until recently, the process of choice for removing chlorate has been ammonia extraction. Now a technique developed by General Electric Co., employing catalytic reduction with another cell byproduct—hydrogen—is available. The process can convert up to 99% of the $NaClO_3$ to NaCl.

In the U.S., virtually none of the chlorine caustic plants completed in the last five years use the diaphragm-cell alternative—mercury cells. To date, about three quarters of U.S. plants use diaphragm cells. However, unlike mercury cells, diaphragm cells always produce, as a byproduct, some sodium chlorate.

The presence of this material severely limits the commercial applications of caustic. The high reactivity of sodium chlorate makes it an undesirable ingredient in the production of many industrial chemicals. Only very low levels can be tolerated in the manufacture of sodium sulfate, sodium hydrosulfite, glycerin and so on (see table).

In a chlor-alkali plant, chlorate removal can best be positioned downstream from the evaporators, so that sodium hydroxide will be purified prior to its use or sale, but if the plant operator wishes to protect his evaporators from chlorate corrosion, he can place the system between the electrolytic cells that produce caustic and the evaporators that concentrate it.

CATALYTIC CONVERSION—Operating continuously since 1977 in General Electric's chlor-alkali plant in Mt. Vernon, Ind., the chlorate removal process has been successfully treating the plant production of caustic. Approximately 85 to 97% of the $NaClO_3$ present is being converted to NaCl.

In the process (see flowsheet), raw caustic is pumped to a feed/product interchanger, an economizer that takes heat from the caustic leaving the reactor, to preheat the feed. Assuming an ambient-temperature feed, about 90% of the heat required by the system can be recovered in this exchanger, with only 10% to be made up through a small feed-heater. Caustic leaves this heater at above 250° F.

The heated caustic enters the top of the nickel-clad reactor. This vessel is relatively compact (e.g., its diameter is less than 6 ft. for a throughput of 170 tons/d).

Caustic trickles over the catalyst, which is composed of a proprietary active ingredient deposited on granular material that is 4 x 8 mesh. Since the bed is not flooded with liquid, there is little pressure drop across it. As caustic percolates through, chlorate reacts with hydrogen off-gas from the chlor-alkali cells.

Originally published February 26, 1979

Low-chlorate caustic is required for the manufacture of industrial chemicals

Chemical	U.S. annual capacity
Sodium sulfate	2,000,000 tons
Glycerin (synthetic)	157,500 tons
Pentaerythritol	100,000 tons
Sodium hydrosulfate	51,000 tons
Other chemicals and applications	
Trimethylol propane	Alkyl sulfonates
Sodium thiocyanate	Cellophane
Ion-exchange-resins regeneration in breweries	

After the concentration of sodium chlorate has been reduced, caustic is passed through the feed/product interchanger for cooling, and then the treated stream enters the caustic flash tank to remove, by venting, any of the gas not used in the reaction. The treated caustic can then be transferred to a bin or shipping container.

Conversion of NaClO₃ to NaCl is typically 75 to 99% in commercial equipment. Chlorate removal will depend on such design parameters as flowrate and catalyst bed size (i.e., liquid space velocity through the bed). Caustic concentration (either 35 or 50%) does not have an effect on removal efficiency.

CAUSTIC QUALITY—Purified product will contain on the average less than 30 ppm of sodium chlorate. In recent operations of this process, treated caustic has contained an average 5 to 10 ppm of $NaClO_3$ for feed concentrations ranging from 280 to 520 ppm.

The total NaCl content of treated caustic will vary, depending on feed concentrations. Whatever is in the feed will be increased by the conversion of chlorates. As an example of the stoichiometry of the conversion: If the reduction of 2,000 ppm of $NaClO_3$ is complete, 1,100 ppm of NaCl will be formed.

For many uses of caustic, the presence of NaCl is not a problem. Typical commercial caustic contains about 1% sodium chloride. For some industries (rayon, for example), the caustic used must be essentially free of NaCl.

ECONOMIC ADVANTAGE—The only other commercial process for the removal of chlorates from electrolytic-cell caustic is based on the continuous extraction of sodium chlorate with liquid ammonia. Although this process removes NaCl as well as sodium chlorate, investment and operating costs tend to be higher.

Ammonia extraction can cost $2 million to $3 million (according to published data) for a system capable of treating 300 tons/d of sodium hydroxide. Operating costs are estimated to be $15 to $16 per ton of caustic treated, and maintenance would be approximately 4% of investment.

In contrast, the capital investment for a chlorate removal system capable of treating 300 tons/d of caustic would be about $1,500,000 (in 1978 dollars). This is an installed cost that includes all process equipment such as exchangers, pumps, tanks and reactor. Operating costs for the same unit are estimated at $1.50/ton of caustic treated, and maintenance at 4% of investment.

Utilities for the chlorate removal system would include steam at a rate of 150 lb/ton of caustic, and power at 3.5 kWh/ton of caustic. Due to the simplicity of the process, no labor in addition to that required by a chlor-alkali plant would be attributed to the operation of the system. Catalyst cost for the 300-ton/d unit is estimated at $85,000/yr. And a small amount of hydrogen from the brine electrolysis (approximately 6% by weight of the amount of sodium chlorate removed) would be consumed in the process.

Due to the severe operating conditions (the corrosive nature of the caustic and chlorate), costs have been reduced by optimizing residence time, enabling the use of more-compact equipment. Nickel and Monel are used only for surfaces that come in contact with hot caustic (i.e., the reactor, the feed/product interchanger and the feed heater).

The Authors

Lawrence C. Mitchell is Manager, Sheet Products R&D, for General Electric Co. (Mount Vernon, Ind.). He holds B.S., M.S. and Ph.D. degrees in chemistry from the University of Michigan.

Michael M. Modan is Manager of Ventures Technology for the Plastics Business Div. of General Electric Co. (Pittsfield, Mass.). He has a Ph.D. in Chemical Engineering from Johns Hopkins University.

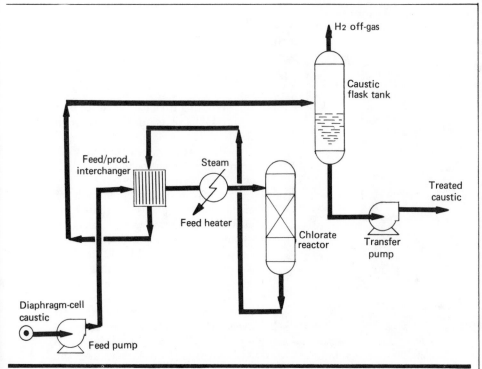

In the reactor, chlorates are reduced to sodium chloride

Membrane cells a hit with chlorine producers

Although no one suggests that they will replace diaphragm or mercury cells, membrane-based processes are increasingly being chosen for small and large chlor alkali units alike.

☐ There's a quiet technological revolution underway in the chlor alkali business worldwide. The two principal electrolytic processes—mercury cell and diaphragm cell—used for making chlorine and caustic soda are being challenged by a versatile new contender, the membrane cell.

Membrane-cell technology, say many of its developers, may revolutionize the structure of the industry because it offers the possibility of decentralizing the production of these vital bulk materials via small but economically viable units at individual consumers' sites. This pares transport costs and bypasses the hazards of moving these commodities via public transportation.

Though some major producers feel that the threat of decentralization is remote, they're evidently keeping a watchful eye on the rapid proliferation of smaller-sized membrane-process plants. They are aware that the technology has developed quickly during the late 1970s. "If you'd asked in 1973 whether the membrane techniques under development would become commercial by 1977, I'd have doubted it, but it happened," says an executive with a major U.K. chlorine maker.

SOME SUPPORTERS—Typical of the membrane-process backers' viewpoint is that of W. Dale Wegrich, vice-president of the Electrolytic Systems Div. of Diamond Shamrock Corp. (Chardon, Ohio), who estimates that a minimum of 20% and a maximum of 50% of all new chlor alkali capacity worldwide will use one of the rival membrane processes (see box) by the late 1980s. Wegrich believes that growing restrictions on the shipping of chlorine will tend to decentralize production, and that cell capacities of less than 75 tons/d will prove a practicable option.

Diamond Shamrock is backing this viewpoint with action: It has assembled a marketing group, currently engaged in discussions with potential clients, to promote a palletized, modular membrane-cell plant ranging in capacity from 1 to 20 tons/d of chlorine. Working from data acquired on a 5-ton/d pilot plant, the firm believes it will be able to scale up or down to suit individual needs. It uses Du Pont Co.'s proprietary Nafion (polyfluorosulfonic acid) membranes.

The rosy marketing prediction for membrane-cell plants is shared by many. Lee Rivers, vice-president of technical matters for Allied Chemical Corp.'s Industrial Chemicals Div. (Morristown, N.J.) agrees that "there is a growing trend worldwide toward the use of membranes, with paper companies and utilities as the most likely users in industrialized areas. In the developing areas, membrane cells will also find use, because the

Photo Courtesy Ionics, Incorporated Watertown, Mass.

Membrane cell and hydraulics skid for Ionics' Model 300 Cloromat unit
Originally published November 5, 1979

Mercury, diaphragm and membrane—pros and cons

Conceptually, the diaphragm and membrane processes have much in common: The former uses an asbestos diaphragm to separate the anolyte and catholyte chambers; the latter a fluorocarbon ion-exchange membrane (in the case of most U.S. processes, one with sulfonic acid functional groups, and in the case of the Japanese processes, carboxylic acid functional groups). In their feedstock needs, the old mercury cell and the membrane cell have much in common—both require a strong, pure brine made up by dissolving solid sodium chloride. Diaphragm cells instead run on a dilute (20%) concentration and relatively impure brine—e.g., one produced by solution mining of salt deposits.

Because of this latter factor, hybrid combinations of mercury cells and diaphragm cells have been common—the diaphragm unit feeds on the stripped brine from the upstream mercury cell. Some experts hold that in the future a similar teaming of diaphragm cells and membrane cells will be common.

The mercury cell's advantages are that the technology is well known, and that design improvements have made mercury losses negligible from such revamped plants. Among the disadvantages are its high electrical power consumption (figured at some 20% more than either diaphragm or membrane processes) and its vulnerability to opposition from environmentalists.

One plus for the diaphragm cell is its ability to deal with dilute and relatively impure brines. Energy consumption, however, is high—the caustic soda produced typically has a concentration of around 20%, and must be concentrated to 50% by means of a large consumption of steam if it is to be shipped. Also working against diaphragms are fears that some of the asbestos they contain may leak into the environment. Also, the product is salt-contaminated, which rules it out for some consumers.

Membrane cells score high on some of these aspects. The product is pure and has a medium (25–30% or higher) concentration. And energy consumption of the technique is low. However, since the membranes are prone to fouling by calcium or magnesium ions in the feed brine, a purification stage is needed upstream of the cell.

membrane plant is cleaner environmentally and can operate on a small scale."

Rivers believes that plants as small as 10 tons/d will prove viable. Allied has demonstrated a membrane process at a scale of 6 tons/d and considers the unit a commercial prototype, though future plans remain undefined at present. Significantly, Allied has decided to get out of the bulk chlor alkali business itself due to low profitability. The firm sold its 184,000-ton/d plant at Baton Rouge, La., to ICI last year and will soon sell three of its remaining four large plants in the U.S.

Another booster of small membrane-cells is Sweden's A. B. Celleco (Stockholm), a subsidiary of the Alfa-Laval AB group. Set up two years ago to market membrane-cell technology using modified Du Pont Nafion membranes developed by Ionics Inc. (Watertown, Mass.), the Swedish firm has already netted one order and is negotiating others. Elkem Spigeverk, a major Norwegian metals group, chose Celleco's process package for a $4.5-million, 25-ton/d chlorine plant at its Bremanger silicon plant. Due onstream next summer, the unit will supplant liquid chlorine deliveries by sea to that remote coastal site.

Celleco, too, will specialize in small units, of 5 to 60 tons/d capacity, says Kjell Rydberg, the firm's marketing director. Investment costs have come down, so that it is now feasible to build small membrane-cell plants," he reasons. Celleco's approach is modular—each module consists of 60 individual cells operating at a current density of 2,700 A/m², and producing at the rate of 4.5 to 5 tons/d of chlorine and 5 to 5.5 tons/d of caustic soda at 5 to 20% concentration.

ANOTHER NORWEGIAN FACILITY—the Tofte Cellulosefabrik A/S bleached pulp mill on the Oslo Fjord—will also be switching to localized chlorine production. The mill will use a 45-m.t./d plant supplied by Uhde GmbH (Dortmund, West Germany). MX-type membrane cells developed by Hooker Chemical Co. (Houston, Tex.), will be employed. Uhde, too, is backing the small-plant concept and believes it will open up new markets.

A membrane supplier, Ionics Inc., has been quietly making inroads in the small-plant field. The firm's Chloromat unit has recorded more than 50 sales (in capacities from 500 lb/d to 25 tons/d) worldwide, including the U.S. Says Floyd Meller, manager of special projects: "Packaged plants will be a growing field." He expects this to be true in the U.S., "especially in the South and Southwest, where there is a trend toward small wastewater-treatment plants and a reluctance to carry chlorine through cities." Hans Hyer, general manager of the Cloromat Div., adds, "there are also smaller users in places where chlorine is either unavailable or extremely expensive."

BIG UNITS, TOO—While U.S. and European firms have zeroed in on the small-plant concept, Japanese developers of membrane cells seem to favor large facilities. Two firms lead the race in Japan (and maybe worldwide, in the view of many European observers): Asahi Chemical Industry Co. and Asahi Glass Co., both in Tokyo. Rival Tokuyama Soda Co. (Tokuyama, Japan) is also a contender, with a 10,000-m.t./yr plant operating since mid-1977.

Asahi Chemical has been running an 80,000-m.t./yr plant at Nobeoka since April 1975 (see *Chem. Eng.*, June 21, 1976, pp. 86-88 for a closer look at the technology). In the meantime, the company has netted a number of license sales: a 10,000 m.t./yr caustic soda plant for Canada's St. Anne-Nackawic Pulp and Paper Co. (Nackawic, New Brunswick), a 60,000 m.t./yr plant for Denki Kagaku Co. (Ohmi, Japan), and a 30,000-m.t./yr unit for Canada's Prince Albert Pulp Co. (Saskatoon, Sask.).

But the biggest catch so far for

Asahi Chemical is a license deal with Holland's Akzo Zout Chemie Nederland BV (Hengelo), which plans to build the first world-scale membrane-cell plant—a 250,000-m.t./yr chlorine, 280,000-m.t./yr caustic unit, to be sited at Botlek, near Rotterdam. Provided that electricity supplies are negotiated and all permits granted, detailed engineering and construction will begin in 1980 for a mid-1982 startup. Instrumental in Akzo's choice of the membrane route was its low-pollution characteristics and favorable operating costs. Choice of the site was influenced by transportation considerations—all the chlorine will be used locally, mainly in vinyl chloride manufacture, supplanting rail or road shipments into the densely-populated Rotterdam area.

The Japanese firms also intend to build large membrane-cell plants at home. In late September, the Japanese government renewed its edict that the remaining mercury-cell plants in that country (accounting for 1.75 million out of the total 4.5-million-m.t./yr capacity) must be converted to other processes by December 1984. It is believed that membrane-cell plants will sweep the board in this scrap-and-build Japanese program.

ITALIAN ENTRY—Another big name in the electrolysis field, Oronzio de Nora Impianti Elettrochimici S.p.A. (Milan, Italy) is set to unveil a new approach in the membrane-cell field early next year. The company plans to offer a process called SPE (for solid-polymer electrolyte) that features an ion-exchange membrane having special catalytic coatings on both sides. These act as the anode and cathode. De Nora says that internal resistance and cell voltage are dramatically lowered in comparison with conventional cells at comparable current densities, and that overall energy savings of up to 20% are achieved in comparison with other membrane processes. (In the U.S., General Electric Co. has been working on a similar concept, and has collaborated with De Nora).

De Nora engineers believe that the predominance of small plants among those ordered thus far reflects the relative novelty of the technology. As the technology improves, they argue, the average size of plants will increase. De Nora also concludes that, by 1983, companies interested in new plants

Modern diaphragm cells have longer life because of improved materials

will find themselves with a straight choice between membrane- and mercury-cell facilities.

BETTER MEMBRANES—According to Alexander MacLachlan, director of the Nafion business group, Du Pont is near a breakthrough that will improve energy consumption and life expectancy of membranes; no further details are yet available.

The company is banking on a 3% annual growth rate in membrane-cell-using plants in the 1980s, from a current worldwide level of 80,000 to 90,000 m.t./yr. It will complete a new membrane manufacturing plant at Fayetteville, N.C., in early 1980.

Du Pont also holds out the hope of another marketing plus for membranes. Manuel Esayian, accounts manager at the Nafion business group, says that retrofitting of diaphragm-cell plants with membrane cells is being studied. If this succeeds and looks economically feasible, it could certainly brighten the picture further for membrane cells.

DON'T WRITE THEM OFF—With so much attention focused on membrane cells, it is easy to overlook how entrenched the two established technologies are. De Nora estimates that 50% of world chlor alkali plants use the mercury cell, while 49% rely on the diaphragm cell. And the Chlorine Institute (New York City) says that total U.S. chlorine capacity is split 74.3% in favor of diaphragm plants, with 20.3% mercury cells. Non-elec-

trolytic routes account for much of the balance.

With the slow market growth (chlorine demand rises no more than 3-4% annually) and general overcapacity, there is little risk of major merchant producers switching overnight to a new process, in view of the huge capital sums tied up in existing plants.

Further, both types of cells are constantly being improved. For example, the introduction of closed-loop operation to enable better housekeeping of mercury or asbestos losses is expected to overcome some environmental objections that have been voiced about mercury cells, while the emergence of such energy-saving items as metal anodes, improved diaphragm materials, and Raney nickel-coated cathodes (from Lurgi Gesellschaft, Frankfurt) will extend the life of diaphragm-cell units.

Plants of both kinds are still being built. For instance, Uhde GmbH recently started up a 150,000-m.t./yr mercury-cell plant in East Germany, while C.I.L. (Montreal, Que.) started up an expansion of a diaphragm-cell plant at Bécancour, featuring a 10% more energy-efficient ICI-Solvay design.

However, even greater advances can be expected in membrane-cell technology, say its supporters. In the development stage are such innovations as air electrodes, said to be better suited to membrane plants than to any others.

Peter R. Savage

Concentrating nitric acid by surpassing an azeotrope

By altering a constant boiling mixture to make distillation possible, this new system manufactures both strong and weak nitric acid from only air and ammonia.

Luis M. Marzo and Jesús M. Marzo, Espindesa

☐ A new process that is able to make up to 99.8% nitric acid has been commercialized in Argentina. The Concentrated Nitric Acid (CNA) system, which is patented in the U.S. and Spain, has been developed by Española de Investigación y Desarrollo, S.A. [ESPINDESA] (Madrid), a subsidiary of Explosivos Río Tinto.

For the most part, nitric acid is manufactured and consumed at concentrations of about 60%. But, concentrated (90% or more) nitric acid is needed for the production of chemicals such as isocyanates and nitrobenzene. Such a concentrated acid, however, is difficult to make. Nitric acid and water form an azeotropic mixture at 68% HNO_3. Distillation of this mixture to generate a higher acid concentration is not economically feasible. In the past, separate systems, even using dehydrating chemicals, have been employed to concentrate acid. Now, the new CNA process integrates the production of weak and concentrated HNO_3 and avoids the use of additional chemicals.

SUPERAZEOTROPE—To manufacture nitric acid, ammonia is oxidized by air over a catalyst, producing NO that is oxidized to NO_2. The nitrogen dioxide is then absorbed in water to make the acid. In the new ESPINDESA technology, gases leaving oxidation are enriched in NO_2 and then absorbed in azeotropic acid to produce a super-azeotropic mixture (about 80% HNO_3) that is then easily distilled.

Compared to methods that have been used to create a concentrated nitric acid, the CNA route offers a low investment cost (see table). Some of the older systems include a route in which liquefied N_2O_4 is combined with oxygen and water in an autoclave, and one that employs an extractive distillation of weak acid with magnesium nitrate or sulfuric acid. The Espindesa technology has a lower investment at any plant size than the autoclaving route, and about 10 to 20% lower investment than the distillation procedures for units over 100 metric tons/d.

A facility using the CNA system can be fully automated—only two operators a shift are required. The process can also make concentrated and/or weak acid (50-60%) in any proportion: the product mix can be changed by adjusting the position of two valves. In addition, most of the towers and exchangers used are made from conventional austenitic stainless steel. And the absorption towers are of austenitic stainless steel that has undergone a nitriding process, permitting a substantial decrease in shell thickness.

This new technology has now been employed in a plant in Río Tercero, Argentina, operated by Dirección General de Fabricaciones Militares. Commissioned in April, and operated continuously since then, this facility has a design capacity of 130 metric tons/d (based on 100% HNO_3) and produces 98.5% acid, with the flexibility to make up to 26 m.t./d of the total in weak acid.

In the ESPINDESA process (see flowchart), liquid ammonia is evaporated and then oxidized with air to form NO at atmospheric pressure and 870°C:

$$4\,NH_3 + 5\,O_2 \rightarrow 4\,NO + 6\,H_2O$$

After reaction, the temperature of the process gas is lowered to its dewpoint in a series of heat exchangers within the oxidizer, which preheat tail gas, produce steam and preheat boiler feedwater. In another exchanger, most of the water formed during reaction is condensed as a slightly acidic stream (about 2% HNO_3). This water is sparged with air to remove nitrogen oxides and eliminated from the plant to maintain the water balance—the water generated is more than that required to make 98-99% HNO_3.

REVERSAL—Now dry, the gas stream is preheated and sent to the bottom of the oxidation tower. Here, 50-60% acid from the absorption sec-

Argentine unit makes 98% and 60% HNO_3

Originally published November 3, 1980

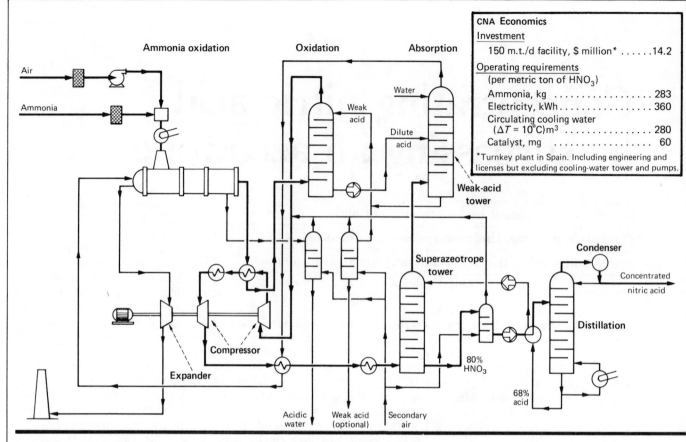

CNA Economics	
Investment	
150 m.t./d facility, $ million*	.....14.2
Operating requirements	
(per metric ton of HNO₃)	
Ammonia, kg	283
Electricity, kWh	360
Circulating cooling water	
($\Delta T = 10°C)m^3$	280
Catalyst, mg	60

*Turnkey plant in Spain. Including engineering and licenses but excluding cooling-water tower and pumps.

Azeotropic nitric acid is forced to absorb more NO_2 to make a mixture that is easy to distill

tion of the system enters the top of the trayed column. As the gas and liquid flow countercurrently, the acid oxidizes NO to NO_2:

$$NO + 2 HNO_3 \rightarrow 3 NO_2 + H_2O$$

This reaction, the opposite of the one conventionally used to make nitric acid, establishes the condition necessary to make a superazeotrope—a high concentration of nitrogen dioxide. Here, three moles of NO_2 are produced per mole of NO, creating a stream very rich in NO_2 and N_2O_4 (formed by the dimerization of NO_2).

Dilute acid from the bottom of the oxidation unit is sent to the absorption section of the route for concentration. The process gas leaving the top is combined with air containing nitrogen oxides, and then compressed to about 10 atm. (Here, heat of compression converts N_2O_4 to NO_2.)

ABSORBING TWICE—The compressed gas is cooled before entering the absorption section of the process. This section actually consists of two columns: one produces the superazeotrope, the other makes a weak acid.

The gases first enter the superazeotrope column and are contacted with azeotropic (68%) acid from the distillation unit. NO_2 is absorbed, and a superazeotrope of 80% nitric acid is formed (concentrations of up to 85-87% have been attained). This is possible only because the partial pressure of NO_2 has been raised to the order of several atmospheres.

This superazeotrope, saturated with nitrogen oxides, is bleached with air, preheated, and then fed to distillation—an operation that is now not difficult. Distilling the superazeotropic acid requires only six theoretical trays.

The normal azeotropic acid mixture used for absorption is taken off the bottom of the distillation column, which can be designed with either packing or sieve trays, depending on the size of the plant (large facilities will tend to use trays). Concentrated nitric acid, from 96-99%, is obtained at the top.

WEAK ACID—Since the process gases leaving the top of the superazeotropic absorption tower are still rich in nitro-

gen oxides, they are sent to another absorption unit. This column is maintained at low temperature by the evaporation of feed liquid ammonia. Here, water is introduced at the top, and low-concentration acid from the oxidation tower enters in the middle. HNO_3 of about 50-60% is removed from the bottom. If this acid is to be sold, it is bleached with air.

The exiting gases have a concentration of NO_x on the order of 300-400 ppm. This tailgas is preheated to about 450°C by process streams in the ammonia oxidizer and by streams leaving the compressor.

The hot gas enters an expander to recover more than 50% of the energy required for compression. After expansion, tailgas is sent to the stack.

Reginald Berry, Editor

The Authors

Luis M. Marzo, General Manager of Española de Investigación y Desarrollo, S.A. holds a Ph.D. in industrial chemistry from the University of Madrid. He managed the team that developed the concentrated nitric acid process.

Jesús M. Marzo holds an M.D. in chemistry from the University of Madrid. He is engaged in the development of fertilizer and inorganic processes at ESPINDESA.

Improved titanium dioxide process keeps plant alive

By replacing an established batch ore-digestion with a patented continuous method in the production of TiO_2, NL Industries reports reductions of particulate emissions by 99.6% and energy costs by about 25%. In addition, the amount of waste acid generated is said to be lowered by 40%.

Leonard J. Kaplan, Assistant Editor

☐ In March of this year, NL Industries, Inc. (New York City) received the National Environmental Industry Award* for the development of an improved method for processing ilmenite, the titanium-bearing ore that the company uses in the sulfate route for manufacturing titanium dioxide (TiO_2) pigment. The new step, called Liquid Phase Digestion (LPD), was the result of a two-year crash program necessitated by the actions taken by federal and state pollution-control agencies to enforce codes for particulate emissions at the company's Sayreville, N.J., plant. According to Fred Baser, director of NL's environmental control department, without the success of the new technology the plant would not have survived.

The LPD process replaces the batch ore-digestion step, standard in the industry for decades, in which concentrated sulfuric acid (90-96%) reacts with titanium ore to produce primarily titanium and iron sulfates. In the old batch method, the reaction of ore and acid was so violent that large amounts of water vapor, containing entrained particulates, SO_x and H_2SO_4, were emitted to the atmosphere. Particulate emissions exceeded the allowable limit (0.02 gr/ft^3) of New Jersey's air-pollution regulations.

According to the company, the new process, which has been used at the

*Sponsored by the White House Council on Environmental Quality and the Environmental Industry Council.

Sayreville plant since last December has had the following effects on emissions and energy use:
■ A particulates reduction of 99.6%.
■ H_2SO_4/SO_3 about 96.8% less.
■ SO_2 reduced by about 99.7%.
■ Steam consumption lowered by about 25%.
■ Oil consumption decreased by 6.8 gal/ton of TiO_2 produced, by the elimination of the ore-drying step.

An added benefit of the LPD process, reports the firm, is reduction of waste sulfuric acid generated by the overall operation. Greater recycling of spent acid, about 1.46 tons/ton of TiO_2 produced, reduces the amount to be barged for ocean disposal.

SCRUBBERS REJECTED—The company says that although scrubbers are employed by other TiO_2 manufacturers using the sulfate process, applying this equipment to solve the pollution problem at Sayreville was studied and found unacceptable because: (1) scrubbers used elsewhere had not shown that the low particulate levels required in New Jersey could be achieved; (2) to handle the large initial volumes of gases given off during the ore digestion, large sizes with high capital costs would be required, and (3) large quantities of scrubbing liquid would require treatment prior to discharge into the river near the plant.

Beginning in 1978, laboratory work had been initiated, at Sayreville, to

change the digestion process from batch to continuous. The latter would use a relatively dilute sulfuric acid, reacting it with the titanium ore over a longer period of time than in the batch method, to eliminate the violent reaction. By doing this, it was hoped, the huge quantities of water vapor that entrain particulate and produce an acid mist would be controlled. A full-scale prototype of the continuous process was installed at Sayreville in mid-1979 and operated for over a year for evaluation. Then the plant operated for a year 50% with batch digestion and 50% with the continuous LPD process until the end of 1981.

PROCESS BACKGROUND—Titanium dioxide pigment is currently made by either of two methods, the chloride or the sulfate process. The chloride process, introduced by Du Pont in the 1950s, chiefly uses mineral rutile—a sand containing about 96% TiO_2. But it is the sulfate process for which the LPD method was developed.

This process (see flowsheet) starts off by reacting dry, titanium-bearing ore, known as ilmenite (the ore used at NL contains 47% TiO_2 and about 30% ferrous oxide) with concentrated sulfuric acid. Water is added to the digester to lower the acid concentration to about 85% and start the reaction. This water, plus that produced in the reaction, vaporizes and this contributes to the pollution-control problem. This exothermic digestion reaction follows the following stoichiometry:

$$FeO + TiO_2 + 2H_2SO_4 \rightarrow FeSO_4 + TiOSO_4 + 2H_2O$$

The resulting product, when made by the batch process, becomes a solid cake that is allowed to cool, and is then dissolved, using either water or dilute sulfuric acid. During the dissolving step, in which the sulfate salts of titanium and iron are solubilized, a suit-

Originally published June 14, 1982

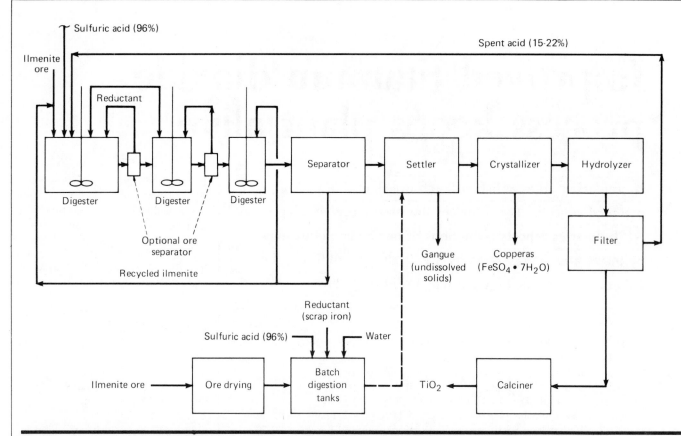

Besides permitting control of atmospheric emissions, the Liquid Phase Digestion process (upper left) recycles about 40% more of the spent acid generated in making TiO₂ than was possible with the previous batch method

able reductant such as scrap iron is added to reduce trivalent iron (ferric) found in the ore to divalent iron (ferrous). This prevents contamination of the titanium hydrate produced further along in the process.

At the Sayreville plant, the batch digestion took place in 22 digestion tanks about (19,000 gal each). When the acid reacted with the ore, a temperature of about 190°C was achieved.

Once digestion and reduction have been completed, the remaining steps are basically the same for the batch and LPD process. The sulfate solution undergoes a step to remove insoluble materials, called gangue. The cleaned solution then enters a crystallizer where the ferrous sulfate (FeSO₄·7H₂O) is removed. The titanyl sulfate (TiOSO₄) can now be hydrolyzed with water to form TiO₂ hydrate, which is separated from the remaining liquid via filtration. Calcining removes water and adsorbed acid from the cake, producing the anhydrous TiO₂ pigment.

LIQUID-PHASE DIGESTION—In the

new process (see flowsheet), the ilmenite does not have to be dried before use, eliminating the drying step and the energy required for it. The ore is added to the first of three (and in some cases four) digesters, where it is mixed with dilute sulfuric acid having a concentration between 25 and 60% by weight. Because the reaction takes place with a much lower concentration of acid than in the batch method, a significantly greater amount of the spent acid generated from the final manufacturing steps can be recycled. The acid solution is now produced from a mixture of fresh acid (96%) and recycled acid with a concentration of 15-22%.

Another important change from the old process, says Brian R. Davis, director of engineering at NL Chemicals, is that more ilmenite ore is used in the digestion step—amounts up to two times the stoichiometric requirements, with the unreacted material being recycled to the reactors.

The ore and dilute acid are agitated continuously at a temperature up to the boiling point of the solution. In the

first digestor, this temperature is preferably maintained between 55° and 140°C, at atmospheric pressure. The residence times and temperatures are set to obtain an acceptable reaction rate and to prevent undesired, premature hydrolysis to TiO₂.

The mixture in the first digester is continuously transported, by gravity, to a series of three or four digestion tanks, with the temperature in the second and third tanks maintained below that of the first. In one of the company's patents, the temperatures for a three-digester sequence are 110°C for the first, 100°C for the second and 75°C for the third. It should also be pointed out that the reductant (scrap iron, usually suspended into the reactor in a basket) was found to accelerate the reaction, in addition to reducing ferric iron.

The solution from the final digester is fed to a separator for the removal of the excess ilmenite and then sent to the remaining process steps described earlier for conversion to TiO₂ pigment.

The company intends to make the process available for license.

Phosphoric acid process proven for large-capacity plants

While hemihydrate routes for making H₃PO₄ have

been developed by other companies, this method used by

Occidental Chemical reportedly boasts a superior

crystallization technique that increases production rates.

Gerald Parkinson, West Coast editor

☐ Occidental Chemical Co. (Houston, Tex.), a subsidiary of Occidental Petroleum Corp., is offering its patented hemihydrate process for license, having proved its commercial viability over the last two years in one of the world's largest phosphoric acid plants, in White Springs, Fla. The company, currently negotiating with several engineering firms to handle licensing agreements, says that the process produces acid at an operating cost of approximately $20-30/ton less than its 1,200-metric-ton/d dihydrate plant at the same location. The acid manufactured by these two different methods finds its way into the production of superphosphoric acid for making liquid fertilizers.

The hemihydrate route for making H₃PO₄ from phosphate rock has the advantage of generating a more concentrated acid (38-44% P_2O_5 than does the commonly-used dihydrate method. This in turn leads to lower energy costs since less evaporation is required to concentrate the finished product to working strengths. According to the company, the use of the hemihydrate process has been limited to smaller-capacity plants (in the range of 40 to 350 m.t./d) because of problems with filtration. These problems are said to be reduced by modifying the crystallization conditions and adding proprietary additives.

OTHER HEMIHYDRATE METHODS— The hemihydrate process is not new; Fisons, Ltd., of London, U.K. (see *Chem. Eng.*, April 30, 1973, p. 62) and Nissan Chemical Industries, Ltd. (Tokyo) have had patented processes available for several years. Both companies offer routes that produce high-strength acid at high yields (up to 98.5%). Sources at Occidental claim that because these other methods still have filtration problems, they are not suited to high plant capacities.

The hemihydrate and dihydrate techniques fall under the classification of "wet processing" methods for making H₃PO₄. Both begin in the same way: phosphate rock (apatite) is dissolved in dilute H₃PO₄ to form calcium monophosphate:

$$Ca_{10}F_2(PO_4)_6 + 14H_3PO_4 \longrightarrow$$
$$10Ca(H_2PO_4)_2 + 2HF$$

followed by a second step in which sulfuric acid is added to the slurry to react with monophosphate to form H₃PO₄ and calcium sulfate.

Depending on the process conditions during the sulfuric acid reaction, the calcium sulfate can be in either of two forms—the dihydrate (gypsum, $CaSO_4 \cdot 2H_2O$) or the hemihydrate ($CaSO_4 \cdot \frac{1}{2}H_2O$). Which of the hydrates will form depends on the temperature and the concentration of the H₃PO₄ maintained during the crystallization step. Higher acid concentra-

Originally published September 6, 1982

59

Hemihydrate cost savings over dihydrate route

Basis: 1,300 m.t./d Occidental plant (54% acid)

Consumption/m.t. of P₂O₅	Hemihydrate advantage	Dihydrate advantage	1982 savings, $
Phosphate rock, tons ($29.70/ton)	–	0.05	–1.49
Sulfuric acid, tons ($66.00/ton)	0.14	–	+9.24
Defoamer & modifier, kg (33.4¢/kg)	–	4.00	–1.79
Electricity, kWh (5.4¢/kWh)	56.00	–	+3.00
Process water, m³ (4.7¢/m³)	45.00	–	+2.10
Steam, kg ($7.865/1,000 kg steam)*	1,425	–	+11.21
Operating-cost saving			22.27
Depreciation, interest & taxes			1.58
Total saving (per m.t. of P₂O₅)			23.85

*Based on Occidental's use of the excess steam for power generation

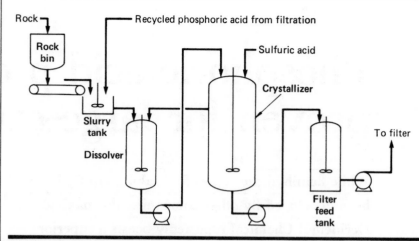

The heart of this hemihydrate process is the crystallizer, where controlled crystal growth takes place

tions (38-52% P₂O₅) and elevated temperatures favor the formation of the hemihydrate crystal form. The following equation shows the routes to both crystal forms.

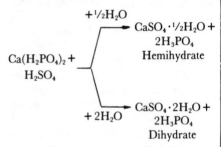

Dihydrate crystals form and are removed when the acid concentration is 28-30% P₂O₅, while the hemihydrate crystallizes at 40-50% P₂O₅. Thus more energy is required to concentrate the acid from the dihydrate process to the so-called "merchant grade" acid at 54% P₂O₅.

CRYSTAL STRUCTURE—Traditionally, the disadvantage of the conventional hemihydrate route is that the calcium sulfate crystals are difficult to filter because of their small size—averaging only 15-30 microns (μm), compared to 40-60 μm for dihydrate crystals (from Florida rock). The reason for the hemihydrate crystal size is reportedly the higher rate of nucleation than with the dihydrate form. As soon as the solution becomes supersaturated, crystal nuclei form and drop out of solution. In addition, the hemihydrate crystals may convert to the dihydrate form during filtration.

The Occidental process is said to overcome the filtration problems by growing so-called "polycrystals"

(about 40-60 μm), whose size and irregular shape permit more efficient separation than the dihydrate form. A specially designed agitation system, in the crystallizer, and a crystal modifier are given as the means for controlled crystal growth.

PROCESS DETAILS—The flowsheet shows the basic steps used in the Occidental hemihydrate process. Wet, unground rock is fed to a slurry tank where it is mixed with recycled H₃PO₄ from the calcium sulfate filters. Vigorous agitation and sufficient retention time allow much of the carbon dioxide in the rock to escape.

The monocalcium phosphate slurry flows by gravity into a dissolving tank where dissolution of the rock is completed. Slurry from the crystallization reactor overflows back to the rock-dissolving vessel to maintain the correct sulfate concentration.

The slurry is now pumped to a single-agitator draft-tube reactor where it is combined with concentrated sulfuric acid. The temperature in this vessel is maintained at 200-210°F under a vacuum of about 14-15 in. of H₂O. Because this reaction is exothermic, control of the temperature is achieved by vacuum evaporation of some of the water. According to the company, fluosilicic acid can be recovered from the crystallizer vapors by conventional equipment.

The hemihydrate slurry now goes from the crystallizer to a filter feed tank that acts as a process surge tank as well as a holding tank for startups and shutdowns. Filtration takes place in rotating-table and belt filters.

One major problem with the hemihydrate route is that about 3-6% P₂O₅ coprecipitates with the hemihydrate crystals. Because of this the standard yield at the Occidental facility is about 92% of the potential P₂O₅. The companies discussed earlier, who offer competitive hemihydrate routes, get around this drawback by offering a secondary recrystallization step, in which the hemihydrate is converted to the dihydrate by mixing with water and sulfuric acid. This additional step is reported to increase the yield into the range of about 98% P₂O₅. While this step is not being used at the Occidental plant, the company says that it has developed the technology for those requiring the higher yield.

ECONOMICS—The above table shows cost figures for Occidental's hemihydrate process compared to the dihydrate method used at its Florida complex. Both processes are evaluated for the production of 54% acid.

The company reports a capital-cost saving of 20% over the dihydrate route. This is attributed to: (1) lower-capacity evaporation requirements, (2) reduced filtration equipment costs, and (3) elimination of the need for rock grinders.

The 1982 operating-cost saving for the 1,300-m.t./d plant is reported to be about $22/m.t. of P₂O₅. The major factors contributing to this figure are: reduced steam consumption required for acid concentration, lower electrical consumption resulting from reduced agitator load, and lower sulfuric acid consumption.

Leonard J. Kaplan, Editor

Section III
Metallurgical Processes

Plasma process is ready for metals recovery

Water-cooled argon gun promotes high-speed reactions in a new method that will shortly undergo demonstration tests in Britain.

☐ Plasma technology is on the verge of a commercial breakthrough in a difficult metallurgical application. A British-born process based on the use of a plasma at temperatures of 10,000°C or more will go on trial this spring at a demonstration unit in Faringdon, Oxfordshire, to recover platinum-group metals from a chromite ore found in South Africa (*Chem. Eng.*, Sept. 11, 1978, p. 107).

Tests conducted with a variety of other materials indicate that the technique has even wider processing potential—e.g., in ferrochrome manufacturing, steel smelting, cement making, remelting and recovery of cast-iron borings or particulate scrap, and recovery of various metallurgical by-products (including a high-value dust arising from stainless steel production). Studies on other, more-esoteric endothermic uses, such as coal gasification, have also been made, at least on paper.

A U.S. CONNECTION—At this point, much of the interest in the new process comes from U.S. firms, partly as a result of a two-year-old agreement between the British developer—a small research-oriented company named Tetronics Research and Development Ltd.—and Foster Wheeler Ltd. (Reading, U.K.), an affiliate of the U.S. engineering and construction company.

Tetronics has already completed successful testing at a 300-kVA pilot plant at Faringdon. The new trial will take place in a 1,400-kVA demonstration unit currently being built by Tetronics in cooperation with Foster Wheeler, which has given advice on instrumentation and other aspects. Testing this spring will take place on behalf of Texasgulf, Inc. (Stamford, Conn.), which has announced that it is considering the plasma process for recovering otherwise-inaccessible platinum-group metals.

Originally published February 26, 1979

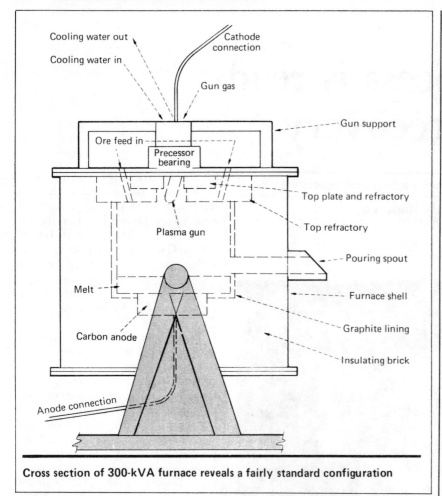

Cooling water out
Cooling water in
Cathode connection
Gun gas
Ore feed in
Gun support
Precessor bearing
Top plate and refractory
Plasma gun
Top refractory
Pouring spout
Melt
Furnace shell
Carbon anode
Graphite lining
Insulating brick
Anode connection

Cross section of 300-kVA furnace reveals a fairly standard configuration

Plans for a commercial-scale plant—probably not much bigger than the 1,400-kVA unit—are currently being formulated, with startup at an undisclosed location slated for 1980. Texasgulf says that it has developed the necessary process steps to recover the desired metals.

Tetronics and Foster Wheeler have also collaborated on the testing performed on other materials. And, in a separate venture, Tetronics has joined forces with Rugby Portland Cement Co. (Rugby, U.K.) in the development of a process that yields a new cement-like material via plasma-induced calcination of lime and coal-mine wastes.

HOW IT WORKS—The Tetronics technique, dubbed the expanded precessive plasma process (EPP), centers on a plasma gun developed at Faringdon. It is a water-cooled, metal-sheathed gun fed with a small flow of argon, and featuring a doped tungsten electrode.

In the 1,400-kVA plant, about 1,500 A of d.c. current is fed to the electrode from a thyristor-controlled power source. The gun is mounted vertically in a hydraulic-motor drive that moves at high speed (1,500 rpm or more) and at an angle to the vertical (typically 15 deg), to produce a "precessing" cone of plasma between the gun and either a counter electrode or the hearth of the furnace.

Material to be processed is fed into the cone of plasma as a falling curtain. In most applications, much of the smelting (or other chemical reaction) occurs during the 300-millisecond time of flight through the extremely-high-temperature zone.

The gun, along with its power and utilities connections, is mounted on a gantry above a refractory-brick furnace shell of fairly conventional configuration (see figure). The process runs continuously, and the vessel is periodically tipped to discharge both refined metal and slag, although it can be adapted to work on a continuous-overflow basis. Larger units are essentially similar in design.

Size of the plasma installation depends on the application. For exam-

ple, the 1,400-kVA unit being completed at Faringdon would be large enough for platinum-group-metals smelting. But ferrochrome or steel manufacturing would call for at least a 6,000-kVA facility. This would cost about $5 million on an f.o.b. basis, according to Peter L. Gulliver, plasma technology coordinator at Foster Wheeler.

Process economics look quite favorable for a ferrochrome operation fed with relatively poor-quality chrome ore and either coke or coal as a reductant. Foster Wheeler estimates that ferrochrome could be made for 25¢/lb of chromium (f.o.b. basis), in comparison with a minimum of 29¢/lb for conventional processes.

SOME ADVANTAGES—Tetronics and Foster Wheeler say that the EPP process has three main advantages:

■ It uses raw materials in powder or particulate form—e.g., less than 0.5-mm dia. This is a plus in nearly all metallurgical applications, because many ores are reduced to particles of such size after beneficiation. So are many "run of mine" ferrous ore concentrates and recovered fines from various other processes. Accordingly, explains Gulliver, "There is a cost bonus before we start; many other processes can use only pelletized or lumpy materials."

■ The ultra-high temperature induced in the plasma allows strongly endothermic reactions in very short periods of time. Charles P. Heanley, managing director of Tetronics, says that reaction mechanisms are hazily understood at present because of the problems involved in monitoring such high-temperature and high-rate processes. Adds Gulliver: "When we first looked at the process, the claims seemed a bit incredible, but it works. The results with ferrochrome were amazing."

■ The EPP technique also allows precise control of the furnace atmosphere. If an inert one is required, it is achieved by excluding air and continuously purging the powder-feed lines with nitrogen. If an oxidizing or reducing atmosphere is needed, air is usually allowed into the furnace, with or without a carbon source.

HEAD START—Both Tetronics and Foster Wheeler believe that they are ahead of the field in metallurgical applications of plasma technology. Among other firms seeking to commercialize such technology are Bethlehem Steel Corp. (Bethlehem, Pa.), which has a patented falling-film process aimed at ferrovanadium production. Bethlehem has conducted extensive tests.

There has also been much enthusiasm for plasma technology in the Soviet Union, and a plasma furnace with a 5-metric-ton hearth has been in operation there since 1972. The Soviets have licensed their knowhow to VEB Edelstahlwerk Freital (Freital, East Germany), which built a 30-metric-ton furnace in 1977.

However, the Soviet and German units are apparently limited in their range of applications, being used primarily for remelting steel scrap, including alloy and high-grade steels. Their technique also relies on water-cooled, argon-fed guns (each has three guns fixed in position, but capable of being adjusted). Last year, East German trade delegations promoted the plasma technology heavily during visits to the U.S. and elsewhere.

Other well-established, plasma-based processes—e.g., plasma spraying and welding—are mostly concerned with highly specialized, small-scale applications. (The road to commercialization of plasma techniques is littered with failures, largely as a result of plasma instability or the limited volume of plasma generated from a fixed gun.)

WHAT'S AHEAD—Success with the Texasgulf trial could trigger a big breakthrough for the EPP process. Already under construction is another 300-kVA unit for tests on remelting and recycling of cast-iron borings for an undisclosed U.S. client.

Ferrochrome manufacturing seems likely to be commercialized soon, too. And Foster Wheeler's Gulliver has his eye on the next stage—the direct remelting of steel and stainless steel. So far, tests have resulted in the manufacture of a high-grade, cast-iron-like material. Work is now aimed at reducing the product's carbon content to obtain a true steel.

One advantage of the EPP process in steelmaking is that coke is not necessarily the only reductant that can be used; if coal works equally well, further cost savings can be made. Says Gulliver, "The process has something going for it in nearly every pyrometallurgical application."

Peter R. Savage

Oil gasification is teamed with iron-ore reduction

Direct reduction of iron ore and gasification of heavy oil are combined to produce hot iron that can be fed directly into electric-arc furnaces.

Henry Johnston, McGraw-Hill World News, Rio de Janeiro

Just in back of the waste heat boiler, the gasification reactor rises through the platform. Scrubbers and stripping columns tower in the rear

Originally published March 26, 1979

☐ In Brazil, the Purofer Thyssen Direct Reduction process developed by Thyssen AG. has been combined with an oil gasification route from Texaco in a 1,060-metric-ton/plant operated by COSIGUA (Companhia Siderúrgica da Guanabara). Here for the first time, a direct reduction plant is making its own gas from heavy oil, which is plentiful in Brazil.

At the COSIGUA plant, oil that has a 4% sulfur content is converted to CO plus H_2. This gas is then used to reduce iron oxide to produce sponge iron, a porous product with a high (92 to 94%) iron content.

Onstream since mid-1977, this facility is located at Santa Cruz near Rio de Janeiro. The operating company, COSIGUA, is owned by the Brazilian Gerdau steel group (51% of the stock) and Thyssen AG. (49%).

Using natural gas, the direct reduction technology has been tested at a 350-ton/d pilot plant operated by Thyssen in Oberhausen, West Germany. Another unit, using natural gas, has been sold to Iranian National Steel Industries. This plant, which has a capacity of 330,000 tons/yr, started up but has been shut down by strikes.

POWER SAVINGS—The Purofer process offers several advantages. The sponge iron product is discharged hot (approximately 720°C) and can be fed directly into electric arc furnaces. This can save steel makers from 170 to 210 kWh per ton of metal as compared to the use of a cold-metal feed.

If the hot iron will not be used immediately, it can be made into briquettes through compression without the need for binders. (Briquettes eliminate an explosion hazard. Porous sponge iron has a large surface area that invites an exothermic reaction between oxygen and iron. Briquetting reduces this area and the probability of reaction.)

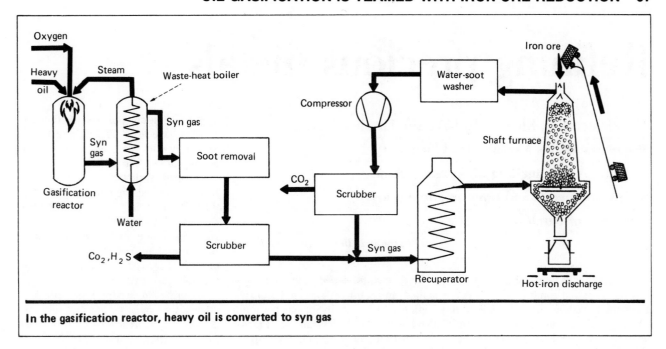

In the gasification reactor, heavy oil is converted to syn gas

An additional advantage is that the process is able to use 100% lump ore (approximately $\frac{1}{4}$ to $1\frac{1}{2}$-in. dia.). The shaft furnace can use up to 1,100 tons/d of lump ore or pellets.

GASIFIED OIL—At COSIGUA, oxygen and steam are reacted with heavy oil (Bunker C). The gas formed is cleaned to remove soot, hydrogen sulfide and carbon dioxide. The remaining components are mixed with a recycled top-furnace gas. This mixture is used to reduce the iron ore.

Steam produced in the waste-heat boiler, as well as oxygen (99.7% pure) and oil, is injected into the reactor (see flowsheet). This reactor has a lining of high-purity alumina bricks; it is divided into two sections: an 8-m³ upper section and a 2.6-m³ spherical section below.

Reactions take place at 1,370°C and 27 bar. They are exothermic in the upper section of the reactor, endothermic in the lower section, where oil and CO_2 form CO. At capacity, this reactor will produce 43,000 m³ of gas having a composition of 48% CO, 45% H_2, 4% CO_2 and H_2S (which is approximately 9,000 ppm), the rest being water.

This syn-gas product is cooled in the waste-heat boiler, which uses the hot gas to produce 70-bar steam for the reaction (a total of 34 tons of steam is produced—5-6 tons are used for the reaction; the rest supplies process heat or generates power).

Soot is then removed from the syn-gas. Hydrogen sulfide and carbon dioxide, undesirable components, are absorbed by a solution of 25% potassium carbonate and 10% diethylamine in a Benfield scrubber operating at 22 bar. Absorbent solution circulates (at 90 m³/h) between the scrubber and a regenerating column operating at low pressure. Here, H_2S and CO_2 are stripped and flared.

The clean syn-gas, after mixing with recycled gas from the top of the furnace (top-furnace gas), is reheated in recuperators and injected at 870°C into the bottom of the shaft furnace.

Ore is charged into the top of the furnace while reductant gas passes upward. In removing oxygen from the ore, the process leaves small holes in the iron. The porous iron (typically at 720°C) is dropped into special containers below and is transferred either to an electric-arc furnace or to briquetting equipment. (Because they resist oxidation, briquettes can be handled like prime-quality scrap.)

Gas exiting from the top of the shaft furnace passes through a water soot-washer, is cooled, and is then compressed to 10.5 bar.

Before it is recycled, this top-furnace gas is fed to another Benfield scrubber to remove CO_2, using a solution of 20% K_2CO_3 and 2.5% diethylamine. The absorbent solution circulates at 500 m³/h.

In normal operation, the Purofer process will have an onstream factor of 90%; due to power failures, this figure has been about 80% for the COSIGUA plant.

ECONOMICS OF METALLIZATION— With a Brazilian ore containing 67.9 to 68.2% total iron, the resulting sponge produced by the Purofer process has been 93% iron, with a metallization factor (the ratio of Fe metal to total iron present) of 90 to 93.5%. Another ore containing 67.6% to 68.1% iron produced a sponge that was 93.6% iron with a metallization of 93.5%.

Investment for the COSIGUA plant has been estimated at 100 million Deutschemarks (1976, approximately $42 million U.S.). The facility consumes about 10,600 m³/h of oxygen and 14,300 kg/h of oil. Electric power is 240 kWh per ton of product. Water that is converted to steam in the waste-heat boiler is used at the rate of 50 m³/h. On the plus side, it may be possible to sell the CO_2 reclaimed from the gas that leaves the direct-reduction furnace.

This process was developed by Thyssen-Niederheim Steel. A development group, Thyssen Purofer GmbH, is responsible for the engineering, building and supply of direct-reduction plants. In October 1978, the direction of projects was placed in the hands of Thyssen AG., which will grant licenses and work as project engineer. Mitsubishi Heavy Industries is the licensee for the Far East. A licensee for North America will be appointed shortly.

Refining precious metals

The elements Ir, Rh and Ru have similar chemical properties and as a result are difficult to separate. A new method for refining these metals saves costs and time by eliminating a number of batch processes.

Reginald I. Berry, Assistant Editor

☐ Ion exchange and solvent extraction has replaced some batch processing steps to reduce the cost, time and labor required for producing and refining the precious metals ruthenium, iridium and rhodium.

Normally extracted during the refining of nickel or copper, Ir, Ru and Rh—platinum group metals—are very rare, very difficult to separate and very valuable (producer prices for rhodium are $700/oz., for Ir about $300/oz. and for Ru approximately $45/oz.). And although uses and markets—electronics and high temperature alloys—for two of them are small, the demand for rhodium for automobile catalytic converters is increasing, and the metal is widely used as a catalyst in the petrochemical and fertilizer industries.

Since Ir, Rh and Ru have similar chemistries and have many different valence states, allowing them to appear in many different forms, it is not easy to refine these metals. The classical method of refining entails many batch operations including leaching and precipitation, with subsequent reprecipitation of salts to achieve the desired metal purities—at least 99.8%.

A new route developed by South Africa's National Institute of Metallurgy (NIM) starts with a reaction between aluminum and platinum group metals forming an alloy, which enables the precious metals to go into solution faster. The method also employs continuous and easily automated separations to eliminate processing steps.

This article is based on a paper delivered in May 1978 in New York at the 2nd Conference of the International Precious Metals Institute.

Originally published June 18, 1979

The NIM-developed route was installed in South Africa by Lonrho Refinery Ltd. at their Springs plant on the Eastern Witwatersrand in Transvaal in 1976, and since then by Impala Platinum Refinery, also at Springs.

Lonrho, which has a platinum production of 80,000 to 100,000 oz/yr, produces from 12,000 to 14,000 oz/yr total of Ir, Rh, and Ru; the Impala Refinery, with a platinum capacity of 800,000 oz/yr, produces an equivalent ratio of the other metals.

ECONOMIC BENEFITS—According to a paper by L.G. Floyd of Lonrho, the new route offers several cost savings when compared to more classical methods of refining:

■ Since single, multi-stage operations replace many batch steps, some equipment can be eliminated and chemical costs and inventories can be reduced. (Instead of reactants being consumed in precipitation, the ion-exchange resin or solvent employed can be regenerated.)

■ Since less equipment is required, the NIM route takes less space, approximately one-fifth of that required by the more conventional method. At Lonrho, it was estimated that the older process would take 7,500 ft²; the new system needed only 1,500 ft². In general, the capital cost of refining has been reduced by 50% or more.

■ Because the number of refining steps can be reduced, processing time of as much as 4 to 6 months has been cut to 20 days. And since more of the process can be automated, labor is only 20 to 25% of that needed for the more conventional route.

■ In addition, purities for the NIM process have been 99.95% and better, and recoveries for each of the metals are typically 97 to 98%.

ALUMINUM ALLOY—At the Lonrho facility, a feed containing 40% Ru, 20% Rh and 5% Ir originates from conventional extraction and the lead smelt technique used in the refining of platinum. The feed undergoes a series of reduction reactions with carbon and then with aluminum.

In order to separate the precious metals, they must be put into solution. The reduction reactions do this faster than conventional techniques in which leaching precedes a Pb smelt (forming a slag that is cooled and then granulated), followed by more leaching.

At first, the carbon reduction (see flowsheet) limits the oxygen content of the process stream to prevent too violent a reaction during the aluminum alloying stage. Then aluminum (as a powder) and ferric oxide, which further controls the oxygen content, are added to the feed in a refractory-lined crucible.

The reaction, reducing precious metals from oxides to their elemental form, takes place at 1,200°C. This temperature is sufficient to drive off volatile impurities and to form a friable alloy of aluminum and precious metals, which—when broken up—has a large surface area, facilitating the wet chemical reaction.

A leach that removes most of the base metals follows. The aluminum/precious metals alloy is mixed with hot hydrochloric acid and streams recycled from solvent extraction. Al, Fe, Ni, Pb and Cu dissolve, leaving finely dispersed solids, of mainly Ir, Ru and Rh. The precious metals are then filtered out of the process stream.

The filter cake is leached in a highly agitated Cl_2/HCl solution. Precious metals are dissolved; Al_2O_3 and SiO are filtered off.

Small amounts of silver and gold are removed by filtration after adding water to lower acidity to precipitate

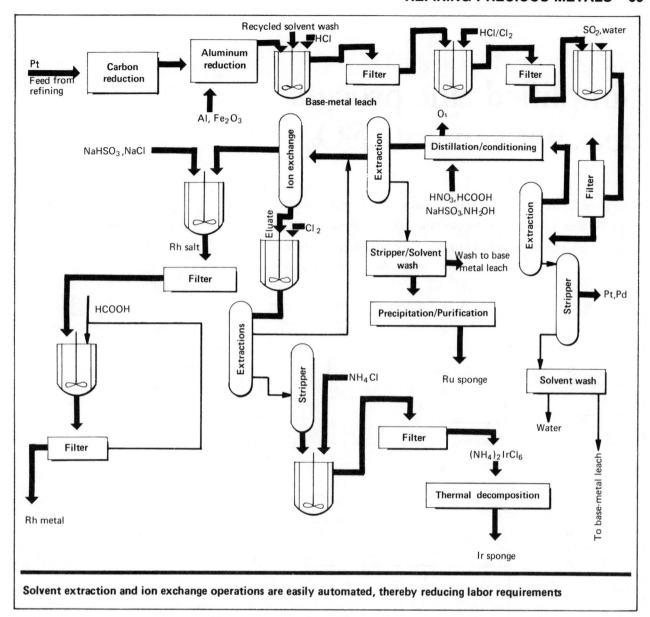

Solvent extraction and ion exchange operations are easily automated, thereby reducing labor requirements

AgCl and sparging with SO_2 to precipitate gold (ion exchange can be used to recover Ag and Au if the concentrations are high enough).

RESINS AND SOLVENTS—The filtered solution enters a packed column where solvent extraction is employed to reduce Pt and Pd levels from 1% to 20 ppm. The solvent used in the liquid-liquid extraction is a secondary amine or its acetic acid derivative, and it is stripped using HCl, and then washed with water to remove extracted acid before it is recycled.

The process stream enters a batch distillation to remove Os, as a tetraoxide. The stream is then conditioned for the extraction of Ru.

Formic and nitric acids convert Ru to a stable nitrosyl species. Sodium bisulphite and hydroxylamine (acting as reducing agents) are added to prevent the coextraction of Ir. Ruthenium is absorbed by a tertiary amine solvent in a packed column; the solvent is stripped in a batch contactor with a 10% caustic soda solution and then contacted with HCl (the acid stream is returned to the first leach and the solvent is recycled).

Ruthenium is purified by precipitation as the hydrated oxide. This material can be directly reduced, by adding hydrogen, or converted to $(NH_4)RuCl_6$, which is thermally decomposed to Ru sponge.

In the NIM route, the recovery of Ir and Rh is the reverse of the classical method in which Rh is obtained first Ion exchange with a strong-base resin, selective against rhodium, is used to separate Ir from the process stream.

Water saturated with SO_2 is used to elute the Ir. This solution is boiled to remove SO_2 and then oxidized with chlorine before a solvent extraction step. Since some base metals such as iron and copper are also absorbed by the resin, liquid-liquid extraction with tributyl phosphate as the solvent is needed to isolate the iridium.

The solvent is stripped with water. NH_4Cl is added to precipitate Ir. The salt formed is thermally decomposed to give the iridium sponge.

Rhodium is recovered by precipitation from a stream containing mainly base metals—predominantly aluminum and nickel—and small amounts of iridium.

Simplified zinc process does not generate SO$_2$

One pressurized reactor has replaced two processing steps in this new zinc process, for which costs may be lower than those for conventional roast-leach methods.

David Whiteside, McGraw-Hill World News, Detroit

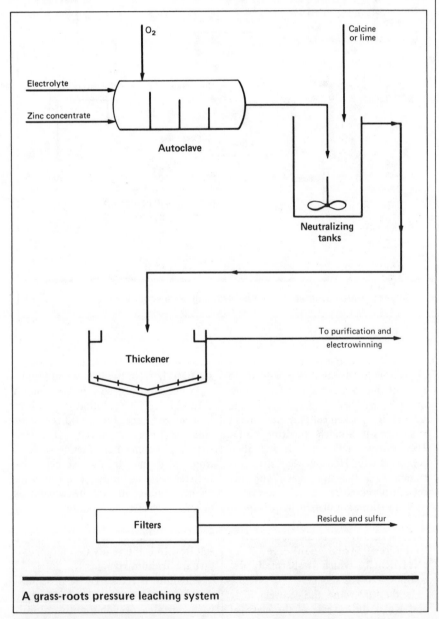

A grass-roots pressure leaching system

Originally published August 13, 1979

☐ For zinc recovery, pressure leaching has several advantages over the conventional roast-leach route: the new technology requires fewer processing steps, improves plant hygiene and does not produce SO$_2$ but instead recovers elemental sulfur.

Sherritt Gordon Mines Ltd. (Toronto), which owns the licensing rights to the new zinc process, developed the system and then invited Cominco Ltd. (Vancouver, B.C.) to join in pilot-plant and commercialization efforts.

In 1977, a pilot plant was operated in Fort Saskatchewan, Alta. And now Cominco has announced that the first commercial application will be at its Trail, B.C., facilities, where pressure leaching will be integrated with an existing conventional plant.

Over an eight-year span, Cominco will spend $370 million to modernize and expand its zinc and lead operations at Trail. Of this amount, $20 million will be allotted for the new system.

Two pressure-leaching reactors, one of which is to be onstream in 1981 and the other in 1983, will replace a suspension roaster and contribute 75,000 tons/yr of zinc to a total plant capacity of 300,000 tons/yr.

WITHOUT SO$_2$—In the conventional method for zinc recovery, concentrates are oxidized in a roaster, transforming the sulfur in the feed to SO$_2$. Zinc oxides are then converted to zinc sulfate in an acid leach. This sulfate is sent to purification and then electrowinning to obtain zinc metal.

A different series of reactions occurs in the pressure leaching system. Here, zinc sulfate is obtained directly in one reactor, and sulfur is produced in elemental form. One reaction replaces two processing steps.

70

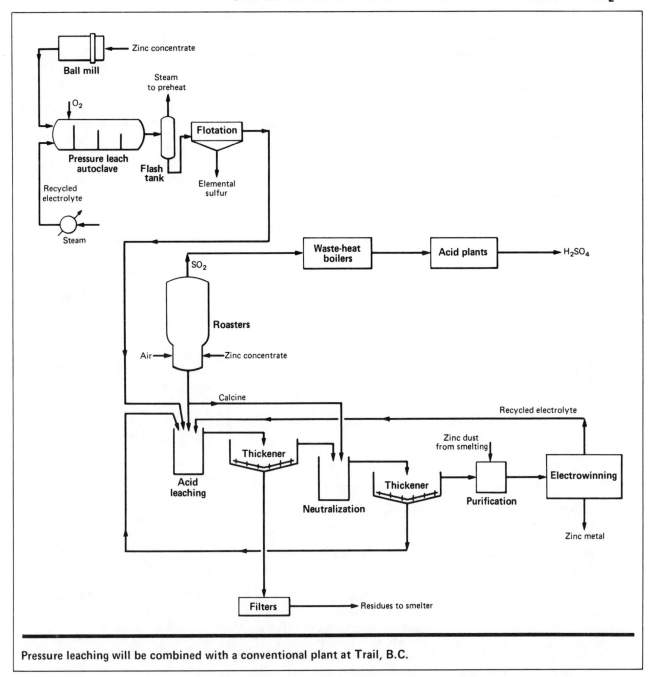

Pressure leaching will be combined with a conventional plant at Trail, B.C.

Cominco points out that the production of elemental sulfur in the pressure leaching route improves and simplifies operations at Trail. Normally, hot gases containing SO₂ leave the roasters and are passed through a waste heat boiler and then sent to a plant for the production of sulfuric acid. But with the new technology, sulfur can be stockpiled and then used when convenient, so that the acid plant's flow is not dictated by the production of the zinc facility.

Also, residue treatment can be avoided. In a conventional system, zinc ferrite ($ZnFe_2O_4$) is formed in the roaster. This compound resists further reaction in the acid leach, and additional treatment is required for the recovery of zinc. Pressure leaching does not form residues that contain zinc, so residue treatment would not be necessary.

In addition, with pressure leaching, plant hygiene is improved. All reactions occur in a slurry within one enclosed vessel. Handling hot gases and dusty solids (both products of the roasters) is eliminated.

SLURRY REACTION—Typically, zinc concentrate is composed of 48–55% Zn, 8–10% Fe, 0–4% Pb, less than 1%

copper and cadmium, 5% water and about 30–34% sulfur. At Trail, the feed concentrate will be split, with a portion going to the new method and a portion going to the roasters.

In the new method, the concentrate goes through a wet ball mill (see flowsheet) that reduces solid particles to less than 325 mesh. This assures that all of the particles react during the time that they are in the autoclave—the pressure leaching reactor.

Following grinding, the concentrate slurry and recycled electrolyte—mainly sulfuric acid—from the electrowinning unit are combined in

**Estimates indicate that there will be economic
incentives to use pressure leaching technology**

Capital spending (U.S. $, millions)*

	Pressure leaching	Roast leach
Roast and acid leach	–	21.7
Leaching and electrowinning		
1. Building and construction	13.1	14.3
2. Equipment	20.9	23.9
3. Miscellaneous (including yards and services)	14.3	14.8
4. Design and management	7.8	7.6
5. Contingency	5.3	4.8
Subtotal	61.4	65.4
Total	61.4	87.1

Operating costs (¢/lb)

	Pressure leaching	Roast leach
Direct labor	2.10	2.34
Materials and supplies		
oxygen	0.95	–
other (reagents, filters, etc.)	2.1	2.43
Utilities		
electricity	0.6	0.6
water and fuel	1.2	1.3
General services	1.8	1.9
Total	8.8	8.6

*For a 60,000-ton/yr zinc plant.

the autoclave. At Trail, the first autoclave reactor to go onstream will be 12 ft dia. and 50 ft long. This cylindrical vessel will have a mild-steel outer shell that is lined with lead and then lined again with acid brick. The internals of the reactor will be divided into four sections with baffles in each.

The slurry and electrolyte will enter the autoclave at approximately 70°C. (The electrolyte is preheated with steam flashed from the process stream.) Oxygen is added.

The reactions in the autoclave are exothermic and take place at 150° C and 150 psi. Zinc sulfide, lead sulfide and iron sulfide react with oxygen and sulfuric acid to form simple sulfates, elemental sulfur and water. The initial reactions are:

$$ZnS + H_2SO_4 + \tfrac{1}{2} O_2 \rightarrow ZnSO_4 + S° + H_2O$$
$$PbS + H_2SO_4 + \tfrac{1}{2} O_2 \rightarrow PbSO_4 + S° + H_2O$$
$$FeS + H_2SO_4 + \tfrac{1}{2} O_2 \rightarrow FeSO_4 + S° + H_2O$$

After a short time, the iron in the slurry is further oxidized, from the ferrous to the ferric state:

$$2FeSO_4 + H_2SO_4 + \tfrac{1}{2} O_2 \rightarrow Fe_2(SO_4)_3 + H_2O$$

As the level of acid in the slurry diminishes, iron precipitates as a number of complex sulfates. Although the precise nature of the compounds formed may vary, two typical reactions are:

$$3Fe_2(SO_4)_3 + PbSO_4 + 12H_2O \rightarrow PbFe_6(SO_4)_4(OH)_{12} + 6H_2SO_4$$
(Plumbojarosite)

$$3Fe_2(SO_4)_3 + 14H_2O \rightarrow (H_3O)_2Fe_6(SO_4)_4(OH)_{12} + 5H_2SO_4$$
(Oxonium jarosite)

After leaving the autoclave, the slurry is flashed (the recovered steam preheats the electrolyte) and then enters a flotation unit for sulfur separation. The process stream is then joined with the conventional plant's at the acid leaching step, since the autoclave reactions would be replacing the roast and leach. Mixing the two streams together at this point will protect the main plant against composition swings and other variations during the initial startup, according to a spokesman from Cominco.

In the conventional operation, concentrate components—zinc sulfide and lead sulfide—react with oxygen to form ZnO, PbO and SO_2. At the same time, ferrous sulfide passes through several reactions, leading to the formation of zinc ferrite. Sulfur dioxide is also a byproduct of these reactions:

$$2FeS + 3\tfrac{1}{2} O_2 \rightarrow Fe_2O_3 + 2SO_2$$
$$2FeS_2 + 5\tfrac{1}{2} O_2 \rightarrow Fe_2O_3 + 4SO_2$$
$$Fe_2O_3 + ZnO \rightarrow ZnFe_2O_4$$

SO_2 passes to a waste heat boiler and then to the H_2SO_4 plant. The metal oxides enter the acid leach, where zinc sulfate is formed:

$$ZnO + H_2SO_4 \rightarrow ZnSO_4 + H_2O$$
$$PbO + H_2SO_4 \rightarrow PbSO_4 \text{ (precipitate)} + H_2O$$

In this way, both the new and the old processes lead to the formation of zinc sulfate and various byproducts (elemental sulfur or SO_2) and precipitates (zinc ferrite, lead sulfate and iron compounds).

After the leach, the combined process stream is thickened (precipitates are removed), neutralized, purified and then sent to electrowinning, where zinc metal is recovered:

$$ZnSO_4 + H_2O + 2e^- \rightarrow Zn + H_2SO_4 + \tfrac{1}{2} O_2$$

POSSIBLE SAVINGS—Cominco estimates that a grass-roots pressure leach unit might offer a roughly 20 to 25% saving in capital investment compared with a grass-roots, full roast-leach plant. If this is accurate, one could expect to save in excess of $35 million for each 70,000-ton/yr facility. Added to that, Cominco says that operational savings can be realized from the reduced labor requirement: fewer operations requiring fewer workers to watch over them. For a 60,000-ton/yr plant, the table compares estimated capital and operating costs for the old and the new technology.

Reginald Berry, Editor

Aluminum: Energy diet pays off

U.S. producers have jumped the gun in meeting the industry's voluntary energy-reduction quota before the 1980 deadline. And a number of new ideas in processing and materials promise significant energy savings over the next decade.

☐ U.S. manufacturers of aluminum—the sixth most energy-consumptive industry*—have reason to celebrate. By the second half of 1978, they had already succeeded in meeting and surpassing an energy-saving target (10%) that the U.S. Dept. of Energy (DOE) had set for 1980 (as part of a voluntary energy-efficiency improvement plan for the ten most energy-hungry industries). And they have done so via changes that leave the biggest consumer in the aluminum-processing chain—the reduction of alumina to primary metal—virtually untouched. The big question now is: What other changes are in store?

Quite a few. Among several developments in the research stage are new materials of construction for electrodes, new electrolysis routes to aluminum, and a direct-reduction ore-to-metal process. Emergence of a more energy-efficient reduction route could conceivably end the reign of the Hall-Heroult method—the industry workhorse for over a century.

Energy, the prime force behind much of the current technological research, is also responsible for other industry changes. For example, cheap-power availability in such nations as Brazil and Australia has lately attracted a handful of grassroots aluminum smelters. Conversely, high power-costs and uncertainty about supplies are precluding aluminum expansions in many industrialized nations. For instance, producers in the Pacific Northwest—home of about one third of U.S. capacity—face serious difficulties linked to the renewal of longterm contracts with the Bonneville Power Administration—*Chem. Eng.*, Dec. 4, 1978, p. 65.

The siting of projects abroad will surely result in the importation by the industrialized world of value-added products (alumina or aluminum) instead of unrefined bauxite, with a corresponding adverse effect on the industrial nations' trade balances.

FINE-TUNING ALTERATIONS—In the 1940s, primary-metal plants used up to 12 kWh/lb of product; this has been cut to less than 6.5 kWh/lb for new plants in recent years (the average for U.S. facilities is about 8 kWh/lb). The saving is one result of several changes made over the years in bauxite refining and alumina reduction operations, as follows:

■ Replacement of rotary kilns for calcining alumina with fluid-bed units. One of these, the Mark III developed by Alcoa (Pittsburgh, Pa.), is said to cut fuel use by 30 to 40% and save on capital and maintenance costs, compared with a conventional rotary kiln.

■ Replacement of mercury-arc rectifiers with silicon diode ones in the conversion of alternating to direct current—as needed in the reduction process. Some plants have managed to reduce power consumption by 3%.

■ Closer control of bath chemistry in the pots. Computerized operations, with sensors developed especially for smelters, have boosted efficiency. And use of such additives as lithium salts has lowered bath temperatures, increased electrical efficiencies and reduced heat loss.

■ Improvements in carbon anode fabrication. Since it takes about 0.5 lb of carbon to make 1 lb of aluminum, this is a prime opportunity for conserving energy. New designs for prebaked-anode furnaces increase efficiency by up to 50%. Better insulation applied to older units can bring up to 13% savings in energy consumption.

"A series of relatively small modifications" is all that Sumitomo Aluminum Smelting Co. (Tokyo) will say about Soderberg-cell changes that it credits with a 12 to 20% cut in power consumption and an overall energy consumption of 6.35 kWh/lb. Sumitomo has licensed its technology to plants in Norway, Brazil and the U.S.

Another Sumitomo development, now being installed in plants in Indonesia and Australia, is said to lower energy use to about 6.14 kWh/lb, and increase cell life to an estimated 6 yr. It is a prebaked-anode technique.

France's Péchiney Ugine Kuhlmann technology, which involves microprocessor-controlled electrolysis, and the use of a continuous feed of high-solubility alumina, can achieve similar overall energy consumption—i.e., about 6.15 kWh/lb. The know-how will be applied in smelters to be built in Australia and the Soviet Union. The Australian facility will be equipped with 180,000-A cells similar in size to those in operation at Aluminum Péchiney's plant in St. Jean-de-Maurienne, France.

In the U.S., Alcoa will showcase its latest cell technology at the plant that Alumax Inc. will build near Charleston, S.C. Scheduled for startup

*Behind steel, petroleum refining, chemicals, paper, and agriculture and food.

Carbon anode is coated with molten aluminum for improved efficiency

Originally published October 22, 1979

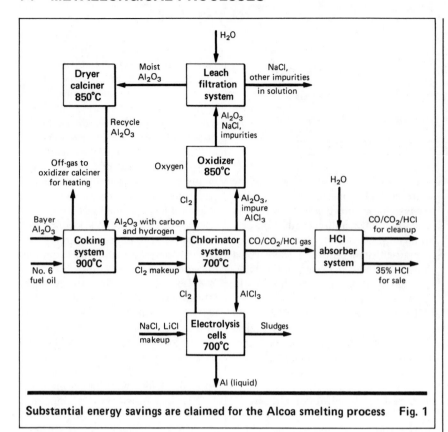

Substantial energy savings are claimed for the Alcoa smelting process   Fig. 1

late next summer, the 197,000-ton/yr aluminum facility will boost pot efficiency from 86 to 90% by lowering current density and reducing swirling.

"Alcoa has alleviated the problem of aluminum falling to the bottom of the pot in a swirl and causing some metal to reoxidize to alumina by entering into contact with evolving oxygen," says J. J. Miller, technical director for the Alumax Primary Div. The result is energy efficiencies close to 6 kWh/lb.

PROTECTING CATHODES—The rationale for developing an inner lining for the carbon cathodes used in Hall-Heroult cells is simple: If the layer of electrolyte in the cell is not kept sufficiently thick, the molten aluminum formed at the cathode (and that does not wet the carbon) can touch the anode and cause shorting. But having thick layers of metal and electrolyte means also having greater anode/cathode spacing, which causes increased electrical heat losses in the cell.

A wettable cathode material that also resists attack from molten metal and the cryolite electrolyte could cut heat losses and boost metal output.

Both Alcoa and Kaiser Aluminum & Chemical Corp. (Oakland, Calif.)

are working on such a development under DOE sponsorship. Much of the R&D focuses on a titanium diboride lining, which allows a reduction in anode/cathode distance—to about 0.5 in., compared with 2 in. for a carbon cell-lining.

The new cathodes boast an improvement of 20% in cell efficiency and also provide operational flexibility. For example, they can either reduce energy consumption by 20% for a constant level of metal production or increase output by the same magnitude at a fixed voltage. The units can be retrofitted into existing facilities.

Kaiser has been running tests in a pilot-plant-sized 15,000-A cell. The next step will be a 200-day run, using a 170,000-A cell in a laboratory operation. The firm expects to complete this test before the end of 1980, after which it will retrofit a potline at one of its plants for 12 to 24 months.

According to a Kaiser spokesman, the main problem of titanium diboride is its brittleness and vulnerability to thermal shock. Cracking occurs, due to chemical instability in the cryolite-aluminum environment.

Alcoa has looked into other metal oxides in addition to titanium diboride. According to Noel Jarrett, assistant

director for the firm's Metal Production Laboratories, "There's still much work to be done. Our biggest problem has been in finding a manufacturing procedure to make the desired shapes stand up to the harsh environment inside the cell, while making economic sense. Our goal is to retrofit all our smelters with these metal inserts if they prove viable."

ANODE ACTIVITIES—Work on anode improvement is proceeding in two directions: (1) use of solvent-refined coal (SRC) to produce the pitch and coke needed for anode manufacturing, and (2) development of a nonconsumable, inert anode.

The first approach would eliminate dependence on petroleum-based raw materials and take advantage of the aluminum industry's large coal reserves (it is believed that U.S. aluminum producers possess sufficient resources to meet 25% of their energy needs). The second avenue of research would yield two advantages: pure oxygen from electrolysis, which is a valuable byproduct; and sealable cells, which would reduce heat losses to the environment.

Alcoa is working on both fronts, but won't reveal which inert anode materials it is exploring. Alusuisse (Zurich, Switzerland), which has some test cells operating with a ceramic-based anode, foresees one problem in switching away from carbon anodes. According to a spokesman: "A carbon anode releases energy to the pot when it burns to carbon dioxide; so with a carbon substitute that doesn't burn, we would have to add makeup energy."

A REDUCTION SUBSTITUTE—A smelting process announced by Alcoa about six years ago is still the subject of an intense research effort. The method involves converting alumina into aluminum chloride, which is then electrolyzed to primary metal. Among the technique's claimed advantages: an energy saving of 30% in comparison to conventional electrolysis of alumina; a completely enclosed system that emits no fluorine or other undesirable fumes; a greater tolerance of power-supply cutoffs, without the problem of metal solidification present in Hall-Heroult cells; and an increased acceptability of power reductions during a utility's peak-load periods.

Alcoa, which has been testing the process in a 30,000-ton/yr plant near Palestine, Tex., completed an evalua-

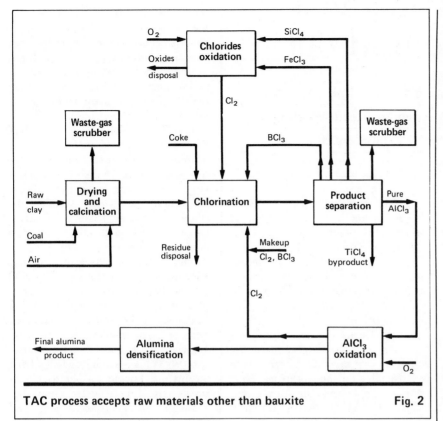

TAC process accepts raw materials other than bauxite   **Fig. 2**

tion of it last May. Further development work will involve a multimillion-dollar redesign of the alumina chlorination operation.

Although Alcoa still keeps processing details under wraps, Arthur D. Little, Inc. (Cambridge, Mass.) has prepared a flowsheet based on available patent literature (Fig. 1).

Overseas, Japan is reportedly interested in developing an aluminum chloride route to primary metal. In Switzerland, Alusuisse is said to have abandoned its aluminum chloride program after conducting some tests.

AWAY FROM BAUXITE—Another development of interest is the TAC process of Toth Aluminum Corp. (New Orleans). Its distinguishing feature is that it can feed on a variety of raw materials, including kaolin clay and other non-bauxitic ores, as well as bauxite itself. An intermediate product is aluminum chloride, which is then oxidized to produce alumina feed for the Hall-Heroult process.

This spring, the Pullman Kellogg Div. of Pullman Inc. (Houston) and Toth Aluminum announced a two-step approach to take the technique from its present bench-scale status to pilot-plant stage and ultimate commercialization (*Chem. Eng.*, May 7,

p. 25). Kellogg is to have sole licensing rights.

Industry sources surmise that the TAC process could be an excellent precursor of the Alcoa smelting process because it produces aluminum chloride. At present, Alcoa is meeting its chloride needs by chlorination of alumina.

When processing kaolin, which contains 1 to 3% TiO_2, the TAC process yields byproduct titanium tetrachloride—at the rate of 1 ton for every 7.5 tons of alumina—suitable for use in making titanium dioxide pigments. This, says Toth Aluminum, is an additional plus for the technique.

THE DIRECT WAY—"There are a number of difficult engineering problems to solve and we're still at least a decade away from realizing any of the advantages the process offers." This is how Alcoa's Jarrett refers to the direct-reduction route to aluminum that his company is working on under DOE sponsorship. On the bright side, DOE believes the technique has the potential to save 0.7 quad/yr of energy by the year 2000, mainly by eliminating the electrolysis step, and by combining a number of energy-intensive operations in one process.

The process, still in the bench-scale

phase (*Chem. Eng.*, Sept. 26, 1977, p. 42), involves mixing ores that contain aluminum and silicon with coke before feeding to the top of a reactor. As the mix falls down the reactor shaft, a number of chemical reactions occur at temperatures of up to 2,000°C. The result is an aluminum-silicon product and a host of valuable byproducts.

One of these is carbon monoxide, which exits from the top of the furnace. "In a hypothetical 300,000-ton/yr plant, CO with a heating value of 49 trillion Btu/yr would be produced," says Jarrett. Other important byproducts are ferrosilicon, and photovoltaic-quality silicon. The end-product alloy can either be used in automotive casting applications or refined further to pure aluminum.

In tests, Alcoa uses a reactor charge containing 50% bauxite and 50% domestic clay. A commercial plant that makes 300,000 tons/yr of end-product "could replace about 750,000 tons/yr of imported bauxite—about 10% of Alcoa's current requirements," says Jarrett. He adds that such aluminum- and silicon-rich materials as coal wastes, fly ash from power plants, and oil-shale tailings (containing dawsonite) could be used as feeds.

The schedule for determining the feasibility of the direct-reduction route calls for a 2,500-ton/yr demonstration reactor by 1986 at the earliest. Scaleup to a 160,00-ton/yr reactor will have to wait until 1990.

BLACKOUT AHEAD?—"As far as I know, this is the only grassroots plant under construction in the U.S.," says an Alumax spokesman. He is referring to the $400-million smelter that the firm is building in South Carolina. Elsewhere, Alumax, as well as other domestic producers, has been thwarted by a lack of power availability that is impeding expansion.

The problem is especially acute in the Pacific Northwest. For example, Alumax's plant there—Intalco Aluminum Corp., in Ferndale, Wash.—has been told by the Bonneville Power Administration that a 20-yr contract that expires in 1984 will not be renewed because not enough power is available.

In a move that may be followed by other U.S. producers, Alumax has decided to build abroad. Its plans include a 197,000-ton/yr smelter in Newcastle, Australia.

David J. Deutsch

New smelting methods aim to get the lead out

Tough proposed U.S. antipollution and workplace-health regulations will be met by new secondary lead-smelting processes, according to a handful of producers that now have projects in the works.

☐ In recent years, government agencies certainly haven't been soft on lead: the U.S. Environmental Protection Agency (EPA) has proposed an ambient standard of 1.5 $\mu g/m^3$, and the U.S. Occupational Safety and Health Administration (OSHA) wants to restrict workplace exposures to 50 $\mu g/m^3$ over 8 hours.

In turn, lead makers, through the Lead Industries Assn. (New York City), have taken a steely line of defense, challenging the standards in federal court on the grounds that they are technically and economically not feasible. (At presstime, a decision by the U.S. Court of Appeals for the District of Columbia is pending.)

At the same time, however, a number of firms have been quietly developing secondary smelting techniques that promise to pare lead emissions. At least three new projects are in the works:

■ In March, Paul Bergsoe & Son A/S (Copenhagen, Denmark) broke ground at Saint Helens, Ore. on a secondary lead smelter that it claims will meet the strict new standards.

■ In February, Cal-West Metals, Inc. (Long Beach, Calif.) brought onstream a 50-ton/d secondary smelter at Socorro, N.M. The firm is counting on its proprietary process.

■ Gould Inc.'s Metals Div. (St. Paul, Minn.) also hopes to use proprietary technology in the replacement for an old secondary smelter in Los Angeles.

On the other hand, Tonolli North America (Toronto) is now trying to get environmental permits for a smelter it plans on the West Coast, but is waiting for the court decision before proceeding in earnest.

DANISH DELIGHT—Paul Bergsoe's $30-million project, a joint venture with East Asiatic Co., another Danish firm, will include a 30,000-35,000-ton/yr smelter and a 30,000-ton/yr refinery. The smelter will accept up to 70,000 tons/yr of used batteries as feed, and the refinery will convert metallic lead from the smelter production and other sources to pure (soft) lead. The complex is scheduled to go onstream in 1982.

Svend Bergsoe, the company's chairman, feels the OSHA standards are too severe and says 100 $\mu g/m^3$ might be reasonable for the workplace. Nevertheless, he believes his firm's new technology will meet even the proposed 50-μg limit.

Bergsoe notes, for instance, that workplace emissions from his company's Copenhagen smelter were measured at 54 $\mu g/m^3$ over 8 hours by Radian Corp. (Austin, Tex.) under a contract with EPA and OSHA. While this is a bit more than the 50-μg ceiling, improvements have been made since the Danish plant started up several years ago.

These improvements will be incorporated in the Oregon plant, which is aimed at proving the Danish technology in the U.S. "Once we meet the Oregon standards we think we could build anywhere in the U.S.," says Thomas S. Mackey, of Key Metals & Minerals Engineering Corp. (Texas City, Tex.), Bergsoe's representative in this country.

Bergsoe's smelter is a shaft furnace, in contrast with the traditional two-stage reverberatory/blast furnace system and the newer rotating furnaces. Also, unlike other systems, the batteries are not broken up but are fed into the smelter whole. This avoids many worker exposure problems, according to the firm.

The Oregon smelter will be completely enclosed. The vehicle unloading and storage area alone will be an enclosed space covering 8,000 m². Batteries, coke and fluxes will be fed into the top of the furnace. In the Bergsoe process, as with others, lime is added to make a liquid slag, and iron is put in to remove the sulfur as iron sulfide; sulfur makes up 25% of the resulting matte. The polypropylene battery cases are completely consumed.

Flue dust is only 2% by weight of the input, compared with about 10% for other processes, according to Bergsoe. This is because the furnace has a large cross section, so that the reducing gases move up slowly through the charge, entraining fewer particulates. Much of the lead-containing steam is condensed on the fresh, cold charge at the top before it leaves the furnace. Exhaust gases go through a 950°C afterburner that destroys organics. Particles that remain are collected in a baghouse, agglomerated in a patented Bergsoe flash agglomerator furnace, and returned to the smelting furnace.

IRONY IN L.A.—Gould's Metals Div., meanwhile, bought an old secondary smelter in Los Angeles from NL Industries in February, 1979, and wants to replace it with a new smelter of more than 50,000-ton/yr capacity, plus a refinery. At presstime, Gould has not received approval from the local South Coast Air Quality Management District. Ironically, the firm has a new process that it says meets lead emissions standards—but the dispute is over emissions of sulfur dioxide and oxides of nitrogen.

The smelter would be the first use of a proprietary process developed over the past two years, says John M. Rossini, director of manufacturing for the Metals Div. In this process, batteries are broken up and separated into their components: metals, plastic, rubber, and "mud" (the lead-bearing component).

Smelting is a two-step reverberatory/blast furnace process. The first step produces molten lead and a lead-bearing slag. The slag is fed to the blast furnace for production of more lead ingot and an inert waste slag.

"We go through two steps because we feel we must get as much pure lead as possible," says Rossini.

The blast furnace has an afterburner for carbon monoxide control, and exhaust gases go through a settling chamber, cyclone, baghouse and scrubber.

Originally published May 19, 1980

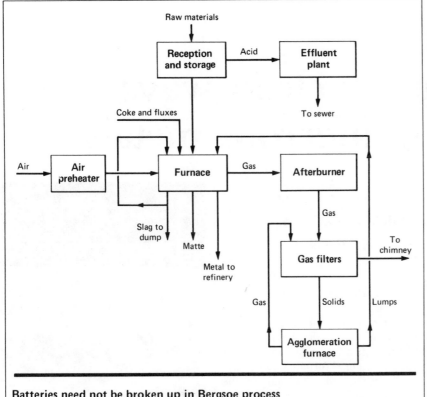

Batteries need not be broken up in Bergsoe process

washed into a separator by a stream of heavy battery liquid and water recycled from the separator.

This liquid serves as a heavy medium in the separator, a rotating drum with no moving parts. An internal separation unit creates a countercurrent flow of liquid and heavy particles to achieve separation, according to LaPoint. Lead metal and fine lead oxides settle and are discharged to a circular, rotating screen. The oxides drop through the screen and are collected separately for feeding to the smelter.

A BID IN CALIFORNIA—And finally, Tonolli North America hopes to build a secondary lead smelter in Fontana, Calif.

A spokesman for the South Coast Air Quality Management District Office in Colton, Calif., says that preliminary information indicates Tonolli will be able to meet ambient air standards, but that the district is still waiting for the firm's final application. He says the key district rule to be complied with is one limiting total emissions of any particular pollutant to 150 lb/day. Therefore, SO_x, NO_x and particulates may be significant problems, even if the lead standards can be met.

Tonolli's plans call for a smelter and refinery that would produce about 40,000 tons/yr of soft lead and lead alloys. The firm's process includes an enclosed battery-breaking system.

Still, Tonolli won't build the unit unless lead environmental and health standards are eased. "We don't believe there is any technology available in the world that could meet the OSHA standard economically," says Vincent Bailini, vice-president, finance.

AN IMPOSSIBLE PROCESS?—Cal-West's New Mexico smelter is a roof-fed, reverberatory type. The firm's Albert LaPoint, who developed the process, says lower lead emissions are achieved because the maximum bath temperature is only 900°C, compared with 1,200°C for a conventional reverberatory furnace. At 900°C, the vapor pressure is much lower, so there is considerably less fume loss, he explains. He declines to give further details. Others in the lead-smelting business say, however, that it is impossible to do a reduction smelt at 900°C.

Cal-West also has a patented separation process for recovering lead from batteries prior to smelting. Batteries are crushed in a stainless steel hammermill, and the fines and particles of lead and plastic (from the case) are

Gerald Parkinson

Search is on for clean ways to produce nonferrous metals

A renewed emphasis on leaching and some interesting smelting twists form the first line of defense for beleaguered copper producers trying to meet emissions standards.

Part I: Copper

☐ World producers of such nonferrous metals as copper, nickel, lead and zinc are in a headlong race to cut sulfur emissions from smelters. And nowhere is this fact more apparent than in the U.S., where metals processors have been notoriously slow in adopting cleaner techniques, and the U.S. Environmental Protection Agency is now clamoring more loudly for compliance.

At least one copper producer—Anaconda Co. (Waterbury, Conn.)—recently opted to close down indefinitely a smelter (in Anaconda, Mont.) rather than spend the money needed for pollution-control curbs. But this processor and others also are looking at cleaner pyrometallurgical (smelting) techniques, or developing hydrometallurgical (leaching) routes that do not emit sulfur.

This article, the first of a two-part series, will examine the most recent copper smelting and leaching developments worldwide. The newest technologies for cleanly recovering nickel (via smelting and leaching), lead (via leaching),* and to a lesser extent zinc, are covered in the next article.

SMELTING DEVELOPMENTS—Some copper producers feel that the U.S. should look more closely at flash smelting—a one-step technique that has been popular abroad for years and is still gaining ground. In flash smelters, the ore acts as its own fuel, making the process virtually autogenous,

*For an update on lead-smelting developments, see *Chem. Eng.*, May 19, p. 93.

Originally published December 15, 1980

especially when the air feed is oxygen-enriched.

Most reactions, including SO_2 formation, occur in the flame in the smelter shaft—instead of in the bath of a reverberatory furnace. Less, and more-concentrated, SO_2 gas is the result, and this allows for easier production of byproduct sulfuric acid, and lower pollution-control costs.

Outside the U.S., flash smelting is the vogue in copper production. One of the leading technologies abroad is that of Finland's Outokumpu Oy, whose president, Lassi Riihikallio, notes that 26 Outokumpu smelters installed worldwide process about a quarter of global copper output.

The Outokumpu smelter consists of a vertical shaft with a horizontal settler and a second shaft for the offgases. Concentrates are kiln-dried by waste heat or an oil burner, then fed into the shaft. Matte tapped from the bottom of the furnace is treated in a converter, then refined in an anode furnace.

"Blister" copper may be produced directly in the flash furnace without the need for conversion, but this requires a very high-grade concentrate, says Riihikallio. Poland's Kombinat Gorniczo-Hutniczy Miedzi Lubin (Glogow) has a commercial plant of this type, and construction of a second one has started in Zaire for La Générale des Carriers et des Mines. Outokumpu is working on improvements to permit blister-copper production from standard concentrates.

The Finnish process is popular in Japan, where Furukawa Co. (Tokyo)

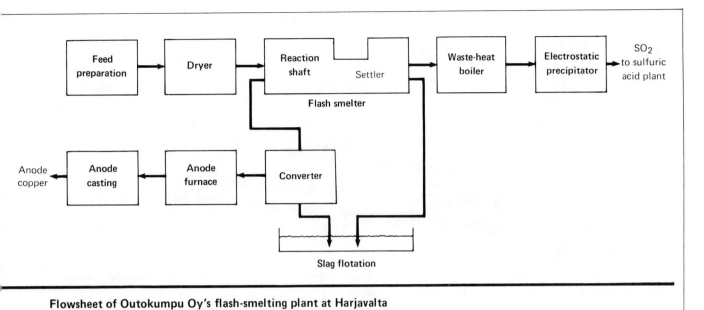

Flowsheet of Outokumpu Oy's flash-smelting plant at Harjavalta

has installed seven smelters for itself and four other copper producers. A Furukawa modification, used at Hibi Kyodo Smelting Co.'s Tamano (western Honshu) smelter, is the installation of graphite electrodes into the bath to reduce the furnace outlet temperature from the normal 1,300°C to about 1,000°C.

U.S. MOVES—The only flash smelter operating in the U.S. is Phelps Dodge Corp.'s 115,000-ton/yr Hidalgo (N.M.) Outokumpu unit, which went onstream in 1976 and has had no trouble meeting pollution-control standards, says a Phelps Dodge spokesman. Although the smelter doesn't use oxygen-enriched air, the SO₂ concentration is still about 11%, compared with only 1½-2% for a conventional reverberatory furnace. Also, about 80% of the sulfur is removed in the smelter, leaving only 20% in the matte, which cuts the capital investment in conversion-furnace equipment.

Phelps Dodge also started testing a new burner last month at its Morenci, Ariz., smelter under an agreement with Dravo Engineers & Constructors (Denver Div.). The sprinkler-burner, as it is called, sprinkles a mixture of concentrate and oxygen into the reverberatory furnace, thereby providing some of the advantages of flash smelting: lower fuel cost, because oxygen substitutes for fossil fuel; higher productivity; and less offgas, with a higher SO₂ concentration.

The burner was developed specifically to convert existing reverberatory smelters to use oxygen. It is designed to fit into the limited space between the roof and the bath, and provides a broad, horizontal dispersion of the concentrate particles. The Phelps Dodge experiment is the first "hot" test of the burner, which was developed by professors Paul E. Queneau and Horst Richter, of Dartmouth College (Hanover, N.H.), and Reinhardt Schuhmann, Jr., of Purdue University (West Lafayette, Ind.).

Dravo has worldwide license rights. It also has a license from Inco Metals Co. (Toronto) for the top-blown rotary converter (TBRC), designed for smelting and conversion of nonferrous metals in small plants (it may be used in plants as small as 5,000 tons/yr). The name is derived from the operation: a mixture of concentrate and process gas—e.g., air—is blown through a lance into the vessel, which rotates to provide rapid mixing and high heat transfer. Smelting, conversion and refining may all be done in the same vessel. Inco uses the process as part of its Sudbury (Ont.) copper and nickel processing operations; it has also been used in the Afton (B.C.) plant of Afton Mines Ltd. since 1978.

CONTINUOUS OPERATION—Kennecott Corp. has invested $280 million to convert its Garfield, Utah, copper-processing facilities to use a continuous smelting process developed by Noranda Mines Ltd. (Toronto). The 270,000-ton/yr plant, which went onstream in 1978, represents the first

commercial use of the process other than Noranda's own operation, and the only fully-integrated facility of its kind, says a Kennecott spokesman.

In the Noranda process, dried concentrate (7-8% moisture content) is belt-fed into a horizontal reactor. A mixture of air and oxygen (Kennecott uses 34% oxygen-enriched air) is blown into the slag bath from the side through tuyeres located below the melt surface. This oxidizes the melt to produce a concentrated SO₂ gas and high-grade matte (about 75% copper), which is subsequently converted to blister copper. The process also yields a slag rich in copper (about 7-8% in this case). This is treated in a slag converter to recover the copper concentrate, which is recycled to the reactor vessel.

Kennecott has three reactors (one standby) and four copper converters. A spokesman says that although there have been operating difficulties, particularly in materials handling, there have been no fundamental problems with the process. The company has had an ongoing dispute with EPA for years over plant emissions, but says this is because of changes in the regulations, and EPA's interpretation of them, since the passage of the Clean Air Act of 1970 and amendments of 1977.

Texasgulf Canada Ltd. (Toronto) is scheduled to complete a 65,000-metric-ton/yr copper plant next year at Timmins, Ont., using a continuous smelting process developed by Mitsu-

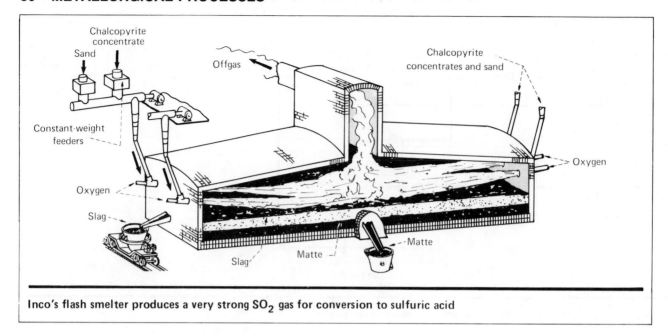

Inco's flash smelter produces a very strong SO_2 gas for conversion to sulfuric acid

bishi Metal Corp. (Tokyo). In Mitsubishi's method, concentrates, fluxes and oxidizing air are top-blown through lances into the furnace. Molten materials move to a second furnace for slag cleaning, then to a third for production of blister copper. Besides producing a concentrated SO_2 gas, the process is said to require only 70% of the capital cost of a conventional smelter installation.

OTHER SMELTING ROUTES—Meanwhile, in Sudbury, Inco operates a proprietary flash smelter for copper and is offering the process for license. The company uses oxygen and produces a very high-strength SO_2 gas (70-80%), which is converted to liquid SO_2 for sale. Output is about 80,000-90,000 m.t./yr of liquid SO_2 and about 800,000-900,000 m.t./yr of sulfuric acid. Inco claims its process involves low capital cost and has high throughput.

An approach completely different from flash smelting is being offered by Amax Base Metals Research & Development Inc. (Carteret, N.J.). In the Amax process, copper concentrates are "dead"-roasted at about 850°C to remove 97-98% of the sulfur (in this case, the sulfur fuels the roasting rather than the smelting process). The calcined material is then agglomerated into briquettes or pellets, and smelted at 1,300-1,400°C in a shaft furnace to produce blister copper directly, instead of a matte.

Key to the process is that the vertical-shaft furnace is only about 8 ft tall.

The short residence time is enough to reduce the copper to metallic form without reducing the iron oxide present. A spokesman says there is only 0.2 to 0.3% iron in the blister copper, adding that while fuel is used for this operation, costly metallurgical coke is not needed. The overall capital cost for a plant of 80,000 tons/yr or less would be 70-80% of that for a flash smelter, and operating costs would be about the same.

So far, the process has not been commercialized, but Amax has reached a worldwide licensing agreement (outside the U.S. and Canada) with the Nonferrous Metals Div. of Davy International (Stockton-on-Tees, England).

LEACHING TECHNIQUES—The only commercial U.S. plant for hydrometallurgical processing of sulfide copper concentrates is operated by Duval Corp. near Tucson, Ariz. (about 90% of the copper mined worldwide comes from porphyry or chalcopyrite ores, which contain sulfur). The plant started up in 1976 (*Chem. Eng.*, Jan. 5, 1976, p. 79), achieved production of 32,000 m.t. of blister-copper equivalent in 1978, and has now reached an annual capacity of about 40,000 m.t., which approximates Duval's goal.

The process, called CLEAR (copper leach, electrolysis and regeneration), leaches copper concentrates in a solution of cupric and ferric chloride. Copper is recovered from solution in continuous-flow electrolytic cells, in which copper crystals form on cathodes. This

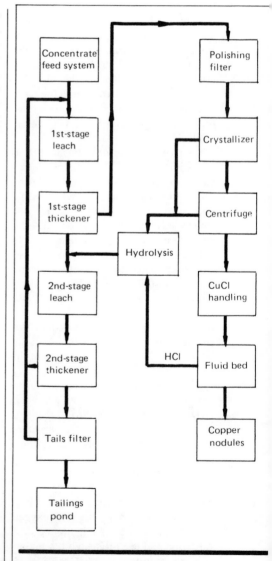

Cymet process uses a chloride leach

Smelting vs. leaching—a general comparison

Despite the emissions hurdle, and over a decade of work on hydrometallurgical (leaching) routes, pyrometallurgy (smelting) is still favored for metals that occur in sulfide form—e.g., copper, lead and zinc. Indeed, on balance, the more modern pyrometallurgical methods, such as flash smelting, still get the nod over leaching for sulfide ores. Smelting is preferred even in the case of nickel, 80% of which occurs in low-grade laterites.

Nevertheless, many hydrometallurgical processes are being developed and used for both sulfide and oxide ores (which do not lend themselves readily to smelting). In general, leaching is expected to become more prevalent in the future, as sulfide ores become depleted or lower in grade.

Smelting requires greater expenditures for pollution-control equipment than does leaching, but this is countered by the greater investment needed for hydrometallurgical processing-system hardware. All told, estimates a leading U.S. copper producer, capital costs generally run on a par for the two technologies for a 50,000-ton-Cu/yr plant. Scaleup economies slightly favor smelting for higher capacities—to around 100,000 tons/yr—and pyrometallurgy's investment advantage becomes more pronounced for still-larger units. Conversely, leaching looks economical for smaller-than-50,000-ton/yr facilities, for which smelting often isn't feasible.

The two technologies also differ appreciably in their operating-cost profiles. Leaching generally requires from two and a half to three times more energy than smelting for each ton of copper produced, says the metal firm, and this is because of the expense of pumping process liquids, and electrowinning (plating out) copper from solution.

On the other hand, hydrometallurgical processing cuts labor expenses by about half, adds the same producer, noting that neither technique holds a clear-cut edge in operating costs. Of course, pyrometallurgy boasts a key plus—it is an established production technique, while hydrometallurgy remains to be proven on an equally large scale.

---

copper, the equivalent of blister, is shipped for melting and electrorefining. Elemental sulfur is recovered as a byproduct.

More work still needs to be done on the process, and it is not yet ready for license, says a company spokesman, who neverthelesss declines to say whether the operation is profitable. However, he asserts that "we believe we have a viable process and we expect that at some time in the future we will have something that is competitive with a new smelter."

Not far from the Duval plant, Cyprus Mines Corp. is taking another look at its Cymet process, following the acquisition of Cyprus by Standard Oil Co. (Indiana) in 1979. The company has just restarted the 12-ton/d (concentrates) pilot plant* and plans to run it for about six months. A company spokesman says a decision will proba-

*Cyprus started up its pilot plant in 1975, but has made extensive process changes in the intervening years.

bly be made next summer on whether or not to build a commercial plant.

In the Cymet process, sulfide concentrates are leached in two steps in a chloride solution and crystallized to obtain cuprous chloride crystals (elemental sulfur is removed during the process). The crystals are reacted with hydrogen in a fluidized bed of sand to form hydrogen chloride, while the copper clings to the sand particles. Copper is obtained by melting the agglomerate in a furnace (*Chem. Eng.*, Nov. 7, pp. 58-60). Potential commercial advantages are relatively low capital and operating costs. Energy consumption is said to be low because the hydrometallurgical steps are done at atmospheric pressure and low temperatures, and hydrogen reduction is used rather than electrowinning.

A process said to produce electrolytic copper for less energy than that needed by conventional processes, and that also generates elemental sulfur as

a byproduct, has been patented by Envirotech Research Center (Salt Lake City). Called ElectroSlurry, it uses a special "stirred ball mill" to grind slurried chalcopyrite concentrate to a fine 2-3 μm, then reacts the pudding-like product with an acidic copper sulfate solution to separate the ore's iron content, which goes into the liquor as iron sulfate.

In a conventional electrolytic process, iron is undesirable because it etches copper from the cathode. In the case of ElectroSlurry, it is part of the standard copper sulfate-sulfuric acid electrolyte.

The iron is oxidized to ferric iron and this oxidizing agent leaches copper from the copper sulfide for deposition on the cathode. Elemental sulfur is recovered simultaneously.

Current density in the cell is 60-100 amp/ft², vs. 15-20 amp/ft² for a conventional process, so copper deposition is faster. Yet energy consumption is said to be lower because ferric iron production takes less voltage. Envirotech is seeking an industrial partner to share the cost of a demonstration plant.

No such plans lie ahead for Anaconda's Arbiter process, which the company has been working on for the past six years. The route, based on leaching with gaseous oxygen and an aqueous-ammonia/ammonium-sulfate solution, was shelved more than three years ago because of high operating costs and assorted mechanical problems. In past test-runs, says a company spokesman, much of the anode copper produced was not pure enough and had to be fed back into the smelter.

However, hydrometallurgy does have a future in heap or vat-leaching of low-grade, oxidized ores. Milton Wadsworth, associate dean of the College of Mines at the University of Utah (Salt Lake City), estimates that 12-15% of the U.S.'s copper production is obtained this way.

Holmes & Narver Inc. (Orange, Calif.) has patented a process called "thin-layer" leaching, which produces cathode copper by leaching thin layers of ore with strong sulfuric acid. This is followed by solvent extraction and electrowinning. Sociedad Minera Pudahuel Ltda. is now starting up the first commercial plant, located near Santiago, Chile.

Gerald Parkinson

Search is on for clean ways to produce nonferrous metals

Routes that offer shortcuts to the end product, reduce steam requirements, or extract such valuable byproducts as cobalt are among the most promising in the latest crop of leaching and smelting methods for nickel, zinc and lead.

☐ Part I of this series covered the latest copper smelting and leaching developments worldwide.

This article will detail similar advances related to the processing of nickel, zinc and lead. Like copper producers, those who process Ni, Zn or Pb ores must meet increasingly tighter emissions standards, so it is no surprise that SO_2 reduction is a prime motivator behind many of the smelting and leaching techniques described in this article.

REDUCTION SMELTER—Pyrometallurgy is the primary method for processing nickel ores and is likely to remain so for the foreseeable future, despite emission problems and the fact that only about 20% of the world's nickel deposits are in sulfide ores, the normal source for smelting. (The other 80% is in lower-grade laterites, for which various hydrometallurgical processes are used or being developed.)

Inco Metals Co. (Toronto), the world's largest nickel producer, is working on a new smelting process to reduce SO_2 emissions problems at its Sudbury (Ontario) smelter, where sulfide ores are processed. Currently, production must be cut back to meet emissions standards when the smelter operates at full capacity. The company spent $14 million over nine years researching cleaner hydrometallurgical processing methods before deciding that the approach was not valid technically or economically.

The conventional pyrometallurgical method is to roast the ore, then smelt it in a reverberatory or electric furnace. Inco's new process, which is being piloted on a commercial scale at its Thompson (Manitoba) smelter, is called reduction smelting. If successful, it could be applied at both the Sudbury and Thompson facilities. Inco declines

Originally published January 12, 1982

to give process details, except to say that it produces a high-strength SO_2 gas suitable for the production of sulfuric acid.

Testing is scheduled for completion by mid-1981 and an economic feasibility study will be available by the end of the year. However, the company points out that even if the results are favorable, one potential problem is the disposal of the large amount of acid that would be produced.

As part of its program to reduce SO_2 emissions at Sudbury, Inco is also installing an improved magnetic separation and flotation process to increase the removal of pyrrhotite from the ore. Pyrrhotite, an iron sulfide, is high in sulfur content, but contains less than 0.5% nickel, against 1-1.5% for the rest of the ore. Inco has been removing some of it over the years, but the improved separation process is expected to reduce the amount of sulfur going to the smelter by 25%.

Inco extracts nickel laterites (oxide ores) in Guatemala and Indonesia, but processes them by smelting rather than hydrometallurgy. Sulfur is added to the ore while it is being reduced in a rotary kiln; this produces a nickel sulfide, which is then smelted in an electric furnace to obtain a matte.

"Most laterites are processed by this method to produce ferronickel," observes Hugh C. Garven, process manager for Inco. (Laterites normally have a high iron content and also contain small amounts of other metals, such as cobalt.)

LEACHING LATERITES—Hydrometallurgical routes for nickel sulfides consume more energy than pyrometallurgical ones. However, if laterite ores are the raw material, smelting consumes twice as much fuel as leaching techniques, according to J. L. Blanco, vice-

president and general manager of Amax Nickel Refining Co. (Braithwaite, La.). This is because pyrometallurgical paths use fuel oil and sulfur to make a synthetic sulfide ore prior to smelting. Garven notes that the cutoff point below which smelting becomes less economical than leaching for laterites is around 2% nickel.

Amax has developed an acid-leach process for nickel laterites that is said to use less energy per unit product than the conventional ammonia leach (Caron) process. Also, extraction of cobalt, often present in laterite, is said to be about twice as effective with the

Amax method—an important factor, since cobalt has been selling for around $25/lb, against about $3/lb for nickel.

Amax's process is a modification of a method developed many years ago at Moa Bay (Cuba), in which sulfuric acid is used to leach ore in a series of steam-agitated pressure vessels. This is said to be more economical because it eliminates the initial roast-reduction step of the Caron process.

Amax claims to have cut steam requirements more than 50%, mainly by increasing multistage flashing and improving steam recovery. Energy losses have been cut by the use of such materials as titanium and high-temperature ceramics. However, the next step in the process—the extraction of nickel and cobalt from the leach solution—has not yet been worked out. Amax is studying various solvent extraction methods to decide which would be most suitable.

Another process that includes selective extraction of cobalt as well as nickel from nickel laterites is being piloted by the U.S. Bureau of Mines' Albany Research Center (Albany, Ore.). This is a variation on the Caron process, but the roast-reduction step is followed by an oxidizing ammonia-ammonium sulfate leach rather than the Caron's ammonia-ammonium carbonate method (*Chem. Eng.*, Nov. 3, 1980, pp. 43-45). The process is designed for the extensive, but low-grade, laterites of southern Oregon and northern California, which contain only 0.8-1.2% nickel.

TWO FOR FERRONICKEL—Hanna Mining Co. (Cleveland), whose nickel division produces ferronickel for steel-making at Riddell, Ore., has developed an improved ferronickel process that will be used on laterites in a 100-ton/h plant at Montelibano, Colombia. Scheduled to go onstream about mid-1982, it is a joint venture between the Colombian government, Holland's Billiton International Metals B.V., and Hanna.

Normally, ferronickel is made by mixing a reductant (coal or coke) with crushed ore, preheating, and then smelting. In Hanna's new process, the ore and coal are pelletized together. The advantages are said to be dust control (it cuts the capital investment for dust collection) and better contact of ore and reductant. This permits more selective reduction of the nickel

and iron in the ore (chemically, iron and nickel are very close) and therefore yields a higher-grade ferronickel.

Another variation on the ferronickel theme will be used by Yugoslavia's FENI, which is scheduled to put a 36-million-lb/yr (contained nickel) plant onstream late in 1981. In FENI's process, the low-grade laterite ore will be concentrated by magnetic separation to remove high-iron particles. The remaining ore, higher in nickel content, will be pelletized, reduced with lignite in a traveling-grate kiln, then smelted in an electric furnace. Davy McKee Corp. (San Mateo, Calif.), the engineering company on the project, says the process is about 25% less expensive than hydrometallurgy.

SULFIDES, TOO—An exception to the "sulfides are for smelting" axiom is the Fort Saskatchewan (Alberta) plant of Sherritt Gordon Mines Ltd., where the company first commercialized its SO_2-free pressure-leach process for nickel sulfide ores 26 years ago. Nickel sulfide concentrates (or matte from a smelter) are suspended in an ammonia-ammonium sulfate solution, then leached under pressure at high temperatures to oxidize the sulfides and extract the metals—first nickel (hydrogen reduces nickel out of solution), then copper and cobalt. Once the metals have been recovered, part of the ammonium sulfate is recycled, but the bulk is evaporated to obtain ammonium sulfate crystals, used for making fertilizer products.

The sale of fertilizer is an important factor in the economic balance of the process. The plant's capacity of 35 million lb/yr of nickel is relatively small, and the Alberta location is good for fertilizer sales.

ZINC RECOVERY—Sherritt Gordon's process is used in Australia, South Africa and the Philippines, and an adaptation of it will be employed to recover zinc at Cominco Ltd.'s Trail (B.C.) operations (*Chem. Eng.*, Aug. 13, 1979, pp. 104-106). Texasgulf Canada Ltd. also will use the process at Timmins (Ont.), where it will handle excess capacity in the refinery's electrolytic tankhouse to produce another 50 tons/day of electrolytic zinc.

Sherritt Gordon and Cominco have also piloted a pressure-leach process for copper [chalcopyrite] concentrate said to recover all the values and convert the sulfide into elemental sulfur.

"We don't have a commercial application for it yet because the economics are not much different from present smelting techniques," says Herbert Veltman, director of Sherritt Gordon's research center at Fort Saskatchewan. However, he says the capital cost of a grassroots zinc pressure-leach plant (none is planned as yet) would be "about 20-25% cheaper than the standard roast-leach approach and the operating costs are basically identical." He adds that the company has discussed the process with a number of zinc producers.

A leaching process for obtaining zinc, lead, copper and other metals from complex ores is being developed by BOM's Reno Research Center (Nev.) in cooperation with Cities Service Co. (Tulsa, Okla.). The firm makes zinc concentrate from a sulfide ore at Copperhill, Tenn., but the declining grade is making it less suitable for smelting. In the BOM process, sulfide concentrates are subjected to a chlorine-oxygen leach at 110°C and 40 psig, then the metals are selectively removed. In the case of zinc, the zinc chloride solution is further purified by cementation with zinc metal, then evaporated to get zinc chloride crystals. Zinc is recovered by electrolysis.

CLEANER LEAD PROCESSING—The lead industry is in a difficult situation because it claims there is no way of meeting the U.S. Environmental Protection Agency's proposed ambient lead standard of 1.5 micrograms/m³, and the OSHA workplace standard of 50 micrograms/m³ over 8 hours. Last year, the Lead Industries Assn. challenged the standards in federal court on the grounds they are technically and economically not feasible. (According to the trade group, enforcement of these rules will force the closure of 45 lead-producing plants totaling nearly 1.5 million tons/yr of refined-lead capacity.)

Two Supreme Court decisions were made last month (*Chem. Eng.*, Dec. 29, 1980, p.12). In one, the Court granted a partial stay of the OSHA standard until it has a chance to act on an industry challenge to the rule. The other decision, however, endorsed a lower court ruling that upheld the EPA standard, scheduled to go into effect in October 1982.

Actually, the latter rule already is being implemented, but compliance timetables for the OSHA regulations have been pushed back because of the legal appeals.

One answer may be a hydrometallurgical process being developed by the BOM Reno Research Center with the support of leading lead producers. Concentrate of galena (lead sulfide) ore, produced by conventional flotation, is leached by a solution of ferric chloride and sodium chloride at 95-100°C. The ferric chloride oxidizes lead sulfide to produce lead chloride and elemental sulfur. The presence of the sodium chloride and the temperature prevent the lead chloride from precipitating. The ferric chloride becomes ferrous chloride in solution.

Elemental sulfur and other unreacted material are filtered out, then the solution is cooled to about 20°C to crystallize the lead chloride. The crystals are washed with water, then put in an electrolytic cell containing molten eutectic salt (lithium chloride-potassium chloride). Electrolysis decomposes the lead chloride into chlorine and molten lead metal (the cell temperature is about 450°C). The chlorine goes to a chlorination or absorber tower, where it is contacted with ferrous chloride to regenerate ferric chloride for the process.

The lead metal is 99.99 to 99.999% pure, says BOM's Morton M. Wong, so no further refining is needed (smelted lead must be refined). Other advantages are: no SO_2 emissions and very low lead emissions, since all unit operations are enclosed or contained. Energy consumption is no more than that of conventional smelting and refining, he says.

BOM Reno has a process development unit that produces 500 lb/d of lead metal. The unit first started up in November 1978, and Wong says tests have proved the feasibility of operating it on an integrated basis continuously, and have confirmed that the materials of construction can resist the corrosive ferric chloride solution.

However, Charles Bounds, of St. Joe Minerals Corp., says that while "in general we are totally in support of the route the Bureau has taken . . . it is very unlikely it will be commercial before the end of the decade because there are too many technical hurdles to be overcome." These include selection of economical materials of construction, the tricky solid-liquid separation step and solid-waste disposal, adds Bounds.

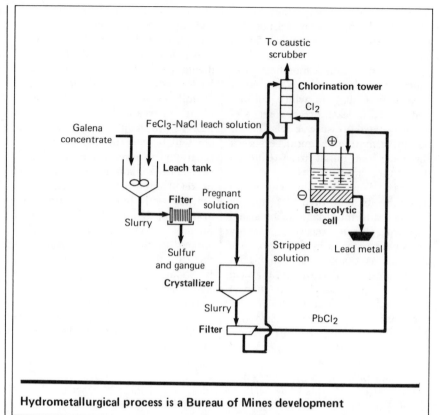

Hydrometallurgical process is a Bureau of Mines development

Another hydrometallurgical process for recovery of lead from galena has been piloted by Minemet Recherche (Trappes, France). This process also uses a solution of sodium chloride and ferric chloride. After leaching, the solution is cemented with electrolytic lead powder so that impurities, including other metals present (e.g., zinc, copper, silver), may be removed. The solution is then treated by an ion-exchange resin to fix remaining copper and silver; lead powder is recovered from the pure lead chloride solution by direct electrolysis.

On the smelting side, a lead smelting plant that uses a new, continuous direct process developed by professor Paul E. Queneau, of Dartmouth College (Hanover, N.H.), and Reinhardt Schuhmann, Jr., of Purdue University (West Lafayette, Ind.), is being built in Duisburg, West Germany, by two German lead producers—Preussag AG Metall (Goslar) and Berzelius Metallhütten GmbH (Duisburg)—with support from the West German government (*Chem. Eng.*, July 28, 1980, p. 9).

The plant, which will handle 240 m.t./d of galena concentrate, is being designed and built by Lurgi Chemie-und Hüttentechnik GmbH (Frank-furt/Main). The firm has exclusive rights to the process and has piloted it in a 2-m.t./h plant in Duisburg. In the QSL process, as it is called, smelting is done in a single reactor, as opposed to the conventional two stages of sintering and blast-furnace smelting.

Lead concentrates are pelletized together with lime, silica, iron oxide (if necessary) and the flue dust generated in smelting. The pellets are charged continuously to the horizontal reactor, into a melt. Oxygen is blown continuously into the melt through submerged injectors to oxidize the sulfides in the lead. The advantages claimed: all reactions are done in a single sealed reactor, fossil-fuel costs are cut by more than a third, and more-concentrated SO_2 is available for cheaper gas clean-up and sulfuric acid production.

The Soviet Union's novel KIVCET flash process will be used for a small lead smelter being built in Bolivia. The KIVCET direct-reduction furnace has a vertical smelting shaft and an electric furnace. The two are divided by a water-cooled wall that penetrates almost to the bottom of the slag bath. The wall permits the melt to pass through into the furnace while SO_2 gases stay behind.

Gerald Parkinson

Copper routes assessed in economics report

A new study tells which are the most energy-efficient processes in a group of 21 proven and unproven hydrometallurgical and pyrometallurgical techniques.

☐ It's no secret among U.S. copper producers that hydrometallurgy (leaching) cannot compete economically with pyrometallurgy (smelting) in the extraction of copper from sulfide ores* (*Chem. Eng.*, Dec. 15, 1980, p. 31). Proof of this fact can now be found in the pages of a new report†— by the University of Utah (Salt Lake City)—that is filled with revealing economics data on 13 pyrometallurgical and 8 hydrometallurgical techniques (see tables p. 86).

The study covers only the energy requirements of the processes, starting with the mining of ore and ending with refined copper, for conventional-size plants (100,000 tons/yr). But the conclusions are the same for the overall economics, says Milton E. Wadsworth, one of the authors, who is with the University's Dept. of Metallurgy and Metallurgical Engineering.

Capital costs are about the same for both types of plants, he adds. Leaching becomes competitive with smelting only when plant size comes down to about 25,000 tons/yr. The obvious conclusion: hydrometallurgy will likely be used only in small facilities, at least for the next ten years.

SOME SPECIFICS—Total energy requirements for the most efficient hydrometallurgical methods range from 87 to 94 million Btu/ton of cathode copper, vs. 80 to 90 million for the currently most efficient pyrometallurgical ones. (Energy-consumption calculations take into account the fossil-fuel energy equivalent of all forms of energy purchased, and that of all major materials consumed—e.g., chemicals.)

Energy consumption is usually greater for hydrometallurgical systems because most of these employ electrowinning for metal reduction. The two exceptions (and two of the most energy-efficient) of those studied were Cyprus Mines Corp.'s Cymet process and the roast sulfite reduction route (an old, unpatented technique that is still in the bench-scale stage), which use chemical reductants.

Also, hydrometallurgical processes have high steam requirements, and make inefficient use of reaction heat. However, the study notes that leading routes are at an early stage of development and offer "considerable potential for improvement." It suggests that energy use may be reduced by developing more-efficient methods of electrowinning (which consumes 21-24 million Btu/ton of cathode copper) or switching from electrowinning to chemical reductants. Other suggested improvements are more-effective heat transfer from the exothermic leaching reactions to the reactants, and better waste-heat recovery.

Pyrometallurgical processes are less energy-intensive because they produce blister copper, which is electrorefined. Also, matte-smelting methods, such as flash smelting and continuous smelting, derive energy from combustion of the sulfides and are more efficient than the conventional reverberatory process, which is "clearly obsolete," the study says.

Flash-furnace technology is described as "modern, transitional, proven technology," and the Noranda Mines Ltd. and Mitsubishi Metal Corp. methods as "modern, continuous, proven processes." Electric-furnace smelting is the most energy-intensive of the pyrometallurgical processes analyzed, but the study says the addition of fluid-bed roasting could cut its fuel requirements by about 3 million Btu/ton of cathode copper.

THE FUTURE WAVE—Of the so-called "newer, unproven processes," INCO Ltd.'s top-blown rotary converter, commercialized for relatively small-scale operations, is described as "limited in size, operates best on high-grade mattes and has the added capability of fire refining."

The Queneau-Schuhmann oxygen-sprinkle smelting process is undergoing a commercial test at Phelps Dodge Corp.'s Morenci, Ariz., smelter, but the firm's continuous-smelting process is still in an early stage of development, although a 50,000-ton/yr lead smelter using the same principle was built recently in Europe (West Germany).

The AMAX dead-roast blast-furnace smelting, and the firm's segregation processes have been piloted, but not commercialized. The Thermo Electron Corp. process, in which concentrates are reacted with chlorine gas to form cuprous chloride, ferric chloride and elemental sulfur, was evaluated from bench-scale tests.

The University of Utah study concludes that the pyrometallurgical reactor of the future, with a production capacity of 100,000 tons/yr, most likely will be a single continuous vessel capable of minimizing convective and radiative heat loss and fugitive emissions, and effective in removal of impurities.

Interestingly, in-situ extraction of copper from a deep deposit by fracturing and leaching the ore body appears to be the most energy-efficient of all, even with only 50% copper recovery. However, the study notes that the process is not proven, that experimental work is needed to demonstrate that 50% recovery is possible, and that it would be applicable only where the geological structure permits containment of the pregnant solution.

Gerald Parkinson

*These account for about 90% of U.S. copper output and most of the world's estimated commercial reserves.
†Done for the U.S. Dept. of Energy's Div. of Industrial Energy Conservation.

Originally published April 6, 1981

Energy use in pyrometallurgical routes, million Btu/ton cathode copper   Table I

Category	Process	Fuel, utilities, chemical reactants	Total consumption, mining to refined copper
Older proven processes	Conventional smelting (green charge)	35.16	93—107
	Conventional smelting (calcine charge)	30.92	
	Electric-furnace smelting	42.97	
	Outokumpu flash smelting	18.92	82—84
	INCO flash smelting	21.25	
Newer proven processes	Noranda continuous smelting	24.00	
	Mitsubishi continuous smelting and converting	19.76	83—92
	Oxy-Fuel reverberatory smelting	28.62	
New unproven processes	Top-blown rotary-converter smelting	23.56	
	Queneau-Schuhmann continuous smelting	22.63	
	Oxygen sprinkle smelting	22.43	83—87
	AMAX Dead-roast blast-furnace smelting	19.58 + 1.5*	
	AMAX segregation	21.03	
	Thermo-Electron chlorination	20.20	

To obtain the total energy required for producing a ton of cathode copper, the energy for mining and concentrator operations must be added. For a 98.7% recovery in the smelting operation, the estimated mining energy is 20.13 and concentrator energy is 42.57 million Btu/ton cathode copper.

*Value reported by AMAX for electrorefining is approximately 1.5 million Btu less than value used for other processes. Also, energy requirement for fugitive-emissions control is not included.

Source: University of Utah.

Energy use in hydrometallurgical routes, million btu/ton cathode copper   Table II

	Concentrate to refined copper		
Process	Fuel, utilities, chemical reactants	Energy consumption range	Total consumption, mining to refined copper
Roast leach electrowin*	30.45		
Cymet ferric chloride leach	30.92	24—31	87—94
Roast sulfite reduction	23.64		
Electroslurry-Envirotech	39.61		
University of Utah/Martin Marietta ferric-sulfate acid leach	49.45	40—50	103—113
Sherritt Cominco	48.13		
Arbiter ammonia leach*	62.05		
Nitric-sulfuric-acid leach	74.45	60—75	123—138
In-situ solution mining			About 80

*Processes that have been used commercially. To obtain the total energy required for producing a ton of cathode copper, the energy for mining and concentrator operations must be added. The same value used in Table I has been added, for 98.7% recovery in the smelting operation.

Source: University of Utah.

Section IV
Organic Chemicals

Ethanol
Other Organics

Low-energy processes vie for ethanol-plant market

Does it cost a lot to convert corn into ethanol? Not really, say a group of firms offering modern routes keyed to an expected alcohol-plant buildup. Reduced energy demand and unusual distillation setups characterize the processes.

☐ Now that a capacity buildup of grain alcohol for automotive fuel seems virtually assured by the Carter Administration program (*Chem. Eng.*, Mar. 10, p. 80), several process developers are ready with new technology aimed at making corn-based ethanol production an economically attractive proposition.

These firms are using a wide variety of lures to attract potential ethanol producers. Some are claiming a reduction in capital costs, but most emphasize energy efficiency. And although the processes for sale remain firmly grounded in the familiar operations of mash cooking, fermentation and distillation, many feature ingenious cost-cutting ideas, mainly in the distillation step.

By these and other means, say the ethanol-plant designers, they have been able to cut down substantially the amount of energy needed to ferment and distill a gallon of corn-derived alcohol—estimated at up to 152,000 Btu for older plants* (*Chem. Eng.*, Feb. 26, 1979, p. 78). The average figure for the new crop of processes is about 40,000-45,000 Btu/gal, but some of the methods lay claim to much lower energy demands.

Several engineering firms now offering ethanol technology save energy by *not* producing distillers dried grain (DDG), which is made by evaporating and drying the distillation bottoms.

One gallon of anhydrous alcohol contains about 84,000 Btu, so older plants sustain in a net energy loss in making ethanol.

Originally published March 24, 1980

Others rely instead on one or more byproducts to improve the overall economic picture. The following processes are a representative sample of available methods.

UNUSUAL AZEOTROPE—Robert F. Chambers, president of ACR Process Corp. (Champagne, Ill.)—a firm with considerable experience in ethanol technology (*Chem. Eng.*, Jan. 29, 1979, p. 41)—is one who believes that an alcohol plant that relies heavily on several non-alcohol byproducts will cost more, partly because a lot of money is spent on additional equipment to process those byproducts. "Besides," he observes, "selling all the nonalcohol items, which are usually food supplements, can take a considerable marketing effort."

Last month, ACR started up the U.S.'s first entirely computerized, continuous grain-based facility—a $3-million, 3-million-gal/yr anhydrous alcohol plant for Big D&W Refining and Solvents, Inc. (Van Buren, Ark.). The ACR design is based on a proprietary process that is the subject of about 13 patent applications covering equipment and all parts of the route except fermentation.

Although Chambers won't reveal specific details, he says the Van Buren plant has a unique distillation setup that uses basically any kind of commercially available gasoline (unleaded is preferable, though) as the azeotrope. This, he claims, saves energy in distillation and avoids the need

for ethanol denaturing, thus shaving an additional 2¢/gal off production costs. "On a 3-million-gal/yr plant, the elimination of denaturing saves about $60,000 a year," he notes.

In addition, the computerized operations reduce labor requirements and optimize efficiency. Overall, says Chambers, capital costs are lower, and production costs are cut by about 30¢/gal. This means that the cost of making grain alcohol is reduced from an average of $1.30-$1.40/gal to about $1 or $1.10—still costlier than the wholesale price of gasoline.

One big saving is in process steam. ACR claims that its process uses less than a third of the energy required by any existing grain-alcohol plant that

Purdue fermenter boosts yields

makes DDG. This translates into a steam demand of less than 30 lb/gal of product, vs. about 109 lb/gal for conventional units.

LOW-PRESSURE TWIST—Inseco Associates Inc. (East Brunswick, N.J.) is offering technology—under a non-exclusive agreement—developed by Grain Processing Corp. (Muscatine, Iowa). Starting with a corn wet-milling operation, the process releases only hydrolyzed cornstarch (minus the oil, meal, gluten, proteins and cellulosics) into the fermenters, and yields a yeast-cell byproduct instead of DDG.

Inseco president Paul P. Eggermann says that the technique con-

Chemapec ethanol process yields several important byproducts

sumes about 40,000-43,000 Btu/gal for distillation. But work is underway to reduce this to possibly less than 20,000 Btu/gal. Eggermann believes that the conventional three-column distillation setup for anhydrous alcohol (in which 190-proof liquid is taken off the second column, and distillation with a benzene azeotrope in a third column yields 199.6-proof product) can be reduced to just one unit.

Eggermann won't disclose any trade secrets, other than to say that the one-column setup would involve operating at sufficiently low pressures. "Not much data has been published on the physical characteristics of water/ethanol mixtures, especially when we get down to very low pressures," he says. Inseco is now designing a pilot plant to test this concept; it will be ready for startup in a few months.

The firm has also been looking into other ways of dehydrating water/ethanol mixtures, and has discussed with Union Carbide Corp. the possibility of using molecular-sieve installations. This kind of dehydration, says Eggermann, is very energy-efficient. But even though it consumes only 5,000 Btu/gal of alcohol, the equipment is too expensive right now. He estimates the total cost for a 3-million-gal/yr alcohol plant at $750,000 ($600,000 for the basic hardware, the balance for the sieve pack).

BYPRODUCT CORNUCOPIA—One process that claims very low energy demands (a total of 20,000-25,000 Btu/gal) and a decidedly rosy energy balance is that of Chemapec, Inc. (Woodbury, N.Y.)—the U.S. subsidiary of the Zurich-based Chemap group, which specializes in fermentation technology.

Crucial to this technique (see figure) is the production of such non-alcohol items as corn oil, corn meal (for animal or human consumption), animal feed from corn residues, fodder yeast, CO_2, and methane.

According to René Loser, chief executive officer of the U.S. company, the energy balance for a 20-million-gal/yr facility that Chemapec will build at Ames, Iowa, works out like this (per gallon of ethanol): 41,000 Btu to grow the corn, plus 20,000 Btu to make the alcohol and all byproducts, adds up to a total consumption of 61,000 Btu. But energy obtained from the alcohol (84,000 Btu) and the

byproducts (67,000) offsets this dramatically. The net energy gain is 90,000 Btu. The firm indicates an ethanol production cost of 50-70¢/gal, after taking credit for all byproducts.

Chemapec, which has filed for patents on its fermentation and distillation technology, employs a dry milling process that doesn't result in DDG production. Ethanol is dehydrated in what the firm calls a "thermo-compression" system. Although Loser won't elaborate on details, he says this is essentially a normal distillation setup under pressure. However, its energy balance is said to be better than that of conventional systems (Chemapec claims a steam requirement of only 18 lb/gal of product).

Some energy actually is recovered within the system. For example, the firm's Anamet process subjects the distillation wastewater (slops) to both aerobic and anaerobic treatment to reduce BOD. This yields methane with an energy equivalent to 7,000 Btu/gal of product—which can be fed to the boilers to generate process steam.

Although the Chemapec route has not been used in any commercial-size U.S. plants, "some of its parts have," observes Loser. For instance, the Anamet system is now producing methane for a beet-sugar plant located in Minnesota. The distillation system is in operation at a Swedish plant that makes ethanol from potatoes.

In addition to the Ames plant, Chemapec is currently doing design work on seven other projects for construction in Nebraska, Indiana, Iowa and Michigan.

SMALL-SCALE PLANTS—Low capital and production costs, plus small size, are some of the attractions of a design offered by Farm Fuels Inc.—an Indiana-based firm owned by Richard M. McGhee, a consultant based in Missouri City, Tex., and a group of Indiana entrepreneurs (*Chem. Eng.,* Dec. 17, 1979, p. 70).

Basically, Farm Fuels is selling the idea of ethanol production to farmers. In order to make 50,000 gal/yr (based on a 300-d schedule), all a farmer needs, says the firm, is $250,000 and 100-200 acres planted with corn or milo (a grain sorghum).

The Indiana company offers proven, if not revolutionary, technology that produces anhydrous alcohol containing 1% fusel oil and ethyl acetate, and slops that contain all the protein and fat present in the original feed. The plants, says McGhee, save energy by not evaporating and drying the distillation wastes into conventional DDG. Instead, the distillation bottoms are fed directly to farm animals.

Ethanol production cost can be as low as 20-45¢/gal, excluding amortization costs and assuming zero cost for raw materials. McGhee's estimate reckons on the farmers' feeding all their corn to the ethanol facility, and letting the slops—along with such non-cash feeds as pasture, silage, and corn stalks and cobs—take over the corn's traditional nutritional role on the farm.

According to McGhee, the Farm Fuels scheme has other attractions. It doesn't, for example, discharge any liquid effluents or pollutants. And the cost picture can be further sweetened if the farmer is able to depreciate the plant in five years as farm machinery, instead of the 17-20 years usually needed for big chemical plants.

McGhee has drummed up some Congressional support for this idea, and indicates that all that is needed is a favorable Internal Revenue Service ruling. Of course, the concept would also benefit from all the other tax incentives that may accrue to ethanol production.

The first Farm Fuels plant is now being completed in Indiana, and start-up is expected by April 1.

GETTING MORE FROM CORN—Elsewhere in Indiana, Purdue University's Renewable Resources Engineering Laboratory has about 50 people working on various ethanol technologies. (This work receives funding from the state, industry, the U.S. Dept. of Energy, the U.S. Dept. of Agriculture, and the National Science Foundation.)

George Tsao, director of the laboratory, considers corn the best material to use from an engineering viewpoint, but calls it expensive at a price of $2.50/bushel.*

Purdue has been operating a pilot plant for the last 4-5 years in order to demonstrate the practicality of a corn-fed process that is available for licensing. Its key feature: the fermenting organism is a mold that feeds on both cornstarch and fiber. Since corn has 70% starch, and 15-20% fiber, which

*One bushel equals 56 lb.

is normally not converted into alcohol, the Purdue process is said to have a good potential for increased yields.

Tsao says that one bushel of corn usually yields 2.4 gal of ethanol; the Purdue method yields 3 gal/bushel—a 20-25% yield increase.

Of course, the corn fiber, which is mostly hemicellulose, must first undergo acid hydrolysis.

Because most of the fiber is converted into useful product, it does not turn up in the distillation wastes (fiber accounts for the bulk of the solids in such wastes). The distillation bottoms, adds Tsao, are a very dilute suspension of mold cells, and this is easily disposed of. Energy that would normally go into evaporating and drying the waste material is saved.

Total energy consumption is about 45,000 Btu/gal, based on conventional distillation. Corn protein and oil are isolated and sold as byproducts. Purdue is now working on a novel distillation twist: The alcohol/water mixture is distilled to about 80-90% ethanol, and a dehydrating agent is employed to remove the rest of the water.

Raúl Remírez

New ethanol route wears a low-energy label

☐ It's an idea of beautiful simplicity—to use a material derived from corn to dramatically reduce the energy needs and capital cost of making grain alcohol for gasohol.

And now, an undisclosed midwestern equipment manufacturer is putting the interesting technique into action: a 792,000-gal/yr anhydrous ethanol plant, scheduled to go onstream by yearend, will employ cornmeal as an adsorbent to dehydrate the aqueous ethanol mixture. The plant will need only one third as many columns and less than one third as much distillation energy as conventional grain-alcohol facilities.

The concept has been developed by Purdue University's Renewable Resources Engineering Laboratory (West Lafayette, Ind.), with support from the U.S. Dept. of Agriculture, and was recently described in a paper presented by Michael R. Ladisch, Juan Hong, and Marcio Voloch at an American Chemical Soc. meeting held in Las Vegas. The University has applied for patents on the process.

CUTTING ENERGY NEEDS—The Purdue development can be seen as part of an industrywide effort to reduce the amount of energy needed to ferment and distill a gallon of corn-derived alcohol. This is estimated at an average of about 40,000-45,000 Btu/gal for newer, energy-efficient methods

Originally published November 17, 1980

(see article on low-energy ethanol processes, *Chem. Eng.*, Mar. 24, p. 57).

In a conventional, three-column distillation setup, 190-proof liquid is taken off the second column, and distillation with a benzene azeotrope in a third column yields 199.6-proof product. Energy requirements, says Ladisch, vary from 16,000 to 40,000 Btu/gal.

In contrast, the Purdue process is said to obtain anhydrous product from an 8-12% ethanol/water mix for approximately 12,000 Btu/gal.

Aqueous ethanol is distilled to 80-90% C_2H_5OH, which requires about 9,500 Btu/gal. Ethanol vapor leaving the top of the column passes through a bed of cornmeal; the result is anhydrous product.

The cornmeal adsorbent, Ladisch says, would cost only 4¢/gal of ethanol, based on 20 cycles of use, and assuming no credit for the spent cornmeal (which can either return to the front end of the process or serve as animal feed).

Because of its low heat of adsorption, cornmeal, which contains 65-75% starch and 5-10% xylan, is easy to regenerate, and has proved superior (in tests) to such other dehydration alternatives as silica gel, lime and molecular sieves. Unlike silica gel, cornmeal shows little or no ethanol adsorption under normal conditions,

according to the Purdue researcher.

Although Ladisch will not now mention specific capital-cost figures, the economics of the Purdue route are expected to be substantially better than for extractive and/or azeotropic distillation processes, mainly because adsorber vessels would replace up to two distillation columns.

The facility now in construction will alternate two adsorber beds—while one is dehydrating the ethanol, the other will be regenerated with hot air, Ladisch says.

Regeneration is said to require about 2,000 Btu/gal of product.

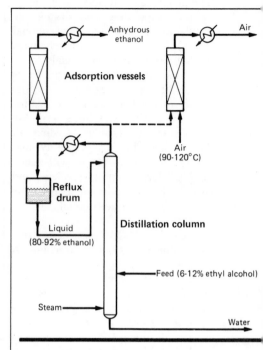

Purdue's ethanol dehydration route

Wood-to-ethanol methods edge closer to fruition

The race to get wood-conversion technologies onstream finds some top contenders scrambling to build and start scaled-up facilities by late 1982 or early 1983.

☐ Critics of the U.S.'s alcohol-for-gasohol program are doubtful that technology based on the conversion of a variety of cellulosic raw materials to ethanol will be ready in time to meet ethanol production goals set for 1990. But developers of cellulose-to-ethanol processes say that it will, and point to experimental work that could culminate in three or more sizable plants (of which at least one can be considered of commercial size) in operation by late 1982 or early 1983.

In an earlier article (*Chem. Eng.*, Oct. 20, 1980, p. 53), *CE* took a realistic look at the U.S. Dept. of Energy's expectations for cellulose conversion, described the leading technologies (raw-material hydrolysis into sugars via acid or enzymatic attack, plus conventional fermentation into alcohol), and discussed some of their strong and weak points. This article will describe some processes that are among the most advanced on the road to commercialization.

BIOMASS REFINING—Geoffrey Noble, vice-president of research and development for Iotech Corp. (Ottawa), says his firm is definitely the farthest ahead among those seeking to commercialize a wood-to-ethanol route, and believes that "we're going to see viable cellulose-conversion technology in existence before a lot of people think it's ready."

Iotech, best known for its development of a cellulose-pretreatment step under the auspices of the Solar Energy Research Institute (Golden, Colo.), is nevertheless into total "biomass refining" aimed at obtaining a wide variety of products—e.g., protein, animal feed, furfural, ethanol. The present emphasis, though, is on ethanol and lignin.

Wood chips of such hardwoods as poplar or aspen are treated with saturated steam at 240-300°C and 500-1,000 psi, in a vessel called a gun reactor. Cooking time can vary between 5 s and 5 min; at some point, the reaction is frozen at the peak of sugar production, and the wood is "exploded" into a fine powder at atmospheric pressure.

The company subjects the exploded product—a mix of lignin, hemicellulose and cellulose—to enzymatic attack, and the resulting C_5 and C_6 sugars are conventionally fermented into ethanol. Tests have recorded sugar concentrations of 10-12%, and alcohol concentrations of 5-6% before distillation.

Lignin can be recovered at the end of the process by filtration. The material, says Noble, is obtained in a very reactive, undamaged form. "In routes that isolate the lignin by extraction with ethanol solvent, the lignin tends to polymerize when it precipitates out of solution. We could take the lignin out this way after the steam-cooking pretreatment step, but then we would need extraction and solvent-recovery equipment, and these are expensive," he adds.

Iotech now uses a proprietary catalyst to speed up its explosion pretreatment, and treated raw material emerges with a texture similar to that of potting soil. The firm says it can now obtain the enzymes it needs directly from this exploded material; normally, the enzyme-producing bacteria can grow only on very pure substrates (high cellulose, low wax and tannin content)—e.g., cotton-gin trash.

The Canadian company plans to build the first commercial cellulose-to-ethanol plant in the U.S.—a 1,000-ton/d wood-processing unit consisting of four 250-ton/d modules. The first of these, says Noble, could be in operation by the fall of 1982, producing 8 million gal/yr of ethanol.

To reach this goal, Iotech has launched an accelerated development plan that involves, among other things, starting the engineering design before pilot-plant testing is completed. The hasty timetable is expected to increase project costs, which Noble estimates at $94 million for the four modules.

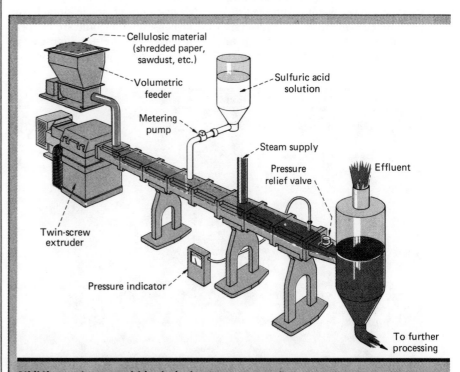

NYU's continuous acid-hydrolysis route uses a twin-screw extruder

Originally published January 26, 1981

Financial support is being sought from DOE, although Iotech claims to be backed by "investment groups in the U.S." The company says it expects to license the construction of as many as 30 plants in the 1980s. "That alone would be enough to supply half of the U.S.'s cellulose-ethanol target," says Noble.

HELP FROM EXTRUDERS—Another process well on the path to commercialization is the much-publicized route developed by New York University (New York City) with the help of the U.S. Environmental Protection Agency. In this process, which has been piloted in Long Island, cellulose waste (sawdust) is fed into the barrel of a plastics extruder, subjected to high temperatures (230°C) and pressures, and hydrolyzed with dilute (1-1.5%) sulfuric acid. Contact times in this continuous hydrolysis method are short (5-25 s).

What comes out of the twin-screw extruder (manufactured by Werner & Pfleiderer Corp., Ramsey, N.J.) is a mudlike material containing unconverted cellulose, decomposition products from the hemicellulose fraction, glucose and degraded lignin. After separation, a liquor containing 8-10% glucose is sent on to fermentation.

Barry A. Rugg, associate professor at NYU's applied science department, says the leftover lignin is sufficient to generate enough steam for hydrolysis and distillation, as well as enough to provide for the electrical requirements of a large commercial plant.

Pilot-plant work has not been able to increase cellulose conversion to glucose beyond the 50-55% level (standard glucose-to-alcohol yield is about 50%). The higher the temperature inside the extruder, the higher the conversion rate to sugars; however, the reaction must be stopped quickly to avoid glucose breakdown. At the moment, R&D work at NYU is focusing on sophisticated monitoring techniques to achieve very close process control, and Rugg believes this will lead to higher cellulose conversion—about 80-90% is possible, he feels.

While R&D continues, detailed engineering work for a demonstration plant—a 4-million-gal/yr module of a larger facility—probably will start this month, to be completed by the end of 1981. Actual construction could begin then, so that the demonstration unit could be ready for startup by late 1982 or early 1983. The participants in this project are Alcohol Fuels of Mississippi (Yazoo City), Vulcan Cincinnati (Cincinnati), Werner & Pfleiderer, and Riverside Energy Technology (New York City)—a consulting firm. Specifically, Vulcan will be involved in design and engineering of the plant, to be located near Vicksburg, Miss., while Werner & Pfleiderer will design the extruding equipment.

No machines of a size suitable for the demonstration module are in operation today, and the 4-million-gal/yr facility will need an extruder with a screw diameter of 480 mm, capable of processing 300 tons/d of feed. Rugg says that such a plant probably won't be economically viable. The cost of alcohol product (NYU has done some economic modeling) is very dependent on feed costs, which Rugg estimates at $30/dry ton.

MUNICIPAL-WASTE FEED—Technology for the Emert process—also known as SSF (for simultaneous saccharification and fermentation)—was developed years ago by Gulf Oil & Chemicals at its facility in Pittsburg, Kans. The company took the technique through the pilot-plant stage and then donated the knowhow to the University of Arkansas (Fayetteville), which is doing follow-up development work.

Since last August, the University has been focusing on detailed engineering and design of a 50-ton/d demonstration unit, and has licensed the process to a company—Cellulose Alcohol Development Co. (Cadco)—formed by several corporations (included are a farm cooperative, a forest-products firm, a chemical company, a waste-handling concern, and an energy company).

Cadco has arranged for DOE financing to build the 50-ton/d facility at Pine Bluff, Ark., and to run it during the first 2½ years. Construction will start in March, with completion set for early in 1982. Total cost will be about $19 million for construction, plus $10.5 million for operating expenses (during the 2.5-yr stint).

Cellulosic waste material, containing about 57% cellulose, is the feed. Typical raw material consists of two-thirds municipal solid wastes (MSW) and one-third pulpmill waste (PMW). About 15% of the MSW is sterilized and employed as feedstock in continuous enzyme production, while the remainder is mechanically treated and mixed with incoming PMW to feed the simultaneous saccharification and fermentation steps.

The SSF process is continuous in four trains of fermenters—each train consisting of three fermenters—with a total retention time of 24 h. The end-product is, of course, ethanol, but the lignin and undigested cellulose can be burned to obtain process steam, and the nonalcoholic soluble materials (protein, C_5 sugars, inorganics) are concentrated to a 60%-solids product that serves as animal-feed supplement. George Emert, former director of

Row of ethanol-producing reactors at the SSF pilot plant

Gulf's biochemical research activities, says that the material is comparable to soybean meal and casein.

BOTH CORN AND BIOMASS—For the past four years, under DOE sponsorship, the University of Pennsylvania (Philadelphia) has been working with General Electric (Fairfield, Conn.) to develop the Penn/GE process. The purpose is a complete conversion of biomass into liquid fuels (primarily alcohol) and byproducts.

Chips from hybrid poplar or aspen, or such pulpmill wastes as waste paper and cardboard, are first passed through a hammer mill for size reduction, and then cooked for an hour in an ethanol/water mixture at about 200°C. Degraded hemicellulose and lignin dissolve, but the latter precipitates upon cooling. Solvent ethanol is recovered by distillation, and the bottoms (stillage) contains degraded hemicellulose, which can serve as substrate for fermentation.

Kendall Pye, the principal investigator at the University, says that GE has developed a thermophilic bacterium that grows at 60-65°C and functions well at such temperatures (the Emert method employs a mesophilic organism that grows at near-ambient—28-40°C—temperatures).

Thermophilic operation, according to Pye, has its advantages: There is less potential for contamination from other organisms, and the rate of enzymatic attack is higher. Also, the enzyme is stable, retaining its activity longer. And the GE bacterium is said to be less inhibited by glucose than mesophilic organisms. "The equilibrium is way over on the side of glucose formation," says Pye.

The GE enzyme is said to be able to yield syrups containing 20% glucose, which translates into 10% ethanol concentrations suitable for distillation.

One way to market this technology, says Pye, is to persuade owners of corn-based alcohol plants to retrofit the Penn/GE process, so that the facilities are able to process both corn and cheap cellulosic biomass. "This alternative will become more attractive as corn prices rise to the level of $4.50/bushel," notes the investigator.

To retrofit, corn-based facilities would need another "front end," consisting of a mechanical breakdown step for chips, a cooking vessel, and provisions for enzymatic saccharification. It is possible, observes Pye, that some of the existing equipment for corn could do double duty and handle cellulose, reducing the added investment. "At present, we are looking at the minimal modifications that might be needed," he says.

A COMBINATION APPROACH—Georgia Tech's Engineering Experiment Station has been working since September 1978 under a DOE contract to develop and design a 3-ton/d process demonstration unit to make ethanol from cellulose. A detailed engineering design has been ready for a year, but Dan O'Neil, program manager at the University, says "bureaucratic snags" have delayed the project.

Now the University is working with several private firms—one of these is Nuclear Assurance Corp. (Atlanta)—and expects to start construction of the demonstration facility in March, with some funding from the Solar Energy Research Institute. The plant is expected to generate engineering data, and will be erected on campus.

Georgia Tech's approach employs a combination of existing technologies that, in O'Neil's words, "the university thinks are the best." Prominent among these are a pretreatment system and an acid hydrolysis reactor system patented in the U.S. by Stake Technology Ltd. (Ottawa). The former already has made it to the commercial stage, being used in plants sited in Florida and Maine to make animal feed from sugar-cane bagasse (*Chem. Eng.*, May 5, 1980, p. 57).

At the University, wood chips first undergo a steam-explosion step, followed by solvent delignification to remove up to 95% of the lignin content. Then comes a dilute-acid, low-temperature prehydrolysis procedure for hemicellulose, and finally cellulose hydrolysis in Stake Technology's tubular, multiple-pass, plug-flow reactor. The resulting sugars (glucose), of course, go through the usual fermentation and distillation to obtain ethanol.

The process yields a "pristine" lignin, now seen as a process fuel although O'Neil says it has value as a chemical feedstock (the economic studies do not take this into account). And the prehydrolysis step yields pentoses that are separated for potential marketing as animal-feed supplement.

Since last summer, Georgia Tech has been testing Stake's pretreatment in a 1-ton/h demonstration unit. O'Neil considers the method superior to batch systems that, he says, consume more steam.

Inside the hydrolysis reactor, material high in cellulose is treated with dilute acid at high temperatures. Residence times are short, and conversion is optimized by recycling any unreacted cellulose several times. O'Neil says the University expects to reach conversion rates of 80-85%, and adds that the program has looked at batch digesters and continuous extruders, but these are costly and consume more energy.

Preliminary economic studies show (in late 1979 dollars) that a commercial plant for treating 1,000 tons/d of oven-dry wood chips can make 25 million gal/yr of ethanol for $1.75/gal—i.e., at a price competitive with that of grain ethanol. (This estimate assumes a byproduct credit only for pentoses, and applies to a facility operating with a fixed-bed reactor. O'Neil says he expects the economics to improve with use of Stake's continuous hydrolysis reactor.) One drawback: Capital costs would be 50-75% higher than those of a grain-fed plant of equal size.

Raúl Remírez

Battelle maps ways to pare ethanol costs

A study indicates that the energy needed to remove water from ethanol can be significantly reduced by a number of nondistillation methods. The most efficient one appears to be solvent extraction under pressure.

☐ A solvent extraction process for removing water from ethanol, now under development at Battelle Pacific Northwest Laboratories (Richland, Wash.), promises to use as little as 13% as much energy as a conventional distillation does. This is significant because more than 50% of the energy used to make ethanol is consumed during distillation, says David E. Eakin, of Battelle's Chemical Technology Dept.

The process, funded by Battelle, was one of eight it recently studied under a U.S. Dept. of Energy contract to try to identify ways to reduce the production costs of gasohol. Battelle notes, in a report due to be released at presstime, that gasohol typically sells for 5-10¢/gal more than unleaded gasoline, due in large part to the required distillation procedure.

Of the eight processes analyzed, seven should use less energy than conventional distillation, but Battelle's approach appears the most energy-efficient, according to Eakin. In this process, still in the laboratory stage, a mixture of solvents is contacted with the ethanol-water mix and extracts the ethanol. This is done under slight pressure, so that when the pressure is subsequently reduced the solvent (not identified) flashes off and is recovered for reuse.

"Right now we are trying to optimize the solvent mixture," Eakin says. "We hope to be ready to go to a pilot plant in six to twelve months."

PARING DISTILLATION—In conventional ethanol production from biomass feed, two ("dual") distillation steps are normally used to extract ethanol from the 90%-water/10%-ethanol mix produced by fermentation. The first ("thermal") distillation step drives off most of the water to leave a solution that is 95% ethanol. This is followed by azeotropic distillation, in which a third component is added to permit complete separation of the ethanol and water.

The eight processes examined by Battelle would replace one or both distillation steps, but the energy savings in each case relate to the overall process—that is, to the 27,400 Btu/gal of normal two-step distillation. Eakin says that the improvement in overall process economics would just about equal the energy savings, since equipment, operation and maintenance costs for the alternative routes would be roughly equivalent to those for conventional distillation.

Only one of the eight techniques—straight vacuum distillation, with a single column—was found to use more energy than conventional distillation.

Other methods examined by Battelle were:

■ Arthur D. Little, Inc.'s carbon dioxide extraction process, in which

---

A number of separation routes promise to trim energy requirements

Type of separation	Ethanol, % Initial	Ethanol, % Final	Process	Energy needed, Btu/gal
Complete	10	100	Conventional "dual" distillation	27,400
Complete	10	100	Extraction with carbon dioxide	8,000-10,000
Complete	10	100	Solvent extraction	3,600[1]
Complete	10	100	Vacuum distillation	37,000[2]
To azeotrope	10	95	Conventional distillation	18,000
To azeotrope	10	95	Vapor recompression	6,400[1]
To azeotrope	10	95	"Multi-effect" vacuum	7,200[3]
Azeotropic	95	100	Conventional azeotropic distillation	9,400
Azeotropic	95	100	Dehydration via adsorption	1,200[4]
Azeotropic	95	100	Low-temperature blending with gasoline	3,000[5]
Azeotropic	95	100	Molecular sieve	4,700-6,270
Other	5	10	Reverse osmosis	500

1. Figure given is the thermal energy required to provide mechanical energy for the process.
2. For single-column distillation.
3. For three-column distillation.
4. For drying with CaO; energy requirements using fermentable grains would be considerably less.
5. Results directly in production of gasohol.

Source: Battelle Pacific Northwest Laboratories

Originally published June 1, 1981

liquid CO_2 is used to extract ethanol, then is depressurized to flash the CO_2. This was found to require only about one-third of the energy needed for "dual" distillation. The A. D. Little process, like Battelle's, replaces both distillation steps.

■ Vacuum distillation, which has not yet been applied to ethanol production, according to Battelle. A study of a three-column system predicted energy consumption at 60% of that needed conventionally.

■ Vapor recompression (yielding an estimated 40% energy saving).

Vacuum distillation and vapor recompression would replace the conventional thermal distillation step.

Three techniques that would replace only the azeotropic distillation step of conventional ethanol production were also examined. One was adsorption, using fresh corn feed as the dehydrating agent. In tests done by Battelle, the corn removed water from a 95%-ethanol/5%-water mix in the vapor phase, for a 30% energy saving. "We did some laboratory work in the liquid phase [also], but it wasn't very successful," says Eakin.

In a second project, Battelle blended a 95%-ethanol/5%-water mix with gasoline and dropped the temperature to −20°F to achieve phase separation of water. This showed a 23% energy saving.

The third substitute for azeotropic distillation was a molecular sieve, using commercially available materials to remove water. The overall energy saving was about 17%.

Gerald Parkinson

Ethanol from wood

Continuous-fermentation knowhow has been combined with a semicontinuous acid-hydrolysis reaction to produce fuel-grade ethanol from cellulosic material. This system also can be energy-selfsufficient.

Henry R. Mendelsohn and Peter Wettstein, Inventa AG

☐ Increasingly, attention has been focused on the development of technology for converting cellulose-containing material to ethanol (see the article on p. 93). Substances such as wood or municipal waste can be employed to make alcohol, suitable for use as fuel or chemical feedstock.

Inventa AG (Domat/Ems, Switzerland) has developed a dilute-acid-hydrolysis process for the conversion of cellulose to ethanol. This technology is based on the wood-saccharification process operated from 1941 to 1956 by Inventa's mother company, EMS Industries, in Domat/Ems; here a plant produced 10,000 tons/yr of fuel-grade ethanol using locally available softwoods (mainly spruce).

The new route offers significant improvements over the older commercial process: The hydrolysis reaction, which was a batch step, is now semicontinuous. Yields of alcohol are higher (now 240 L/dry ton of wood versus 220 based on spruce). Energy consumption has been reduced; steam consumption is now 14-16 tons/ton ethanol versus 23.5. And consumption of chemicals has been reduced (from 735 kg sulfuric acid/ton of ethanol to 500-600 kg).

Since 1980, Inventa has operated a demonstration plant, also at Domat/Ems, for the development of its new technology. This unit has three hydrolysis reactors, the largest of which is over 30 ft high with the ability to process 2 tons/d of wood. This plant has also been used to test a number of other cellolosic materials— including bagasse, straw, and waste cellulose from both the wood industry and municipal sources—which pose special processing problems due to the absence of lignin, or due to a low cellulose content.

WOOD AND ACID—In Inventa's technology, dilute sulfuric acid is used as a catalyst to depolymerize cellulose to monomer sugars (pentose and hexose); the hydrolysis reaction products are then neutralized and the hexose sugars are fermented to form ethanol. The ethanol is then recovered and purified by distillation.

Before processing, wood logs are fed to debarking machines, because the bark contains compounds that will adversely affect fermentation. So this material is used as fuel, and the logs are converted to chips.

Wood chips then enter the hydrolysis reactor. The plant may require several reactors, depending on the wood types used and their cellulose content. These units will operate at about 10 atmospheres and at 140-180°C, starting at 140°C to convert the hemicellulose to pentose sugars.

The 140°C temperature is chosen to avoid excessive decomposition of pentose to furfural. The C_5 sugars pass through the process unchanged, to be used to produce methane by anaerobic fermentation. The temperature of reaction is then raised to 170-180°C for the cellulose constituents that provide the hexose sugars required for conversion to ethanol.

Before the hydrolysis reaction, the wood is contacted with hot water without acid, which allows the chips to swell. To catalyze the reaction, the wood is then contacted with dilute (0.6 wt %) acid. This concentration provides acceptable reaction times: 40-60 min for the hydrolysis of the hemicellulose, and 150-170 min for the hydrolysis of the cellulose. This acid concentration also provides the best yields (see Table I).

Because little was known about the corrosion during hydrolysis, tests were done to find the proper materials of construction for the reactor. The Inventa vessel uses three different steels to guarantee a plant life of 15 years and corrosion rates of 0.2 mm/yr.

PLUG FLOW—Filling of the reactors takes place in a staggered time cycle. Wood chips are exposed to dilute acid in small successive plug flows, each plug being removed from the reactor before the next is introduced.

Acid concentration affects yields of fermentable sugars   Table I

Basis: Wood tropical eucalyptus; Wort ratio kg/kg WDS ≅ 10:1; Temperature 177°C.

H_2SO_4 conc. (%)	Cellulose conversion (%)	Yield hexose kg/kg WDS* (%)	% of theory	Yield lignocell residue, kg/kg WDS (%)
0.45	76.5	24.7	51.7	39.0
0.60	89.5	30.6	64.2	32.5
0.80	96.0	29.7	62.2	28.9

*WDS: Wood dry substance

Originally published June 15, 1981

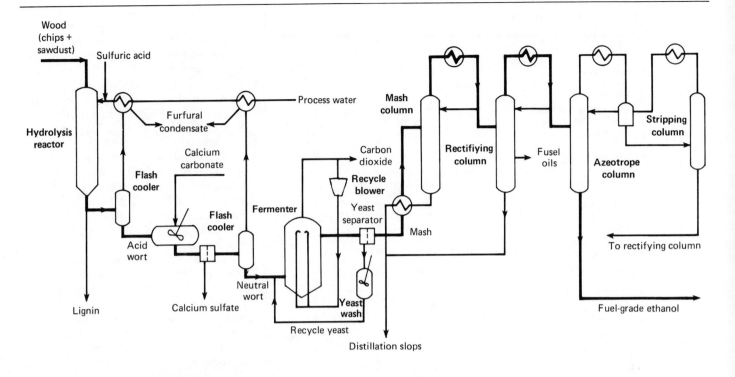

Ethanol produced by continuous fermentation is purified by azeotropic distillation

The interval between the plugs, the operating conditions of the reaction, and the concentration of the acid are all determined by kinetic and diffusion considerations. Basically, the wood chips are subjected to a series of cycles in which they are exposed to the acid. Attack by the solution occurs after the liquid diffuses into internally exposed surfaces via capillary pores in the wood. The released sugar diffuses to the outside of the chips by way of the same capillaries, and finally these sugars are washed out of the reactor by the next plug of acid solution. This method results in a conversion of well over 90% of the cellulose.

To limit raw-material costs, the use of acid and extraction water must be minimized. Using too much results in a dilute sugar solution and increased production costs. Therefore, the ratio of wood dry substance (WDS) to dilute acid is fixed at about 1:10, yielding a solution containing from 3 to 4% fermentable sugars.

FLASHED—The solution of acid and sugars, the wort, is removed from the base of the reactor and collected in a flash tank (where pressure is reduced from 10 bar to 0.7 bar). The lignocellulose cake leaving the reactor is removed and used for energy.

Flashing the wort accomplishes a number of things: The temperature of the solution is reduced; the flash system serves as a buffer between the semicontinuous reactors and the rest of the process, which is continuous; also, furfural is removed, and energy is recovered as waste heat.

When the solution is flashed, furfural is vaporized. The vapor stream is condensed in exchangers that use the heat of condensation to preheat the process water. Extra heat is used to produce low-pressure steam for the distillation section. The condensate, rich in furfural, is collected for processing in a furfural recovery section.

The cooled wort is then neutralized with limestone ($CaCO_3$) in a stirred vessel (CO_2 produced during the neutralization reaction is vented). The reaction is controlled so that sugars do not decompose and the solubility of calcium sulfate is limited. Almost all the gypsum ($CaSO_4 \cdot 2H_2O$) is formed as fine crystals.

A small percentage of organic acids that are present in solution must be preserved, since they are essential to the proper course of the later fermentation reaction. Neutralization is therefore terminated at a specific pH, and the gypsum is separated for disposal or conversion to ammonium sulfate and $CaCO_3$ for recycle. The wort is then passed to a set of flash vessels for cooling to 30°C, the temperature required for fermentation.

CONTINUOUS FERMENTATION—The C_6 sugars in the wort are converted to ethanol by a yeast strain specially adapted by Inventa to wood hydrolysates. The conversion takes place in a series of reactors split into two parallel lines for operating flexibility. The entire system is a continuous operation. Wort enters the plant and is mixed with nutrients that supply the necessary nitrogen and phosphates. Recycled yeast is then added to enable the production of ethanol to begin.

The fermenters are stirred by gas. The gas used in this system is carbon dioxide, scrubbed, compressed and recycled to the fermenters at a rate to ensure good distribution of the yeast. Complete conversion of the sugars to ethanol has occurred by the time the wort leaves the last fermenter.

The fermented wort or mash then passes a sieving unit in the yeast separators, where it is split into two streams; the yeast-free mash ready for

The acid hydrolysis system can be entirely energy self-sufficient	Table II

Basis: 10 million gal/yr of 200-proof ethanol

Investment $46 million

Raw materials
Wood (cellulose 58%)   35 lb/gal
Sulfuric acid, 100%   2.5 lb/gal
Calcium carbonate, 95%   3.0 lb/gal

Utilities
Steam and electricity are cogenerated in a power plant by the
combustion of lignin and methane produced in battery limits.

Net fuel wood required   zero

Personnel
Management   3
Supervisory   10
Operators   60

Credits
Furfural   0.43 lb/gal
Methanol   0.30 lb/gal

the separation of ethanol, and a concentrated yeast stream that is recycled to the front end of the process.

Before mixing with fresh wort, however, about 20% of the yeast stream undergoes a special wash to inhibit the growth of any unwanted organisms that might be present and to remove dirt. Both of these conditions could inhibit the efficiency of the yeast metabolism.

It has to be mentioned that sterilization is a major factor in maintaining a continuous fermentation. The design ensures that contamination does not occur even under tropical conditions. Yeast propagation, to replace inactive or dead cells in the system, is kept to a minimum (balanced against loss of cells in the fermenters) to maximize alcohol yields. Operating experience and continuous tests have shown that only about 3% of the yeast can be inactive. The mash stream leaving fermentation has an alcohol content of about 2% ethanol.

AZEOTROPE—Fuel-grade ethyl alcohol produced from the fermented wort is fed to a four-column distillation system designed to minimize steam requirements.

The feed to the mash column is preheated by the slops from the column bottoms. The cooled slops are then fed to the methane digestion system for boiler fuel production. Overhead product from the mash column is then fed to the rectifying column, where it is concentrated to an alcohol content of 94 vol. %.

Removal of the fusel oil (higher-boiling alcohols) is accomplished by withdrawing a side-stream having 60% alcohol from the center of the rectifying column and passing it to an extractive distillation unit (not shown) where higher-boiling components are stripped off. Ethyl alcohol is returned to the rectifying column. The fusel oil recovered is used as boiler fuel, if it is not sold as a byproduct.

In the azeotropic distillation column, the 94% alcohol solution obtained as overhead in the rectifying column is distilled with an entrainment agent, cyclohexane, to remove the residual water as an azeotrope and obtain a bottom product containing 99.3 vol. % fuel-grade alcohol.

The azeotropic mixture leaving the column as overhead is passed to a stripping column for recovery of the entrainment agent, which is then recycled to the azeotropic distillation column as reflux.

Methanol can be obtained from a methanol-water mixture recovered from the azeotrope column. This product can be added to the fuel-ethanol or burned in the steam boiler.

SELF-SUFFICIENT—The efficiency of this wood-to-ethanol process is defined by the ratio of the energy input (wood and process energy) to the energy output (in the form of produced ethanol plus the energy credits of the byproducts). The Inventa process can be energy-autonomous, with wood the only resource required.

The heating value of the wood feedstock introduced into the system amounts to 19,200 megajoules (MJ)/ton of dry softwood. This energy is split into the following products: fermentable sugars 6,300 MJ; other sugars 1,150 MJ; resins, waxes and decomposed sugars 3,100 MJ; and residual lignin 8,650 MJ. Moist lignin can be mechanically dewatered to some 50% moisture and directly fired in a boiler.

Unfermented organic substances are withdrawn from the process with the distillation slops. This material, due to its content of pentoses and organic acids, can be easily converted to methane, which can be used to generate steam. Steam produced from these byproducts (at a conservative boiler efficiency of 60%) is 2.84 tons for each ton of wood processed.

The energy requirements of the process are principally in the hydrolysis and distillation sections. The hydrolysis is carried out in a way to extract sugars as soon as they are formed. This leads to a dilute ethanol solution. And the energy consumption of the process is a function of this dilution. Yet, because of the availability of steam produced by burning the lignin and methane, this dilution is far less important a factor than one might normally suppose. If bark removed from the wood before chipping is burned as well, the energy requirements of the plant can be met without using additional firewood.

The energy efficiency of the process can be summarized as follows: One ton of dry wood results in 195 kg of 200-proof ethanol (65 U.S. gal); or 19,200 MJ feed energy in the form of wood results in 5,300 MJ of energy in the form of ethanol—an energy efficiency of 27%.

Reginald I. Berry, Editor

The Authors

Henry R. Mendelsohn was process coordinator in the Process Dept. of Inventa AG, responsible for the design of the ethanol-from-wood process. He has a B.Sc. (hons.) degree from the University of Cape Town, South Africa.

Peter Wettstein is a project manager in the Project Dept. of Inventa AG, Domat/Ems, Switzerland. He is responsible for the planning and analysis of wood alcohol projects. He has a Dipl. Ing. degree from the Swiss Federal Institute of Technology (Zurich).

Bright future for CO feed

Vinyl acetate plant
under construction by
U.S.I. Chemicals relies
on acetic acid made from
CO and methanol

**With natural gas and its derivative products
becoming scarcer and more expensive, the
chemical process industries are looking at
carbon monoxide's potential as a basic
building block in organic chemistry.**

☐ The simple carbon monoxide molecule is not as convenient a building block as ethylene for the more complex organic compounds. But someday, CO may take over ethylene's role as a key starting material or intermediate in many organic syntheses.

Some movement in this direction is already taking place. Both Monsanto Co. (St. Louis, Mo.) and BASF AG (Ludwigshafen, West Germany) have routes to acetic acid based on CO instead of ethylene. Both firms have licensed their technology to various users around the world. Halcon Chemical Co. (New York City) is developing technology that could make acetic anhydride or vinyl acetate monomer (VAM) directly via carbonylation, bypassing output of intermediate

products. Japanese firms are working on methods for producing CO-based toluene diisocyanate (TDI) and terephthalic acid (TPA). And Union Carbide Corp. (New York City) and others are investigating the possibility of a direct route to ethylene, and its derivatives, from synthesis gas.

BEHIND THE ACTIVITY—Two factors are the prime movers of these developments:

■ CO is only about half as expensive as ethylene (though its availability right now in feedstock purity and quantities is somewhat limiting). And projections indicate that CO's price should increase much more slowly than that of ethylene.

■ Continued development of coal gasification and liquefaction technolo-

gy, which many observers indicate should be commercially viable in the mid or late 1980s, will increase the availability, and should lower the unit cost, of synthesis gas—primarily a mixture of CO and H_2. (For more details on coal-conversion developments, see *Chem. Eng.*, Nov. 6, 1978, pp. 73-75.)

But what if coal-conversion technology does not develop within the timetable envisioned? Not to worry: there is plenty of CO around. In fact, the U.S. Dept. of Energy (DOE), in a report issued in November 1977, states that "almost 50% of the carbon monoxide potentially usable for chemical syntheses could be obtained from industrial byproduct gases."

Capitalizing on this resource is Tenneco Chemicals, Inc. (Saddle Brook, N.J.), whose Cosorb process, agree industry spokesmen, is by far the leading technology for extraction of feedstock-purity carbon monoxide from gas streams. Tenneco has five licensees already signed up, with four more waiting in the wings.

Despite this activity, many observers urge caution against being overly opti-

Originally published January 29, 1979

mistic about the immediate possibilities of CO-based technology. They claim that many process breakthroughs are really still only on the drawing-boards, and that marketing and economic factors could relegate CO-based routes to the distant future.

BYPRODUCT CO—According to the DOE report on chemical production from waste carbon monoxide, the estimated potential requirement for chemicals manufacture in the U.S. is greater than 51 million tons/yr. The report estimated that annual CO production as byproduct gas is approximately 142 million tons, of which about 63 million tons are used as fuel, and 21 million tons are exhausted or flared. The remainder is vented as dilute gas or oxidized to CO_2. (DOE figures date from 1974-76.)

The greatest amounts of byproduct CO are produced by the iron and steel industry—an estimated 117 million tons/yr. Large amounts are also produced by petroleum refiners (some 14.5 million tons/yr), ferroalloys makers (about 3 million tons/yr), and the coke, carbon black, phosphorus and aluminum industries (around 8 million tons/yr, combined).

Another source of byproduct CO is the array of organic chemical processes that involve partial oxidation. But, in its own words, the DOE report was "unable to identify any large-scale sources of high-concentration waste CO in the organic chemical industry, partly because industry is reluctant to discuss sources of CO, and partly because organic chemical producers are already well aware of the potential of waste CO as an intermediate."

In an effort to detect whether a trend is developing in capturing this byproduct CO for use as feedstock, *CE* polled several leading firms in each of the producing categories, and was unable to detect any moves in that direction.

A typical comment was that of a spokesman for a major integrated steel company: "We have looked at economic studies. There are well-defined, established processes for extracting CO and using it potentially as a chemical feedstock. But it hasn't made any economical sense to us, because if you take the CO from an existing fuel use, then you have to replace that [gas] with a higher-cost fuel. So, the economics just are not very favorable at present."

Other spokesmen note that CO in process offgases is often quite diluted, and that cleanup to the point where it could be used as feedstock might be quite complex and costly.

Another problem, pointed out by Charles A. Rohrmann, resident consultant at Battelle Northwest Laboratories (Richland, Wash.) and one of the authors of the DOE report, is the geographic location of suppliers in relation to potential users. In most cases, says Rohrmann, they are far apart, so the economy of recovery and use varies widely from case to case.

Two areas where capture and use of byproduct CO would seem to make most sense are in petroleum refining and organic chemical processes. Both would certainly be near potential users. But, according to the DOE report, although 90% of the byproduct CO from petroleum refining comes from fluid catalytic crackers, 70 to 80% of these units burn the CO to recover its heating value and minimize pollution problems.

The elimination of petroleum refining as a likely source thus leaves the offgases from organic chemical processes as the sole imminent resource for byproduct CO. And this is where Tenneco's Cosorb system is finding its niche.

CO-WINNING—Cosorb (*Chem. Eng.*, Dec. 5, 1977, pp. 122 − 123) recovers CO from gas streams at low pressure (about 1 atm) and ambient temperatures. It is effective on lean, as well as rich, gas mixtures, and can handle nitrogen-bearing streams that pose difficulties for competitive cryogenic routes.

The process relies on selective absorption of CO, complexing the compound with cuprous aluminum tetrachloride ($CuAlCl_4$) in a toluene solvent. Most other components in the offgas streams, such as nitrogen, hydrogen, carbon dioxide, oxygen and methane, being only slightly soluble, are not complexed. (Water, hydrogen sulfide and sulfur dioxide must be removed prior to Cosorbing, however, because they poison the solvent mixture.) The absorbed CO is then stripped out of the enriched solvent at 99 + % purity by a slight elevation of temperature.

Paul A. Lobo, Tenneco's director of corporate planning, sees carbon monoxide-based organic syntheses becoming far more prevalent as a result of

Cosorb because rising hydrocarbon prices make CO more attractive as a feedstock and because Cosorb can economically recover CO that might otherwise have to go out the stack.

Dow Chemical Co. (Midland, Mich.) was the first commercial user of Cosorb, with a 75-million-ton/yr unit recovering CO from acetylene-plant offgas for TDI and acrylates production. Israel's Koor Chemicals Ltd. has installed a smaller Cosorb unit to purify, for phosgene production, CO generated by naphtha reforming. The Yugoslavian firm Prva Iska Baric is involved in a TDI project that will include a Cosorb unit producing about 20 million lb/yr of CO. And in Cosorb's most recently announced application (*Chem. Eng.*, Nov. 6, 1978, p. 65), Romania's Romchim will tie the process with BASF's route for reacting CO and methanol to yield acetic acid. The Cosorb unit will produce about 5,000 normal m^3/h of CO from acetylene-plant offgas.

CO-BASED CHEMISTRY—The granddaddy of CO-based syntheses is the making of methanol, either via reforming of natural gas, or from synthesis gas produced by partial oxidation of hydrocarbons or coal gasification.

Both BASF and Monsanto have chosen methanol as the starting point for their acetic acid routes. (In conventional technology, acetic acid is made by conversion of ethylene to acetaldehyde, which then goes to acetic acid.)

BASF reacts methanol with CO in a high-pressure, gas-phase reaction to form acetic acid directly. Licensees for the technology include the Romchim plant, and Borden Chemicals in the U.S. Because of the "brute force" nature of this technology (reaction takes place at pressures of over 100 atm), the economics of the route are presently being questioned in some quarters because of high capital costs for high-pressure equipment.

Indeed, more interest is being paid today to Monsanto's liquid-phase technology that encompasses more-moderate reaction conditions. BP Chemicals Ltd. (London) confirms that it has picked Monsanto's process for its next acetic acid plant. The reason: BP's existing plants use "light distillate fuel," i.e., naphtha, which is getting short in supply and long in price. According to a spokesman, the switch to a methanol/CO route makes sense, since prices of these materials are

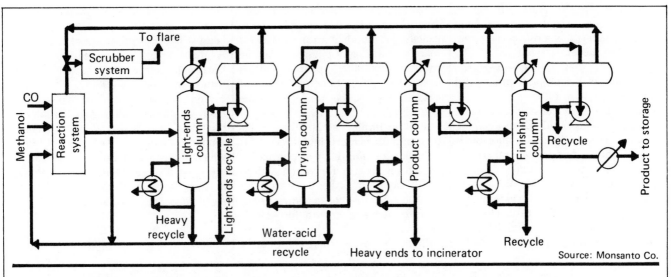

Monsanto's liquid-phase acetic acid route has been chosen for plants to be built by British and Japanese firms

dictated by natural-gas prices, which are relatively low, in Europe, compared with naphtha costs.

Monsanto technology will also find use in two acetic acid projects in Japan. Both facilities—one planned jointly by Nippon Synthetic Chemical Industry Co. (NSC), Osaka, and Mitsubishi Chemical Industries, Ltd. (MCI), Tokyo, and the other by Daicel, Ltd. (Osaka)—will use heavy residual oil as a raw material for CO production. NSC/MCI are scheduled to start up their 200,000-metric-ton/yr plant at the Kurosaki, Kyushu Island, works of Mitsubishi by 1981. The same year, Daicel will start operation of its 150,000-metric-ton/yr unit at Himeji City, near Osaka. Full-capacity operation of the two acid facilities will require 175,000 metric tons/yr each of carbon monoxide and methanol.

Du Pont Co. (Wilmington, Del.) and USI Chemicals Inc. (New York City) are engaging in a joint venture that would make use of both the Monsanto acetic acid route and a Texaco, Inc. plant that will make synthesis gas from high-sulfur crude oil bottoms. Du Pont will convert the syngas into methanol, part of which it will supply to USI and part of which it will sell or use internally. USI, in turn, will then react the methanol with CO from the Texaco facility to produce around 600 million lb/yr of acetic acid. The acid will then be reacted with ethylene to make VAM.

A BETTER WAY—John L. Ehrler, director of development for Halcon

Chemical Co. (New York City), points out that most acetic acid finds its way into two end uses: VAM and acetic anhydride. So, Halcon has been developing processes that could make either product, or both, by direct carbonylation of methanol. Ehrler reports that Halcon's moderate-pressure and -temperature catalytic route is very close to commercialization, having been demonstrated in continuous pilot operations. The technology is presently being "fine-tuned," according to Ehrler, and Halcon feels that the route will be more cost-effective than a new plant using conventional technology.

Ehrler indicates that the mostly likely application of Halcon technology would be a tie-in with a methanol facility, similar to the Du Pont/USI arrangement. The syngas for methanol and the CO for carbonylation would come from the same source—likely a coal or lignite gasification unit.

Intermediate-Btu gas (300 to 350 Btu/ft^3)—with some upgrading likely—meets Halcon process requirements, according to Ehrler. Halcon envisions that a syngas cost of $4 to $5/million Btu would be attractive for use in carbonylation, and that pipelining of intermediate-Btu syngas will be commercially viable—though not before 1985, because of the high capital cost of gasification and related environmental considerations. In the long term though (by 1990), Halcon is bullish about carbonylation processes being serviced by specific CO-generating facilities.

OTHER EFFORTS—Halcon does not

stand alone in its belief in the strengths of syngas. A spokesman for Exxon Chemical Co. (Houston) says that the firm is not doing anything that it could specifically talk about at present, but indicates that Exxon is taking a closer look at CO and H_2 chemistry for the long-term future. The company's interest, he says, is in the fact that syngas can be made from a wide range of feedstocks—and this is a great hedge against future peak energy shortages.

Celanese Chemical Co. (Dallas) is also believed to be studying the possibilities of syngas chemistry, though the firm would not comment.

Union Carbide is promoting the idea of a CO-plus-H_2 route to ethylene and downstream products, and BASF and Ruhrchemie, both in Germany, have also been doing work in this area. Knowledgeable observers feel, however, that commercialization is far off.

In the meantime, Chevron Research Co. (Richmond, Calif.) has applied for patents on a CO/formaldehyde route to ethylene glycol (*Chem. Eng.*, Jan. 15, p. 67). Du Pont had operated such a process until 1968, when it was shut down. Chevron will use a different catalyst, and expects to get around the pollution problems that DuPont had.

NOT ALL ROSES—A number of involved experts caution against being overly optimistic about CO technology in the immediate future.

Seaton Hunter, director of technology for petrochemicals for Monsanto's Chemical Intermediates Co., says that CO won't have a clear economic advantage for 20 years. "It has to happen," says Hunter, but he notes that the high cost of a conversion plant, and the moderate volume of CO needed by CPI firms, make the economics of the proposition doubtful for the near-term future.

A U.K.-based consultant concurs. "There's no doubt that CO is a coming thing," he says, "but the question is really when." And although there are a number of processes under development, he warns that not all of them can be considered new breakthroughs. For example, he believes that replacing phosgene ($COCl_2$) with CO, as in the TDI process proposed by Tokyo's

Mitsui Toatsu Chemicals Inc. (*Chem. Eng.*, Mar. 14, 1977, p. 73), is not exactly revolutionary, since CO is merely an analog of phosgene. And he cautions that other apparently novel techniques remain to be proven.

Other factors, too, can dampen the outlook for CO-based technologies.

For example, a 50,000-metric-ton/yr TDI plant using direct CO-reduction of dinitrotoluene, proposed by Mitsui Toatsu for completion in 1980, has been pushed back to 1982 or 1983 because of a slower-than-expected demand growth for TDI in Japan.

Similarly, sources there rule out any

near-term possibility of ethylene production from CO in Japan, partly because of the large volume of relatively unavailable CO required, and also because present ethylene capacity will suffice until the mid 1980s.

And depression of the polyester fibers market is discouraging Mitsubishi Gas Chemical Co. (Tokyo) from commercializing its TPA process, in which toluene and carbon monoxide are reacted to *p*-tolylaldehyde, which in turn is oxidized to TPA. (*Chem. Eng.*, Aug. 16, 1976, pp. 27–28).

Philip M. Kohn

Supercritical fluids try for CPI applications

The chemical process industries are seeking to exploit the special characteristics of fluids beyond their critical points. Regeneration of adsorbent materials and various extraction uses look promising.

☐ Compress carbon dioxide beyond its critical pressure, and heat it above its critical temperature, and the resulting supercritical fluid becomes an excellent solvent for removing organic impurities from spent activated-carbon filter beds. The CO_2 itself can be cleaned and regenerated; activated carbon is spared the heat degradation caused by thermal regeneration because CO_2's critical temperature (31°C) is only slightly above ambient; and there is no pollution or toxicity problem.

This application, now being readied for commercialization by Arthur D. Little, Inc. (ADL), Cambridge, Mass., exploits just one of several chemical process industries (CPI) end-uses for which supercritical fluids are currently being considered. Among the others:

■ Extraction and separation of asphaltenes and resins from distillation residua in oil refineries. Kerr-McGee Refining Corp. (Oklahoma City, Okla.) has already commercialized a

process that uses aliphatic hydrocarbons at supercritical conditions as a solvent. Installed at one of the Kerr-McGee refineries, the ROSE process (for residuum-oil supercritical extraction) has been licensed to Pennzoil Corp. (Houston).

■ Coffee decaffeination. Hag AG (Bremen, West Germany) is building a plant based on supercritical-fluid technology. The company has also patented use of the technique to reduce the nicotine content of tobacco, and obtain extracts from spices and hops.

■ Extraction of anticancer drugs and other substances from plants. ADL is currently working on these applications.

■ Extraction of light components of coal, leaving a gasifiable char residue. This work is being conducted by researchers at Britain's National Coal Board Research Establishment (Stoke Orchard, U.K.).

Further down the development trail are: (1) work by the Massachusetts

Institute of Technology (MIT), aimed at using supercritical water to gasify and reform coal; (2) research at West Germany's Nuremberg-Erlangen University, targeted at the separation of mixed compounds via supercritical extraction aided by an "entraining" substance; and (3) studies to replace azeotropic distillation with supercritical extraction—by researchers at the University of Manchester Institute of Science and Technology (UMIST), Manchester, U.K.

IDEAL SOLVENT—The ADL process (see flowsheet), which was actually developed at MIT, makes use of a desorption column to regenerate spent activated carbon with CO_2 fed from a storage tank at 300 atm and 35°C.

Regenerated material exits through the bottom of the column, whereas the CO_2 goes through a turbo-expander, in which pressure drops to about 80 atm. As the CO_2 expands, its temperature declines to considerably below 35°C, but a subsequent heat exchanger adjusts it to just below 35°C—a point at which CO_2 becomes a non-solvent for the adsorbates it carries. The now-two-phase system moves on to a cyclone- or hold-tank-type separator for adsorbate removal. Clean CO_2 is recompressed, cooled and stored for reuse.

Originally published March 12, 1979

According to Richard P. de Filippi, manager of ADL's supercritical-fluids unit, activated carbon is by far the most popular adsorbent. CO_2 was chosen as the regenerant because, at $60/ton, "it is cheap in comparison with other organic solvents, and is totally nonhazardous and nontoxic."

The process, says ADL, can operate in both batch and continuous fashion. And multiple-step separations of various adsorbate components are possible by dint of successive increases or decreases in temperature. This happens because the supercritical fluid may be a nonsolvent for one component at a particular temperature, while remaining a solvent for all other components.

Process economics compare favorably with those of thermal regeneration, ventures Michael Modell, of MIT's Dept. of Chemical Engineering. Lower operating costs are expected because the adsorbent is not subjected to cycles of heat-related expansion and contraction, so adsorbent replacement is lessened. As for capital costs, the major items of a supercritical process, notes Modell, are fluid-circulation compressors and high-pressure desorption vessels, and these would cost less than multiple-hearth furnaces or the associated air-pollution-control equipment used in conventional techniques (*Chem. Eng.*, Sept. 12, 1977, p. 95).

Modell explains that supercritical fluids make ideal solvents because their density is only about 30% that of a normal fluid. This is high enough to provide for good solvent capability, but low enough for high diffusivity and rapid mass transfer. A slight drop in temperature or pressure precipitates dissolved material.

Says de Filippi: "We're presently considering our commercialization strategy. Several varied-interest groups have approached us and we want to introduce the process as soon as possible, but we want to do a good, careful job on technical development first."

WEAPON AGAINST CANCER?—Encouraged by previous ADL work that indicated that supercritical fluids have potential in the extraction of drugs from plants, the consulting firm is getting ready to seek funds from the National Cancer Institute (NCI) to continue development studies.

Allan Branfman, senior staff member of ADL's Biomolecular Sciences Section, says that tripdiolide—a diter-

pene triepoxide believed to be useful against certain leukemias—has not yet entered into clinical trials in the U.S. because NCI researchers have had trouble obtaining it from a plant (*Tripterygium wilfordii*) via standard solvent-extraction techniques.

Branfman believes that the difficulty stems "possibly from the basic instability of triepoxides, but also because of interaction with the organic solvents used during extraction." Supercritical extraction with carbon dioxide, he notes, would provide an organic-solvent-free alternative. And there would be no need for evaporation to remove the solvent.

DEASPHALTING ET AL.—Elimination of the need for heat to evaporate the solvent is also the major strength of Pennzoil's application of Kerr-McGee's ROSE process. Pennzoil will use supercritical propane to deasphalt oils for use as feed in lubricating-oils manufacturing. The company will convert to the supercritical route a conventional, 2,700-bbl/d solvent extraction unit at its Rouseville, Pa., refinery. The revamping is scheduled for completion late this year.

A spokesman for Pennzoil says that the ROSE process "is expected to offer savings in utility costs of about $100/h of operation." An added bonus for new installations, says Kerr-McGee, is that capital cost for the ROSE process is about 20% less than for a conventional deasphalting route.

Kerr-McGee's own use of the supercritical technique is different. At the company's Wynnewood, Okla., refinery, the process extracts lubricating-oil stock and cat-cracker feed from atmospheric and vacuum-distillation residua.

Resid and a light, aliphatic hydrocarbon are contacted in a reactor, and then pressure and temperature are raised. At this point, heavy asphaltenes separate and are removed. Another temperature increase causes the separation of a resin fraction, and, after a final temperature boost, the solvent passes its critical point, leaving behind the desired oil product. Upon cooling, the solvent is ready for reuse.

A spokesman for Kerr-McGee says that the firm continually evaluates customers' feedstocks for applicability of the ROSE process. He indicates "active interest" in the process by other refiners, and adds that the company is also looking at non-petroleum uses.

DECAFFEINATING COFFEE—Supercritical extraction technology licensed from the Max Planck Institute for Coal Chemistry (Mülheim/Ruhr, West Germany)—the acknowledged birthplace of this kind of knowhow—will be used in Hag AG's decaffeination plant.

Beyond the fact that carbon dioxide will be the solvent, not much else is known about the company's plans. Hag has declined to comment on capacity, investment and operating costs, and onstream date.

Interest in this technique among U.S. food companies has been reported by a knowledgeable source. But *CE* has not been able to confirm whether the mentioned firms—General Foods, Inc. and The Nestlé Co., both in White Plains, N.Y.—are active in this area.

CHEMICALS FROM COAL—Britain's National Coal Board Research Establishment is planning to use its supercritical-gas solvent-extraction (SGSE) process in a 1-metric-ton/h development plant. The British researchers have been operating a 5-kg/h pilot plant since the fall of 1977.

SGSE relies on a light organic solvent (presently toluene gas) at 350 to 450°C and 100 to 200 atm to remove coal's light components; the resulting char residue can be gasified. Although up to 40% of coal weight can be extracted, economics are more favorable with less extraction (at least at this stage of development), according to the coal-board researchers.

The process is said to have several advantages over solvent-refined coal: (1) Gas/solid separation is easier than liquid/solid, (2) solvent recovery is high, (3) hydrogenation of the coal extract to a distillate oil is simple, and (4) the char is easy to gasify. The coal board hopes to bring onstream a 50-100-metric-ton/h SGSE demonstration plant by 1990.

NO-CHAR GASIFICATION—MIT's Modell believes he has found a process based on supercritical water for gasifying and reforming coal and organics. (Water's critical temperature is 374°C and its critical pressure 220 atm.) A patent on the technique was issued to MIT last fall. The university hopes to interest the U.S. Dept. of Energy in funding further work that would help define the exact chemistry of the reaction and characterize the fractions.

Modell explains that water is a poor

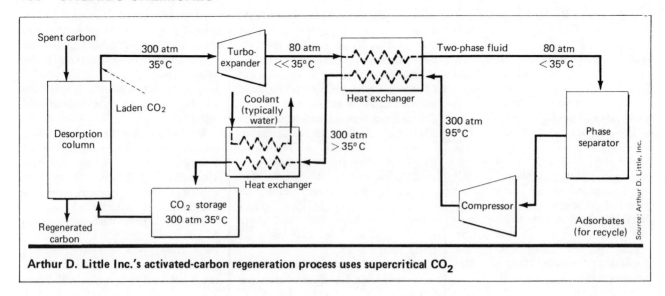

Arthur D. Little Inc.'s activated-carbon regeneration process uses supercritical CO$_2$

organic solvent at room temperature, primarily because of hydrogen bonding within the molecule. But because at its critical temperature water has only about one third of its normal density, most of the hydrogen bonds (whose strength is distance related) break, and the liquid exhibits an affinity for, and an ability to react with, organics.

Using glucose in a glucose/water system as a model for study, Modell has concluded that supercritical water prevents charring during gasification by keeping organic molecules from clustering and entering into condensation-polymerization reactions.

OTHER EFFORTS—Researchers at Britain's UMIST report that, under certain conditions, a liquid mixture contacted with a gas at supercritical temperature and high pressure exhibits liquid-liquid phase separation. The resulting three-phase equilibrium provides the basis for a separation process that, when tested with binary mixtures that show azeotropic behavior at normal pressures, may provide an alternative to azeotropic processes.

A research team at West Germany's Nuremberg-Erlangen University is working on a process that uses supercritical fluids and an "entrainer" to achieve separations. In a 1-kg/h bench-scale facility, the group has demonstrated—with supercritical propane, and acetone as entrainer—the separation of mono-oleic glycide from a mixture of glycides.

The entrainer is credited with enhancing separation and extraction, but its necessary removal adds a process step to the overall procedure.

Philip M. Kohn
Peter R. Savage
Silke McQueen, World News, Bonn

New technology sparks ethylene glycol debate

The unveiling of a giant U.S. facility that employs a direct, liquid-phase route is provoking fresh comparisons with the traditional vapor-phase methods. Experts have yet to pick a clear winner. Meanwhile, glycol producers are trying to improve the yield of vapor-phase oxidation.

☐ Ethylene glycol manufacturers, forever on the search for a direct route to the alcohol, will be closely watching the progress of a showcase plant in Channelview, Tex., currently going through startup motions. It is an 800-million-lb/yr facility owned by Oxirane—a subsidiary of Atlantic Richfield Co. (Los Angeles) and Halcon International, Inc. (New York City)—which will allow a comparison between the merits of novel, liquid-phase acetoxylation and those of older, conventional techniques based on vapor-phase oxidation of ethylene and further hydration to ethylene glycol. The latter routes now account for almost all U.S. production*.

It is difficult to gauge which will be the winning concept. In the U.S., four companies—Dow, Union Carbide, Scientific Design (a Halcon subsidiary) and Shell—have basic oxidation

*The old chlorohydrin process for ethylene oxide has been almost universally replaced by basic oxidation, although Dow still operates a small chlorohydrin unit at Freeport, Tex.

technology that uses either oxygen or air. Those favoring acetoxylation point to the traditional routes' poor yields in the ethylene-to-ethylene oxide step. But their opponents say that this problem is being alleviated by recent process improvements (including use of better catalysts). Their acetoxylation rival, they further note, suffers from corrosion problems and has large utility requirements.

Union Carbide has done a comparison of energy requirements for the acetoxylation process vs. conventional basic oxidation (see table), which shows the former to be a bigger consumer of energy.

CONJECTURES ON CHANNELVIEW— The new plant, which is twice the size of conventional units, represents the biggest single chunk of ethylene glycol capacity built in the U.S. As can be expected, the facility is still having startup difficulties (the process was never tried out in a pilot plant).

While, according to a Halcon spokesman, "the plant is running and producing fiber-grade material," Robert Malpas, Halcon's new president, freely admits that "the startup is rightly claiming a great deal of effort from Oxirane and Halcon/Scientific Design people."

Neither Oxirane nor Halcon has said much about process problems, but industry sources speculate that two of these could be corrosion and catalyst containment. Acetoxylation patents indicate that ethylene glycol is produced by reacting ethylene with oxygen in an acetic acid solution in the presence of a tellurium oxide/hydrogen bromide catalyst.

"On paper, the Halcon process is simple," says A. M. Brownstein, manager for new-ventures technology with Exxon Chemical Co. (Florham Park, N.J.). "The initial step of olefin bromination is followed by acetic acid displacement of the bromine to yield ethylene diacetate, plus monoacetate due to some partial hydrolysis. Displaced bromine is reoxidized by tellurium, which itself is reoxidized by oxygen."

Brownstein adds, "The trouble with this route is that it is costly to recycle acetic acid. You must distill to get the water out, and this takes a lot of energy. And running the reaction in glacial acetic acid creates a corrosion problem, so that investment cost is high."

Celanese Chemical Co. believes that corrosion problems at Channelview may parallel those of an acetic acid-consuming vinyl acetate process it had in operation until recently. Says David Medley, vice-president of marketing for Celanese: "We have had a lot of experience in liquid-phase chemistry, but from a corrosion standpoint, our process was in a league by itself. We had to use titanium equipment, and this added substantially to capital cost."

Oxirane won't confirm that it may be using titanium at Channelview, but admits that the new plant is "a giant of stainless steel and exotic metals."

COST PICTURE UNCLEAR— George Intille, program manager for Chem Systems, Inc. (New York City)—a company that periodically publishes in-depth analyses of ethylene glycol technology—believes that the Halcon route has some advantages over conventional basic oxidation. But he notes that the increasing cost of energy and

Ethylene use and energy balance of competing processes		
	Conventional	Acetoxylation
Ethylene use, lb/lb of glycols	0.646	0.486
Energy consumed by process,* Btu/lb of glycols	7,280	11,600
Byproduct energy,[†] Btu/lb of glycol	5,660	2,200
Net energy supplied by utility systems, Btu/lb of glycols	1,620	9,400
Gross energy used by utility systems[‡] Btu/lb of glycols	2,310	13,400

*Process heat losses, steam for separations, etc.
†Available from reaction heats and residue-fuel values
‡At 70% overall efficiency in the utilities area
Source: Union Carbide Corp.

Originally published January 15, 1979

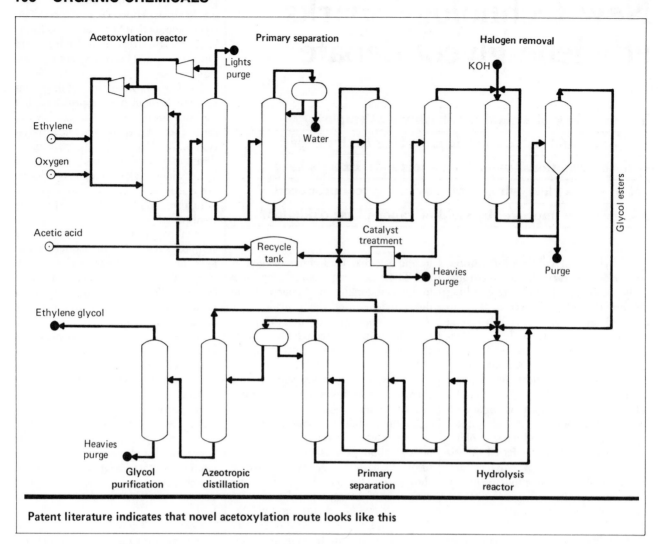

Patent literature indicates that novel acetoxylation route looks like this

raw-material ethylene could work against acetoxylation. Says Intille, "Clearly, the liquid-phase process has considerable utility costs, which can be offset by the higher yields. But as energy costs get higher, the new route's advantages begin to fade."

B. F. Smith, Union Carbide's engineering manager for ethylene oxide and ethylene glycol at South Charleston, W.Va., is more specific: "Rising energy costs render acetoxylation old before its time," he asserts.

One West European glycol producer, who disagrees with published appraisals of ethylene glycol technology, notes: "The size of the plant at Channelview makes me think that process economics are very much dependent on scale. The process may be suited only to the largest markets; if so, it is of potentially limited commercial importance."

In the U.K., ICI's Petrochemical Div. says it has developed a direct route to the glycol, but has no plans to

commercialize it. In fact, the company will stay with the Shell vapor-phase technique for a new 80,000-metric-ton/yr plant that is due for startup next year.

NEW CATALYST FOR OLD METHODS—While trying to assess the new techniques, many experts are keeping an eye on improvements of the old processes.

"If any of the present research efforts are successful, then the economics would improve tremendously for the vapor-phase routes," says E. H. Ivey, president of E. H. Ivey & Associates (Houston)—a firm specializing in ethylene technology. "Catalyst improvements such as that claimed by D. Bryce-Smith at the University of Reading [Reading, U.K.] could remove restrictions imposed by heat transfer and recycle problems, which lead to low conversion."

Bryce-Smith reports the development of a new oxidation catalyst, in which silver deposition is controlled to

create a unique structure that improves selectivity. Side reactions are reduced, and ethylene oxide yield is higher.

Catalyst development is still in the laboratory stage. But, says Ivey, "If it can be established, it changes the whole economic picture. A few percentage points in selectivity can add greatly improved capacity to an existing plant merely by changing the catalyst. The reaction is controlled to keep out of the explosive range, using real low ethylene concentrations [approximately 30%] and keeping the temperature down to acceptable limits."

REVAMPING THE OLD—Of the four U.S. companies that have basic oxidation (vapor phase) technology, only Scientific Design and Shell license their processes. The first offers air or oxygen routes, whereas Shell focuses on oxygen technology.

Robert Merims, a process engineer with Scientific Design, says that most ethylene oxide plants use oxygen rather than air. Some air-based facilities are actually being converted to oxygen use. A Texaco spokesman at Port Neches, Tex. (where the Scientific Design process is employed), confirms the conversion of two such units in order to "increase yields and further reduce air pollution emissions." Many experts believe that as producers start looking for pollution offsets, this kind of switch will increase.

Three of the U.S. exponents of basic oxidation technology claim that their processes have been considerably improved.

Reactor stability, according to Scientific Design's Merims, is the forte of his company's process. "The reactor operates with high gas recycle and has small-diameter tubes to provide a very high ratio of heat transfer to catalyst volume. With this good heat removal, temperature control of the catalyst surface is excellent and only trace quantities of inhibitor are required to maintain a high selectivity," he notes.

Shell claims that improvements in the catalyst and process have boosted yields. Recent Shell patents covering alkali-metal (cesium, rubidium and potassium) catalysis indicate ethylene-oxide selectivity of over 80% in some cases.

Union Carbide also reports improvements in basic oxidation. Says Smith, "Modern, large-scale basic oxidation units can take advantage of advanced machinery design and plant-wide integration of steam and power balances to optimize the use of byproduct heat from the reaction, remove heat and power from residual organics, and recover power from pressure swings. For example, we can recover an enormous amount of the mechanical energy used in getting the air up to reaction temperature. Use of multiple-effect technology enhances heat utilization."

OTHER PROCESSES—While U.S. producers busy themselves with comparisons between Halcon's acetoxylation route and conventional basic oxidation, two Japanese firms are developing techniques similar to Halcon's. And Chevron Research (Richmond, Calif.) is looking into a variation of an old ethylene glycol process.

In Japan, both Kuraray Co. and Teijin Ltd. have patents on liquid-phase schemes. Both of these are at the bench-scale level, although Kuraray hopes it will be able to license its knowhow to local and foreign glycol producers. This, say experts, is not likely to happen soon. For one thing, Kuraray has yet to scale up its technology. Also, the Kuraray catalyst is based on palladium—a very expensive component, so catalyst losses would have to be minimal.

In the Kuraray process, the olefin reacts with oxygen in acetic acid, in the presence of a mixed catalyst, to produce yields of up to 95%.

Teijin's patents indicate that the company has examined at least two processes. One of these is based on a copper-iron salt/bromine catalyst. U.S. sources suggest that Teijin's yields are not very good.

At the moment, neither Japanese company has plans for commercialization, partly because their polyester-fiber (an important end-use for glycol) capacities are too small to absorb the output of a commercial-scale ethylene glycol plant. The two companies also believe that it is impossible for them to sell ethylene glycol in open markets.

In the U.S., Chevron Research has applied for patents on a process based on the reaction between carbon monoxide and formaldehyde, using a hydrogen fluoride catalyst. This is a different version of an old Du Pont technique that uses the same reactants, but employs sulfuric acid as catalyst. In 1969, Du Pont shut down such a facility because of pollution problems.

PPG is also reportedly studying a modification of the carbon monoxide/formaldehyde process.

Meanwhile, Celanese's Medley says there could well be an ethylene glycol shortage in the U.S. this year. The Oxirane plant has not yet reached full capacity, PPG has shut down a 400-million-lb/yr plant in Puerto Rico, and a 230-million-lb/yr plant in Lake Charles, La., is being rebuilt after a fire last year.

Guy E. Weismantel

Phthalic anhydride made with less energy

A new reactor and new catalysts have reduced energy requirements for phthalic anhydride production. Heat produced during reaction has been increased, and as a result enough steam is produced to run air compressors.

Otto Wiedemann and Walter Gierer*

☐ The von Heyden process, which accounts for 45% of the world's phthalic anhydride (PA) production, partially oxidizes *ortho*-xylene or naphthalene to phthalic anhydride. This process, which has been used in over 60 plants with a total capacity of 1.1 million tons/yr, has been modified to reduce energy requirements. Electrical consumption has been slashed from 850 to 150 kWh per ton of PA, and fuel-oil requirements have been reduced to zero.

Reaction heat is used to generate steam, so that air compressors can be

*Mr. Wiedemann is with PSA Lizensverwertungsgesellschaft Chemische Fabrik von Heyden GmbH und Wacker-Chemie GmbH. Mr. Gierer is with Chemische Fabrik von Heyden GmbH.

run without electrical power. New catalysts have been developed, and construction of the reactor has been modified so that the ratio of *o*-xylene or naphthalene to air can be increased. This change in ratio has boosted the amount of heat produced during the exothermic reaction, and this additional heat is applied to the generation of steam for use in turbines and as process heat for distillation.

The low-energy process went onstream in October 1978 in a 31,000-metric-ton/yr installation for Veba Chemie AG at Bottrop, West Germany. Three other units are under construction, with startups scheduled for 1979. They are: a 23,000-metric-ton/yr unit at Northfield, Ill., for Stepan Chemical Co.; a plant in Taiwan for Union Petrochemical Co.; and one in Venezuela for Oxidaciones Orgánicas C.A.

In order to modify the von Heyden process, hydrocarbon-to-air loadings used in the reactor were brought into the explosive range. The lower explosive limit for *o*-xylene or naphthalene air mixtures is 47 g/normal cubic meter of air. For the new low-energy process, loadings were increased from 44 to 60 g/Nm³ of air, and the new catalysts, which are of the vanadium pentoxide type, have been tested with concentrations as high as 100 g/Nm³.

SAFE OPERATION—The new catalyst and the reactor construction work together to minimize the risk of explosion—the catalyst deters an explosion, and the reactor design minimizes the effects should one occur.

It has been determined that an explosion starting in the catalyst area of the reactor would be quenched by the heat capacity of the catalyst. To further control temperature and remove heat, the reactor is cooled by a recirculating salt bath.

For this process, a fixed-bed vertical-tube reactor is used. Tubes with an inner diameter of 25 mm hold the catalyst. To produce 31,000 tons of PA, 15,000 tubes would be used.

Although explosions are not known to occur in the upper reactor space, because of the risk the thickness of the reactor cover is 10 to 20% greater than what had been used in the older method. The cover also uses a number of rupture disks for pressure relief, should pressure exceed the permissible level. Tests performed by the Bundesanstalt für Materialprüfung (the federal agency for material testing in Berlin) have established specifications for such pressure relief. The agency determined explosion characteristics

---

Economic data

Plant costs	$, million
a. Engineering	$1.84
b. Equipment	8.75
c. Construction and steel	1.06
d. Assembling	2.12
e. Catalyst	0.303
f. License (lump sum)	0.437
Operating costs for 1,000 kg PA	
Electrical energy	150 kwh
Cooling water	40 m³
Surplus steam	0.8–1.0 metric ton
Onstream time (minimum)	8,400 h/yr

Basis: 30,000-metric-ton/yr PA plant. *O*-xylene feed. December 1978 costs. (1.9 DM per dollar). Steam turbine for compressor.

Originally published January 29, 1979

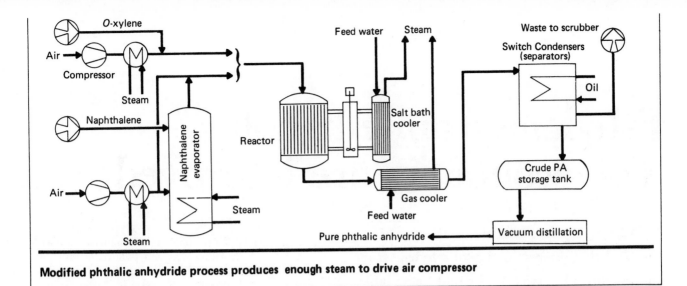

Modified phthalic anhydride process produces enough steam to drive air compressor

(of maximum pressure rise per second and maximum explosion pressure) for o-xylene and naphthalene air mixtures.

In addition, there is instrumentation to monitor hydrocarbon-to-air mixtures during startups and shutdowns.

OXIDATION AND DISTILLATION—In the low-energy process, air is used in the oxidation of o-xylene or naphthalene to phthalic anhydride. The crude PA formed is recovered by condensation and is then purified by vacuum distillation.

O-xylene or evaporated naphthalene is injected (see flowsheet) into a stream of compressed (1.5 bar), filtered and preheated air. This gas mixture then enters the reactor holding the catalyst.

During this reaction, and depending on the feedstock, either 1,293 kJ/mole of o-xylene or 1,742 kJ/mole of naphthalene is released. This reaction heat is removed by the salt bath, a mixture of potassium nitrate and sodium nitrite, with a melting point of 140 to 145°C. As this salt bath recirculates through the reactor and back to the salt-bath cooler, its temperature ranges between 375 and 410°C.

As the salt is cooled, medium- or high-pressure steam is generated. This steam can replace some of the electrical requirements of the process if it is used to drive a turbine for the air compressor. Additional steam is generated in an exchanger that cools the gas leaving the reactor.

Crude phthalic anhydride is then condensed in special separators, which are switch-type condensers.

With a melting point of 130°C, crude PA is separated as a solid. Hot and cold oil—or water and steam—are used alternately in the separators. As one section of the separator is condens-ing PA, the other is in a heating cycle melting the crude PA for recovery.

Air leaving the separators contains organic byproducts. This air is purified by water scrubbing or by thermal or catalytic incineration.

Crude phthalic anhydride collected from the switch condensers is typically 98.5% PA, with trace amounts of phthalic acid, maleic anhydride, monocarboxylic acids and phthalide. This PA is gathered in a storage tank, and then subjected to vacuum distillation for purification.

After distillation, the final product will be 99.9% pure, with trace amounts of maleic anhydride, benzoic acid and phthalide.

YIELDS AND ECONOMICS—Between 107 and 109 kg of pure phthalic anhydride are obtained from 100 kg of o-xylene. And 90-96 kg of pure PA are obtained from 100 kg of naphthalene (the exact amount depending on the quality of the naphthalene).

Costs and utilities for a phthalic anhydride plant with an annual production of 30,000 metric tons of PA are provided in the table. Costs for construction and steel refer to concrete and brick work, i.e., foundations and buildings; assembly costs refer to installation. All the figures are based on an o-xylene feed and the use of a steam turbine to drive the air compressors.

Use of process steam reduces electrical requirements and eliminates the need for fuel oil. The table presents costs based on the use of a condensing turbine (in which superheated steam would be reduced to condensate). Other steam produced in the process is used to supply heat needed for vacuum distillation. Still more steam would be surplus.

In contrast, although a steam-driven turbine does increase the investment cost, utility requirements without one would be greater. For each metric ton of PA (1,000 kg), 40 kg of fuel oil would be required; also, electrical consumption would be higher (530 to 600 kWh).

With a naphthalene feed, operating costs are slightly higher. (Energy is needed to heat and evaporate solid naphthalene.) Using the same economic basis, electrical demand would be 200 kWh per metric ton of PA. Also, onstream time would be 8,280 h/yr.

The recently developed catalysts are expected to have a three-year lifetime (they are now guaranteed for two years).

An added benefit: byproduct credits might be claimed. If the air leaving the phthalic anhydride separators is scrubbed, the wastewater will contain maleic acid. This can be used to produce either maleic anhydride or fumaric acid. For each metric ton of PA produced 40 to 50 kg of maleic anhydride can be reclaimed.

Licensor of the new low-energy process and the traditional von Heyden process is PSA Lizensverwertungsgesellschaft Chemische Fabrik von Heyden GmbH (Munich, West Germany). World licensees are the engineering firms of Davy Powergas GmbH (Cologne) and Lurgi Kohle und Mineralöltechnik GmbH (Frankfurt).

The authors

Otto Wiedemann is President of PSA Lizensverwertungsgesellschaft Chemische Fabrik und Wacker-Chemie GmbH. He holds a Ph.D. in chemistry from the University in Freiburg/Breisgau. **Walter Gierer** is technical Director of the Regensburg plant of Chemische Fabrik von Heyden GmbH. He has a degree in chemistry from the Polytechnikum in Nuremberg.

Ethylene from ethanol: The economics are improved

As petroleum prices increase, some countries may find that the production of ethylene from an agriculturally based feed could be a viable alternative to hydrocarbon cracking.

N. K. Kochar and *R. L. Marcell*, *Scientific Design Co.*

☐ A new catalyst—offering increased productivity and operating life—and a new reactor design have cut the cost of ethylene derived from the dehydration of ethanol.

Before 1945, a significant portion of the world's production of ethylene was based on ethanol dehydration. Later, the large-scale thermal cracking of relatively cheap hydrocarbons made ethanol dehydration uneconomical, and this route to ethylene disappeared almost overnight in western countries.

However, some developing nations eager to begin petrochemical production found the ethanol dehydration route very attractive. And in the 1960s, a number of plants based on the technology were established in Brazil, India, Pakistan and Peru, providing the first ethylene produced in these countries.

Today, several factors that govern the ethylene business again emphasize the advantages of ethanol dehydration. Petroleum prices are increasing; the size of an economical hydrocarbon cracker is increasing; immense capital (sometimes involving foreign currency) and skilled manpower are needed to build and to operate ethylene facilities based on petroleum. All told, from planning to plant operations, a decade-long effort is involved for a hydrocarbon cracking system. Now, ethanol dehydration may offer an alternative:

- It provides ethylene at the lowest investment per unit of product.
- It does not produce any byproducts (avoiding the burden of their processing and market development); consequently, even small ethylene markets can be serviced.
- Since ethanol is a feedstock available from an agricultural economy, this route increases domestic employment and avoids use of petroleum feedstock, which frequently has to be imported.
- Also, the process uses conventional, relatively unsophisticated technology and equipment.

Scientific Design Co. (New York) became involved with ethanol dehydration technology in the early 1960s when it designed, engineered and built two plants: the first (in Brazil, for Eletro Cloro, a Solvay subsidiary) has a capacity of 20 metric tons/d of ethylene with a purity of about 98%; later, Scientific Design built a 6,400-m.t./yr ethylene plant for Valika (now Synthetic) Chemicals in Pakistan to produce 99.9+%-purity material to be used for the manufacture of polyethylene. Both plants operated successfully, with actual yields and purities exceeding design projections.

However, the catalyst used in these plants suffered from some basic shortcomings. Its productivity was low, necessitating large, expensive reactors for even small plants; though the ethylene selectivity exceeded 90%, the relatively high price of ethanol resulted in a large economic penalty; further, the catalyst became fouled easily and needed regeneration about once every four weeks.

CATALYST ACTIVITY—Recently, one of the commercial ethanol-to-ethylene facilities has run pilot-plant tests and is scheduled to put into commercial operation (in May) a new catalyst developed by the Halcon International/Scientific Design group.

In contrast to the old catalyst, the new one has a greater activity, and in the region of optimal operation is six times more productive. The ethanol selectivity to byproducts has been reduced from about 6% to about 3%, resulting in significant savings in capital, utilities and quantity of waste products.

The new catalyst has a greater resistance to deactivation, and operating cycles exceeding four months each should be attained. This increases reactor onstream time by about 5%. (And extended cycles result in further cost savings.)

Ethylene yield is also significantly improved. With an almost complete conversion of ethanol, ethylene selectivity is about 97 mol %. The resultant reduction in ethanol consumption is very important, because even at the subsidized Brazilian price of $170/m.t., ethanol contributes about 75-80% of the net operating cost.

Even the reactor system has been cheapened, with savings amounting to as much as 15-20% of the battery-limits capital cost. Due to the increased activity of the catalyst, reactor volume can be substantially reduced; and because the catalyst has excellent selectivity over a broad temperature range, it is possible to expand the hot-oil temperature range across the reactor—resulting in a significant decrease in pumps and piping required for the hot-oil system.

In the process, ethanol is vaporized and dehydrated to ethylene:

$$C_2H_5OH \longrightarrow C_2H_4 + H_2O$$

The endothermic reaction occurs at

Originally published January 28, 1980

112

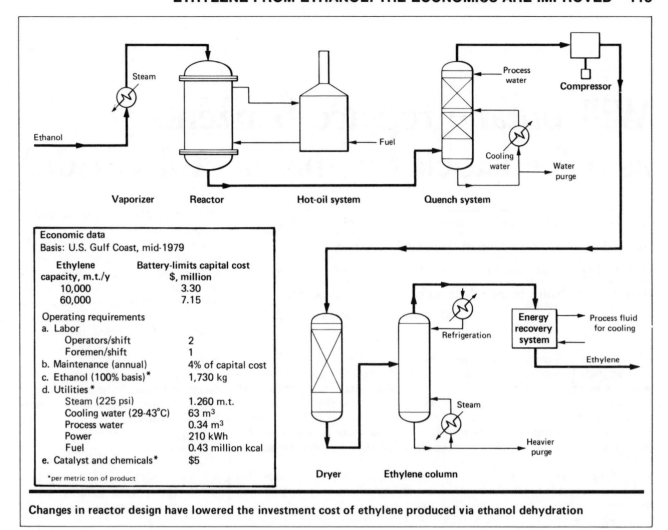

Economic data
Basis: U.S. Gulf Coast, mid-1979

Ethylene capacity, m.t./y	Battery-limits capital cost $, million
10,000	3.30
60,000	7.15

Operating requirements
a. Labor
    Operators/shift   2
    Foremen/shift   1
b. Maintenance (annual)   4% of capital cost
c. Ethanol (100% basis)*   1,730 kg
d. Utilities *
    Steam (225 psi)   1.260 m.t.
    Cooling water (29-43°C)   63 m³
    Process water   0.34 m³
    Power   210 kWh
    Fuel   0.43 million kcal
e. Catalyst and chemicals* $5

*per metric ton of product

Changes in reactor design have lowered the investment cost of ethylene produced via ethanol dehydration

an optimal temperature of about 330-360°C and is conducted at essentially atmospheric pressure in shell-and-tube reactors (see flowsheet). The catalyst is within the tubes, and a hot oil circulates in the shell to provide reaction heat.

An ethylene-steam product contains a small amount of unreacted ethanol, methane, carbon oxides, ethane, propylene, butylenes, acetaldehyde and diethyl ether. This hot reactor effluent is quenched in a tower with circulating water. Product water, unreacted ethanol and some heavy organics are removed. (For large-scale plants, it may be economical to recover heat from the reactor effluent.)

The cooled stream is then compressed to a pressure of about 17 atm, and then dried in a molecular-sieve dryer system to produce approximately 98% C_2H_4, which is purified in a low-temperature distillation column where the heavier-boiling impurities are removed. Depending on the amount of ethane rejected with the bottoms, the ethylene product, obtained overhead, can have a purity of 99.8 mol % or higher.

PURE PRODUCT—The process takes as feed 190-proof fermentation alcohol and turns out ethylene having a purity of about 99.8 mol %, with ethane as the other major component. The product at the battery limits is about 4.4 atm. Economic data for such a plant based on Scientific Design's process are given in the table.

Typically, modern polyethylene plants require ethylene of about 99.95% purity. To make such a product, only minor process design changes and marginal additional equipment would be required.

The quench tower could incorporate a section for removal of carbon dioxide by scrubbing with aqueous caustic soda solution. The ethylene column can be designed to reject sufficient ethane to the bottoms to produce the required overhead produce purity.

Also, the overhead product can be treated further in a stripper to obtain the desired specifications with respect to lights.

As a result, it is estimated that the economic data presented in the table will change as follows: capital cost would increase by about 10-15%, ethanol consumption by 10 kg per metric ton of C_2H_4, power by about 50%, and catalyst and chemicals by about half a dollar per metric ton of product.

Reginald Berry, Editor

The Authors

N. K. Kochar is a process manager in the Process Development Dept. of Scientific Design Co. In addition to his U.S. experience, he has handled process design, development and licensing responsibilities in India. Kochar is a chemical engineering graduate of the University of Bombay and has a Ph.D. from Rensselaer Polytechnic Institute.

R. L. Marcell is vice-president of process development at Scientific Design. He has worked in process design and development at Chemico and for the last 20 years at Scientifc Design. Mr. Marcell is a chemical engineering graduate of Johns Hopkins University and is a licensed engineer in the State of New York.

Will butane replace benzene as a feedstock for maleic anhydride?

A new catalyst has made it possible to use a cheaper feed for maleic anhydride. But in the event that raw material costs vary in the future, it may be best to have the ability to switch from one feed to another.

D. A. DeMaio, Halcon Catalyst Industries

☐ In Western Europe and North America, where most of the world's maleic anhydride (MAN) is produced (see Table I), there remains little doubt that benzene is no longer the most economical feedstock for this important petrochemical (maleic anhydride is used for polyester resins, alkyd resins, agricultural chemicals and lubricating additives). And with the current growth rate of maleic anhydride estimated at 7% per year, producers are trying to determine the feedstock of the future.

For a period of approximately forty years, starting in 1933 when MAN was first produced commercially, the feedstock of choice has been benzene. Despite this chemical's ever-increasing price, minor improvements in engineering, coupled with major improvements in catalyst efficiency, have succeeded in maintaining its economic· advantage.

But beginning in 1970, maleic anhydride technology has been subjected to scrutiny both here and abroad, motivated by the availability of low-cost C_4 fractions. Petrotex Chemical Co. has built a plant in Houston (now owned by Denki Kagaku Kogyo K.K., Denka) and Mitsubishi Chemical Co. has built a plant in Japan, both based on the oxidation of butanes to maleic anhydride. Within the past year, both Monsanto Co. (St. Louis, Mo.) and Denka have announced the development of a proprietary catalyst

Originally published May 19, 1980

for the oxidation of butane to maleic anhydride. And this year will see the startup of a butane-fed plant that will use a catalyst developed by Halcon Catalyst Industries (HCI), a division of Halcon Research & Development, a sister company of Scientific Design Co. (New York).

The main driving forces behind this current movement to a butane feed are:

- The benzene-emission control standards that are now being proposed by the U.S. Environmental Protection Agency.
- Progressive increases in the price of crude oil, which trigger increases in the price of benzene.
- The theoretical advantage of converting a four-carbon feedstock into a four-carbon product.

For these reasons (and for the first time in decades), each MAN producer must evaluate feedstocks to remain competitive. Two companies in the U.S. have already discontinued operation of their benzene-fed plants, probably as a result of one or all of these forces significantly affecting the economics of their operation.

OXIDATION CATALYST—During the past twenty years, the Halcon/Scientific Design route for the oxidation of benzene has accounted for thirty-eight facilities that have been put onstream in sixteen countries. And this technology is responsible for approximately 60% of the world capacity of MAN,

representing about 77% of the world's supply via benzene.

Early in 1979, HCI revealed the development of a viable commercial catalyst for butane oxidation. After prototype trials and laboratory evaluation by another company, HCI later in the year announced the construction of a plant for producing the catalyst. Now this catalyst will be used in a major facility in the U.S.

MAN is produced by the partial oxidation of butane or benzene with air over a promoted vanadium oxide catalyst. The reactions are:

For benzene:

$$C_6H_6 + 4.5O_2 \rightarrow C_4H_2O_3 \text{ (MAN)} + 2CO_2 + 2H_2O$$

(where heat of reaction is 5,400 kcal/kg benzene)

The principal secondary reaction is the complete oxidation to carbon dioxide and water:

$$C_6H_6 + 7.5O_2 \rightarrow 6CO_2 + 5H_2O$$

(Heat of reaction is 10,000 kcal/kg benzene)

For butane:

$$C_4H_{10} + 3.5O_2 \rightarrow C_4H_2O_3 + 4H_2O$$

(Heat of reaction is 5,500 kcal/kg butane)

The secondary reaction is:

$$C_4H_{10} + 6.5O_2 \rightarrow 4CO_2 + 5H_2O$$

(Heat of reaction is 11,800 kcal/kg ·butane)

TUBULAR REACTOR—The Halcon/SD route is divided into three principal sections: reaction, recovery and refining. Following the flowsheet, ambient air is filtered, compressed and mixed with vaporized feed. This mixture passes through a tubular reactor.

The reactor is equipped with a pump that circulates molten heat-transfer salt through its shell—removing the heat of reaction and thereby

114

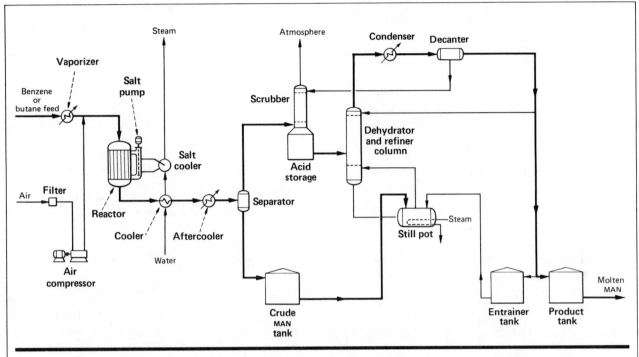

Route to MAN will be the same with either a butane or benzene feed, but equipment sizes will vary

controlling temperature (340-360°C). Other reaction variables, such as air rate, feed concentration, and pressure, are adjusted to balance catalyst performance with catalyst life.

The gases leaving the reactor are cooled (here steam is produced) to below the dewpoint of MAN. About 50% of the product (using a butane feed) is condensed in the separator and collected in the crude tank.

The off-gas from the separator is fed to the water scrubber, where essentially all the residual maleic anhydride is hydrolyzed to maleic acid. Overhead gases, which contain low concentrations of the feedstock, are passed to a system for the decomposition of hydrocarbons.

Maleic acid is dehydrated in the column; the resultant crude maleic anhydride is collected in the still pot. The MAN from the crude tank is then also pumped into the still pot, and distillation is started. The distillation results in a maleic anhydride stream that is recycled, a product that is transferred to a tank, and residues or bottoms that accumulate in the still pot until they are periodically removed by water washing.

PRODUCTIVITY DIFFERS—The flowschemes and equipment for both benzene and butane oxidation plants are identical; only the equipment differs in size. For peak efficiency, the butane catalyst must be operated at a productivity rate somewhat lower than that used for a benzene-fed plant, and therefore it is necessary to increase the number of reactor tubes to maintain productivity.

Also in the butane system, the concentration of maleic anhydride in the reactor product is lower than that in the benzene plant, which provides a 60% condensation of crude in the separator. Thus the butane system

At present, benzene is the predominant feedstock for maleic anhydride production

Table I

Maleic anhydride is produced from different feedstocks in different parts of the world. In the Far East, the oxidation of C_4 streams is and will continue to be the dominant practice; in Europe, Latin America, and North America, benzene has been the major feed.

World maleic anhydride production (tons)

Location	Benzene feed	C_4 feed
North America	246,500	50,000
Latin America	28,500	--
Western Europe	232,200	17,300
Eastern Europe	67,000	--
Far East	34,400	55,000

At present, butane seems to be the most economical feed for maleic anhydride   Table II

Basis:
60 million lb/yr MAN
Butane at 10¢/lb
Benzene at 21¢/lb

	Butane	Benzene
Battery-limits capital cost (1979 costs)	$15,700,000	$13,000,000
MAN Transfer Price	¢/lb	¢/lb
Raw materials		
Butane or benzene	11.65	23.87
Catalyst and chemicals	2.00	1.12
	13.65	24.99
Utilities	0.20	(0.33)
Labor related	0.59	0.59
Capital related (19.5%)	5.12	4.22
	19.56	29.47
At 20% return before taxes	5.24	4.32
Total transfer price	24.80	33.79

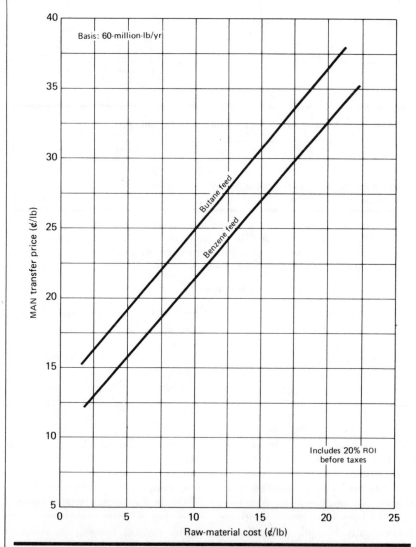

If the gap between MAN raw-material costs narrows, benzene may remain the feedstock of choice

requires somewhat larger dehydration equipment.

Comparing plants of equal productivity rates, the butane-based process will require slightly higher capital costs in the reaction, recovery and dehydration sections. This fact is apparent from Table II, which compares the economics of oxidizing both butane and benzene.

Of great significance is the lower transfer price for butane-derived MAN, which indicates that the higher capital costs are more than offset by the raw-material cost advantage of *n*-butane. If the projected rate of increase in price for benzene exceeds that for butane (see graph), the transfer price for butane-derived product will be affected to a lesser extent.

ALTERNATIVE, NOT SUBSTITUTE— Nevertheless, despite rising prices for benzene, and the eventual imposition of benzene-emission standards, it is reasonable to assume that acceptance of butane oxidation will not be universal. Where a benzene feed is captive and less costly, the maleic anhydride producer can install systems for the control of benzene emissions and continue to use the more-efficient benzene oxidation system.

What should be considered, however, is whether for the long term the difference in price between benzene and butane will widen. If the government prohibits the use of benzene in gasoline because of toxicity regulations, its price could decrease. In addition, there are indications that major companies are directing their research and development efforts toward using butane as a feedstock for other purposes. The combination of these two factors could affect the economic scales enough to create a better, or at least a different, balance between the viability of the two feedstocks.

The solution to the problem of which feedstock to choose probably lies in the convertible-plant concept. New plants can be designed to accept and operate with either a benzene or a butane catalyst. Existing plants can be made convertible at a relatively low capital cost.

Reginald I. Berry, Editor

The author
D. A. DeMaio is currently Product Director for maleic anhydride at Halcon Catalyst Industries. Formerly Director of Development, he has been involved in the maleic anhydride effort for Halcon for more than twenty years.

Aniline: phenol feed chosen

A U.S. firm will bypass the traditional nitrobenzene route to aniline and try a single-step process that offers economic and environmental benefits.

Ian McKechnie, Frank Bayer and *John Drennan,*
The Halcon SD Group

☐ For the first time in the U.S., phenol and not nitrobenzene will be used as a feedstock for aniline. U.S. Steel Corp. (Pittsburgh, Pa.) is building a 200-million-lb/yr phenol-based facility at its USS Chemicals Div. complex in Haverhill, Ohio. This plant, which is expected to be completed in December 1981 for startup in the first quarter of 1982, will use technology developed by The Halcon SD Group Inc. (New York).

Since the 19th century, aniline has been produced by nitrating benzene and then hydrogenating the nitrobenzene formed. It was not until the 1960s that Halcon developed the route based on the ammonolysis of phenol. This process was first commercialized in 1970 by Mitsui Petrochemical Co. at its Chiba works in Japan; this unit is now the only one in the world that does not use nitrobenzene technology.

NO BENZENE—These days, conventional nitrobenzene technology is being reevaluated. The U.S. government has considered placing restrictions on the handling of benzene because of its alleged carcinogenicity. And the market for aniline is growing. Traditionally, aniline has been used for sulfa drugs, pesticides, rubber chemicals, hydroquinone and a myriad of dyes. But mainly due to its use for polyurethanes (not a factor before 1960), U.S. consumption has grown 9.9%/yr in the 1960s and 7.4%/yr in the 1970s. Worldwide consumption (see table) is 680,000 metric tons.

In the conventional aniline process, benzene is charged to a nitrator and mixed with an aqueous solution of nitric (37%) and sulfuric (55%) acids. Crude nitrobenzene that is formed is separated from the spent acid, vaporized and catalytically reacted with hydrogen. The reacted gases are condensed and the aniline product is purified (water is removed) by distillation.

In contrast, the Halcon process is based on a one-step, reversible ammonolysis reaction with phenol as a feed:

$$\text{C}_6\text{H}_5\text{OH} + \text{NH}_3 \rightleftharpoons \text{C}_6\text{H}_5\text{NH}_2 + \text{H}_2\text{O}$$

Here, phenol conversion to aniline is favored by high ammonia-to-phenol ratios and low reaction temperatures. The forward reaction is mildly exothermic. Heat of reaction is calculated at 2.4 kcal/g-mole. This small exotherm at high ammonia/phenol ratios results in a low temperature rise. This helps to control reactor temperature, which must be limited to prevent the dissociation of ammonia to N_2 and H_2. (The catalyst also helps to control dissociation by allowing a short residence time.)

Nevertheless, a second reaction occurs: equilibrium exists between ammonia, aniline and diphenylamine (DPA). This reaction is also exothermic, reversible, and dependent on the phenol-to-ammonia feed ratio. However, high ammonia concentrations retard the formation of DPA, so that only a small percentage of it is formed.

In the system (see flowsheet), phenol is first vaporized, mixed with fresh and recycled NH_3 and then fed into the reactor, which contains a unique catalyst (several Lewis-acid-type catalysts have been developed by Halcon).

Gases leaving the reactor are cooled and fed to the first column in the purification train to recover ammonia for recycle. This distillation column recovers ammonia as the overhead product with aniline, water, phenol and the small quantity of diphenylamine as the bottoms stream.

The overhead ammonia is recycled (the system is also equipped to remove small quantities of N_2 and H_2 that may eventually build up). The bottoms are fed to the drying column, where water is removed overhead.

AZEOTROPE BROKEN—In the third purification step, specification-grade aniline is recovered. However, this purification can be difficult because aniline and phenol have almost identical vapor-pressure curves. In addition, aniline-phenol mixtures are nonideal, as evidenced by the formation of a

Polyurethane accounts for the largest share of world aniline demand

Aniline consumption (1979), thousand metric tons/yr	
U.S.	
Polyurethane	171
Rubber chemicals	83
Dyes	11
Hydroquinone	9
Drugs, pesticides, misc.	35
	309
Western Europe	
Polyurethanes	182
Rubber chemicals	62
Other uses	36
	280
Japan	61
Eastern Europe	25
Rest of world	5
TOTAL	**680**

Source: U.S. Intl. Trade Commission, *Chemical Marketing Reporter,* Feb. 28, 1980. Halcon Chemical Co. estimates.

Originally published December 29, 1980

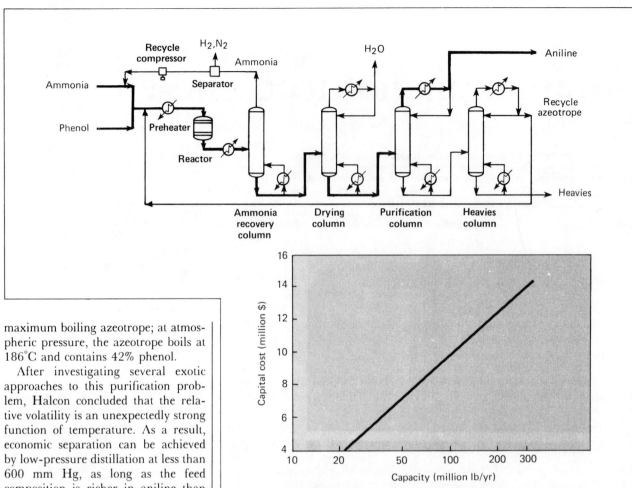

One-step phenol-based route to aniline has a lower capital cost than traditional nitrobenzene technique

maximum boiling azeotrope; at atmospheric pressure, the azeotrope boils at 186°C and contains 42% phenol.

After investigating several exotic approaches to this purification problem, Halcon concluded that the relative volatility is an unexpectedly strong function of temperature. As a result, economic separation can be achieved by low-pressure distillation at less than 600 mm Hg, as long as the feed composition is richer in aniline than phenol—a situation that occurs when one of the new Halcon catalysts is used. This technique results in an energy saving of approximately 40% compared with an extractive distillation, which had been studied for this separation.

In the purification column, aniline is recovered overhead. The bottoms enter the final (heavies) column, where diphenylamine is removed, and a small quantity of aniline-phenol azeotrope is withdrawn and recycled to the reactor.

ONE-STEP BENEFITS—The Halcon phenol-based system offers a number of advantages over the traditional nitrobenzene process. It is a one-step system that is simpler to operate. The raw materials, ammonia and phenol, are articles of commerce and are transportable, in contrast to nitrobenzene, which is not and is generally produced on-site by available nitric acid and nitrobenzene facilities. Capital cost of the Halcon system, taking into account that the dedicated nitrobenzene and hydrogen facilities are not needed, is

about one-fourth that of the older route (see fig.). Problems posed by regeneration of sulfuric acid or the handling of now-controversial benzene are avoided. Waste is minimized in the newer process since yields are almost stoichiometric—and, at that, 80% of the organic effluent is diphenylamine, which has market value.

Also, unlike the catalyst used for the nitrobenzene synthesis, the Halcon catalyst does not require regeneration. And the color stability of the Halcon product is superior to that obtained by the conventional technology, which discolors faster with exposure to air.

Even rising natural-gas prices result in an advantage for the new system, since its production costs will be affected to a lesser degree. Ammonia costs will rise, but so will costs for nitric acid and hydrogen required for the older route. And the nitrobenzene method requires an extra three moles of hy-

drogen per mole of aniline produced.

In addition, the phenol-based technology opens the possibility of developments leading to production of a wide spectrum of amines using either the same or an analogous feedstock. For example, in addition to aniline, *m*-toluidine is commercially produced with *m*-cresol as feedstock.

Reginald I. Berry, Editor

The Authors

Ian M. McKechnie is Vice President, Market Development, at Scientific Design Co. (Two Park Ave., New York, N.Y., 10016), and has been involved in the licensing, sales and business development of Halcon SD technology.

F. K. Bayer is Technical Sales Coordinator in Scientific Design's international sales and licensing department. A chemical engineering graduate of Cooper Union, he holds an M.S. in that field from the University of Pennsylvania.

John M. Drennan, Manager of Business Analysis for The Halcon SD Group, is responsible for opportunity definition and analysis. He holds a B.S.Ch.E. from Newark College of Engineering and advanced degrees from Lehigh University and Rutgers University.

Separating paraffin isomers using chromatography

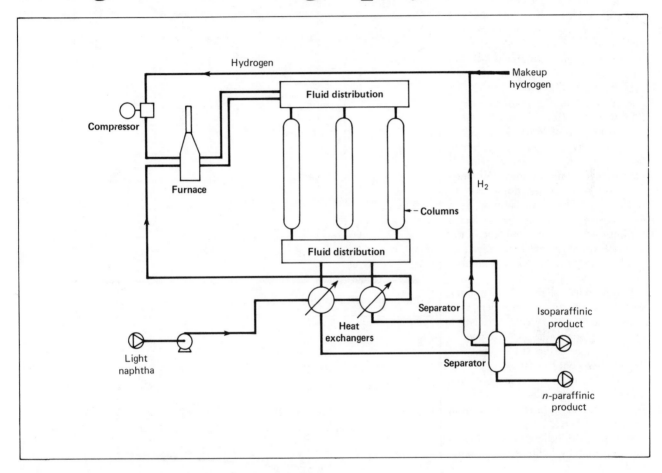

Molecular sieve columns operating at constant

pressure and temperature are separating normal and iso

paraffins. Since operating conditions are stable,

heat integration is possible and utility requirements are low.

Jean René Bernard, Jean-Paul Gourlia
and Michel J. Guttierrez, Elf Aquitaine

☐ Using chromatography, the new N-ISELF process can separate normal paraffins from a light naphtha feed. This system, which uses molecular sieves to make the separation, operates at a constant pressure and temperature. It produces a normal paraffin product that can be used as a petrochemical feedstock, and a high octane fraction, rich in isoparaffins, that can be used in a gasoline pool (see *Chem. Eng.*, Oct. 20, 1980, p. 17).

The technology, which can handle a range of material from C_4 to C_{10}, was developed by Elf Aquitaine (Paris). And by the end of 1981, it will see its first commercial application in a 100,000-metric-ton/yr demonstration unit at Donges, France. Here the process will produce a fuel base, and an *n*-paraffin cut that will serve as a feedstock for an isomerization unit (such a facility converts normal paraffins to isoparaffins).

OCTANE AND OLEFINS—The new

Originally published May 18, 1981

process, which is an outgrowth of Elf's work in chromatography (see *Chem. Eng.*, Mar. 24, 1980, p. 70), offers advantages to refining and petrochemical operations.

It can help solve one of the problems that will be facing the refiner over the next ten years—the production of high-octane gasoline without the benefit of lead. Light naphtha (C_5-80°C), has a low research octane number, RON (typically about 64), but a high susceptibility to the addition of lead (the RON would be 78 with the addition of 0.4 g/L of lead).

N-paraffins, accounting for about 50% of the naphtha, are responsible for the low octane number. *N*-paraffins have lower octane numbers than isoparaffins. The octane number of a light naphtha stream can be considerably improved, even without lead, if the *n*-paraffins are removed (see Fig. 1).

The refiner can benefit in a number of ways by using N-ISELF:

■ It permits a slight reduction of the severity of operation of reformers, thus improving overall yield, because it provides octane to the gasoline pool.

■ It increases the susceptibility of the gasoline pool to lead.

■ It improves the motor octane number of the gasoline pool, particularly when the refinery has a fluid catalytic cracker.

■ It provides octane to a light fraction and thereby lightens the gasoline pool.

Light naphtha is also used as a cracking feedstock for the production of olefins, such as ethylene. And olefin yields increase with *n*-paraffin content (see Fig. 2).

So, if the light naphtha has been divided into two streams, one will be high in isoparaffins for the gasoline pool and the other will provide a high ethylene yield when cracked. In this way, the process has pluses for the petrochemical manufacturer:

■ When cracked, a feedstock rich in *n*-paraffins yields more ethylene, propylene and butadiene; at the same time, it produces slightly lower amounts of pyrolysis gasoline.

■ When an *n*-paraffin-rich feed

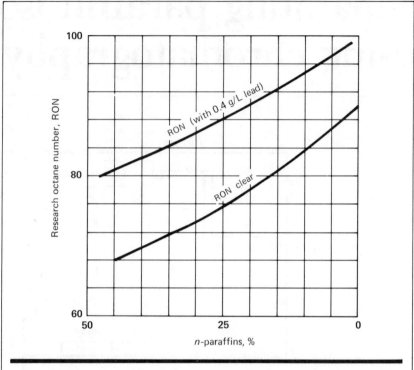

Octane of light naphtha increases as *n*-paraffin content is reduced   **Fig. 1**

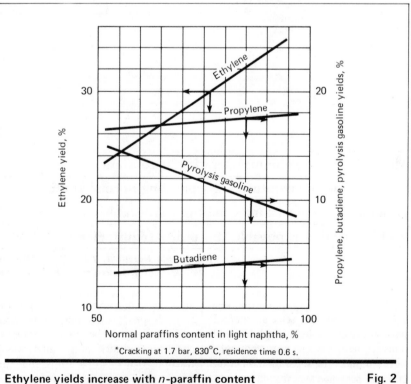

Ethylene yields increase with *n*-paraffin content   **Fig. 2**

Yields from N-ISELF can be varied by adjusting operating variables

Composition, wt. %	Typical Yields		
	Feed	Iso product	Normal product
i-C	14.7	20.3	9.1
n-C$_5$	33.6	13.5	53.7
i-C$_6$	30.3	52.7	7.9
n-C$_6$	17.4	7.1	27.7
Cycloparaffins, aromatics	4	6.4	1.6
Normal paraffins	51	20.6	81.4
Specific gravity	0.645	0.645	0.645
Yield		50%	50%
Octane			
RON, clear	69	81	
RON, 0.4 gPb/L	82.5	93.5	
MON,* clear	68	80	
MON, 0.4 g Pb/L	81.5	92.5	

*Motor octane number

is used, there is also less coking in the cracking tubes.

SEPARATING ISOMERS—Although there are advantages in separating paraffin isomers, the separation can be difficult and energy-intensive. Boiling points of isomers are close (e.g., n-pentane has a boiling point of 36.07 °C, isopentane, 27.85°C).

Distillation could be used, but a large nunber of trays would be required. Also, different columns would be needed for each separation (one for pentanes, one for hexanes, etc.); and distillation requires a great deal of energy.

N-ISELF makes use of a molecular sieve adsorbent that because of its crystalline structure can only adsorb significant quantities of molecules having an equivalent diameter less than 5A. This adsorbent, developed by Elf, permits the adsorption of n-paraffins, while isoparaffins, aromatics and cycloparaffins are not adsorbed.

In the process (see flowsheet on p. 92), a carrier gas—usually reformer hydrogen of 57-70% purity—is compressed and circulated through the system. Feed hydrocarbons are preheated by product gases leaving the columns and vaporized in a furnace. The hydrocarbons are then injected periodically into the carrier gas.

The mixed gas comes into contact with a bed of molecular sieves. During the operation of the process, the columns containing the sieve are maintained at a constant pressure and temperature. The compounds that can be adsorbed (the n-paraffins) are retained. The other compounds, leave the bed. The hydrogen carrier gas, constantly recirculating in the system, continues to enter the bed. As it does so, the adsorbed compounds desorb.

Therefore, for each cycle, at first a mixture rich in isoparaffins leaves the bed; this stream is followed by a product rich in n-paraffins. The separated hydrocarbons are condensed and the hydrogen is removed and recycled.

During the operation, the columns of molecular sieves are maintained at a constant pressure and temperature. Although the ability of the sieves to perform a given separation does decrease over time, they can be regenerated by burning off carbon deposits. This regeneration must be done after 2,000 h of use. Typically, the molecular sieves can be regenerated 15 to 20 times. They have a life of approximately 5-8 years.

In an industrial application, three or four sieve columns will be placed in parallel so that the operation is practically continuous. (Injection of hydrocarbons and switching between columns are accomplished by a valve system controlled by an electronic controller.) While one column is injected with the hydrocarbon/hydrogen mixture, another column will discharge isoparaffins and the third will discharge normal paraffins.

LOW OPERATING COSTS—Since the product flow is continuous and the operating conditions are constant, flow through the heat exchangers is continuous and the load is constant. This allows heat integration—a high heat recovery between hot effluent from the column, cold naphtha feed and recycled carrier gas. This heat integration leads to low utility requirements.

For each barrel of feed processed, only 1,700 Btu/h of fuel oil, 0.075 kW of electricity and 1.6 standard ft^3 of makeup H$_2$ are required. Molecular sieves would cost $45/bbl of feed.

Total fixed costs (using 1980 dollars) would be $300,000/yr for a unit that would process 300,000 metric tons/yr. This would include labor (representing 0.75 of a shift worker and supervision) at $150,000/yr, and taxes, maintenance and overhead expenses at $150,000/yr.

The investment for a 300,000-ton/yr facility would be $700-800/bbl per stream day of feed. This is a battery-limits cost including start-up, working costs, engineering and product stabilization.

FLEXIBLE OPERATION—For any given feed, the yields of the isoparaffin cut and, consequently, the yields of the normal paraffin product may be selected by adjusting a number of operating variables. And the selection of product yields determines product characteristics such as the octane number of the isoparaffin product.

A typical yield is given in the table. However, process variables can be adjusted to change this. It is possible to achieve paraffin purities from 80 to 99.7%. Temperatures may be adjusted in the range of 150-300°C; pressure may be from 8 to 30 bar; the time cycle may be from 1 to 6 min; the ratio of moles of hydrogen to moles of hydrocarbons from 2 to 5.

Reginald I. Berry, Editor

The Authors

Jean René Bernard is a research project chief in the Centre de Recherche Elf (Solaize, France, Telex 842 300591) where he specializes in separation by adsorption and reactor chemical engineering. He has an engineering degree from the Ecole Supérieure de Chimie Industrielle de Lyon and a Ph.D. in catalysis from the Institut de Recherches sur la Catalyse.

Jean-Paul Gourlia is Manager of Petrochemical Research, Elf Aquitaine. He received the degree in chemical engineering from Ecole Nationale Supérieure des Industries Chimiques and a doctorate from Institut National Polytechnique de Lorraine.

Michel J. Guttierrez is Manager of Development, Elf Aquitaine. He received a degree in chemical engineering from Ecole Nationale Supérieure des Industries Chimiques and a degree in Science Economique (Grenoble).

New catalyst revises dimethylaniline production

Now dimethylaniline can be made via a low-pressure, vapor-phase reaction instead of the traditional liquid-phase reaction that required an acid catalyst.

L. K. Doraiswamy, G. R. W. Krishnan,
and S. P. Mukherjee, National Chemical Laboratory

□ Once again, the development of a catalyst has allowed significant change in process technology. Via a new system, the National Chemical Laboratory, NCL (Poona, India), has created a low-pressure, continuous, vapor-phase process for the methylation of aniline to dimethylaniline, DMA (see *Chem. Eng.*, Sept. 8, 1980, p. 17).

In the past, DMA, an important intermediate used in the dyestuffs industry, has been manufactured by the liquid-phase methylation of aniline under pressure and in the presence of an acid catalyst. This traditional route suffers from the disadvantages of high capital cost, corrosion in the reactor, and the formation byproducts that cannot be recycled.

The new vapor-phase technology avoids corrosion problems since acid is not needed, and minimizes the formation of byproducts. This technique was first tested in a 400-ton/yr pilot plant operated at the Poona site of M/s Sahyadri Dyestuffs & Chemicals Ltd. (Bombay). After the successful operation of the pilot unit for over a year, Sahyadri Dyestuffs & Chemicals, which participated with NCL in the development of the new route, installed a 3,000-ton/yr commercial plant at Dewas, India. This unit was started up in 1978.

Now a second plant based on the new technology has been proposed, by Hindustan Organic Chemicals Ltd. (Maharashtra). NCL will be involved in the design of this facility, which is to have a capacity of 3,000 tons/yr of DMA and will be located in Rasayani, India.

ACID NEEDED—In the past, DMA—which is also used as an intermediate for vanillin, and as a solvent, alkylating agent, and an activator for polyesters—was produced by the methylation of aniline. Either sulfuric or phosphoric acid was needed as the catalyst in the liquid-phase reaction that took place at about 200°C in the range of 30 to 50 kg/cm².

A large excess of methanol is required in this process. Byproducts N-methylaniline ($C_6H_5NHCH_3$) and dimethylether (CH_3OCH_3) are formed in substantial quantities. In addition, the crude product would also contain some unreacted aniline.

Due to the closeness of the boiling points of aniline (184°C), N-methyl aniline (194-196°C) and DMA (192-194°C), it is extremely difficult to separate these compounds into pure components by distillation. The normal method of separation is the acetylation of the crude product, which fixes aniline and N-methylaniline chemically. In this way, however, the recovery of the byproducts and purification of the DMA can consume substantial amounts of expensive chemicals.

And since the reaction is carried out in the presence of an acid catalyst, the material of construction of the reactor has to be resistant to dilute acid. Typically, lead-lined mild steel is used.

CATALYST IMPROVED—The new NCL process offers a catalyst system that provides a very high selectivity and conversion to DMA. This system uses a commercially-available pelleted alumina having the desired porosity and surface area, combined with a proprietary additive.

The additive, a volatile liquid compound, is introduced into the reactor with the feedstream. Here it forms a complex with the high-boiling-point substances, preventing their deposition on the catalyst surface. This helps to enhance the life of the catalyst and ensure uniform performance over a long period of time. Additional advantages of the system:

■ Nearly complete conversion of the aniline feed (about 95-96%) to DMA.

■ Less feed is required per ton of DMA product. The NCL route requires 670 kg of methanol and 795 kg of aniline versus requirements of 725 kg of methanol and 820 kg of aniline for the older route.

■ Formation of dimethylether is significantly reduced. Only 95 kg/ton of DMA is formed versus 140 kg in the liquid-phase process.

■ Formation of nuclear-substituted products and of N-methylaniline is kept at very low levels. With each ton of DMA produced, only 30 kg of N-methylaniline is formed versus 65 kg in the older system.

■ The reaction is carried out at low temperature (200-250°C), which helps reduce the cracking of aniline and as a result minimizes the deposition of carbonaceous material on the catalyst surface.

■ The yield of DMA is about 100 kg/h per ton of catalyst.

■ The catalyst in the presence of the additive retains its initial activity for over 5,000 h, with an average selectivity of 92% to DMA.

Originally published July 13, 1981

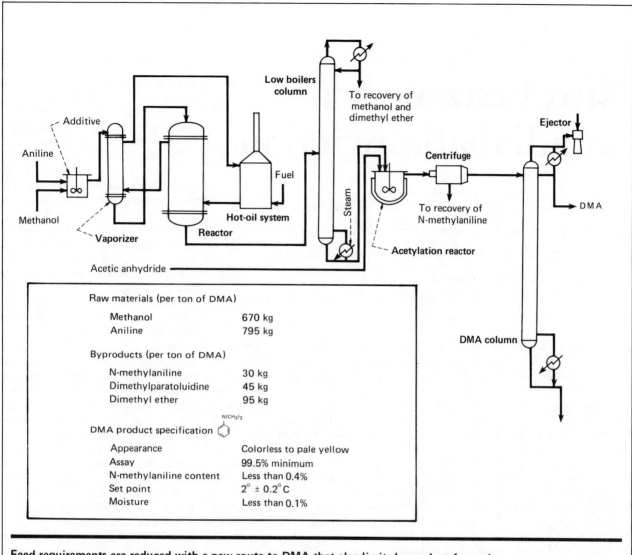

Raw materials (per ton of DMA)

Methanol	670 kg
Aniline	795 kg

Byproducts (per ton of DMA)

N-methylaniline	30 kg
Dimethylparatoluidine	45 kg
Dimethyl ether	95 kg

DMA product specification

Appearance	Colorless to pale yellow
Assay	99.5% minimum
N-methylaniline content	Less than 0.4%
Set point	$2° \pm 0.2°C$
Moisture	Less than 0.1%

Feed requirements are reduced with a new route to DMA that also limits byproduct formation

TUBULAR REACTOR—In the newly developed NCL process, liquid aniline and methanol are mixed together in predetermined ratios (see table) along with the additive, and this stream is pumped to a vaporizer (see flowsheet). The gas generated is preheated before it enters the tubular reactor, which holds the fixed-bed alumina catalyst.

For the system, the reactor operates essentially at atmospheric pressure. And since the process is noncorrosive—no acid is used—carbon steel is the main material of construction in the plant.

The heat required for the reaction is provided by a hot-oil system. The overall reaction, including the generation of byproducts, is mildly endothermic, even though the main reaction, the conversion of aniline to DMA, is exothermic.

The crude product obtained is sent from the reactor directly to a fractionating column, where substances with low boiling points, such as dimethyl ether, methanol and water, are separated overhead. This stream is then processed for the recovery of pure components in a set of continuous columns. The dimethyl ether and methanol recovered can be recycled into the feed.

The bottoms from the low boilers column, containing DMA, N-methylaniline and other heavier components, is sent to a stirred reactor, where the mixture is treated with acetic anhydride. The acetylated products formed are insoluble in the liquid stream and are then separated by centrifuge. The filtrate is sent to a vacuum distillation where pure DMA is recovered as the overhead product.

A facility using this technology to produce 6.6 million lb/yr of DMA would cost approximately $3.9 million (in mid-1980). Also, since the plant can be instrumented to a very large extent, the operating personnel required would be only three per shift, with one foreman.

Reginald I. Berry, Editor

The Authors

L. K. Doraiswamy is currently the Director of the National Chemical Laboratory (telex: Poona 266). He received his M.S. and Ph.D. from the University of Wisconsin. He had been head of the chemical engineering and process development division of NCL.

G. R. V. Krishnan is the head of the process development division of NCL. He holds a Ph.D. in chemical engineering from the Indian Institute of Science, Bangalore.

S. P. Mukherjee is the assistant director of process development at NCL. He holds a B.S. in chemical engineering from Jadavpur (West Bengal) University and has been involved with the development and commercialization of a number of major processes from NCL.

Nitrobenzene via an adiabatic reaction

Unlike in isothermal systems, the heat of reaction is not removed but is used to reconcentrate spent acid. This eliminates problems of acid storage and disposal, while reducing energy requirements.

A. A. Guenkel, H. C. Prime, and *J. M. Rae, Chemetics International Ltd.*

☐ In a joint effort, American Cyanamid Co. and C-I-L Inc. have developed and commercialized an adiabatic process for the nitration of benzene. This route costs 30% less than, and requires only about 10% of the energy needed for, traditional nitration technology, which is isothermal in nature. The new process, which avoids sulfuric acid disposal problems and limits benzene emissions, can provide a yield of nitrobenzene well above 99%.

The adiabatic process was developed at C-I-L's Explosives Research Laboratory in McMasterville, Quebec. Based on this work, two commercial plants were put onstream in 1979: one unit with a capacity of 380 million lb/yr for Rubicon Chemicals Inc. at Geismar, La., and the other a 110-million-lb/yr unit for American Cyanamid at Bound Brook, N.J. (this plant has now been shut down). Now a third unit has been designed and equipment has been ordered. This technology is licensed by Chemetics International Ltd. (Vancouver, B.C.), which is wholly owned by C-I-L.

In the U.S. more than 90% of all nitrobenzene is converted to aniline, which is used mainly in the manufacture of methylene-diphenylene isocyanate (MDI) and as an intermediate for the manufacture of rubber chemicals, pesticides and dyes. In recent years, demand for aniline has followed the growth pattern of MDI and polyurethanes, and in view of the increasing range of applications of polyurethane foams, a growth in aniline demand can be expected. Nitrobenzene should follow this growth even though a new phenol route to aniline (see *Chem. Eng.*, Dec. 29, 1980, p. 26) will offer some competition. In addition, a direct route to MDI from nitrobenzene is under development by Arco Chemicals and this should have a significant affect on nitrobenzene demand.

ACID AND WATER—Nitrobenzene is obtained by reacting benzene with nitric acid in the presence of sulfuric acid:

$$\text{C}_6\text{H}_6 + \text{HNO}_3 \xrightarrow[\text{H}_2\text{SO}_4]{} \text{C}_6\text{H}_5\text{NO}_2 + \text{H}_2\text{O}$$

Sulfuric acid acts as a dehydrating agent to remove the water produced and provides a strong acid medium so that nitric acid can act as a base and form the nitronium ion.

In the conventional continuous-nitration process, this exothermic reaction takes place at 60°C—a temperature maintained by the use of cooling coils on the reactors, which remove the heat of reaction.

Also, in the conventional system the spent-acid concentration is maintained at 70% H_2SO_4 and 1% HNO_3. If the sulfuric acid strength falls below 70%, the nitration rate falls off rapidly. In contrast, higher strengths of acid generally lead to the formation of unac-ceptably high levels of dinitrobenzene and other byproducts. The flowrate of the feedstock—typically 64% nitric acid and 93% sulfuric—is adjusted to maintain the proper concentration.

The use of 64% HNO_3 introduces 1.97 moles of water into the system for each mole of product. Since the reaction produces 1 mole of water, the sulfuric acid must absorb a total of 2.97 moles of water. This requirement, together with the constraints on spent-acid strength, sets the amount of sulfuric acid feed required for the system. If the sulfuric acid is to be recycled, then 0.435 ton of water must be evaporated per ton of nitrobenzene produced.

The use of strong nitric acid (99%) can reduce the requirements for sulfuric acid, and thereby cut down the spent-acid discharge by about two-thirds. However, the premium charged for high-strength nitric acid usually makes this approach uneconomical.

Manufacturers have had the option of recycling the spent acid by reconcentration (which is energy-intensive) or disposing of the acid in another facility, such as a fertilizer plant.

TEMPERATURE RISE—In the new adiabatic process, benzene is nitrated in the presence of a large volume of H_2SO_4. The temperature rises as the heat of nitration and dilution is absorbed by the acid as sensible heat. Now it is possible to reconcentrate the acid by flashing it under vacuum—significantly reducing the energy needed for the removal of water.

Also, since the heat of reaction is absorbed by the acid, cooling coils are not needed for the reactors. Overall in the system, the required heat-transfer area of the equipment used is about 10% that used in a conventional plant of similar capacity.

In the process (see flowsheet), recycled H_2SO_4 is pumped from the sulfuric acid pump tank. This acid is mixed with feed nitric acid to create a stream with a composition of 3-7.5% HNO_3, 58.5-66.5% H_2SO_4 and 28-37% water. The mixed acid enters the reaction vessels (the nitrators), which can be stirred tanks in series. The

Originally published August 10, 1981

number and size of the nitrators are based on required plant capacity.

Benzene is fed to the nitrators through an exchanger that recovers heat from crude nitrobenzene product. The feedrate is typically 10% in excess of the stoichiometric rate, to ensure the complete conversion of HNO_3.

In the nitrators, the temperature rises from 90°C to 135°C as the reaction proceeds. The temperature rise increases the reaction rate and allows operation with a more-dilute spent acid than in the conventional system.

The crude nitrobenzene, after being separated from the acid phase, provides heat to the benzene feed. The product is then washed to remove mineral and organic acids and stripped to recover unreacted benzene, which is recycled into the process.

EVAPORATING—The spent sulfuric acid leaves the separator and enters a glass-lined flash evaporator containing a tantalum bayonet heater. Here acid is reconcentrated under vacuum.

Although the heat of reaction provides most of the heat required for the reconcentration, some low-pressure steam is needed. Operating data have shown that the sulfuric acid concentration requires about 0.085 ton of steam per ton of nitrobenzene, or that one ton of water is evaporated using only 0.2 ton of steam.

The condensate from the concentrator is used to wash the crude nitrobenzene, reducing the need for process water. The reconcentrated acid at a strength of about 70% returns to the sulfuric acid pump tank.

This acid strength helps to protect the process. Because of their exothermic nature, nitration reactions, in general, are potentially hazardous. The dangers associated with the adiabatic process were identified as secondary reactions that would occur at 190°C, a temperature well above the normal operating maximum of 135°C.

As a precaution, rupture disks on the separator are set to burst before the temperature reaches 190°C. The process is inherently safe when vented to the atmosphere because any energy release would be dissipated through boiloff of water and benzene, made possible by specifying the acid strength at 66-70%.

EXPERIENCED—Two plants have used the adiabatic technology now for about two years. After initial mechanical problems, an onstream factor of

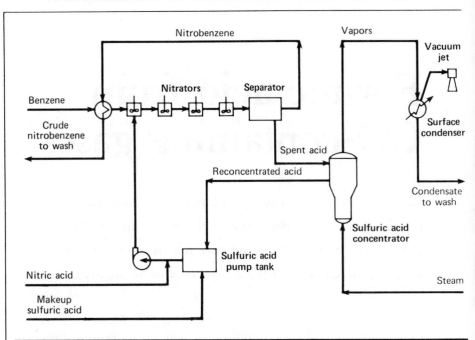

In the adiabatic nitrobenzene process, sulfuric acid accepts the heat of reaction

greater than 90% has been sustained for over a year. It has been shown that plants can be brought to full operating rates in less than an hour. Shutdown can be achieved simply by stopping the nitric acid and benzene feeds.

Nitrobenzene yields based on benzene and nitric acid feeds have been 99.5% and 99.1%. Sulfuric acid losses are on the order of 0.004 ton/ton of nitrobenzene. (The low acid loss is achieved by reconcentrating the acid to a strength, 70%, at which the partial pressure of H_2SO_4 is low.)

Emissions from the system are limited. The sealed nitration system and vacuum acid concentration permit easy collection and treatment of gases. Benzene and nitrogen oxides are controlled by scrubbing the vent stream. For a 15,000-lb/h (of nitrobenzene) plant, a total benzene emission of less than 1 lb/h is achieved; for NO_x, total discharge is less than 0.05 lb.

A facility using the adiabatic process would cost $8.8 million for 200 tons/d of nitrobenzene and $12 million for 500 tons/d. This capital cost includes engineering, equipment, and installation and facilities for washing and stripping the nitrobenzene. Maintenance, for a North American installation, is estimated to be 1 to 2% of the capital cost in the early years, rising to 3 to 4% as the plant ages. To operate the plant, one and a half operators

Utilities and feedstock requirements have been minimized by recycling sulfuric acid

Operating requirements (per ton of nitrobenzene product)	
Raw materials	Metric tons
Benzene	0.64
Nitric acid (100%)	0.515
Sulfuric acid (100%)	0.0033
Caustic soda	0.008
Utilities	
Cooling water	14,200 gal
Steam	800 lb
Electricity	20 kWh
Compressed air	180 scfm

would be required per shift, with one part-time foreman and one part-time supervisor. Utility and raw-material requirements are given in the table.

Reginald I. Berry, Editor

The Authors

A. A. Guenkel, Chief Process Engineer for Chemetics International Ltd., 1770 Burrard St., Vancouver, B.C., Canada, holds a Dipl. Ing. degree from the Technische Hochschule, Aachen, West Germany, and a Ph.D. in Chemical Engineering from McGill University in Montreal, Canada.

H. C. Prime, Manager of Technology and Engineering for Chemetics International Ltd., holds an M.Ch.E. degree from Nova Scotia Technical College.

J. M. Rae, Engineering Group Manager—Nitration, is a graduate in mechanical engineering from the University of Saskatchewan and is currently responsible for project execution and design for Chemetics' nitration plants.

Formic acid from CO-containing gases

New processes have been developed to manufacture formic acid from carbon monoxide, water and methanol, instead of via the major conventional route of obtaining it as a byproduct from the production of acetic acid.

Leonard J. Kaplan, Assistant Editor

☐ Two similar processes have been developed for manufacturing formic acid (HCOOH) from the carbon monoxide (CO) found in various waste vent-gases that are generated from the production of chemicals such as ammonia and methanol, as well as from the off-gases from basic-oxygen furnaces. Both methods combine: a carbonylation step, in which CO and methanol (CH_3OH) react to form methyl formate ($HCOOCH_3$), and a patented hydrolysis reaction between the ester and water to generate formic acid and liberate recyclable methanol.

The Leonard Process Co. (Englewood Cliffs, N.J.) and Kemira Oy (Helsinki, Finland) are joint licensors of a process now employed at a 20,000-metric-ton/yr plant operated by Kemira in Oulu, Finland since the beginning of the year. According to Jackson Leonard of Leonard Process Co., licensing agreements for facilities in Malaysia and Korea have been signed.

And The Halcon SD Group Inc. (New York City) and Bethlehem Steel Corp. (Bethlehem, Pa.) have technology that, while not proven commercially, is of interest to companies in Japan, Malaysia, the Soviet Union and India, reports Alan Peltzman, the SD Group's technical director.

The feed to the Kemira plant comes from an ammonia reformer vent-stream that contains N_2, CH_3 and H_2 as well as CO (47 vol%). Water is removed from the gas before the latter enters the formic acid system. According to Leonard, the concentration of CO in the gas stream is to be increased—and along with it the system product capacity—by installing hollow-fiber membrane separators (Prism units made by Monsanto, see *Chem. Eng.*, Feb. 25, 1980, p. 54).

The SD Group/Bethlehem process is said to economically use synthesis gas streams that contain as little as 50% CO, as long as contaminants such as H_2O, CO_2, O_2, and sulfur compounds have been reduced to low-ppm levels. H_2 and N_2, meanwhile, present no problems. In addition, the CO concentration of the feed stream can be upgraded by removing impurities or selectively recovering CO by using methods such as Union Carbide's pressure-swing absorption; Monsanto's Prism separators; and Tenneco's Cosorb process.

CURRENT SOURCE—The major source of formic acid in the U.S. has been as a byproduct of the butane oxidation process for making acetic acid. But experts feel that this traditional route will not be able to keep up with future demands for the chemical. Leonard says the oxidation process is rapidly becoming obsolete due to the increase in the cost of butane, butane shortages, and increased use of the Monsanto acetic acid process, which uses methanol and produces no byproducts. Echoing this belief, Andy Bayne, a process manager at the SD Group, does not expect to see any new butane oxidation systems built in North America.

Both of these processes hope to take advantage of the availability of low-cost CO from other chemical operations as to make formic acid more cheaply than has been possible. This would make the chemical attractive for uses not considered in the past, says Leonard.

OLD AND NEW USES—Major applications for formic acid today are as a reactant in chemical synthesis; as a coagulant for natural rubber; and for the treatment of cattle silage (stored green fodder) to prevent it from spoiling. The latter use is a major consumer of formic acid solution in the Soviet Union and Scandinavian countries such as Finland, where short drying seasons prevent adequate processing of animal feeds. In the U.S. and Canada, reports Peltzman, there is not a big demand for formic acid for this purpose because the drying season is longer and farmers dry with fuels to obtain high water removal. But he and Leonard both see treatment of silage as a potential use for the acid in the U.S. and Canada if fuel costs begin to increase again.

Leonard anticipates increased demand for formic acid in applications where its cost had been too high compared with other chemicals. He feels that using formic acid to pickle steel plate and produce pulp for making paper will eventually become alternatives to the conventional technologies. He reports that a major Japanese company is looking at the possibility of making formic acid, from CO-containing gas, to use as a replacement for the hydrochloric acid used in pickling steel. (Formic acid has not been considered for this because it cost considerably more than HCl.)

While no pulp manufacturer would comment on the potential for using formic acid to compete with the established Kraft process, Leonard insists that pilot work is being done. In fact, several processes have been developed by Robert Jordan, director of the Engineering Research Institute at Gannon University (Erie, Pa.).

PROCESS DETAILS—Both new pro-

Originally published July 12, 1982

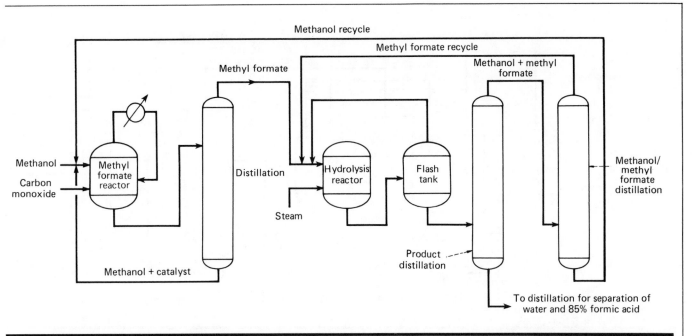

In the Leonard process, recycled methanol from the methanol/methyl formate distillation is recycled directly back to the methyl formate reactor (a maximum 4 lb/100 lb of product is lost in the reactor's vent gas)   **Fig. 1**

Economics: Leonard process, 20,000-m.t./yr facility		Table I
Product		
Formic acid, wt%	100	85
Raw materials		
Methanol, lb/lb of product	4	4
Carbon monoxide (97 vol%), lb/lb of product	67.6	57.5
Catalyst, lb/lb of product	0.007	0.007
Water, lb/lb of product	40	49
Utilities		
Steam (150 psig min.), lb/lb of product	970	650
Cooling water ($\Delta 10°C$), gal/lb of product	6,500	4,500
Electricity, kWh/lb of product	7.2	6
Labor		
Operating labor/shift	1	1
Capital investment		
Battery limits, $ million	7.265	7.265

cesses follow the same basic reaction steps, but differ in how their recycle streams are handled, the type of catalysts used, and the component separation schemes. The chemical reactions are:

Carbonylation step:

$$CH_3OH + CO \xrightarrow{\text{Catalyst}} HCOOCH_3$$

Hydrolysis step:

$$HCOOCH_3 + H_2O \rightleftarrows HCOOH + CH_3OH$$

The carbonylation process is a well-established route that generally uses sodium methylate as catalyst, but both of the developers say they use improved versions of it. The Leonard catalyst reportedly contains a proprietary additive that improves yield and lowers the pressure required during the methyl formate reaction.

During the carbonylation step (see Fig. 1 and 2), compressed CO-containing gas enters a reactor where it reacts with methanol in the presence of a catalyst. In the case of the Leonard

process, this takes place at a pressure of about 600 psig and temperature that is called moderate. Temperature and pressure information is not available for the SD Group/Bethlehem technology.

A continuous stream of liquid containing methyl formate, methanol and a small amount of the catalyst is withdrawn and fed to a distillation column, where the methyl formate is taken off overhead, and methanol and catalyst are removed from the bottom for return to the reactor. The SD Group/Bethlehem process recycles methanol and methyl formate from the hydrolysis separation back to this distillation column (Fig. 2), while the Leonard process recycles purified methanol directly back to the initial reactor.

HYDROLYSIS—It is the hydrolysis technology that each of the developers has modified (and patented) to achieve its own processing objectives. But in general the methyl formate stream enters a reactor, where it reacts with water at what is again referred to as moderate temperature and pressure. Because of this reaction's reversibility, each process uses slightly different techniques to prevent too much of the formic acid from reacting with methanol to form methyl formate and H_2O. Leonard says this reverse reaction prevented commercialization in the past

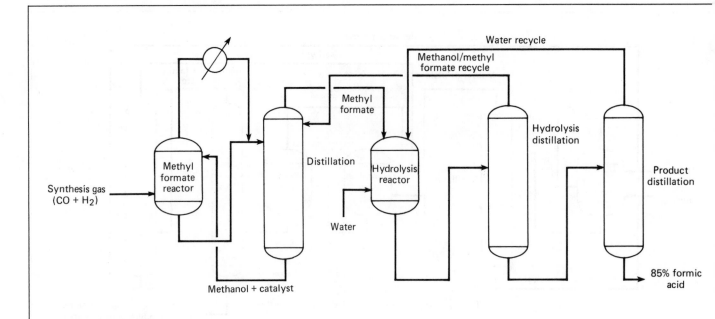

In the SD Group/Bethlehem process, a methanol/methyl formate stream from the hydrolysis distillation is returned to the column associated with the methyl formate reactor for separation into the two components　　**Fig. 2**

because at least 50% of the acid produced could be lost.

The discharge liquid from the hydrolysis reactor in the Leonard process enters a flash tank, where cooling takes place to stop the reverse reaction. From the flash tank, the liquid flows continuously into a distillation column, where the methyl formate and methanol are removed overhead, while the formic acid and water leave through the bottom. The contact time between the formic acid and methanol is said to be reduced to a minimum, and operating conditions established to prevent the reverse reaction.

The formic acid solution is fed to a second column, where a maximum boiling azeotrope of 85 wt% acid is taken off the bottom, and the excess water taken off the top is recycled to the hydrolysis reactor.

The methyl formate and methanol mixture from the first column is separated by distillation, with the methanol from the bottom being recycled to the first reactor and the methyl formate from the top returned to the hydrolysis reactor.

The SD Group/Bethlehem process uses no flash tank. The hydrolysis reactor effluent is fed to a distillation column, where methanol and methyl formate leave overhead for recycle to the carbonylation section via the separation column. At the same time, the

formic-acid water stream from the bottom enters a purification column, in which 85 wt% formic acid is produced. The water removed from overhead is recycled to the hydrolysis reactor. These distillation columns are said to be optimized to prevent losses from re-esterification and from thermal decomposition.

ECONOMICS—One of the major objectives of both new processes is to produce formic acid at lower cost. According to Leonard, the current selling price of about 29¢/lb makes it too

expensive to use in the new applications mentioned earlier. He expects that the Leonard process will eventually produce formic acid at 5-6¢/lb. While Halcon SD Group sees the cost of formic acid being reduced, it is not sure this could be lowered to the range claimed by Leonard.

Tables 1 and 2 give the process economics for the technology currently available from the developers. While the same-size facilities are compared, the feed-gas pressure and component concentrations may not be the same.

Economics: SD Group/Bethlehem process, 20,000-m.t./yr plant		Table II
Product		
Formic acid, wt%	85	85
Raw materials		
Carbon monoxide, vol%	97	50
Feed gas, m^3/m.t. of product	660	1,530
Vent gas, m^3/m.t. of product	67	945
Catalyst and chemicals, $/m.t. of product	8.05	9.95
Utilities		
Steam (18 kg/cm^{2}g), m.t./m.t. of product	4.5	4.5
Cooling water ($\Delta T = 6°C$), m^3/m.t. of product	600	617
Electricity, kWh/m.t. of product	155	303
Process water, m^3/m.t. of product	0.6	0.6
Labor		
Operators/shift	2	2
Foremen/shift	1	1
Capital investment		
Battery limits, (3rd-quarter 1981), $ million	9.7	10.6

Ethylene: Feedstocks and their impact on by-products

Constantly changing ethylene production economics caused

by the shifting of feedstock costs and availability

and by-product revenues will require manufacturers

to crack a range of materials. But propane is expected

to be the feed of choice for the future in the U.S.

Tim B. Tarrillion, The Pace Co.

☐ Over the past year, several petroleum companies in the U.S. have taken the time and money to add feedstock flexibility to existing ethylene plants that have used primarily heavy liquids such as naphtha and gas oil. Taking advantage of the low ethylene demand, Exxon (Baytown, Tex.), Shell Chemical Co. (Deer Park, Tex.) and Corpus Christi Petrochemicals Co. (Corpus Christi, Tex.) shut down their plants to make the necessary modifications. The reason: to be able to take advantage of the most economical feedstock at any given time.

Ethylene is the basic building block for many petrochemicals, serving as the primary feedstock for over 40% of this industry's products. Among the primary petrochemicals, ethylene is unique in that it is produced almost entirely in plants constructed for its manufacture, not as a by-product or coproduct, and is used entirely as a chemical building block. There are essentially no direct uses for ethylene (such as ammonia for fertilizer), nor is it used as a fuel to any appreciable extent. In addition, significant amounts of by-products such as propylene, butadiene, aromatics and other petrochemicals are generated during its manufacture. But the amounts of these materials produced vary by the feedstock used (see box).

During the 1950s and 1960s in the U.S., ethylene was supplied primarily from cracking ethane and propane. The reasons for this included the price and availability of natural gas and natural gas liquids (NGL), the strong growth for ethylene relative to other petrochemicals, and the early technology, which favored the use of feedstocks with a high selective yield of ethylene.

The situation changed, however, in the early 1970s. Although ethane remained the dominant feedstock, refined products such as naphtha and gas oil were assessed as the feedstocks of the future. This happened because all energy products looked to be less available and much more expensive, and since the industry felt crude oil would be imported anyway, the incremental amount required for petrochemical feedstocks would be minor. The petrochemical industry believed that price controls on natural gas would lead to decreased production and a scarcity of NGL. So with the demand for other petrochemicals (e.g., propylene, butadiene and benzene) looking bright, refined products were in favor for their greater by-product yields.

By-products are important to economics

By-product production from ethylene plants can vary considerably because ethylene is produced by "cracking" different hydrocarbon feeds, each having its own mix of by-products. Thermal cracking technology has developed to a point where the feedstock for making ethylene can be ethane, propane, normal butane, natural gasoline, raffinate, naphtha, or gas oil (see *Chem. Eng.*, Jan. 26, 1981). Thus, the choice of feedstock affects the amounts of by-products available. Typical ethylene yields for these feedstocks follow:

	Weight %
Ethane	78
Propane	41
n-Butane	37
Naphtha	30
Gas oil	25

As the heavier materials are used, the yields of by-products increase significantly. The more important of these by-products are propylene, butylene, butadiene, isoprene, benzene, toluene and xylene; and ethylene facilities that are designed to use the heavier feedstocks have fractionation systems to recover these chemicals. The relative value of the by-products becomes very important to the total net cost (total costs minus by-product revenues) of producing ethylene.

Originally published December 27, 1982

Now the outlook again seems to be changing. The political disruptions of petroleum crude supplies in 1979 led to sharp increases in the prices of naphtha and gas oil. In addition, such feedstock supplies were restricted and the large plants designed for heavier feedstocks were forced to reduce operating rate or crack lighter feeds. It now seems that both natural gas and NGL will be available at attractive prices.

FLEXIBILITY IMPORTANT — This change in the perceived feedstock outlook has caused flexibility to become very important in the ethylene industry. Pace expects it will remain important for the following reasons:
■ The volatile relationship between crude and natural gas as price decontrol approaches could severely change feedstock economics and/or limit supplies of NGL.
■ The variations in demand and pricing of key olefin by-products can also swing economics.
■ The high potential for future political disruptions of feedstock supplies will force producers to vary sources.
■ Given an expected slow recovery of ethylene demand and the continued overcapacity situation, each producer must have the flexibility to use the best feedstock (determined by the cost of the feed less by-product revenues, plus the production cost) if it expects to maintain a reasonable margin.

A study by our firm shows that there is currently sufficient flexibility inplace. Thus, the industry could consume significantly more NGL. But this shift to NGL as a feedstock from the previously expected naphtha/gas-oil slate will mean that less total by-products will be produced.

FORECASTING FEEDSTOCKS — To predict which feedstock slate will be optimum throughout this century, we calculate the variable cost of producing ethylene from each feedstock, using its own price forecasts for the feeds and by-products. The variable-cost analysis is applicable because all existing plants in the U.S. have enough flexibility to choose the lowest-cost feedstock. The company's forecast of ethylene manufacturing cost for various feedstocks is shown in Fig. 1.

The figure shows that propane is expected to remain the lowest-cost feedstock. Early in 1982, its low price and the high propylene price gave it almost a 5¢/pound advantage over the other feedstocks in production costs. In the current market, much of this advantage has been eliminated. If propane remains priced at 85% of the Btu value of No. 2 fuel oil or below, it will continue to maintain a significant advantage over other feedstocks. We feel it will be priced at the 85% level.

Given an ethylene demand as forecast by this firm, the flexibility of the industry to crack a range of feedstocks, and the relative economics of the various other feedstocks, the company has determined the feedstock slate through the year 2000 as shown in Fig. 2. The key trends obtained from this forecast are:
■ Ethane usage as a feedstock will decline primarily because higher gas prices will limit the profitabilty of extracting ethane. Limited availability will make ethane a marginally more expensive feedstock than propane.
■ Propane will be the feedstock of choice in the long term in the U.S., because its price is determined by its value as a fuel.
■ Normal butane will increase in use as a petrochemical feedstock because of its suitability as a feed to heavy-liquid crackers. There should be no availability problems.

Production of ethylene from heavy liquids will increase as demand grows, but most of the growth will be in the form of gas oil, raffinates, paraffinic naphtha, natural gasoline, and

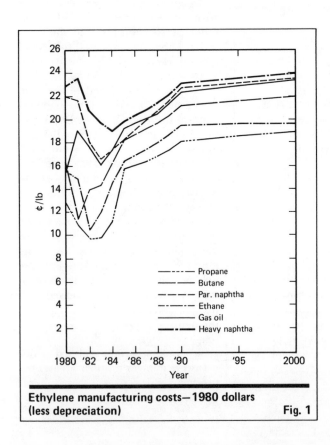

Ethylene manufacturing costs— 1980 dollars (less depreciation) — Fig. 1

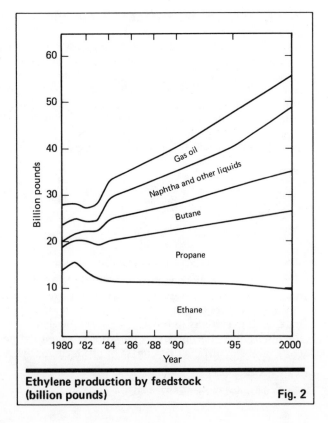

Ethylene production by feedstock (billion pounds) — Fig. 2

other low-value refinery liquids.

BY-PRODUCT SUPPLY—The continued emphasis on gas liquids as the primary feedstock for the ethylene industry will have a major effect on the supply of key by-product. However, should there be disruptions of natural gas liquids supplies from foreign or domestic sources, due to either political or economic considerations, there could be times when the order of feedstock preference is changed and the relative use of different feedstocks is altered. Such periods could last for a month, a quarter, or years, and the yield of by-products would vary accordingly. To analyze what impact such changes would have on the need for incremental supplies of the key by-products, the potential supply/demand balance has been done for such products, including propylene (benzene and butadiene have also been analyzed but do not appear in this article).

Capacity to produce propylene cannot be stated in the same manner as for most other petrochemicals where size of production units is essentially the only consideration. Propylene supply could be limited by extraction ("splitter") capacity in olefins plants and refineries, but availability and composition of feedstock streams are usually the controlling factors. In ethylene plants, propylene production is influenced by feedstock, operating rate, and cracking severity. In refineries, it is affected by overall operating rate, crude-oil characteristics, catalytic cracker operation, and size/composition of the gasoline pool.

Coproduct propylene from olefins plants has been an increasingly important contributor to total supply. Presently over 50 percent of domestic propylene supply comes from this source, and the share will continue to grow as olefins plants continue to consume more propane and heavier liquid feeds. The choice of feedstock

	Weight %	Propylene per pound of ethylene
Ethane	3.1	0.04
Propane	16.8	0.41
Normal butane	21.8	0.59
Naphtha	16.5	0.66

to an olefins plant significantly impacts coproduct propylene. The table above shows typical propylene yields from various feedstocks under high-severity cracking conditions.

Even with the rapid growth of propylene from ethylene plants, propylene from refinery production is still necessary to balance demand. The bulk of refinery propylene is produced as off-gas in fluid catalytic cracking (FCC) units. Less significant quantities of propylene come from thermal cracking and coking operations. Propylene concentration from refineries is usually 40 to 60 percent and therefore requires a splitter tower to achieve the desired chemical or polymer-grade quality.

The supply/demand balance for propylene, given the validity of our forecast for ethylene feedstock slate discussed previously, is shown in Fig. 3. The need for refinery propylene grew (from 5.9 billion pounds per year in 1980 to 7.3 billion in 1990), but actually it declined as a percentage of the total supply.

This need for refinery propylene could be increased if use of ethane as an ethylene feedstock continues to grow. The consequent decreased supply of propylene is depicted by line "A" in Fig. 3. Such a scenario would increase the need for refinery propylene to 9.5 billion pounds per year in 1990. This amount could easily be supplied by the industry with only minor additions to existing refinery extraction capacity. Thus, no long-term shortages of propylene should occur.

On the other hand, increased use of butane and heavy liquids as ethylene feedstocks could increase the amount of by-product propylene, as shown by line "B" in Fig. 3. This change would decrease the need for refinery propylene to below its present production level and could produce a potential "glut" of propylene on the market.

Leonard J. Kaplan, Editor

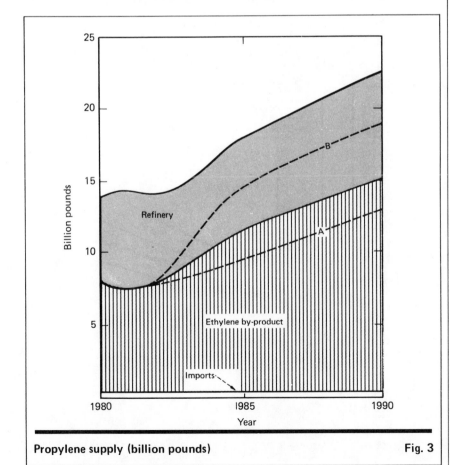

Propylene supply (billion pounds)　　　　　Fig. 3

The author

Tim B. Tarrillion is Manager of Market Analysis for Pace Company Consultants & Engineers, Inc. (Houston, Texas). He has a Master of Chemical Engineering degree from Rice University as well as an M.B.A. from Harvard University. He is primarily concerned with market and strategic analysis for companies operating in the petrochemicals, refining, crude oil, and energy industries.

Section V
Petroleum Processes

General Processes
Rerefining Used Oil

Gas oil becomes feedstock for SNG plant

Either integrated with a refinery or used by itself, this noncatalytic process can gasify high-sulfur oils to produce a high-Btu substitute natural gas

John C. Tao and *Joseph Klosek*, Air Products and Chemicals, Inc.

☐ Existing SNG plants in the U.S. are based on catalytic processes that require a low-sulfur, light-naphtha feedstock. But the policy of the federal government, as administered by the Dept. of Energy (DOE) is to preserve such light-petroleum fractions for use as chemical feedstocks.

Air Products and Chemicals, Inc. (Allentown, Pa.) has applied British Gas Corp.'s Gas Recycle Hydrogenator (GRH) technology to produce substitute natural gas from high-sulfur gas oil—a lower-cost feedstock that is not allocated by DOE.

In contrast to catalytic processes, GRH is a thermal route in which oil reacts with hydrogen to form gas and aromatics. Since GRH depends solely on temperature and pressure, it can tolerate heavier, high-sulfur feeds that are detrimental to catalysts.

In Europe, GRH units have been used to gasify light-naphtha feeds to produce a town gas of 700-750 Btu/std ft³. Now, this technology has been adapted and combined with a cryogenic separation to produce higher-Btu gas from a heavier oil.

The first commercial plant to use the GRH/Cryogenic Recycle process is now being constructed by Air Products and Chemicals alongside Sun Co.'s refinery at Marcus Hook, Pa. This system will produce 4.3 million scfd of gas having a heating value of 1,300 Btu/scf. The product will be mixed with 8.0 million scfd of refinery off-gas from the Sun Co. facility, to produce a total flow of 12.3 million scfd of SNG having a heating value of 1,050 Btu/scf. The SNG will be distributed by the Philadelphia Electric Company.

Originally published October 8, 1979

The gasification facility will also produce a small byproduct stream of benzene-toluene-xylene (BTX), which will be returned to Sun Co. for use in its refinery.

At Marcus Hook, the gasification process is being integrated with the refinery. Feedstock, hydrogen, utilities and sulfur-recovery facilities are supplied by Sun Co. The GRH/Cryogenic Recycle system has been designed to operate with four different feedstocks (see Fig. 1) ranging from 330°F end-point naphtha to 750°F end-point gas oil. (In comparison, existing U.S. SNG plants have been designed to operate with feedstocks having end-points of 330°F or less.)

RECYCLING REACTOR—As liquid hydrocarbon feed enters the process, it is mixed with hydrogen recycle from the cryogenic separation (see flowsheet). The combined stream is then vaporized by means of preheat supplied by hot gas leaving the GRH reactor and by additional heat supplied by process heaters.

The vapor then enters the reactor, which operates at 1,400°F and 600 psig. The reactions are exothermic, forming a gas that contains mainly methane, ethane, unreacted hydrogen and aromatics. Sulfur in the feed is converted to H_2S.

Temperature in the reactor is kept uniform by introducing the reactants at high velocity. This induces a recycle around the concentric draft tube in the vessel (see Fig. 1).

The primary reaction is the hydrogenation of aliphatic hydrocarbons:

$$(-CH_2-) + H_2 \longrightarrow CH_4$$
$$2(-CH_2-) + H_2 \longrightarrow C_2H_6$$

The aromatic constituents of the feed are relatively stable. However, dealkylation reactions do occur:

And organic sulfur compounds are converted to hydrogen sulfide:

Partly through heat exchange with the incoming feed, hot gas leaving the reactor is cooled to 400-500°F. Heavy aromatics that contain unconverted feed sulfur are condensed from the gas, separated, and sent to the refinery.

Cooled product-gas is combined with crude reformer hydrogen from the Sun Co. facility. H_2S and CO_2 are removed from the process stream in a monoethanolamine absorber.

The process stream then enters an oil-scrub column where light aromatics are absorbed in heavy oil. Direct steam stripping separates the heavy oil and the aromatics, which are cooled and condensed to produce a BTX condensate, a low-pressure gas, and water.

Following the oil scrub, the main product gas is cooled to condense water. Before the dried gas goes to the cryogenic section of the process, alumina is used to absorb any residual water.

In the cold box, the gas is chilled to below −200°F, condensing SNG and leaving a vapor rich in hydrogen. The hydrogen is recycled to the GRH reactor; the liquefied SNG is flashed to provide refrigeration via the Joule-Thomson effect (a decrease in pressure will cause a decrease in temperature). The low-pressure gas stream from the aromatic separation is mixed with the SNG. The combined product is compressed and sent to a pipeline.

By removing hydrogen from the SNG, the cryogenic step makes the product gas compatible with pipeline natural gas. Also, process economics are improved by recycling hydrogen, a gas that is costly to produce.

STANDS ALONE—The GRH/Cryogenic Recycle process does not have to

135

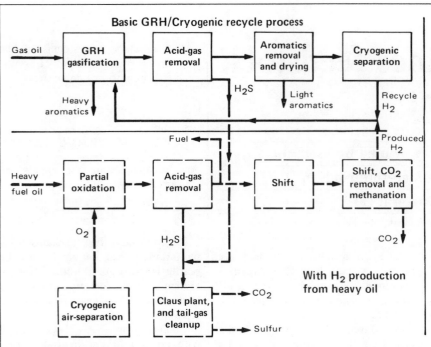

Basic GRH/Cryogenic recycle process

The GRH/Cryogenic process can be integrated with a refinery or it can stand alone

SNG heat balance (Marcus Hook plant)

Btus, input	%	Btus, output	%
Oil feedstock	56.1	Product gas	86.1
Hydrogen	39.3	Light aromatics	4.0
Fuel for fired heaters	1.1	Heavy aromatics	3.5
Electricity	3.5	Process heat loss	6.4
	100.0		100.0

Estimate for 150-million-scfd SNG plant

	\$/ million, Btu	
	CRG	GRH/cryogenic recycle (stand alone)
Hydrocarbons (net)[1]	3.93	2.80
Operating costs[2]	0.20	0.58
Capital charges[3]	0.54	0.66
Total	4.67	4.04

1. Current feedstock costs based on *Platts Oilgram*—March 1979
   Light naphtha at \$3.70/million Btu
   High-sulfur gas oil at \$3.00/millo million Btu
   High-sulfur residual oil at \$2.20/M million Btu
2. Includes 4% of investment for annual maintenance
3. At 19% of plant investment

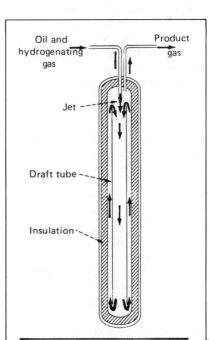

Feedstock specification

	Straight-run gasoline	Light naphtha	Heavy naphtha	Gas oil
Gravity, °API	68-73	48-52	39-43	32-37
Distillation, °F				
50% maximum	220	320	450	580
90% maximum	260	400	500	675
End-point, maximum	330	450	550	750
Aromatics, vol. % maximum	10	17	25	30
Sulfur, % maximum	0.01	0.05	0.1	1

The GRH reactor can accept a variety of oils

---

be integrated with a refinery. Hydrogen for the system can be produced by the partial oxidation of residual oil (see flowsheet). In such cases, oxygen can be provided by a cryogenic-air-separation plant.

Partially oxidized residual oil will produce a synthesis gas composed mainly of H_2 and CO. This gas would be desulfurized. A portion of the syngas would be used for fuel. The remainder would undergo a shift to increase hydrogen concentration. Carbon dioxide would be removed, methanation would eliminate any other carbon oxides, and the H_2 that remained would be suitable for the GRH/Cryogenic Recycle system.

COST AND EFFICIENCY—The cost of a stand-alone plant using the system can compare favorably with a competitive catalytic route such as British Gas Corp.'s Catalytic Rich Gas (CRG) process. The table provides comparative economics for a 150-million-scfd SNG facility. For the estimate, investment costs have been based on late-1978 figures, but current (March 1979) feedstock prices have been used.

Thermal efficiency for Air Products' Marcus Hook facility is given in the table. As shown, most of the input Btus are in the oil feedstock and in the hydrogen. The output Btus are mostly in the product gas and in the aromatic streams. Because the facility does not generate its own H_2, the overall thermal efficiency is very high, about 94%. With a stand-alone plant where hydrogen will be generated within the facility, the overall thermal efficiency will be in the low to middle 80s.

Reginald Berry, Editor

The authors

John C. Tao is Energy Technology Development Manager for Air Products and Chemicals, Inc.'s Energy Systems Dept. He received his B.S. and Ph.D. degrees in chemical engineering from Carnegie-Mellon University and is a member of Tau Beta Pi, Sigma Xi, AIChE and the Combustion Institute.

Joseph Klosek is a Process Design Manager in Air Products' Process Engineering Dept. He received B.S. and M.S. degrees in chemical engineering from Lehigh University and is a member of AIChE.

Partial oxidation in comeback

With light petroleum feedstocks often in short supply and expensive, companies throughout the world are increasingly turning to partial oxidation of heavy liquids for petrochemicals manufacturing.

☐ On both sides of the Atlantic and in Japan, petrochemical producers seeking flexibility in feedstock choices are taking a hard look at the bottom of the barrel—i.e., at cheap, abundant residual oil. This novel interest in resid is generating a minor revival of a decades-old technology—partial oxidation (PO$_x$)—that feeds on this kind of material.

In the U.S., a PO$_x$-based plant was completed last May in Louisiana, and another will be built in Texas. And in Europe, a number of PO$_x$ facilities are either planned or coming onstream, especially for ammonia manufacturing. Japan, China and India also are committed to the technology, with plant startups scheduled for the early 1980s.

According to Patrick E. Baggett of Chemical Market Associates (Houston)—a firm specializing in feedstock analysis—"partial oxidation has a wonderful potential because it provides feedstock flexibility." This is good news for the U.S., where, despite the current existence of a natural-gas "bubble," future raw-material-gas availability is iffy, and refiners continue to compete for the lighter cuts of crude oil to process into high-octane gasoline blending components. It is good news also for Europe and Japan, where commonly-used naphtha is in tight supply and costs more every day (since late last year, the spot price of naphtha in Europe has climbed on occasion to a top price of about $350/metric ton).

PO$_x$ technology, which yields a syngas that can serve as a starting point for downstream synthesis, also can count on entering into more process-choice considerations as refiners turn to heavier, more-sour crude oil of higher metals content.

AVAILABLE PROCESSES—Today, nearly all the PO$_x$ units operating use methods developed by either Shell Oil or Texaco. Both firms' processes have two main steps: (1) reaction of liquid hydrocarbons and oxygen to make a syngas and (2) carbon recovery for recycling as a reactor feed.

Partial oxidation with high-purity oxygen produces a syngas consisting mainly of carbon monoxide and hydrogen, with the ratio varying with feed type and quality and with operating conditions. The typical 300-Btu/scf synthesis gas may be used as is, or shifted to adjust the hydrogen-carbon monoxide ratio to levels required for methanol or oxo-alcohol synthesis, or to make pure hydrogen for ammonia manufacture.

Recently, both Shell and Texaco

Partial listing of PO$_x$ plants built in recent years					Table I
Company	**Country**	**Feed**	**Startup date**	**Output, Nm^3/stream day**	**Products**
Veba Chemie AG	West Germany	Vacuum resid	1973	4,000,000	Ammonia and methanol
SIPM	Netherlands	Full range	1971	50,000	— —
Oxo-Chem Enterprise	Puerto Rico	Refinery off-gas and naphtha	1975	300,000	Oxo chemicals
National Fertilizers Corp. of India	India	Heavy fuel oil	1977	2,100,000	Ammonia
Fertilizer Corp. of India	India	Heavy fuel oil	1977	2,100,000	Ammonia
National Fertilizers Ltd.	India	Heavy fuel oil	1978	2,100,000	Ammonia
Neyveli Lignite Corp.	India	Heavy fuel oil	1979	800,000	Ammonia
Veba Chemie AG	West Germany	Heavy fuel oil	1979	4,000,000	Ammonia
Exxon	U.S.	Naphtha	1979	Not available	Oxo alcohols
Mitsubishi/Nippon	Japan	Heavy resid	1980	45,000	Acetic acid
CNTIC	China	Vacuum resid	1981	715,000	Ammonia
Quimigal	Portugal	Vacuum resid	1982	2,400,000	Ammonia

Originally published October 8, 1979

Natural-gas processing: PO$_x$ vs. steam reforming

Partial oxidation is not often considered an alternative in natural-gas processing, because the method requires investing in a high-purity oxygen plant that eats up about a third of the capital cost. But if reasonably priced oxygen is available from an outside source, PO$_x$ can compete with the traditional steam-reforming process, depending on the end-product desired.

For example, if pure CO is needed—to run an acetic acid unit—PO$_x$ of natural gas yields a more-desirable H/CO ratio (1.83) than steam reforming's 3.00 (see Table II).

One disadvantage of steam reforming is that having a syngas richer in hydrogen content increases handling costs as well as size (and capital cost) of the hardware. This is particularly true if there is downstream cryogenic processing.

Also, in PO$_x$ schemes all heat is internally generated, and although a waste-heat boiler is needed, the methods generally handle lower gas volumes than reformers, which require external heating and heat-recovery systems.

Choice of feedstock has a big influence on PO$_x$ economics. For example, partial oxidation of sour resid requires a much larger capital investment in pollution control (for S recovery) than does partial oxidation of natural gas. This is one reason why resid is not considered a feasible raw material for small PO$_x$ units.

have made new sales. Exxon Chemical Co., for instance, has turned to Shell's technology at its oxo-alcohol plant in Baton Rouge, La. A major revamp, completed in May, has been designed to reduce the facility's dependence on natural gas. Among other things, an old syngas unit has been replaced by a modern PO$_x$ unit that can feed on various liquids. The revamp has also enabled Exxon to shut down five lower-efficiency furnaces that fired natural gas to make syngas needed for alcohols manufacture.

This is the first partial-oxidation unit in the Exxon family, according to the firm. The changes at Baton Rouge, coupled with compressor and furnace improvements, provide not only feedstock flexibility but a 40% increase in production with only a 5% increase in required energy, the firm says.

Texaco's technology will get its latest workout in Texas, where it will be used in a joint venture between Du Pont and U.S. Industrial Chemicals Co., a division of National Distillers and Chemicals Co.

Built at the property line between the companies' plants in Deer Park, the PO$_x$ unit will feed a portion of the manufactured syngas to Du Pont to produce 200 million gal/yr of methanol. USI will use CO from the PO$_x$

unit, combining it with methanol supplied by Du Pont to make acetic acid via a Monsanto process (*Chem. Eng.*, Jan. 29, p. 49). The acid is further reacted with ethylene to make vinyl acetate monomer by USI vapor-phase technology.

USI had anticipated feedstock problems in its VAM expansion from 375 million to 600 million lb/yr, but the partial-oxidation route assures an adequate supply of syngas for both it and Du Pont.

Industry sources report that the companies are using heavy, sour resid from Exxon's Baytown, Tex., and Shell's Deer Park refineries. The material, a solid at room temperature, requires heating to 300-350°F to maintain pumpability. This doesn't create any major problems, but does put certain limits on PO$_x$ processing.

For example, an engineer with a large construction firm notes: "You have to heat like heck to pump the resid, so PO$_x$ will be used only near the source of supply. This probably limits partial-oxidation facilities to areas close to refineries, such as on the Gulf Coast."

He emphasizes, however, that "partial oxidation of resid definitely has a place in the petrochemical-feedstock picture. This is because for years, the

U.S. in particular has selectively consumed the light fractions of the barrel, but now we are relying more and more on reserves of heavy, dirty crude. PO$_x$ is one way to utilize the resid."

Easy resid availability isn't always the case, however. USI and Du Pont shopped at length to assure adequate supplies for the Deer Park partial-oxidation unit. Norman E. Kraus, manufacturing manager of the methanol products division of Du Pont, cautions that "one must talk to the petroleum companies to get a good answer to this feedstock question. There are competing processes for the bottom of the barrel—for example, Flexicoking and resid hydrogenation—and one has to look hard at feedstock availability and economics before making a decision."

"In addition," he warns, "partial-oxidation facilities should not be dependent on one source of supply, because in case of emergencies or planned shutdowns, the resid raw-material would not be available."

In Europe, however, there is certainly a resid glut. "There's a trend toward partial oxidation," a Lurgi representative says, "because it takes care of the resid problem." Some companies in Europe view partial-oxidation technology as a way to add petrochemical feedstock without necessarily having to build more oil-refinery capacity.

There's a caveat for European PO$_x$ planners, though. Local refineries are rapidly installing conversion capacity to alleviate the great resid surplus so by the early mid-1980s, this excess may well have disappeared.

CHOOSING PO$_x$ IN EUROPE—After taking a hard look at this and other factors, Veba AG turned to partial oxidation for two facilities: one for 1,250 metric tons/d of ammonia and 450 m.t./d of methanol, started up four years ago at Gelsenkirchen-Buer in the Ruhr area; the other for 1,650 m.t./d of ammonia, inaugurated last fall at Brunsbuettel, on the North Sea coast. (These facilities are now operated by Chemische Werke Hüls, a Veba subsidiary.)

PO$_x$ was chosen because, unlike a steam cracker, it doesn't require precious naphtha. In Europe today, ammonia sells for about half the price of naphtha, so the latter is an uneconomic choice as raw material.

The Brunsbuettel plant suffered about six months of startup troubles caused by trying to make all six major units (air separation, gasification, hydrogen sulfide removal, carbon monoxide conversion, hydrogen purification, and ammonia synthesis) work together as a big, one-train facility. Hüls points out that heat recovered in the ammonia plant and recycled to the front-end PO_x unit cuts fuel consumption for heating to only 10% of the oil-feedstock consumption.

The Gelsenkirchen plant uses resid to extract carbon soot from the process to make $\frac{1}{4}$-in.-dia. pellets (10-20% soot content) that are fed to a power plant. Some design engineers, however, question the efficiency of handling the soot this way, unless a mammoth power plant is close at hand.

A number of other PO_x plants are planned for Europe (see table).

In Portugal, for instance, government-owned Quimigal has awarded a contract to Lurgi for a 900-m.t./d ammonia plant based on partial oxidation of heavy fuel. The Shell-licensed unit will use a Haldor-Topsoe synthesis loop, and is projected for startup in 1982.

Greece, too, plans a PO_x-based ammonia plant. So far, ten chemical engineering companies have participated in a pre-qualification round, and five companies are to be selected for the actual bidding. The project, still in a state of flux, started as a naphtha-fed one, progressed through natural gas as a feedstock, and is now being tendered for design with a basis of partial oxidation of 80% fuel oil and 20% natural gas.

PO_x-ING ELSEWHERE—Partial oxidation is winning converts outside Europe and the U.S., as well.

For example, Gujarat State Fertilizer Co. has chosen Linde AG to build a 1,350-m.t./d ammonia plant near Bombay, India. Based on heavy fuel oil and scheduled for 1980 startup, the project will incorporate a new process sequence that includes gasification at 85 atm, instead of at the usual 30-40 atm. This not only saves space (smaller components) but, more importantly, saves about one-third of the compression energy required for the subsequent ammonia synthesis that takes place at about 200 atm, Linde explains. This is a plus over steam-reforming, which operates at atmospheric pressure.

H/CO ratios: matching feed and process with end-product needs				Table II
Partial oxidation of:		**Methane steam-reforming**	**End-products**	
Propane asphalt	0.87	3.00	Methanol	2.00
Heavy fuel oil	0.97		Oxo alcohols	1.00
Light naphtha	1.22		Acetic acid	Pure CO
Natural gas	1.83		Aldehydes	Pure H_2
			Ammonia	Pure H_2

In Japan, meanwhile, heavy-resid partial oxidation will be applied commercially in acetic acid production by early 1980. Also, three PO_x-based ammonia plants built there will be exported to China in the last half of 1980.

Until recently, the Japanese have shown little interest in the technique, as exemplified by the fact that only a few PO_x units are onstream in that country. Ube Industries, Ltd. (Ube City and Tokyo) currently uses vacuum residual oil for ammonia production. Its 329-m.t./d plant, operated at Ube City since 1960, is based on Texaco technology. Small Shell units operate at Mitsubishi Petrochemicals, at Yokkaichi.

But in Japan, too, rising costs for petroleum and naphtha have led to increasing interest in partial oxidation of resid. The first major commercial applications will be for two 200,000-m.t./yr acetic acid plants that Daicel Ltd. (Osaka City) and the Mitsubishi Chemical Group (Tokyo) separately plan to complete early next year. The Mitsubishi project, at Kurosaki, Kyushu, is designed to produce 45,000 normal m^3/h of syngas from 124 m.t./d of resid, using Shell technology. The Daicel project, at Aboshi, near Osaka, is designed to use Texaco technology to produce 100,000 m.t./yr of carbon monoxide for acetic acid feedstock (hydrogen will be consumed internally as a fuel).

Japan's Ube Industries will also export three 1,000-m.t./d ammonia plants to China, which will enable the Chinese to produce urea from asphalt. One plant is to be erected near Shanghai; the other two locations have not been revealed.

China also has plans for other PO_x installations. It has signed a contract with Lurgi, which will use Shell's partial-oxidation process in a 715,000-m^3/d syngas plant ordered last December. The project is due to start up in 1981/82 at an undisclosed location.

SOME DRAWBACKS—Despite its powerful plus as a process capable of handling all sorts of feedstocks, PO_x has some disadvantages. For example, one expert for a London-based design, engineering and construction firm feels that engineers evaluating PO_x should pay especially close attention to the interface between the technology and downstream plants. Traces of carbonyl sulfide present a special problem. "If you don't get rid of this, it causes all sorts of hassles with urea production because the carbonyl sulfide is a near analog to carbon dioxide and has very similar solubilities in the most commonly used solvents."

Sulfur removal, too, is critical in processes based on partial oxidation, especially those for ammonia synthesis. As little as 1 ppm in the syngas, in fact, can seriously disrupt catalyst activity (*Chem. Eng.*, Oct. 24, 1977, p. 79).

The dirtier crudes upon which current PO_x plants often feed can cause other problems as well. For instance, some older PO_x units were able to totally recycle unconverted carbon soot (which amounts to about 0.5-1.5% of the feed) because the crudes' metals content was low enough. But now, with higher-metal-content feeds, the metals must be removed to avoid buildup in the system.

In the Texaco process, the metals end up in a sludge stream or metal concentrate that normally goes to a reclaimer for reprocessing to recover metal values. In the Shell scheme, the metals pass through the reactor bottom as part of the friable ash, and also exit in the water effluent—

Guy E. Weismantel, Larry Ricci

Gasoline or olefins from an alcohol feed

A new catalyst makes it possible to produce hydrocarbons in the right range for gasoline. And fluid-bed reactors promise to offer even more product selectivity.

Reginald I. Berry, Assistant Editor

☐ Mobil Oil Corp.'s methanol-to-gasoline, MTG, process can turn a crude methanol (as much as 30% water) feedstock into high-octane gasoline or into olefins. This technology, based on the development of a new catalyst, will be commercialized in a New Zealand facility that will produce 12,500 bbl/d of synthetic gasoline when it goes onstream in 1985 (see *Chem. Eng.*, Apr. 7, p. 43).

In a sense, the route catalytically dehydrates methanol, obtaining a stoichiometric yield of 44% hydrocarbons and 56% water. The reaction, which is highly exothermic, at first produces dimethyl ether (CH_3OCH_3) and water. The dimethyl ether and methanol are transformed to light, then heavier, olefins. With enough recycle and the proper catalyst selectivity, the olefins rearrange to paraffins, cycloparaffins and aromatics without generating hydrocarbons higher than a C_{10}:

$$2\,CH_3OH \rightleftharpoons CH_3OCH_3 + H_2O$$

$$\downarrow$$

Light olefins $+\ H_2O$

$$\updownarrow$$

$C_5{}^+$ olefins

$$\downarrow$$

Paraffins, cycloparaffins, aromatics

It is the catalyst, the basis of the

Originally published April 21, 1980

MTG route, that limits the production of heavier hydrocarbons. This catalyst, a synthetic zeolite, has a structure different from a wide-pore faujasite (with openings 9-10 Å) and a narrow-pore zeolite (5-Å opening). The new catalyst is composed of straight and angled channels with openings of about 6 Å—a size just wide enough to produce compounds in the gasoline range ($C_5{}^+$). Hydrocarbons larger than a C_{10} can be produced but they will not be able to leave the catalyst and will therefore undergo further reaction that will reduce their size.

CHOICE OF REACTORS—A number of different designs can be used to take advantage of the new catalyst. The New Zealand facility will employ a fixed-bed reactor system. But in 1983, a $35-million, 100-bbl/d pilot plant incorporating a fluid-bed design will come onstream in Wesseling, West Germany, through cooperation between the U.S. Dept. of Energy, the German Federal Ministry for Research and Technology, Union Rheinische Braunkohlen-kraftstoff (Wesseling), Uhde GmbH (Dortmund) and Mobil.

Since 1975, the process has been tested on a 4-bbl/d pilot unit at Mobil's Paulsboro, N.J., facility. Here both fluid-bed and adiabatic fixed-bed reactor systems have been studied. In addition, independent investigators are examining a tubular reactor design.

The type of reactor used for this route is influenced by the desire to

limit temperature rise during a reaction that will produce about 750 Btu/lb of methanol converted. This heat must be dissipated, since the adiabatic temperature rise could be as high as 1,100°F. Also, olefins production, if that is desired, is favored at lower temperatures.

Tubular reactors can be used to maintain temperatures; and fluid-bed reactors offer excellent heat transfer and may be operated almost isothermally; however, the fixed-bed system, which uses two different reactors to distribute the heat, is the simplest design—the easiest to scale up—and for that reason has been chosen for the first commercial system.

In the fixed-bed system, 20% of the heat of reaction is generated in a reactor that produces only dimethyl ether, and the other 80% in another reactor that carries out the conversion to heavier hydrocarbons. This design minimizes the amount of recycle needed to control temperature during the reactions that produce gasoline components.

In a commercial design (see flowsheet), crude methanol, which may come directly from the methanol plant, is heated and vaporized by conversion reactor-effluent to about 570°F. This vapor then flows down the DME reactor, where some of the methanol is converted to dimethyl ether and water over a proprietary Mobil catalyst that differs from the new zeolite one.

The stream leaving the DME reactor is mixed with gas recycled from the product separator. The recycle is typically 7 to 9 moles to every one of feed. This combined stream enters the conversion reactors, where at 300 psig and 650 to 765°F olefins, aromatics and paraffins are formed. The temperature rise in the reactor, operated adiabatically, is about 100°F.

Four conversion reactors are shown, but the actual number would vary, depending on plant capacity and on the frequency set for regenerating the zeolite catalyst. (During normal operation, at least one vessel would be going through a regeneration cycle.)

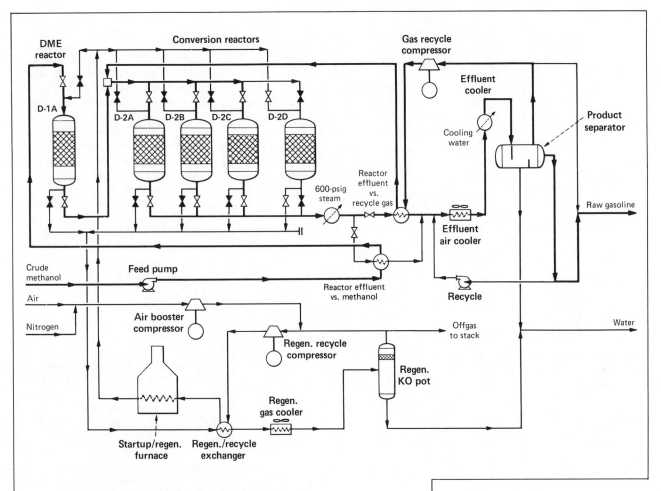

Fixed-bed system divides the methanol-to-gasoline reaction   Fig. 1

Estimated economics for fixed-bed system

Basis
December 1979, U.S. Gulf Coast costs

Methanol required	
Crude, 17% water	57,000 bbl/stream-d
100% pure	50,600 bbl/stream-d
Gasoline product	20,300 bbl/stream-d
C_3/C_4 LPG product	4,500 bbl/stream-d
Capital investment	$100-140 million
Catalyst fill, lb	368,000

Gasoline Costs (¢/gal)	
Pure methanol at 40¢/gal	100
Operating costs	3
LPG byproduct credit at 30¢/gal	(6)
Capital charges (incl. 12% DCF return)	8
Cost at plant, ¢/gal	105

Typical properties of finished gasoline

Components, wt. %		Composition, vol. %	
Butanes	3.0	Paraffins	53
Alkylate	3.0	Olefins	12
C_5^+ gasoline	94.0	Naphthenes	7
	100.0	Aromatics	28
			100

Physical properties		Octane		
			Motor	Research
Reid vapor pressure, psig	9.0			
Specific gravity	0.728	Clear	83	93
Sulfur, wt %	Nil	Leaded	90	101
Nitrogen, wt %	Nil			
Corrosion, copper strip	1A			

Regeneration is necessary because coke formation deactivates the catalyst. At any one time during operation, only a narrow band of the catalyst bed is actually being used. As the catalyst deactivates, this band moves down the bed, and unconverted methanol will eventually begin to break through into the product stream. When this occurs, the reactor vessel is put on a regeneration cycle (see vessel D-2D), where coke is burned off with a heated air-nitrogen mixture. The time between regeneration cycles is approximately 20 days. (In contrast, the catalyst used in the DME reactor can operate for something like a year without having to be regenerated.)

The product stream leaving the conversion reactors is cooled: by generating steam, by heating the methanol feed, by recycle gas, and by air and water. The cooled stream then enters the product separator, which isolates water, and gas and liquid hydrocarbons. The water is removed. Some of the gas is compressed and recycled to the reactors.

GASOLINE AND COSTS—The remain-

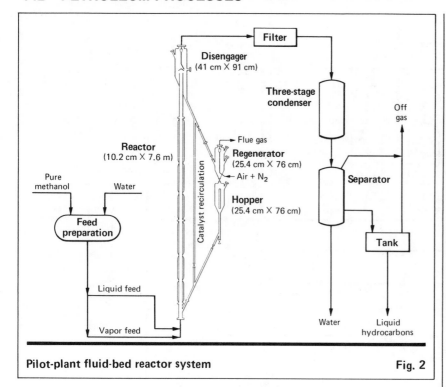

Pilot-plant fluid-bed reactor system   **Fig. 2**

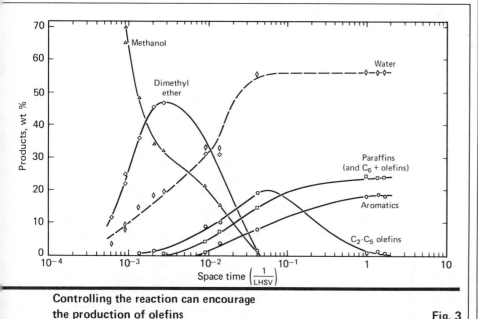

**Controlling the reaction can encourage
the production of olefins**   **Fig. 3**

indicated that engine performance is satisfactory with durene concentration less than 5%.

Economics for a fixed-bed MTG plant are given in Figure 1. These costs would apply to a plant adjacent to, but not integrated with, a methanol plant. The investment does not include working capital, land or insurance. Gasoline costs do not include royalties or catalyst costs (the zeolite catalyst has approximately a one-year life).

According to Mobil researchers, the MTG process will provide essentially 100% conversion of methanol when fixed-bed reactors are used. Studies have also shown that over 85% of the hydrocarbon yield is gasoline. In addition, the overall energy efficiency of the process, including processing energy, is said to be 92-93%.

FLUID POTENTIAL—Instead of using two reactors as in the fixed-bed scheme, the entire process can be carried out in one fluid-bed reactor. Fig. 2 illustrates Mobil's pilot plant. Here, to simulate a crude methanol feed, water was mixed with pure methanol. This feed was typically vaporized, but the reactor could be fed with vaporized and liquid methanol.

After reaction in the fluid bed, the disengager separated catalyst and product. The catalyst was regenerated and recirculated; the product was filtered to remove catalyst particles and then condensed. A light-gas stream was removed from the condenser for recycle to the reactor. In the separator, water was removed from the hydrocarbon product.

Although the fluid-bed system must be developed further before it is commercialized, this technology offers certain advantages: With a fluid-bed system, catalyst activity can be controlled via continuous catalyst regeneration; recycle will be significantly lower, since it would be required only for conversion purposes and not for heat removal; more reaction heat can be turned into high-pressure steam; and gasoline yield may be improved.

With the fluid-bed system, the product may be more accurately selected. Dimethyl ether and methanol are more rapidly converted to olefins, then paraffins, and finally aromatics (see Fig. 3).

This article is based in part on a paper presented at the 1980 NPRA annual meeting in New Orleans; and on a paper from "Large Chemical Plants," Elsevier Scientific Publishing Co., Amsterdam.

ing gas and the liquid product are mixed and fractionated. LPG is removed. Propenes and butenes may be alkylated and blended with the C_5 fraction. The gasoline is pressurized with butanes.

The resulting synthetic gasoline (see table) does not contain impurities such as oxygenates. Its boiling range is similar to that of premium gasoline. With Mobil's standard additive package, the gasoline has passed tests for carburetor detergency, emulsion formation, filterability, copper attack, metal corrosion and storage stability. Vehicle tests have shown that this product is comparable to unleaded premium gasoline.

The synthetic gasoline does contain more (about 3-6 wt %) durene (1,2,4,5-tetramethylbenzene) than is normally present (about 0.2-0.3%) in gasoline. Although durene has a high octane, it has a freezing point of 175°F. However, driving tests have

Polygas spells relief from alkylation ills

Reeling from the spiraling price of isobutane, which is needed to alkylate olefins, refiners are now considering olefin polymerization as a supplement to alkylation.

☐ Tight isobutane supplies have sent prices soaring over 90¢/gal in some parts of the U.S. If you are a refiner who relies on this paraffin to produce high-octane blending components (alkylate) from olefin streams, you may find the answer in an old technique formerly used as an alkylation alternative—olefin polymerization. High tags for isobutane, plus several other factors (see box) are bringing back polymerization, but this time as an adjunct, not a rival, of alkylation.

"Polygas fits well into the unleaded-gasoline pool," says process engineer James G. Eckhouse, of Universal Oil Products, Des Plaines, Ill. (UOP sells polymerization technology), adding that "a refinery saves on utilities with polymerization because there is no isobutane recycle and other costs associated with alkylation."

And these are not the only economies. According to Ken L. Comontofski, a market analyst with The Pace Co. (Houston), "an investment in polymerization is peanuts compared with the normal refinery investment." Indeed, industry sources say that some poly processes cost only about one-half as much as alkylation hardware and can be installed almost overnight.

In addition, the real cost of alkylation raw material is actually higher than its market price. "When you take into account that isobutane suffers a 20% reduction in volume during processing, the actual cost of isobutane raw material is over $1.10 per gallon. So when you add the cost of converting the paraffin into alkylate, you are actually losing money," says Edwin K. Jones, president of International Energy Consultants Inc. (IECI), Sarasota, Fla.—a company that claims it has developed low-cost polymerization technology.

AVAILABLE TECHNIQUES—So far, no major announcements of IECI polygas

facilities have been made, even though four refineries are said to be close to making a decision. On the other hand, at least five Dimersol units (Dimersol is a polymerization technique developed by Institut Français du Pétrole, Paris) are operating or will come online in the U.S. by the end of 1981, and six more are in the planning stage. Dimersol isolates and polymerizes propylene streams to C_6 (dimate compounds), with special emphasis on high olefin conversion.

In addition to the IECI and IFP routes, there is UOP technology that processes mixed C_3/C_4 streams. It, too, shoots for high conversion rates.

Perhaps better suited to polymerization's emerging role as an alkylation supplement is the IECI process. It

polymerizes basically only excess refinery olefins, enough to balance the facility's onsite supply of isobutane.

The design (see flow diagram) consists of a reactor, a charge pump, heat exchangers, and a small rectifier (about 10 trays) added to the cracking unit's debutanizer. All the polymerization effluent is returned as reflux to the debutanizer.

The big attraction of the IECI scheme, says Jones, is capital cost savings. No fractionators (depropanizers or debutanizers) are needed, except for the small rectifier, and this does not require additional energy because it uses the heat of reaction as a heat source. Having no fractionators also means the elimination of associated hardware—e.g., condensers, piping, reflux units, reboilers—and operating costs.

Adds Jones, "We are not pushing for high conversion. We'll only need about 50% conversion, vs. about 90% for other poly routes. The goal is to get just enough olefins reacted so that the charge to alkylation will match whatever isobutane is available."

While other polymerization techniques also can run at lower conversion rates, "one still has to purchase the additional equipment and contend with the disadvantage of higher utility

Polymerization: Why it's coming back, and how it fits in

Both alkylation and polymerization transform refinery olefins coming from catalytic crackers, cokers and other units into high-octane blending components.

In the past, the former process was preferred by refiners because it yields components that have higher octane numbers and respond better to lead additions. But the phaseout of lead has robbed alkylate of some of its octane-response glamour. Also, use of alkylation requires a steady supply of isobutane, which combines with the C_3 and C_4 olefins to produce alkylate, and isobutane supplies have been getting scarcer and more expensive.

True, refineries usually produce isobutane onsite—as a coproduct from hydrocrackers, catalytic crackers or reformers, or from C_4 isomerization—but often in insufficient quantities to fill the needs for alkylation. Some facilities must purchase isobutane on the open market. And prices have climbed so rapidly that for some refiners alkylate production has already become a money-losing proposition.

Further boosting the cause of olefin polymerization is a general improvement in polymerization techniques. The old conventional poly units, designed as an alternative to alkylation for plants lacking isobutane-producing equipment, are high-severity units that process the entire olefin stream. Modern polymerization routes are less capital-intensive and more flexible. Operating at lower severities, they supplement—rather than compete with—alkylation, because they can be tailored to handle just enough of the olefin feed to balance a refiner's alkylation-isobutane supply.

Originally published June 16, 1980

Alkylation plus polymerization means no purchased isobutane

	Alkylation only, bbl/d		IECI process, bbl/d			
	Alky unit		Alky unit		Poly unit	
	Alky feed	Alky product	Alky feed	Alky product	Poly feed	Recycle to cat cracker debutanizer
Propane	500	610	500	547	321	321
Propylene	1,100	0	465	0	706	71
Isobutane	1,400	20	1,400	20	899	899
Normal butane	450	481	450	463	289	289
Butylenes	1,550	0	654	0	995	99
Alkylate	0	4,676	0	1,975	0	0
Polymer	0	0	0	0	0	1,172
Purchased isobutane	1,888	—	0	—	0	—
Octanes						
F-1 clear		95		95		96
F-2 clear		93.5		93.5		83

Source: IECI

costs around the fractionators [depropanizer and debutanizer]," explains Jones.

LOWER OCTANE—Of course, there are some drawbacks for users of polymerization. Harvey Olsen, principal manager of process engineering at Fluor Corp. (Irvine, Calif.), notes a definite volume shrinkage in the manufacture of polygas. Also, "there is a greater spread [higher sensitivity] between research octane number [RON] and motor octane number [MON] with polygas, vs. alkylate."

IECI's Jones agrees. Referring to the table, in which F-1 represents RON and F-2 stands for MON, Jones points out that "with polygas, the F-2 is about 13 numbers lower, so that the familiar F-1 plus F-2 average you see on gasoline pumps results in a 6.5-point drop when you use polymer."

The low F-2 values were recently confirmed by William J. Benedek and Jean-Louis Mauleon, of Total Petroleum Inc. (Alma, Mich.), in a paper presented at the annual meeting of the National Petroleum Refiners Assn., held last March in New Orleans, La. While describing the operation of the first U.S. commercial Dimersol unit—a 2,000-bbl/d unit at the Alma refinery—the authors included graphs showing low F-2 values. (They noted, however, that the presence of dimate in a no-lead gasoline pool is a significant plus because dimate can replace light straight-run naphtha, due to its excellent front-end distillation.)

A low F-2, however, will not deter Good Hope Refining Inc. (Good Hope, La.) from making polygas; the company says it will boost the F-2 by adding methyl *tert*-butyl ether.

THE OLEFIN PICTURE—Aside from whether polygas will make inroads in refineries, the isobutane-supply situation will remain tight for some time. One reason is that refinery olefin-streams will swell in the near future.

Even though new, improved catalytic crackers have reduced the olefin-to-paraffin ratios, roughly 500,000 bbl/d of catalytic-cracking capacity will be installed during 1979-83, according to Gregory A. Lester, a market analyst with McClanahan Consultants (Houston). And, notes Lester, "when a refinery adds a cracker, it normally builds an alkylation unit."

So, unless refiners find other ways of handling the excess olefins—such as sending them through poly units, selectively removing isobutylene for MTBE feed, or selling olefins as petrochemicals—isobutane will remain scarce. Lester believes that in order to obtain all the isobutane needed for alkylation, "it will be necessary to split some mixed butanes from sources outside the U.S. This will flood the market with normal butane and maintain the approximately 30¢/gal differential that now exists between isobutane and normal butane."

According to Chem Systems (New York City), the value of such refinery olefins as propylene and butylenes for gasoline use will drop with the lead phaseout. However, the value of at least propylene as a petrochemical may rise. After all, ethylene crackers in the U.S. have been running on lighter feeds (which means less co-product propylene), so propylene supplies are tight.

The situation calls once more for flexibility in refinery operations. Says IECI's Jones: "Refiners realize the tradeoffs, and many of the poly units being designed call for olefin flexibility. The olefins may go into either polymerization, alkylation, or for sale as petrochemicals."

Guy E. Weismantel

Few extra pieces of equipment are needed for the IECI process

Feedstock flexibility: Cracking butanes for ethylene

Price and availability of raw materials have become unpredictable, so many industries are examining alternative feeds. Yet, changing is not always simple and many processing and economic factors are involved.

Vincent J. Guercio, CTC International

☐ These days it has become very important to choose the right feedstock. In the past, the feed that was used to manufacture a particular substance was set by predictable economics. Now this situation is more complex. Raw materials may not always be available; the price of one feed relative to that of another may be changing so rapidly that a decision today may be invalid next year; and the cost of building a facility that can accept more than one type of feed may be hard to justify in the current market, even though a commitment to a specific feed may be disastrous.

Several industries have faced this problem. For example, ammonia, hydrogen and methanol manufacturers are choosing between natural gas and coal or oil; maleic anhydride producers are looking at a choice of butane or benzene.

Recently, Monsanto Co. and Conoco Inc., in announcing the startup of their joint petrochemical operation at Monsanto's Chocolate Bayou facility (Alvin, Tex.), pointed to "feedstock security" as an impetus for the project (see *Chem. Eng.*, Dec. 15, 1980, p. 25). Here, new ethylene crackers operating on naphtha and gas oil will complement existing natural-gas-condensate furnaces, and both parties in the venture will be responsible for obtaining feed.

ETHYLENE EXAMPLE— Ethylene may serve to illustrate many of the problems associated with achieving feedstock flexibility. It is feasible to initially design an ethylene plant for wide variations in feedstock type. Unfortunately, most of the world's ethylene plants have been built for a specific type of raw material; therefore, use of more readily available and cheaper feedstocks is limited, unless the plant is revamped or there is a relaxation of product quantity and/or quality.

Around the world there is a desire to obtain more feedstock flexibility from existing ethylene plants. In Europe and Japan, 90% of ethylene capacity is based on naphtha. Now that this naphtha is no longer a "surplus" product, European and Japanese ethylene producers are searching for alternatives. Here, in the U.S., there is more diversity, with capacity divided between ethane (37%), propane (13%), n-butane/refinery gas (4%), naphtha (24%), atmospheric gas oil (16%) and vacuum gas oil (6%). But almost all of the naphtha/gas-oil capacity has been installed since the mid-1970s, and these plants have been generally designed with no flexibility to crack lighter feedstocks.

Now, while the U.S. trend toward naphtha/gas oil should continue, new plants will be more flexible, and in existing units some opportunistic cracking of ethane, propane and normal butane (see box) will occur. In addition, within the last two years, 21 facilities worldwide (7 in Japan, 8 in Europe, and 6 in the U.S.) that have been designed for naphtha/gas

---

Butanes are more suitable than propane

Butane from LPG may be a better feed for ethylene than propane because it it much more easily cracked in existing naphtha furnaces and the coproduct distribution is similar to that of original naphtha feed.

The butanes stream recovered from gas processing is 20-35% isobutane, with the remainder being n-butane. Butanes from refinery units can be up to 70% isobutane.

Most butanes are currently used by refineries for gasoline production. Isobutane can be used to alkylate propylene and butylenes to a high-octane product, or the butanes can be blended directly into the gasoline to supply volatility.

Since gasoline demand is seasonal, peaking in the summer, the demand for butanes for blending into gasoline is also seasonal, but this demand peaks in the winter because then more butanes are added to gasoline to give it high volatility. Butanes are also heavily used in the winter as residential and commercial fuel in warm climates such as Spain, Algeria, Turkey, Egypt and Mexico. Perhaps the optimum ethylene plant should crack more naphtha in the winter and more butanes in the summer.

Originally published January 26, 1981

Comparing ethylene feeds: single-pass cracking yields (wt%)			Table I
	Naphtha*	n-butane	Isobutane
Conversion, wt%	–	96	90
Hydrogen	1.0	1.0	1.3
Methane	14.2	23.0	23.0
Ethylene	26.0	34.5	9.0
Ethane/acetylene	4.0	5.0	1.5
Propylene	15.0	16.5	20.0
Propane/MA/PD[†]	1.3	1.0	2.5
Butadiene	4.4	3.0	1.7
Isobutylene	2.7	0.2	17.0
N-butylenes	2.3	1.8	1.0
Butanes	0.5	4.0	10.0
Benzene	7.0	3.0	3.5
Toluene	4.0	1.4	1.5
C_5/400° F remainder	14.0	4.6	6.0
Fuel oil	3.6	1.0	2.0
	100.0	100.0	100.0

*Medium-severity cracking
[†]MA=methylacethylene
  PD=Propadiene

oil have taken in supplemental butane feeds.

ANOTHER FEED—Liquefied petroleum gas (LPG) is being increasingly considered as an alternative cracking feedstock for ethylene. (The term LPG covers any mix of propane, isobutane and normal butane). In the next few years, increased quantities of LPG will be available from Saudi Arabia, the North Sea, Indonesia and Australia. And sharp increases in the price of

energy, as well as the nationalization of oil companies, are encouraging the gathering, conditioning and also the transportation of LPG to the industrialized countries.

Prices of LPG relative to crude oil have been traditionally set by residential fuel use, but the large quantities of LPG projected will overwhelm the residential market. At that time, the value of LPG as a cracking feedstock will be a factor in setting its price.

That price will also be determined by other feedstock prices, coproduct prices, and the investment required to feed LPG to the cracker. LPG must be priced low enough on a delivered basis to compete with naphtha/gas oil in order to encourage investment in LPG receipt, storage and use in the ethylene plant.

LOGISTICS—Many times, planners tend to emphasize the ability of process units to handle a feedstock change, and do not pay enough attention to offsite facilities. LPG transportation, terminaling, and storage must be considered early in feasibility studies from a political as well as a technical viewpoint. Because it is shipped as either a pressurized or refrigerated liquefied gas, LPG has been opposed in many instances by local groups concerned about the safety of installations.

For inland transportation, butanes from LPG offer a significant advantage over propane. Unlike the latter, butanes can more easily use existing pipelines by being batched with crude oil, gasoline, gas oil or naphtha. Some butanes could be blended into naphtha and stored in existing naphtha storage tanks. But blending butanes with naphtha may not be desirable because the butanes are best cracked separately from the naphtha.

So when butanes must be segregated, small amounts might initially be stored in pressurized tanks or in some of the C_4 product tanks (used for raw C_4s, butadiene and butylenes/butanes). Refrigerated tanks will be used in most locations for large storage requirements. If the geology of the area is good, salt-dome or mined underground storage would be more economical than refrigerated storage.

PROCESS LIMITS—Many studies on feedstock flexibility have concentrated on drastic changes, such as conversion from all-naphtha to all-ethane or all-propane. But a total conversion may be limiting. A much more likely case would be the cracking of some LPG along with naphtha, in order to maximize the amount of LPG cracking, while minimizing equipment modifications of the existing plant. The LPG market is extremely erratic, so the ethylene plant should be able to go back to all-naphtha feedstock if the price of LPG gets too high.

The amount of butanes from LPG that can be accommodated by a naph-

Ethylene unit material balances (1,000 metric tons/yr)				Table II
Case	I	IIA	IIB	III
Feedstocks				
Naphtha	1,712	1,537	1,407	1,424
Mixed butanes	–	171	298	298
Products				
Ethylene	500	500	500	500
Propylene	259	262	265	253
Butadiene	77	74	71	71
Butylenes/butanes	94	106	115	–*
Benzene/toluene	188	177	168	167
Fuel oil	62	58	55	54
Fuel gas	291	306	317	302
Pyrolysis gasoline	241	225	214	212
Alkylate[†]	–	–	–	163
Butanes cracked	–	171	298	225
Net naphtha requirement	1,471	1,312	1,193	1,049
Fuel gas consumed	292	293	294	293
Isobutylene production**	46	50	54	39

*Butylenes/butanes to alkylation unit: 93,000 m.t./yr.
[†]At 8 psia Reid Vapor Pressure. Contains about 5% n-butane.
**Contained in butylenes/butanes stream. For Case III, all butylenes are
  consumed in the alkylation unit.

tha cracker depends upon how the butanes are cracked. The use of butane/naphtha mixtures is generally not economical, because it requires high-severity cracking of the naphtha and low butane conversion. (The highest conversions attainable are 90% for n-butane and 82% for isobutane.) A conversion this low causes processing problems because of the large quantity of C_4s that must be handled.

Segregated-butanes cracking is optimum especially in a medium-severity naphtha cracker. Table I compares the single-pass cracking yields for normal and isobutane with those from medium-severity naphtha cracking.

The yields point out potential trouble spots when the alternative feed is used. Since the product is different, existing equipment may not be sufficient. Large amounts of methane will overload demethanizer columns, and cause lesser problems in compressor and refrigeration systems. The large amounts of C_4s present when isobutane is cracked will cause problems during C_4 recovery and butadiene extraction.

ALL THE PRODUCTS—Strictly speaking, it is incorrect to call a naphtha/gas-oil cracker an ethylene plant. The propylene, butadiene and benzene coproducts are significant contributors to plant revenues, and the impact of an alternative feedstock on coproduct yields must be considered.

Table II compares the material balance of a medium-severity naphtha cracker (Case I) with the balances obtained as increasing amounts of a 30/70 iso/normal butane mixture are cracked. Case IIA corresponds to 10% butanes cracking and Case IIB to 17.5%. (Net naphtha requirements for Cases IIA and IIB are reduced by 11 and 19%, respectively.) Here, propylene production increases slightly, while butadiene and benzene/toluene decrease moderately.

The decrease in butadiene production is no cause for alarm, for unless naphtha/gas-oil cracking decreases, there will be a world surplus of butadiene by the end of the decade. This surplus would cause a drop in the butadiene price relative to ethylene. But the decrease in benzene/toluene will have to be made up (e.g., by increased naphtha reforming).

Another coproduct is the pyrolysis gasoline remaining after benzene and toluene are recovered. The most eco-

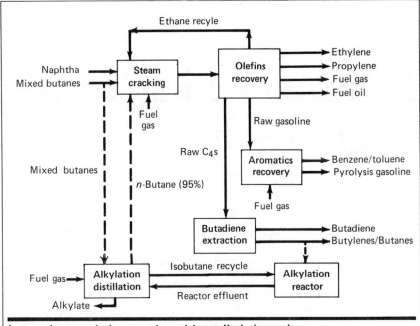

Integrating an ethylene cracker with an alkylation unit solves processing problems associated with a change in feed

nomical use of this product is as a high-octane gasoline blending stock. It can be traded off for virgin naphtha on at least a 1/1 weight ratio and effectively decrease the "naphtha vulnerability" of the cracker.

INTEGRATION—Combining a plant with another processing unit may alleviate equipment problems associated with a change in feed. Cracking up to a 10% butanes mixture (as in Case IIA) will cause some processing problems, but these might be moderated by minor revamping of equipment. However, butanes cracking at the 17.5% level (Case IIB) will require major equipment changes if the ethylene capacity is to be maintained: C_4 and methane removal facilities would be overloaded by about 10%, and butadiene extraction by 22%.

Integrating a refinery alkylation unit with the ethylene unit could solve the problem. Alkylation is a reaction of propylene and/or butylenes with isobutane (in the presence of a hydrofluoric or sulfuric acid catalyst) to form a high-octane gasoline component called alkylate. The isobutane/olefin ratio in the reactor is kept high by recycling isobutane. Normal butane does not participate in the reaction, and it is fractionated out. Since an alkylation unit will do the separation and use isobutane, integration with a cracker will either eliminate or greatly reduce many of the equipment

throughput restrictions associated with a mixed butanes feed.

The flowsheet illustrates the possible integration. In Table II, Case III provides a material balance when the same amount of mixed butanes (as in Case IIb) is first fed to the distillation section of the alkylation unit. The cracker feed is then 95% n-butane. The alkylation unit reacts all of the isobutylenes produced in the furnace with essentially all of the isobutane in the mixed-butanes stream.

The alkylate produced can be used with the pyrolysis gasoline from the cracker to reduce the naphtha demand. Because of the alkylate's low density and very high octane, it can be traded off for virgin naphtha at better than 1/1 on a weight basis. (Case III shows a 29% reduction in naphtha requirements over the all-naphtha case.)

Integration, therefore, can make it possible to use existing equipment with little or no revision and it provides an important way to lessen dependence on a feed without major capital spending.

Reginald I. Berry, Editor

The Author

Vincent J. Guercio is president of CTC International (Montclair, N.J.), a new affiliate of Chemical Technomics Co., which provides advice in the areas of natural-gas liquids, petrochemicals and oil refining, with ethylene, propylene C_4s and C_5s as particular specialties. Mr. Guercio received his B.Ch.E. at Catholic University of America and did graduate work in chemistry at Princeton University.

Hydrocracking without catalysis upgrades heavy oil

Heavy crude oil, atmospheric or vacuum bottoms
can be turned into lighter, more valuable, products
via a hydrogenation system that avoids coking problems
and offers high conversion and desulfurization.

Michael A. Menzies and *Adolfo E. Silva*, Petro-Canada, and
Jean M. Denis, Dept. of Energy, Mines and Resources, Canada

□ CANMET is a new hydrocracking process for the upgrading of heavy oils (see *Chem. Eng.*, Nov. 3, 1980, p. 19). This route, unlike other hydrocracking techniques, does not use a catalyst but employs a low-cost additive to prevent coke formation and allow better than 90% conversion of high-boiling-point hydrocarbons into lighter products.

This process was developed by the Energy Research Laboratories of the Canadian Federal Government (Ottawa), which has granted Petro-Canada (Calgary) the exclusive rights to commercialize the system. And now, Petro-Canada, after operating two 1-bbl/d pilot units, is negotiating to build a demonstration plant of around 5,000 bbl/d capacity. The demonstration unit will be employed within a conventional refinery to upgrade vacuum bottoms, or to upgrade a heavy crude. Construction of this facility is slated to begin in mid-1982.

CRUDE QUALITY—As world reserves of conventional crude oil decrease and the price of oil increases, interest in heavier crudes and in oil obtained from oil shale and tar sands continues to grow (see *Chem. Eng.*, July 30, 1979, p. 25). However, while huge deposits of these materials exist in Canada, Venezuela and the U.S., this heavy oil has a high density, viscosity and pitch content (more than 50 wt % of the material has a boiling point over

524°C). The oil also contains impurities such as metals, sulfur and nitrogen, and it has a high carbon residue (as measured by Conradson carbon testing), which indicates a tendency to coke.

Most present-day catalytic reactors cannot economically upgrade these heavy oils, due to their high metals content and coking tendency. But at least a minimum of upgrading is required to produce an oil that is transportable by pipeline and acceptable to refineries.

The usual approach to upgrading is by thermal cracking, where carbon is removed as coke, or by hydrogenation where hydrogen is added to convert more of the oil to lighter products. Carbon removal processes, such as commercially available Fluid Coking, Delayed Coking, and Flexicoking, operate at low pressure and without a catalyst but have low liquid yields and offer no desulfurization. Hydrocracking processes, represented by routes such as LC-Fining and H-Oil, make use of a catalyst but produce a high yield of light products and achieve some desulfurization.

NON-CATALYTIC—CANMET technology offers the advantage of hydrocracking without the aid of a catalyst. The process uses an additive of pulverized coal impregnated with iron sulfate or other metallic salts. This additive, which represents about 0.5 to 5.0 wt % of total feed material, prevents the

deposition of coke in the reactor and allows operation at 66% lower pressure than conventional hydrocracking routes. The additive also attracts metals, e.g. vanadium and nickel, removing them from the liquid products.

In tests with Cold Lake vacuum residuum (boiling point above 404°C), conventional hydrocracking could not achieve more than 50% conversion of heavy material into lighter components without reactor coking problems. With the use of the CANMET additive, conversions of over 90% were attained without operating difficulties. The process has also been tested on Athabasca bitumen and residues from Lloydminster, South Saskatchewan and Boscan (Venezuela) heavy crudes as well as residues from conventional light crude.

The CANMET process has been tested with a variety of crudes and residues that vary in boiling points, and in concentrations of sulfur (up to 5.5 wt %), nitrogen (up to 0.6 wt %) and metals (up to 1,300 ppm). The system provides yields close to 100 vol % of distillates (from C_4 to 524°C boiling-point components). The liquid products exhibit less than 5 ppm of metals in the heaviest fraction. Desulfurization is better than 60%. And the route offers low hydrogen consumption (less than 2 wt % of the feed).

In the process (see flowsheet), additive is slurried into the crude-oil feedstock. This oil slurry, recycled gas and makeup hydrogen are fed to the reactor through separate heaters (reactor temperatures are less than 470°C).

An empty upflow reactor is used for the hydrocracking. Here, hydrogen and oil rise through a bed of particles composed of the additive residues and deposited feed metals. During operation, solids are continuously removed from the reactor vessel: fines go overhead with the product, and the rest

*This article is based on a paper delivered at the 30th Canadian Chemical Engineering Conference, held at Edmonton, Alta., October 1980.

Originally published February 23, 1981

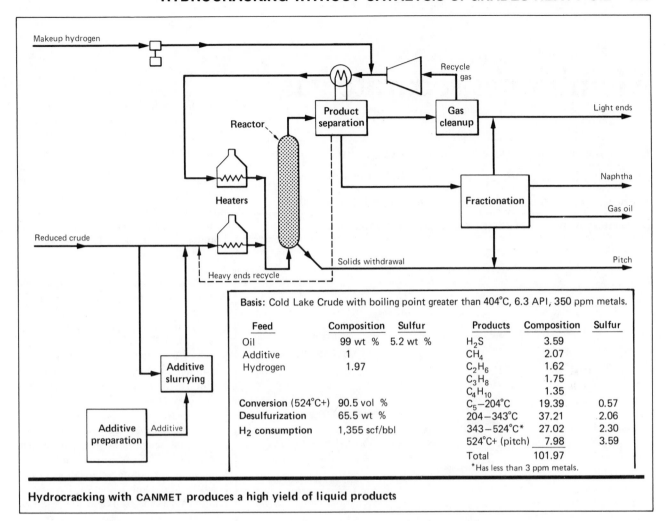

Basis: Cold Lake Crude with boiling point greater than 404°C, 6.3 API, 350 ppm metals.

Feed	Composition	Sulfur		Products	Composition	Sulfur
Oil	99 wt %	5.2 wt %		H_2S	3.59	
Additive	1			CH_4	2.07	
Hydrogen	1.97			C_2H_6	1.62	
				C_3H_8	1.75	
				C_4H_{10}	1.35	
Conversion (524°C+)	90.5 vol %			C_5–204°C	19.39	0.57
Desulfurization	65.5 wt %			204–343°C	37.21	2.06
H_2 consumption	1,355 scf/bbl			343–524°C*	27.02	2.30
				524°C+ (pitch)	7.98	3.59
				Total	101.97	
				*Has less than 3 ppm metals.		

Hydrocracking with CANMET produces a high yield of liquid products

head with the product, and the rest exits through the bottom.

Leaving the reactor, the mixed liquid product is flashed (pressure is reduced) to separate the lighter and heavier components. Gas is scrubbed to remove H_2S and recycled to be mixed with makeup hydrogen. Heavier components are separated by atmospheric and vacuum distillation. (CANMET, as now developed, is a liquid once-through system but capabilities for heavy-ends recycle are provided for additional flexibility.)

SYNTHETIC CRUDE—An estimate has been prepared showing the application of CANMET technology to the upgrading of 100,000 bbl/d of Cold Lake crude (10.6° API, 4.3 wt % sulfur). In this application, atmospheric and vacuum distillation would precede the CANMET unit, which would be fed 72,500 bbl/d of material with a boiling point above 404°C. Downstream hydrotreating would be added to obtain maximum desulfurization of distillates. Propane, lighter gas compo-

nents and heavy oil would be used for fuel balance.

In this application, the distilled products would be blended to form a synthetic crude of 31.6° API and 0.21 wt % sulfur. Total yield is calculated to be 95.6 vol % of feed. Sulfur produced would be 587 long tons/stream-d; pitch, 4,823 bbl/stream-d. Utilities for this plant would be: 125,000 gpm of cooling water, 49,300 kW, and 37.5 million standard ft³/d (scf) of natural gas. Fuel requirements would be (in bbl of fuel oil equivalents): 5,786 for oil plus 5,552 for gas.

The economics of such an operation based on first-quarter-1980 Canadian dollars shows a discounted cash flow rate of return of 21.8%/yr, with a payout time of 3.8 yr for a stand-alone operation. Other components of the basis of this estimate are a plant operating life of 20 yr, a capital cost allowance of 50%/yr, tax rate of 41%, and prices (Canadian $) of $2.15/thousand scf of natural gas; $16.25/bbl crude feed; $10/bbl of pitch; zero

value for the sulfur, and $29.38/bbl for the synthetic crude.

The CANMET additive would cost $10,200/d. Catalyst for hydrotreaters would be $3,400,000 for an initial charge and $6,300/d for consumption. Direct cost (materials and labor) for the CANMET unit alone would be $127 million, not including fractionation and sulfur removal. Direct cost for the entire upgrading facility would be $344.4 million.

Reginald I. Berry, Editor

The Authors

Michael A. Menzies holds a Bachelor of Engineering degree from the University of Canterbury, New Zealand, and a Ph.D. in chemical engineering from McMaster University in Hamilton, Ontario. He is the Project Manager for development of CANMET at Petro-Canada.

Adolfo E. Silva received his Bachelor's degree in chemical engineering in Chile and his M.A.Sc. and Ph.D. from the University of Waterloo, Canada. He is now Senior Engineer at Petro-Canada, P.O. Box 2844, Calgary, Alberta T2P 3E3. He is responsible for all technical matters related to CANMET technology.

Jean M. Denis holds a B.A.Sc. in chemical engineering from Ottawa University. He is now manager of the Synthetic Fuels Laboratories at the Department of Energy, Mines and Resources in Ottawa.

Membranes separate gas

Use of membranes to sweeten natural gas is finally going commercial, and other applications are not far behind. Though high product purities seem beyond the reach of present systems, there are many advantages.

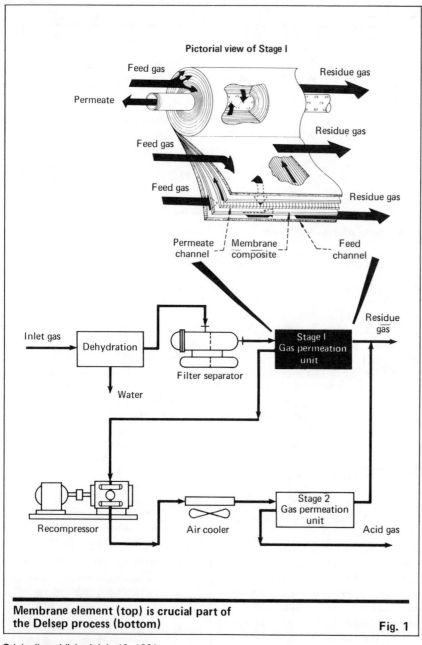

Pictorial view of Stage I

Feed gas

Permeate

Residue gas

Feed gas

Residue gas

Feed gas

Residue gas

Permeate channel / Membrane composite / Feed channel

Inlet gas → Dehydration → Filter separator → Stage I Gas permeation unit → Residue gas

Water

Recompressor → Air cooler → Stage 2 Gas permeation unit → Acid gas

Membrane element (top) is crucial part of the Delsep process (bottom)

Fig. 1

Originally published July 13, 1981

150

□ The many attractions of membrane systems (see box)—e.g., simplicity, low capital and production costs—seem to be suddenly within reach of chemical process industries (CPI) firms attempting gas-separation jobs. Several such systems, mainly for natural-gas sweetening, are nearing the commercial stage in the U.S. or have already reached it. Others are under development for such tasks as air separation and separation of olefin streams.

At least one of the techniques—Monsanto's hollow-fiber Prism method for obtaining purified hydrogen—has been in large-scale use for several years. The firm installed it in 1977 in a commercial-size plant, but waited until last year (*Chem. Eng.*, Feb. 25, 1980, p. 55) to announce its commercial availability. (Since then, nine Prism units have gone up in the U.S., and Monsanto has orders for twelve more plants worldwide.)

Delta Engineering Corp. (Houston) says that its process to separate acid gas (CO_2, H_2S) from natural gas will be ready for commercialization in late 1981. And Dow Chemical Co. (Midland, Mich.) is confident that the first full-scale plant to use its CO_2 separation route will go onstream next year. Not as close to the marketplace are an air separation technique that would produce low-purity oxygen, and an olefins-separation system. The former is being piloted by the Fluid Systems Div. of UOP Inc. (San Diego, Calif.); the latter, a development of Amoco, is in a dormant state right now because of uncertain economics.

Overseas, work on membranes' gas-separation capabilities is also intense, expecially in Japan and Italy. Japan's Ministry of International Trade and Industry and the chemical industry are planning a 10-year, $45.6-million development program aimed at such applications as water/alcohol separation, uranium recovery, separation of materials with close boiling points, and others. In Italy, an air-separation route is under development at the University of Naples, with joint funding from the government and Montedison. Commercialization is thought to lie at least two years away.

LIKE ROLLED PAPER — Delta's Delsep process reached the pilot stage in 1977, and has been extensively tested in a sour-gas field in Alberta, Canada, and in a CO_2 enhanced-oil-recovery project in western Texas.

The sour-gas stream, with concentrations ranging from 5-50% H_2S and/or CO_2 at inlet pressures from 800 to 1,200 psig, flows through membrane modules arranged in spiral-wrap fashion reminiscent of a roll of paper (see Fig. 1). The spiral wrap consists of alternate layers of a feed-gas channel, permeable membrane, and acid-gas channel. The acid gas passes through the membrane to the low-pressure side, where it collects in a perforated internal pipe, and exits. Purified residue gas continues to flow along the surface of the membrane, and exits at the opposite end, with little loss of pressure.

Delta's president, A. B. Coady, lists several impressive advantages over conventional sour-gas processing: reduced operating costs (fuel consumption alone is about 30% less), lower space requirements, and ease of expansion (via additional modules). Just as important is the capital-cost edge. According to Coady, "Where the acid concentration is 20% or higher, and the inlet pressure is in the 800-1,200-psig range, we expect that this process would save nearly half of the capital cost of a conventional system." (H_2S and CO_2 removal usually is handled by amine-absorption units.) Total life of the membrane elements is estimated at 3-5 yr.

Coady sees a promising potential for Delsep in the separation of CO_2 contained in sour-gas reservoirs previously considered uneconomic. The CO_2 can be used in enhanced oil recovery.

MORE SWEETENING. — Carbon dioxide separated from natural gas via Dow's hollow-fiber membranes (believed to be of a modified triacetate) may also wind up used in tertiary oil recovery. Cynara Inc. (Houston), a joint-venture company that Dow has formed with Production Operators Corp. (Houston), is in charge of developing and commercializing the separation processes.

General Electric Co. also is using membranes to sweeten methane—but from such sources as landfills, which emit gas containing about 40% CO_2. The firm has been operating a pilot unit that can separate both CO_2 and

Membranes—what they can and cannot do

Gas separation via membrane technology works on the principle that different components in a gaseous mix have different rates of permeation. The driving force for the separation is differential pressure—the operation can take place over a wide range of temperatures and pressures and is restricted only by the membrane's physical limits. Such membrane materials as polysulfone, polystyrene, Teflon and various rubbers have different separation characteristics, which qualify them for specific jobs.

In its most basic form, a membrane separation system consists of a vessel divided by a flat membrane into high- and low-pressure sections. Feed gas entering the high-pressure side selectively loses the fastest permeating substances to the low-pressure side.

The flat-plate design generally is not used in commercial applications because it doesn't provide enough membrane surface area to achieve the desired mass transfer.

In the capillary or hollow-fiber design, the separator modules resemble shell-and-tube heat exchangers. Anywhere from 10,000 to 100,000 capillaries, each less than 1 mm dia., are bound into a tubesheet surrounded by a metal shell. Feed gas is introduced into either the shell or tube side of the module. If the permeability rates of the gaseous components are close, or if higher product purities are desired, separator modules can be arranged in series, and feedstreams can be recycled.

Compared with other gas-separation routes, membrane systems offer these advantages:

■ Mild operating conditions. Air separation, for example, can be done at atmospheric pressure and room temperature, instead of at cryogenic conditions.

■ Large reductions in electricity and fuel consumption. Membrane systems usually do not require utilities unless compression is needed.

■ Little space required.

■ Economic viability at both low and high system-capacity. Conventional techniques can be expensive at small flowrates, but membrane systems can be easily scaled down. And membrane modules can be added in stages to accommodate higher capacities. Says Amoco's Robert Hughes, "This is not the case with distillation, where the separation train would be a bottleneck."

■ Low capital cost.

There are, of course, limitations. Membrane technology generally is not indicated in cases requiring very high purities and recovery rates. Says professor Sun-Tak Hwang of Iowa State University, "Membrane processes generally provide only partial enrichment, and are not economical in obtaining very high purities. This is because of the number of stages and compression costs involved."

For this reason, many experts see membrane separation systems as working in combination with conventional techniques. In air separation, for example, cryogenic separation could be preceded by processing with membranes.

H_2S from 10,000 ft³/d of sewage gas from a digester. The system employs a flat membrane; the high-pressure side is at 1-7 atm; vacuum is applied at the low-pressure side. The end-product is a pipeline-quality gas (96-98% methane, 2% CO_2); with multistage units, gas yields can be as high as 90%.

AIR SEPARATION — Other researchers are developing membrane processes for low-purity oxygen that could be used in medicine and to increase the efficiency of furnaces, annealing ovens and smelters.

Working under a U.S. Dept. of Energy contract that funds two-thirds ($924,000) of the cost of the project, UOP has started operation of a pilot unit that employs membranes to separate air and obtain the equivalent of 2 tons/d of pure oxygen. Pressure difference across the membrane is about 0.5 atm. The goal is to supply a copper-tube annealing furnace with an enriched stream containing 50% O_2.

A system under development at the University of Naples seeks to provide oxygen enrichment for small combustion systems used to make glass, or used in gas turbines. Employing a hollow-fiber membrane made from polysulfone and butadiene, Italian researchers have achieved enrichments of 30-35% oxygen.

MARGINAL FOR OLEFINS—Amoco Research Center (Naperville, Ill.), which recently revealed details of a membrane process that separates such olefins as ethylene and propylene from paraffins up to C_5 (*Chem. Eng.*, Apr. 20, p. 20), has had to shelve the technique because of marginal economics. Several membrane stages are needed to achieve polymer-grade-purity compounds (polymer-grade propylene is 99% + pure), and this makes the system uncompetitive in comparison with conventional distillation. (Amoco initially had hoped to be able to replace the distillation train in olefin plants with membranes.)

The Amoco technology employs a cellulose acetate hollow-fiber impregnated with an aqueous solution of silver nitrate. The silver ions form a complex with olefins, and this facilitates their transport across the membrane.

To demonstrate the process, the firm has operated a commercial-size separator, fed with a vent-gas stream (containing 75% propylene) from a propylene plant. Fitted with 200 ft² of membrane surface area, the separator did succeed in yielding a 98% + propylene product.

But the firm has shifted directions, and is now pursuing a membrane route that would separate CO_2 from methane, for use in enhanced oil recovery.

MEMBRANE COLUMN—At a recent (Apr. 5-9) meeting of AIChE in Houston, Iowa State University (Iowa City) researchers described new equipment—dubbed a continuous-membrane column—that they have used to separate methane from mixtures containing CO_2 and N_2.

Design of this unit (see Fig. 2) is similar to that of a packed column for distillation or absorption, and it combines several individual membrane stages. The column has stripping and enriching sections; the highest concentration of the most permeable component can be found near the compressor in the latter section. The degree of separation can be increased by increasing the reflux (recycle through the compressor.).

Such a continuous column can be used to separate a binary mixture or one component of a more complex mixture. And several columns can be arranged in series to isolate the substances in a multicomponent feed. In fact, the university researchers have

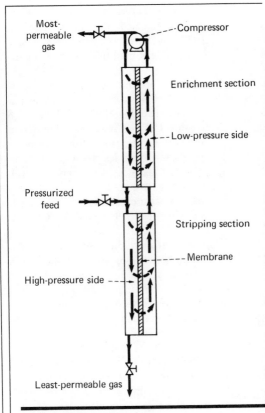

Most-permeable gas
Compressor
Enrichment section
Low-pressure side
Pressurized feed
Stripping section
Membrane
High-pressure side
Least-permeable gas

Packed column integrates several stages   **Fig. 2**

used this arrangement (two columns in series) to separate ternary mixtures.

In one case—a combination of methane, carbon dioxide and nitrogen—researchers were able to separate methane by using silicone rubber capillary membranes.

Reginald I. Berry

New lease on life for enhanced oil recovery

☐ The scenario for enhanced oil recovery (EOR) in the U.S. has changed considerably in recent years. On the plus side, oil-price decontrol, which became fully effective in January 1981, considerably brightened the dismal outlook of the late 1970s (*Chem. Eng.*, Nov. 19, 1979, p. 102). On the debit side of the ledger, the current oil glut and depressed prices now make it less attractive to recover heavy oil. Even so, for the long term many experts see a steady growth for EOR, fueled by improved recovery techniques.

Originally published June 28, 1982

The current methods—according to Harry R. Johnson, director of the U.S. Dept. of Energy's Energy Technology Center (Bartlesville, Okla.)—will produce 1.1 million bbl/d by the year 2000, while more-advanced processes now under development may recover another 2.7 million by then*. U.S. oil output via EOR is now about 400,000 bbl/d (roughly 4-5% of total production, which averaged 8.56 million bbl/d last year).

Even more optimistic are the predictions of a study by Colin A. Houston & Associates, Inc. (Mamaroneck,

N.Y.), which anticipates EOR production by 1991 of at least 1.3 million bbl/d, and perhaps as much as 2.5 million if new technology now being tested proves successful. (The company notes that the windfall-profits tax favors investment in EOR because it lowers the tax bracket for EOR oil from 70% to 30%.)

*Johnson made his predictions at a conference on advances in petroleum recovery and upgrading technology, held in Calgary, Alta., on June 10-11.

The expected boost in heavy oil output, adds Houston, will increase the market for recovery chemicals to $800 million by 1986 and to as much as $4.7 billion by 1991 (in constant 1980 dollars).

STEAM HEAT—Development of new EOR techniques is being spurred by DOE, whose Bartlesville center manages 14 field test projects in a cost-sharing program with industry, as well as research work in university and DOE laboratories.

Bartlesville and two other DOE centers are not included in the agency's proposed budget for 1983, but a DOE spokesman says that efforts are being made to have the facilities operated by a private institution and funded by industry. DOE spending on EOR was about $18.6 million in 1981.

So far, the only commercially successful EOR technique has been steam injection. In California, where most of the recovery activity is taking place, about 325,000 bbl/d was produced last year by that method. All others combined—including carbon dioxide injection, in-situ combustion, polymer flooding, and caustic injection—recover only about 50,000 bbl/d.

"These other methods just haven't been proven," says Todd Doscher, a petroleum engineering professor at the University of Southern California (Los Angeles) who heads his own consulting firm. Doscher is not very bullish about the near-term prospects for heavy-oil recovery. "The market is bad now because Alaska crude is rather cheap, and it won't improve until North Slope production starts falling off about 1985-87. Then steam-enhanced production will go to about 500,000 bbl/d by the second half of the decade."

Whether by steam or other methods, only EOR will be able to recover at least part of more than 300 billion bbl of oil that neither primary nor secondary techniques are able to wrest from its formations. DOE believes, for example, that EOR processes now in use or under development can reclaim up to 18-53 billion bbl from that reserve total.

ALL ABOUT FOAM—Steamflooding, which involves injecting steam continuously into a formation, can recover 50-55% of the oil in place, says a spokesman for Getty Oil Co. Among the method's limitations are heat loss and the steam's tendency to form chan-

nels and gravitate upward, which leaves a lot of oil untouched.

Various projects now underway aim at reducing these deficiencies. One way around the channel problem is to inject chemical foaming agents that block the conduits and divert steam to unreclaimed zones of the oil formation. Houston & Associates estimates that chemicals consumption for steamflooding could reach 1.6 billion lb by 1991 (from less than 10 million lb in 1981), of which 25% will be steam-diverting agents; the rest will be pollution-control chemicals for the steam generators. Various families of surfactants are used for steam diversion.

Chemical Oil Recovery Co. (Corco), Bakersfield, Calif., under a DOE contract, is doing field tests in a Petro-Lewis Corp. oilfield near Bakersfield, and has shown through injection profiles that foam does divert steam within the formation. Corco formulates its own foaming agents and applies them through a process licensed from Phillips Petroleum Co. (Bartlesville, Okla.). The company has used a standard liquid surfactant that takes a few hours to react completely in the oil formation, and has encapsulated the material in a cross-linked polymer gel. This penetrates farther into the reservoir before foaming (it reacts between 48 and 72 h after injection).

"Use of foam costs $3 to $5 per incremental barrel of oil produced on average," says Rod L. Eson, vice-president of engineering and administration at Corco. "This is a very low investment for a barrel of oil." Eson adds that it is more effective and cheaper to inject "slugs" of surfactant (55-165 gal) about once a week than to inject small amounts continuously.

A foam injection process being developed by Stanford University Petroleum Research Institute (Stanford, Calif.) is being tested in another Petro-Lewis oilfield near Bakersfield. In the Institute's process, large slugs (30,000-40,000 gal) of aqueous solution containing 15% active material will be injected every 3-6 mo to determine whether it will permit a better steam sweep of the oil formation. The material is Suntech IV, a long-life foam made by Suntech Group of Sun Oil Co. (Marcus Hook, Pa.).

CLD Group, Inc. (Ventura, Calif.), claims to have increased the ratio of oil to steam in a California oilfield from 0.1 bbl per bbl steam (as water) to

more than 0.2 bbl/bbl steam, using proprietary foams under a DOE contract. Todd Doscher, who is a member of the group, says the improvement makes production economical.

UNDERGROUND GENERATOR—Heat losses in steam injection may be reduced 50% or more by insulated, as compared with noninsulated, tubing, says Alan J. Leighton, a program manager for thermal recovery methods in DOE's San Francisco Operations Office (Oakland, Calif.). A typical insulation method is to place an outer tube around the injection tube and fill the annulus with ceramic fibers, calcium silicate or sodium silicate.

Another way to tackle heat loss is to use a downhole steam generator, especially for deep zones, as steam injection is not practical below about 2,500 ft. This also avoids pollution problems: In California's Kern County, where steam generation requires the burning of about 1 bbl of crude to recover 3 bbl, oil companies have had to install scrubbers to control sulfur emissions, and now methods are being tested (including ammonia injection into fluegas) to control oxides of nitrogen.

In the case of a downhole generator, fuel (usually diesel), air or oxygen, and water, are pumped downhole to the underground combustor, then steam and combustion products are injected into the formation.

Sandia National Laboratories (Albuquerque, N.M.) has conducted three field tests of downhole steam generators, and concluded that aboveground steam generation with insulated injection piping is marginally more efficient than the other systems. However, Rod Eson of Corco, which claims to have the first commercial downhole system generator, says that this type of unit is more efficient at depths greater than 4,000 ft, and also notes that the combustion gases help pressurize the reservoir and avoid pollution problems.

Tests of a process that combines steam injection with mining techniques are scheduled to start next month on property leased from Shell Oil Co. near Bakersfield. Patented by Cornell Heavy Oil Process, Inc. (Dallas, Tex.), the concept is to excavate a shaft to the bottom of an oil formation, then drill lateral wells radially into it. These wells serve alternately for steam injection and oil collection. The advantage of this approach, says Cornell, are that it exposes about six times as much steam-injection (and oil-collection) casing to the oil formation as does conventional steam injection, and it takes advantage of steam's natural tendency to rise.

Cornell has sunk a 490-ft, 5-ft-dia shaft for its tests, and eight radial steam/oil pipes, four of them 700 ft long and four of 430 ft (the test parcel is rectangular). Initially, steam will be injected for 3 mo, then oil will be pumped for 6-9 mo.

BURNING IN PLACE—*In-situ* combustion, or fireflooding, has the advantage that it removes essentially all the oil from the traversed portion of a reservoir, and it can be used at substantial depths. A good fireflood yields about 50% of a reservoir, says an oil company spokesman. Combustion is started by a downhole burner or spontaneously by compressed air, then air is injected as necessary to control burning. The high temperature and steam (from water in the formation) drive out oil ahead of the combustion front, leaving a coke residue that is burned to continue the process.

Fireflooding has not found extensive use, however, because it is difficult to control. Also, the corrosive combustion gases may require the use of high-nickel steel for well casings, and air compression is expensive. Gulf Oil Exploration and Production Co. (Houston), which has a lot of experience with fireflooding, plans to cut compression costs by using a company-developed LHV (low heat value) combustor for two fireflood projects near Bakersfield. The catalytic combustor will be fueled by well exhaust gases (it can function on 50 Btu/ft^3 gas).

CARBON DIOXIDE USE—Of the techniques used for light-oil recovery, carbon dioxide flooding has the best potential for the near future. Royal J. Watts, a project manager at DOE's Morgantown, W. Va., Energy Technology Center, expects CO_2 to account for about 100,000 bbl of oil recovery by about 1988 (current production is around 22,000 bbl/d). Houston & Associates predicts that miscible-gas flooding (CO_2 and some nitrogen) could produce as much as 800,000 bbl/d by 1991. The firm notes that a number of major oil companies are building or planning several CO_2 pipelines in the Southwest to supply CO_2 to EOR projects.

CO_2 is miscible with oil; when injected under pressure, it causes the oil to swell and decreases its viscosity, so that it may be driven out by waterflooding (or nitrogen). The gas is normally used at depths of more than 2,000 ft to avoid fracturing the formation (injection pressures are 1,500 psi or more, depending on the depth). Some general criteria for use of CO_2 are that the oil should have an API gravity of not less than 25-30 deg, and a viscosity of no more than about 10 cP. The recovery efficiency of CO_2 is 15-19% of the original oil in place, according to DOE's Johnson. DOE's total recoverable target is 21 billion bbl, of which 7 billion could be produced by current technology.

While CO_2 flooding is inexpensive compared with chemical flooding, it is marginally economical because the enormous amounts of CO_2 needed are normally not readily available. A study done for DOE by Science Applications, Inc. (Morgantown, W. Va.), determined the potential CO_2 demand for four major basins at 12 billion ft^3/d, based on a 15-yr life.

Natural CO_2 and high-purity sources (from hydrogen and ammonia plants) could satisfy less than 20% of the demand, the report says. Secondary sources, such as power plants, could supply the rest, but the cost of recovery alone would be around \$2.00-\$2.50/1000 ft^3, without considering pipelines. Harish Anada, of Science Applications, says it takes about 6,800 ft^3 of CO_2 for secondary recovery of a barrel of oil (i.e., in a field that has not been waterflooded) and 7,800-13,900 ft^3/bbl for tertiary and heavy oil (assuming 50% recirculation of CO_2).

FLOODING METHODS—Polymer flooding (the addition of polymer to water to increase viscosity) recovers an average of only about 4% additional oil over a plain waterflood, according to DOE, and has the potential to recover only about 1 billion bbl. However, surfactant/polymer flooding, still in an early stage of development, and expensive, seems to have long-term potential because of its efficiency.

In surfactant/polymer flooding, the surfactant (typically a petroleum sulfonate) is injected first to disperse the oil as fine globules in the water. This is followed by a polymer flood (polyacrylamides and polysaccharides are commonly used). However, petroleum sulfonates are affected by salinity and will react with calcium that may be present in oil reservoir brines. Polyacrylamides lose their effectiveness if the brine's salt content is much over 1% (which is the case in 90% of U. S. oil reservoirs, says one oil company spokesman), and polysaccharides are subject to biodegradation. A water preflush may only partially alleviate these problems, say industry people, and a long-term resolution would require better chemicals.

A successful surfactant polymer-flood pilot test in a high-salinity reservoir in Fayette County, Ill., was described by Exxon Co. U.S.A. (Houston) at the Soc. of Petroleum Engineers/DOE third joint symposium on EOR in Tulsa, Apr. 4-7.

Exxon used a proprietary microemulsion surfactant, designed to withstand the 10%-salinity brine, and a xanthan biopolymer, which is unaffected by salt (but suffered some bacterial degradation). The firm said that about 55% of the oil in the reservoir had already been removed by waterflooding, and the test project recovered about 60% of the remainder.

At the same meeting, Goodyear Tire & Rubber Co. reported promising laboratory tests of a new water-soluble acrylic polymer, said to retain its viscosity after extended aging at elevated temperatures (up to 86°C) in various media, including under high-brine conditions.

Alkaline or caustic solution injection for EOR has the advantage of low cost, as it uses relatively inexpensive sodium hydroxide or sodium orthosilicate to activate natural surfactants in the oil and thereby reduce surface tension between the oil and reservoir water. Several large-scale field tests are underway in California, but it will be at least a year before meaningful data are available, says E. H. Mayer, chief staff engineer for Thums Long Beach Co., which is working on a DOE-supported field test.

Some hazards of alkaline flooding: reaction with divalent ions (e.g., magnesium and calcium) to form salts, absorption of the alkali by reservoir clays, and dissolution of silica from reservoir rock.

At some time in the future, microbial methods may find a place in EOR. DOE is sponsoring a number of research projects, but they are still in early stages of investigation.

Gerald Parkinson

Rerefining waste oil

In the U.S. and Europe, new techniques are being applied to rerefine waste oil—obtaining high recoveries and avoiding waste-disposal problems.

Reginald Berry, Assistant Editor

☐ Recently-developed methods may make rerefining waste oils more attractive, particularly since the EPA plans to classify used lubricating oil as a hazardous substance.

Corrosion and waste-disposal problems associated with older acid-clay or solvent-extraction processes have been lessened by the new routes, which avoid caustics, clays, sulfuric acid and solvents. These processes offer high recoveries (over 90%), high lube-oil yields, and costs comparable to conventional acid-clay rerefining. A significant amount of the fuel required by the systems is supplied by the waste oil itself. And the product is comparable to virgin oil.

RECENT APPLICATION—In Raleigh, N.C., Phillips Petroleum Co. (Bartles-ville, Okla.) will be starting up a 2-million-gal/yr plant based on its new Phillips Re-refined Oil Process (PROP). This facility will be operated by the state of North Carolina. Another PROP unit, rated at 5 million gal/yr, will be delivered in late 1979 to the Mohawk Oil Co. at a site near Vancouver, British Columbia. And it has been announced that Lubricants Inc. (Pewaukee, Wis.) will use this system to treat 2 million gal/yr of waste oil.

In Europe, Kinetics Technology International, B.V. (Zoetermeer, the Netherlands) has received an order from West Germany's Haberland & Co., to convert a plant now based on the acid-clay process to a new process developed by Kinetics Technology International (KTI) in close cooperation with Gulf Science and Technology Co. (Pittsburgh, Pa.). Located near Hanover, West Germany, the unit for Haberland will treat 120,000 metric tons/yr of used oil; startup is expected by the end of 1980.

With experience gained through a 1-metric-ton/d plant operated in Switzerland, Leybold-Heraeus GmbH (Hanau, West Germany) is negotiating a commercial application of its Recyclon process in Germany, the U.K., Yugoslavia and Switzerland.

REREFINING WASTES—Lubricating oil that has been drained from engines contains water, broken-down additives, metal sludge, antifreezes, brake fluids, impurities drawn from the atmosphere, and gasoline. In the traditional acid-clay process (see flowsheet), after water is removed waste-oil is contacted with strong sulfuric acid, which extracts impurities and forms an acidic tar that settles out. The slightly acidic oil that remains is mixed with active fuller's earth (a clay) to adsorb additional contaminants and improve color. Finally, the oil is neutralized and distilled. The rerefined product is removed as overheads; the spent clay is separated from the bottoms by filtration.

This method has several problems.

Typical compositional range of waste-oil feedstock		Table I
Viscosity, cSt @ 100°F	41.3 —	149.7
Water, wt.%	0 —	15
Volatiles below 600°F, wt.%	2 —	15
Nitrogen, wt.%	0.03 —	0.54
Sulfur, wt.%	0.13 —	0.60
Chlorine, wt.%	0.03 —	0.25
Sulfated ash, wt.%	0.50 —	1.60
Metals, ppm		
Lead	1 —	11,000
Calcium	600 —	3,720
Zinc	560 —	1,500
Barium	2 —	1,630
Magnesium	3 —	500
Iron	10 —	600
Phosphorus	600 —	1,410
Copper	1 —	120

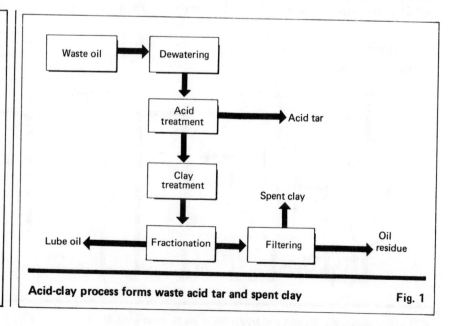

Acid-clay process forms waste acid tar and spent clay   Fig. 1

Originally published April 23, 1979

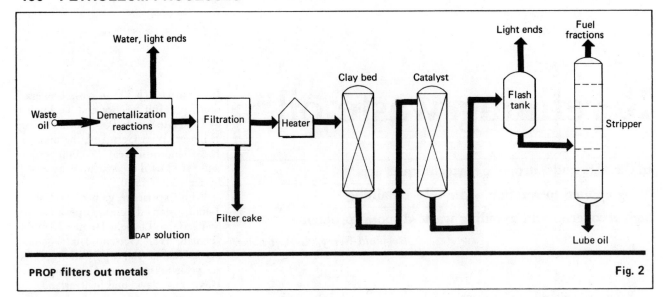

PROP filters out metals   **Fig. 2**

Both acid and clay are corrosive; the acidic tar and spent clay pose disposal problems; and a sizable fraction of the oil is lost with the discharged residues. Today, multigrade oils have a large number of additives, so more acid and more clay than ever are required.

Solvent-based routes, such as phenol and furfural extraction, also pose disposal problems. Wastewater from such processes may contain phenol or other solvents and require special treatment.

NEW TECHNOLOGY—Phillips' continuous process, PROP, begins with a demetallization step. An aqueous solution of diammonium phosphate (DAP) is mixed with heated waste oil. Through a series of reactions—conducted at temperatures and pressures not exceeding 300°F and 20 psig—metallic phosphates are formed. These compounds have a low solubility in water or oil and are filtered out, removing most metals. During the reaction, water and light ends are taken off as overheads.

The oil that remains is heated, mixed with hydrogen, percolated through a bed of clay (to remove trace inorganics) and passed over a nickel-molybdate catalyst. This hydrotreating removes sulfur, nitrogen, oxygen and chlorine compounds and improves color. Treated oil is then flashed and cooled. A final stripping step removes any remaining fuel from the lubricating oil.

Phillips claims that (1) the process recovers over 90% of the waste oil, (2)

the filter cake produced during demetallization (about 100 lb/h for a 2-million-gal/yr plant) can be used safely in landfills, (3) in some areas, resulting wastewater can be sent to municipal disposal systems without pretreatment and (4) the metals content of the rerefined oil is less than 10 ppm. The process also produces a sidestream of heavy gasoline (in the diesel-fuel range) that can be sold.

The cost of process materials, which include H_2, catalysts and DAP, is estimated at about 11.5¢/gal of feed. Phillips estimates the cost of skid-mounted PROP units that can treat 2 million, 5 million, or 10 million gal/yr at $1.8, 3.1 and 4.3 million, respectively. Utilities for a 2-million-gal/yr plant are given as 900 lb/h of

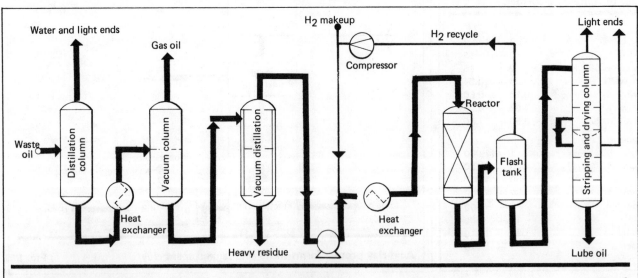

KTI uses vacuum distillation and hydrofinishing to treat used oil   **Fig. 3**

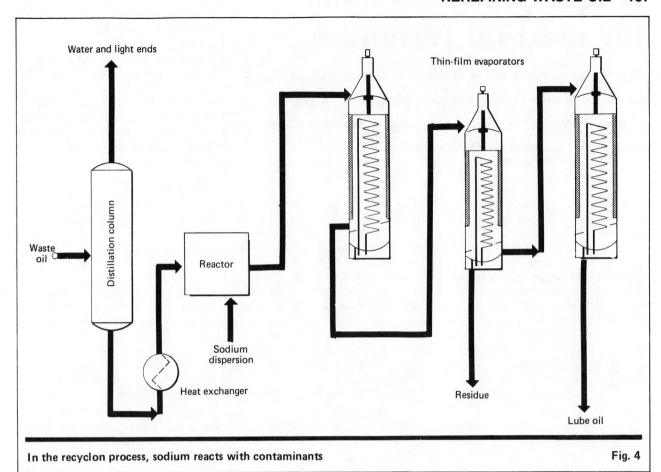

Water and light ends

Thin-film evaporators

Waste oil

Distillation column

Reactor

Sodium dispersion

Heat exchanger

Residue

Lube oil

In the recyclon process, sodium reacts with contaminants

Fig. 4

150-psi steam, 85 kWh/h of electricity, 300 gal/h cooling water and 0.85 million Btu/h fuel.

DISTILLATION AND HYDROFINISHING—In the KTI Process, distillation removes water and gasoline. Vacuum distillation produces an overhead in the lube oil range and a heavy residue containing metals, polymerization products and asphaltenes.

KTI uses Gulf's refinishing process to improve the color and odor of the overheads. The lube oil is mixed with a hydrogen-rich gas, heated and passed through a reactor holding a fixed catalyst-bed. Little or no cracking occurs. Excess gas that has not been used in the reaction is recycled and some makeup hydrogen is added. To obtain a product with the right specifications, the treated oil is either stripped with steam (any remaining light ends are removed)—or fractionated into different lube cuts—and then dried in a vacuum column.

Lube-oil yield for a European spent-oil feed averages 82% on a dry basis. The heavy-residue bottoms, which represents about 8% of the yield, can be used to make asphalt products and special lubricants, or it can be fired in power plants or cement kilns. The remaining 10% represents lighter fractions (from gasoil mixed with waste lubricating oil in Europe). KTI states that the process can vary product specifications by changing hydrofinishing operating conditions.

Based on a feed of 10 metric tons of waste oil, utilities would be: 2.5 tons of medium-pressure steam, 300 kWh of electrical power, 600 m^3 cooling water, approximately 250 m^3 of makeup hydrogen (100% pure), and 3.5 million kcal of fuel (based on 10% water in the feed).

RECYCLON—Sodium plays a key role in removing oil contaminants in the process offered by Leybold Heraeus. Waste oils are first distilled to remove water and light ends. Sodium particles sized from 5 to 15 μm are dispersed into the dry oil in amounts less than 1% metal to oil.

In a few minutes, impurities in the oil are transformed into compounds that cannot be distilled, such as salts or polymers. Typical reactions would include: polymerization of unsaturated olefins, conversion of halogen compounds to sodium salts, and conversion of mercaptans into materials that remain in the residue. To a certain extent, this desulfurizes the product.

Following this, low-boiling-point compounds are flashed (in thin-film evaporators at pressures lower than 1 millibar). Residues obtained from this flash can be mixed with light ends obtained in the early stages of the process to form a fuel of medium viscosity. Approximately 25 to 30% of this fuel will be needed for the rerefining; the rest can be sold.

The Recyclon process yields over 70% refined lubricating oil; the rest is fuel. Water from the first processing stage will be slightly acidic, but after neutralization it can be sent to a public sewerage system.

In specification and performance, the rerefined oils produced by these routes are comparable to virgin oils, developers claim. Phillips is operating a test fleet of 19 passenger cars and pickup trucks (so far accumulating over 500,000 miles) with rerefined oils. KTI has had its rerefined product tested by Gulf Science and Technology Co. (Pittsburgh).

Bright prospects loom for used-oil rerefiners

The U.S. government's emphasis on conservation bodes well for the used-oil recycling business. Meanwhile, a group of new rerefining techniques claim significant advantages over conventional acid-clay processing.

☐ A combination of new technology and a favorable attitude by the U.S. government is doing much to revitalize the used-oil rerefining business. Several federal agencies, responding to directives contained in both the Energy Policy and Conservation Act and the Resources Conservation Recovery Act (RCRA), have been revising their policies and encouraging the use of recycled oils. And several processes that have been under development for some time seem ready to challenge the traditional acid-clay treatment technique.

The following measures reflect the government's interest in boosting utilization of used oil:

■ Under a regulation proposed last December by the U.S. Environmental Protection Agency (EPA), waste oil would be considered a hazardous waste. If adopted, the measure would divert more used oil to rerefiners, because untreated waste oil would no longer be allowed in such applications as road oiling (for dust suppression) and paving.

■ The National Bureau of Stan-

dards is working on test procedures for rerefined oil, and the Federal Trade Commission (FTC) will use these as the basis for standards that will compare the recycled product with virgin oil for a given application. Use of oil that is "substantially equivalent" to the virgin product will be encouraged.

■ FTC is writing new labeling procedures for rerefined oil. (In 1965, the agency had ruled that all oils made from recycled material had to bear a "made from previously used oil" label.)

■ The U.S. Dept. of Energy (DOE) is funding state programs aimed at creating collection centers for waste oils. This will, it is hoped, encourage do-it-yourselfers—e.g., those who change their own crankcase oil—to recycle waste material.

■ RCRA is promoting federal purchasing of recycled materials, even at prices slightly higher than those of competing virgin products, and calling for changes in specifications that exclude rerefined products. (For the last 20 years, the U.S. Dept. of Defense has had a policy of excluding

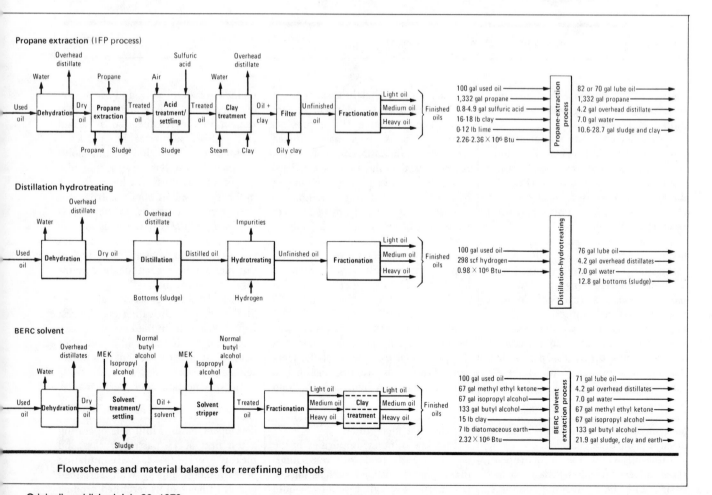

Flowschemes and material balances for rerefining methods

Originally published July 30, 1979

rerefined oils from use in military vehicles.)

According to a recent DOE report, "federal utilization of recycled oil would do much to stimulate its market." As DOE sees it, a governmental stamp of approval would have a beneficial impact on industrial applications, since many industries rely on federal specifications when buying such products.

PROCESS SHORTCOMINGS—Buoyed by the government actions, refiners have stepped up research on waste-oil processing. Until now, the acid-clay method—which involves dehydration, sulfuric acid treatment, and clay clarification—has reigned supreme. But refiners using the process have had to increase the severity of treatment in recent years to remove difficult-to-handle oil additives, and this has led to reduced yields and higher production costs. In addition, acid-clay processing leaves an acid sludge that presents disposal problems.

"Waste disposal from rerefining is a major issue in successful operation," says Michael Kerrin, vice-president of Double Eagle Refining Co. (Oklahoma City, Okla.). "Without the availability of an appropriate disposal site, you simply can't operate—this factor is responsible for driving some refiners out of business."

SOLVENT EXTRACTION—After several years of development work, DOE seems ready for a full-scale test of its BERC rerefining scheme. The patented solvent extraction/distillation process uses a solvent mixture of isopropanol, butanol, and methyl ethyl ketone in a 1:2:1 ratio, by volume. Three volumes of solvent are needed for each volume of used oil.

Dewatered, fuel-stripped oil is mixed with solvent in an inline mixer at 155°F. This pretreatment step removes coking and fouling compounds; the resulting sludge is separated from the oil/solvent mixture in enclosed (to prevent solvent loss) vertical solid-bowl centrifuges. Solvent is recovered in an atmospheric distillation column. Bottoms temperature in this tower is set at 550°F to minimize solvent loss. (Total solvent losses do not exceed 0.2%.) For a 10-million-gal/yr plant, 1-2 drums/d of makeup solvent would be required.

Vacuum distillation of the treated oil is at 5 mm Hg absolute to minimize cracking. However, to maximize yield, the bottoms temperature is maintained at 600°F, so some cracking does occur. Blending-stock lube is the major product of vacuum distillation.

Finishing can be accomplished by clay contacting or hydrotreating. For hydrotreating, the lube is heated to 600-650°F and pumped into the top of an 800-psi ASME Code, low-alloy-steel vessel. For a 10-million-gal/yr plant, 160 ft³ of nickel/cobalt catalyst is required. Lube is flashed in two stages down to 100 psi, cooled, filtered and stored. Off-gases are scrubbed with a caustic solution and flared.

The yield is 71%—which means that 71 gal of lube-oil product can be expected from a used-oil feed of 100 gal containing 7% water and 4.2% fuel dilution (light ends or naphtha). Fleet testing of oil being produced by the BERC process has been completed by the state of Iowa, which used it in 46 vehicles over a 2-yr period. Results from an independent testing lab "were very positive," according to DOE.

The agency is now in the process of contacting refiners and others for selection of a site to demonstrate the process. In a departure from its original intent to build a new 10-million-gal/yr unit, the present program calls for a retrofit in an existing rerefinery or other closely related facility. Finished design drawings are expected by the end of the year. Construction of the project would require some financial involvement by the participant.

PROPANE ENTRIES—Over the years, many firms have studied the use of propane as solvent, but the first commercial process, according to DOE, was developed by the Institut Français du Pétrole (IFP). The technique is being used in two plants—a 35,000-metric-ton/yr unit (in operation for 10 yr) for Viscolube, Lodi, Italy (near Milan) and a 20,000-m.t./yr plant that started up a year ago at Belgrade, Yugoslavia.

The IFP process does not totally replace acid-clay treatment, but it reduces the quantity of processed materials—and this results in less waste.

Although the reported process yield is 82%, the plant at Lodi, according to a DOE source, shows a much lower yield—about 70%. Propane addition is between 6-13 times the volume of used-oil feed.

At the Third International Conference on Waste Oil Recovery and Reuse, held last October in Houston, Tex., Edward T. Cutler, president of Pilot Research & Development Co. (Merion, Pa.), compared his PVH (propane-vacuum-hydrogen) process with other methods. His firm is seeking financial backing to develop the patented process through the pilot-plant stage and into full-scale commercial production.

Consisting of dehydration, propane treatment, vacuum fractionation and hydrofinishing, the PVH process has a reported yield of 73% of high-quality lube stock. PVH is said to require considerably less of chemicals and energy than many other commercial processes, which means lower costs, says Cutler.

Pilot's dehydration and propane treatment steps are claimed to work without heat-treating the used oil to excessive temperatures. Propane requirements are only four times used-oil feedrate—two to three times less than other conventional propane-based processes.

In the first commercial use of another propane extraction process, construction has begun at Ceccano, Italy, for a 50,000-m.t./yr rerefining plant for Clipper Oil Italiana S.p.A. Developed by Snamprogetti, the process is not suitable for such emulsified oils as waste oils from lathes and rolling mills.

The rerefining cycle begins with a flash distillation that removes 97-98% of the water and light hydrocarbons. The two-stage propane extraction is said to provide high yield (up to 97-98% of recoverable product). In the first stage, a propane oil ratio of 7:1 is maintained at a low temperature; a high-temperature second step requires a solvent addition of 10:1. Fractionation by vacuum distillation and hydro-treating provide SAE 10, SAE 30 and SAE 40-quality oils and a bright stock base. Yields are expected in the 85-90% range, on a dry basis.

With completion scheduled for before the end of 1980, the new facility will replace Clipper's 10,000-m.t./yr acid-clay processing unit. Construction costs are said to double those for a comparable acid-clay plant, but operating costs are lower because the acid-sludge control system is not needed. Snamprogetti says no new contracts have been signed, but negotiations are underway with a West German firm and East European countries.

An application using propane at supercritical conditions has been successfully tested at Friedrich Krupp GmbH's research center at Essen, West Germany. A patent has been applied for—but not yet granted—for the Krupp process, which uses countercurrent propane extraction as its key step. Usable oil products are extracted from a dehydrated waste-oil feed by propane.

Yields are said to be 90% (of drier oil after atmospheric distillation removes water and gas oil). and costs are comparable with those for conventional acid-clay technology, according to a Krupp spokesperson. Propane requirements are said to be slashed to only a 1:1 ratio of waste-oil flow. Plans are underway to build a 2-2.5-m.t./d pilot plant.

PROGRESS OF THREE ROUTES—Three other rerefining processes are getting closer to commercialization (*Chem. Eng.*, Apr. 23, p. 104). At presstime, the state of North Carolina was preparing a site at Garner, N.C., for installation of a skid-mounted facility based on Phillips Petroleum Co.'s PROP (Phillips Re-refined Oil Process). This continuous method starts with a demetallization step; the oil that remains is heated, mixed with hydrogen, percolated through a bed of clay (to remove trace inorganics) and passed over a nickel-molybdate catalyst. After hydrotreating, the oil is flashed, cooled, and subjected to a final stripping step.

Phillips, which claims a 90% recovery from waste oil, says it will deliver a 5-million-gal/yr PROP facility to Mohawk Oil Co. (Vancouver, B.C.) in late 1979.

Sodium plays a key role in removing oil contaminants in the Recyclon process, offered by Leybold Heraeus GmbH (Hanau, West Germany). Micrometer-size sodium particles dispersed into the dry oil transform impurities into such undistillable compounds as salts or polymers. Process yield is over 70% rerefined lube oil; the rest is fuel. The West German Research Ministry is expected to place an order for a 1-metric-ton/h plant, to be located in West Berlin. Tentative startup date: mid to late 1981.

In the KTI process—developed by Kinetics Technology International, B.V. (Zoetermeer, The Netherlands) in close cooperation with Gulf Science and Technology Co. (Pittsburgh,

Evaporator is used for dehydration or solvent stripping in BERC method

Pa.)—distillation removes water and gasoline. Vacuum distillation produces an overhead in the lube-oil range, and a heavy residue containing metals, polymerization products and asphaltenes. KTI uses Gulf's refinishing process to improve the color and odor of the overheads. Lube-oil yield for a European spent-oil feed averages 82% on a dry basis.

The Dutch firm will build a 120,000-m.t./yr plant—for Haberland & Co., near Hanover, West Germany—that is expected to start up by the end of this year.

U.S. METHOD GOES COMMERCIAL—Motor Oils Refining Co., a division of Estech, Inc. (Chicago) is already using its own technique at plants in McCook, Ill., and Flint, Mich. The latter facility, which started up last year, is billed as the largest rerefinery ever built in the U.S.

The process involves an undisclosed pretreatment technique to remove low-boiling-point materials, followed by vacuum distillation of the lube-base cut, and final treatment using clay filtration. Presently, the company is trying to patent this process while gearing up to license the technology to other rerefiners in the near future. The oily clay waste that is generated is a "fairly dry product, disposed of at a controlled landfill," says Motor Oils president John O'Connell. He adds, "the shift to the new technology was to yield higher-quality products, to improve process efficiencies, and to steer clear of acid-sludge problems."

WORLD PROCESS—EkoTek Lube, Inc. (Oakland, Calif.), a division of A. Johnson & Co., Inc. (New York City) has developed a rerefining technique called the WORLD (for Waste Oil

Reclamation through Lube Distillation). It consists of a two-stage stationary thin-film vacuum distillation followed by conventional clay contacting. The key unit's nonrotary design differs from that of other thin-film distillation equipment available, says technical manager Peter Mehiel.

EkoTek's equipment is said to be designed so that internal fouling of the distillation units is minimized—without requiring a pretreatment step, as used by other distillation processes. In the first stage, used lube oil is stripped to remove water and light hydrocarbons—which are used as fuel. The dehydrated oil is then fed to the high-vacuum-distillation second stage. Finishing is done via the clay-contacting process, hydrotreating is being looked into.

The distillate oil produced is said to be a light neutral lube, comparable in quality to virgin oil. Residue from the vacuum distillation unit is asphalt flux, which is marketed as an additive for asphalt and roofing tar.

Claimed yields (based on dehydrated feed) are 7% light ends, up to 65% lube oils, and 28% asphalt flux—with yields of the latter two adjustable, depending on marketing needs.

EkoTek purchased two existing acid-clay rerefining plants late in 1978, in order to obtain an existing market and collection system. One of these—a 10-million-gal/yr facility in the San Francisco area—has been converted to the WORLD process and started up recently. The other—a 5-million-gal/yr Salt Lake City, Utah, plant—is currently operating with the acid-clay treatment; conversion is expected sometime in 1980.

David J. Deutsch

Oil re-refining route is set for two plants

Using only distillation know-how, a new system for recovering lubricants from used oil avoids problems of hazardous-waste disposal and offers high yields.

Reginald I. Berry, Associate Editor

☐ Resource Technology, Inc. (Kansas City, Kans.) has developed a new process for the re-refining of used lubricating oils. According to the firm, this method does not require acids, solvents or additional chemicals and does not produce hazardous wastes as do traditional acid/clay re-refining methods (see the previous article for details).

In contrast, the new technology uses a series of vacuum distillation steps involving equipment of a unique design that minimizes coking during operation. And according to Timothy F. Sparks, president of Resource Technology, the route will recover 97% of a gallon of dehydrated used oil as marketable products.

Now, Resource Technology will see its process used for the first time in the U.S. in a 10-million-gal/yr facility that is to be built for Lakewood Oil Service, Inc. (Santa Fe Springs, Calif.), a used-oil collector. Construction on the $5-million plant, to be located in Fontana, Calif., will start this month, and is expected to be completed in 15 months.

The first contract for the use of the technology was signed in September 1980 between Resource Technology and Regionol A/S (Oslo, Norway). Here Regionol's existing 2,000-L/h (about 3 million gal/yr) acid-clay re-refinery has been retrofitted to accept the new system. This unit is to start up this month.

Trials started in 1977 formed the basis for the new route. An acid/clay plant in Kansas City was retrofitted in order to try the new distillation techniques. This demonstration/pilot unit had a capacity of 435 gal/h.

MORE THAN WASTE—Re-refining of used oil dates back to 1915. And re-refined oil was used successfully in both World Wars I and II in military aircraft. The used-oil industry peaked in the 1960's when over 150 companies were producing over 300 million gallons of re-refined lubricants—nearly 18% of the U.S. lubricating oil demand. However, today re-refiners contribute less than 5% of the total lube oil market.

The change in demand was attributed to a number of factors:

■ A cost/price squeeze generated by the application of used oil as a fuel and as a dust suppressant, raising the price of the oil to the re-refiner.

■ A Federal Trade Commission rule that required the labeling of re-refined oil as previously used—thereby hurting the marketing of the product because it appeared inferior.

■ The traditional acid/clay technology caused environmental problems associated with disposal of acid sludge. Also, the older processes were inconsistent in their refining of oils with complex additive packages, and as a result re-refined oils were excluded from military and government purchase lists.

Now, these problems are not significant. The "Used Oil Recycling Act of 1980" prohibits labeling restrictions and encourages state participation in used-oil recovery. Re-refined oil has been shown to be of the same or better quality than product made from crude. Also, the re-refining of the nation's estimated 1.4 billion gallons of used crankcase oil would produce a net energy saving of 42,000 bbl/d of crude oil, reveals a Congressional report. And Sparks points out that at current oil prices the U.S. "could be saving more than $500 million/yr in trade deficits to oil producing countries."

IN A VACUUM—In the Resource Technology process, used oil first enters one of three cone-bottom settling vessels (see flowsheet). Free water is separated and sent to wastewater treatment. The oil is then heated before going into an atmospheric flash unit where emulsified water and a fuel cut with a boiling range up to 425°F are vaporized. The fuel, which is sold as a number 2 fuel oil, is condensed and separated from the water.

The dehydrated used oil is then heated and transferred to a vacuum flash column operating at 100 mm Hg. Here, a spindle oil fraction with an atmospheric boiling range up to 620°F is vaporized. This recovered oil, after clay polishing, is sold as a base blending oil (100 Ssu at 100°F).

The remaining used oil is transferred to a cyclonic vacuum distillation tower. This column, operating at 20 mm Hg, has an internal design that eliminates the need for conventional internals such as trays or packing. In this vessel, oil is injected at high velocity to generate high centrifugal force, which aids in separation while minimizing the formation of coke.

Using an ordinary design, this distillation unit, which vaporizes oils hav-

Originally published October 5, 1981

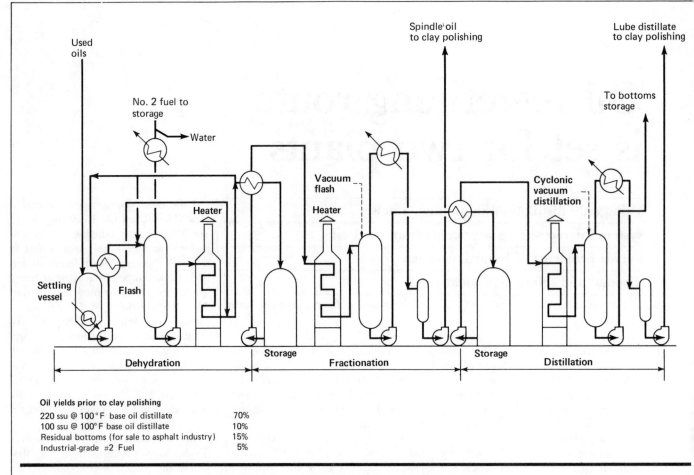

Used oils

No. 2 fuel to storage

Water

Spindle oil to clay polishing

Lube distillate to clay polishing

To bottoms storage

Heater

Vacuum flash

Heater

Cyclonic vacuum distillation

Settling vessel

Flash

Dehydration | Storage | Fractionation | Storage | Distillation

Oil yields prior to clay polishing

220 ssu @ 100°F base oil distillate	70%
100 ssu @ 100°F base oil distillate	10%
Residual bottoms (for sale to asphalt industry)	15%
Industrial-grade #2 Fuel	5%

New re-refining process is said to recover 97% of used oil as marketable products.

ing an atmospheric boiling range of 950 to 1,200°F, might be fouled by the formation of coke and force plant shutdowns for maintenance. According to Resource Technology, the proprietary design avoids this problem.

In this column, more than 80% of the feed stream is vaporized and sent overhead. The distilled lube oil, which is a 220-Ssu base blending oil, is condensed, collected and sent to clay polishing. The bottoms from the tower consist of the spent additive package and contaminants. This material is marketable, without further processing, to the asphalt industry.

CLEAN PRODUCTS—Clay polishing is used to remove any impurities remaining after the distillation steps. Here lube oil is mixed with diatomaceous earth and activated clay. The oil/clay slurry formed is passed through a plate-and-frame filter press where the base blending oil is recovered. The oily clay filter cake that is generated may be disposed of in a conventional landfill, states the company.

The clean product oil has been shown to contain less than 5 ppm metals. Heavy metals, typically found in used oil, are removed in the residual bottoms from the cyclonic vacuum distillation tower. In asphalt, where the bottoms are used, the metals are encapsulated, thereby prohibiting leaching. Also, according to the developers, the process has no significant effect on air quality, and the wastewater produced requires only minimal treatment.

ECONOMIC RECOVERY—The process has a throughput cost of less than $0.30/gal, according to Resource Technology. The firm projects that, given an oil feedstock cost of $0.32/gal, a 5-million-gallon facility will produce a before-tax earning of $2.5 million/yr; a 10-million-gallon plant should produce an estimated $5.3 million. The system requires the following utilities per gallon of dehydrated used oil feed: 8,660 Btu of fuel, 8.70 lb of 100-psig steam, and 0.197 kW of electricity.

A grassroots re-refining facility using the process is estimated to cost, respectively, $3.5 million or $5.2 million for a 5-million or 10-million gal/yr facility. This cost does not include land and tankage but it does cover a site-constructed turnkey unit.

Resource Technology also can provide a skid-mounted unit with a capacity of 500,000 gal/yr. Cost of this type of plant is estimated to be $650,000—but this does not include land, tankage, and transportation of the unit to the re-refining site.

It is also possible to retrofit the technology to an existing plant. This would consist of the addition of a cyclonic vacuum distillation tower, which would replace the acid-treatment phase of an acid/clay process. Cost of skid-mounted equipment with a capacity of 3 million gal/yr would be $525,000. The firm estimates that a retrofit can result in a net process saving of nearly $0.34/gal, and at the same time eliminate the problems of hazardous waste.

Section VI
Plastics and Elastomers

New catalyst cuts polypropylene costs and energy requirements

Process steps have been eliminated via a better new catalyst that improves yields and eliminates impurities.

Cipriano Cipriani and *Charles A. Trischman, Jr.,*
El Paso Polyolefins Co.

☐ The El Paso Co. (Houston) has derived a new efficient polypropylene technology by combining its liquid pool process with a new catalyst that was jointly developed by Montedison S.p.A. (Milan, Italy) and Mitsui Petrochemical Industries, Ltd. (Tokyo, Japan). The new catalyst increases polymer yield and allows the elimination of process steps. With these advantages, El Paso has been able to make significant process improvements: In comparison to its conventional system, process steam has been reduced by 85% and electricity by 12%; also, the capital cost of a new plant is 30% lower (see *Chem. Eng.,* Mar. 9, p. 17).

The surge in feedstock and energy costs, starting in 1974 with the formation of OPEC, forced polymer producers to reassess their technology. In 1978, El Paso Polyolefins Co. began bench-scale studies of high-yield catalysts. By late 1978, parameters for the production of typical polypropylene grades (homopolymer, ethylene-propylene random and block copolymers) were developed in pilot plant facilities. Soon after, scaleup runs began in El Paso's operating units, and last year the new technology was commercialized in two plants, one in Odessa and the other in Bayport, Texas. These facilities, which are operated by Rexene Co., El Paso's subsidiary, have a combined capacity of 150,000 metric tons/yr.

And now two licenses have been granted—one to the Nigerian Oil Co.,

which plans to build a 35,000-metric-ton/yr facility at Warri to be completed by 1983, and the second to the Azzawiya Refinery Co. of Libya, which will build a 68,000-m.t./yr plant at Ras Lanuf to be onstream by 1984.

OLD AND NEW—To appreciate the improvement in energy savings and in monomer utilization achieved with the new technology, it must be compared to El Paso's conventional liquid-pool process that was patented in 1963, commercialized in Odessa over 15 years ago, and licensed to producers in Europe, Japan and the U.S. The conventional route uses liquid propylene monomer as the reactor medium, and a Ziegler-type catalyst—based on titanium trichloride activated with alkyl aluminum halide. However, yields are low and impurities must be eliminated from the product.

The Ziegler-type catalyst used in the conventional route provides a relatively low polymer yield—on the order of 500-1,000 lb/lb of catalyst. In addition, the polymer product leaving the reactor contains 5-7 wt % atactic material* and an excessive amount (0.3-0.5 wt %) of catalyst residues. Both of these byproducts must be extracted in order to generate a commercially viable resin.

In the older system (see flowsheet), propylene is purified and excess pro-

*When polymerized, propylene forms stereoisomers (spatially different molecules): atactic, syndiotactic and isotactic forms. The atactic polymer is disordered and not crystalline. The syndiotactic and isotactic forms are symmetrical. The isotactic form, the most symmetrical, is highly crystalline and is what is normally considered commercial polypropylene.

pane and other impurities accumulated in the process are removed. Then the liquid monomer together with catalyst and hydrogen (used for molecular weight control) is introduced to an agitated reactor that already contains some liquid propylene. The reaction takes place at 130-180°F and 400-600 psi. Propylene homopolymer forms as particles, and this product leaves the reactor in a slurry with unreacted liquid propylene.

The slurry enters a flash vessel; here, the monomer is separated and recycled to be mixed with the feed. The product polymer-powder is treated with a solvent mixture (an aliphatic hydrocarbon and alcohol) that is acidified with HCl to solubilize the catalyst residues. This operation, called deashing, in addition to removing the catalyst residue also extracts the atactic material.

After deashing, the polypropylene product-powder is separated from the solvent, dried, stabilized and finally pelletized. The deashing solvent is neutralized, and then rectified by distillation. Makeup solvent is added. The atactic material and the catalyst residues are discarded, and the solvent is recycled.

Ethylene-propylene random copolymers can be and have been produced in this process by introducing ethylene as a comonomer in the polymerization vessel. Additionally, ethylene-propylene block copolymers are produced by reacting the homopolymer powder, still containing some active catalyst, with ethylene in an auxiliary reactor.

BETTER CATALYSIS—With the new high-efficiency catalysts, which are supplied to the U.S. market by Montedison, the liquid pool technology has been significantly improved. Polymer yields have been increased and the

Originally published April 20, 1981

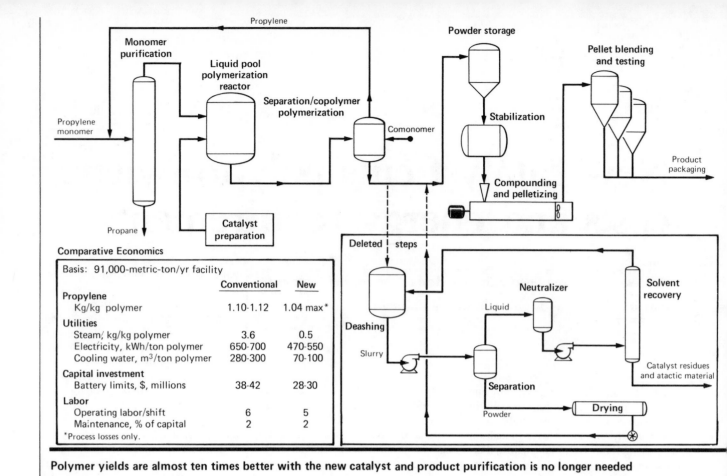

Comparative Economics		
Basis: 91,000-metric-ton/yr facility		
	Conventional	New
Propylene		
Kg/kg polymer	1.10-1.12	1.04 max*
Utilities		
Steam, kg/kg polymer	3.6	0.5
Electricity, kWh/ton polymer	650-700	470-550
Cooling water, m³/ton polymer	280-300	70-100
Capital investment		
Battery limits, $, millions	38-42	28-30
Labor		
Operating labor/shift	6	5
Maintenance, % of capital	2	2
*Process losses only.		

Polymer yields are almost ten times better with the new catalyst and product purification is no longer needed

deashing step has been eliminated.

Also, the new technology benefits by not requiring a change in the purity of the propylene monomer. Excellent results (i.e., resin with a high crystallinity) have been obtained with the same sources of supply of chemical-grade propylene.

With the new supportive catalyst, polymer yields are on the order of 5,000-7,000 lb/lb of solid catalyst. And the polymer produced contains minimal catalyst residues, so deashing is unnecessary.

In addition, this catalyst is also highly stereospecific—almost no atactic (heptane-soluble) material is formed. Now in the liquid pool reactor, monomer is directly polymerized to 94-95% crystalline (heptane-insoluble) polymer. Since this is the quality of the product attained after the deashing operation in the older route, atactic extraction is also not needed.

In the new process, the polymerization section is essentially unchanged and operates in the same manner and at the same conditions used with the old catalyst. Yet some mechanical modifications of the catalyst preparation and feed systems were made to accommodate lower catalyst feed-

rates—made possible by higher yields.

SAVING ENERGY—The basic difference between the old and new routes is the elimination of the deashing operation. This portion of the process is the most energy-intensive, so significant savings are possible (see table). Elimination of solvent extraction saves 15% of the total process steam (used here to maintain temperature).

Also, the ancillary solvent recovery and recycle systems are no longer necessary. Formerly, 65% of the total process energy was used to distill and recover solvent. And since the polymer product does not now undergo deashing, it does not have to be dried in order to remove solvent.

Elimination of these operations also saves 12% of electricity use. And since the new route uses less equipment, capital cost has been reduced. A new grass-roots plant constructed on a Gulf Coast site is estimated to save 30% in capital.

A NEW RESIN—The revised technology produces a broad range of homopolymer, random copolymer and block copolymer grades called A Resins. Physical and mechanical properties of these new products are comparable to

those of earlier materials. But because the molecular weight distribution of the products is broader than for earlier polypropylenes, the new resins have lower apparent melt viscosities and therefore can be converted at lower temperatures.

A reduction in injection molding cycles of 6 to 20%, processing temperatures 30° to 50°F lower, and polymer-conversion energy savings of 5 to 10% are obtained. Increased line speeds have been demonstrated in high-tenacity-tape production for carpet backing and cordage. High extrusion rates combined with high orientation have been achieved in the production of textured yarns and staple fibers for carpet and upholstery uses.

Reginald I. Berry, Editor

The Authors

Cipriano Cipriani, Director of Product Development of El Paso Polyolefins Co. (P.O. Box 665, Paramus, NJ 07652), holds a Ph.D. in industrial chemistry from the University of Bologna. His specialty is R&D in the area of free-radical, anionic and condensation polymerizations, polymer processing and applications.

Charles A. Trischman, Jr., a chemical engineering graduate of Virginia Polytechnic Institute and State University, is currently Director of Process Development for El Paso Polyolefins Co. He has also held positions in marketing, product development, customer technical service, and plant process engineering with the company.

British route to polymer hinges on bacteria

The process, now under development, is a spinoff of the ICI path to single-cell protein. Current work focuses on purifying the product and improving its properties.

ICI's Pruteen plant at Billingham provided knowhow for polymer work

☐ ICI's latest candidate among the microorganisms it wants to put to work at making "everyday bulk products" is the bacterium *Alcaligenes eutrophus*. So says Peter King, research manager for the firm's agricultural division, headquartered at Billingham, U.K.

Speaking at the Second European Congress of Biotechnology, held on Apr. 6-10 in Eastbourne, U.K., King revealed that the company has launched a serious development effort to make use of the bacterium in a process to produce the polymer polyhydroxybutyrate, whose chemical formula is: $[-CH(CH_3)CH_2COO^-]n$.

PRUTEEN SPINOFF—According to King, the process is based largely on knowhow gained by ICI during the evolution of its new single-cell protein plant at Billingham, which became fully operational late last year. There, bacteria are grown in a pressure-swing fermenter containing a mix of methanol, ammonia and inorganic nutrients. The plant produces up to 70,000 tons/yr of Pruteen, marketed as a protein supplement for animal feedstuffs.

Operating conditions of the new PHB process "have been greatly altered from those of the Pruteen process," says King, with respect to such fermentation parameters as temperature, pressure and pH. Although King would not give further process details, he said that the yield of PHB in the cell mass of the fermentation broth is 80% by weight (dry basis).

The new process also differs from the Pruteen route in that it is being designed to avoid dependence on methanol and other hydrocarbon feedstocks derived from oil and natural gas. Glucose from corn is the most likely choice among alternative feedstocks, although gaseous mixtures of carbon dioxide and hydrogen, produced from coal or electrolytic solar cells, also are being looked at.

While King declined to say how much was being spent to develop the process, he indicated that the effort was begun three years ago, and now involves 19 researchers. The program is still at the bench-scale level (10 kg of product per week) and is being carried out in collaboration with Hans G. Schlegel, of the Institute for Microbi-

ology at the University of Göttingen, in West Germany. Current work focuses on steps required to purify the polymer.

BETTER PROPERTIES SOUGHT—A look at PHB properties shows that the polymer is roughly comparable to polypropylene homopolymer in several categories, such as melting point and tensile strength. ICI has already tested PHB and found it satisfactory as a material for extruding such small items as golf tees, says King.

However, in a number of important property characteristics, PHB appears to get poor marks. Its high glass-transition temperature and low extendability make the material prone to brittleness, and its solvent-resistance is poor.

ICI's King is confident that these drawbacks can be largely offset by the addition of plasticizers, and fillers such as glass. He adds that by purifying the polymer to a high degree, ICI already has overcome the "crumbliness" of PHB, a roadblock that obstructed previous investigators—e.g., W. R. Grace & Co., which looked at the polymer during the 1960s.

TAILORED GENES?—Another avenue to improved PHB properties—genetic engineering of the *Alcaligenes* bacterium—was not ruled out by King. Recombinant-DNA techniques would afford another means of altering the fermentation step so as to provide higher yields of PHB or a more desirable polymer structure.

"We could consider modifying the organism to get a lower molecular weight and shorter chains," says King. The result would be lower density (advantageous because plastics are sold by volume) and better rigidity and toughness.

FEEDSTOCK HURDLE—At the moment, the most serious barrier to scaling up the PHB process is feedstock economics. "Oil is still too cheap," says King, to justify near-term commercialization of fermentation polymer routes based on alternative feeds. Corn-derived glucose, the cheapest nonfossil feed, is still nearly three times as expensive as oil-derived raw materials.

"We've simply got to get our product costs down," says King. "They are still about twice as high as they ought to be." He estimates that to be economic, commercial-scale plants will have to have product capacities on the order of 10,000-100,000 tons/yr.

James H. Mannon

Originally published May 4, 1981

Plastics producers search for paths to linear LDPE

A healthy demand outlook and the potentially attractive economics of linear low-density polyethylene are luring a growing roster of international manufacturers.

☐ Everyone, it seems, wants to get into the linear low-density polyethylene (LDPE) act. Attracted by the lure of substantial capital- and production-cost savings over conventional high-pressure technology, at least 17 firms in the U.S., Europe and Japan are developing their own linear low-pressure routes (some are near commercialization), and others are licensing established processes (see table).

Another powerful reason for acquiring or developing low-pressure capability is the bright outlook for linear LDPE—at least in the U.S. According to a study released in July 1980 by New York City-based consultant Chem Systems Inc., U.S. demand for the polymer will reach 1.2 million metric tons in 1985, and double in the ensuing 5 years, reaching 2.5 million m.t. by 1990. In Europe, the demand for linear LDPE is expected to reach 1.1 million m.t. by 1990, but development of this market, because of current overcapacity, will not be as rapid as in the U.S.

A common thread running through upcoming PE projects is flexibility—producers want to be able to make the linear polymer in existing high-pressure LDPE—or HDPE—facilities. (One newcomer to LDPE manufacture will even make the resin in a polypropylene plant.) Here's a rundown on the latest linear LDPE developments in the U.S., Europe and Japan.

FROM PP TO PE—This year, two U.S. firms—El Paso Co. (Houston) and Arco Polymers, a subsidiary of Atlantic Richfield Co. (Philadel-phia)—have announced linear LDPE processes.

El Paso's route combines new polypropylene knowhow announced last year by a subsidiary (Rexene Co., Paramus, N.J.) with the use of a high-yield catalyst developed by Italy's Montedison. The technology will enable Rexene to adapt a train to linear LDPE production at the firm's 25-million-lb/yr polypropylene plant in Odessa, Tex.

Referring to the facility's polypropylene capacity, Rexene president R. T. Kelley says, "We can make that much or more polyethylene without sacrificing any polypropylene capability." This is because the company's new PP process will boost throughput in the trains that will still be making PP.

Kelley adds that "our linear LDPE unit at Odessa is not intended to swing back and forth between the two polyolefins, but with a short shutdown and some fairly simple modifications, the changeover is possible. Some equipment simply will sit idle, depending on the material being manufactured."

Rexene and Montedison intend to license the process, once the operating experience at Odessa proves to be satisfactory.

Not much is known about the Arco process. The firms says that it is now entering into large-scale pilot-plant testing at a site operated by an unidentified joint partner. The route, which uses a proprietary catalyst system, will be able to run on existing high-pressure LDPE equipment. The resin product, says a spokesman, is comparable in price and performance to similar resins already commercialized. Full commercialization is expected before yearend.

A MODIFIED TECHNIQUE—Last year, Phillips Chemical Co., a division of Phillips Petroleum (Bartlesville, Okla.), announced the construction of a small plant (8 million lb/yr) in Pasadena, Tex., that will be able to make both HDPE and linear LDPE. It will use a modified version of the company's particle-form process—a technique employed throughout the world for HDPE production.

The feed consists of ethylene, an alpha olefin comonomer, a hydrocarbon diluent, and powdered chromium oxide catalyst. These are sent to a loop reactor operating at 70-100°C or higher to maintain a liquid phase. A slurry of polymer and liquid hydrocarbon is discharged from a settling zone in the reactor into a flash tank, in which the diluent and ethylene are recovered for recycle. The polymer crumb is purified and pelletized. Catalyst yield is high, so no catalyst removal step is needed.

Phillips is now expanding its 1-billion-lb/yr HDPE plant in Pasadena by 450 million lb/yr, and observers speculate that the addition will allow the company to produce some linear LDPE, especially since this July it announced that it had developed a new polymerization catalyst for this purpose.

Phillips will not be the only one to benefit from its knowhow; according to P. T. Swoden, the company's manager of R&D licensing, Phillips technology is used by 17 licensees to make nearly 6 billion lb/yr of HDPE worldwide. "This volume," he says, "exceeds one third of the combined worldwide capacity, both in place and planned,

Status of various worldwide LLDPE Projects

Company	Process	Comonomer	Capacity, thousand metric tons/yr
Commercial technology			
Du Pont (Canada)	Solution	C_5 and higher	227 online in Ontario; 227 planned for Alberta
Union Carbide	Gas phase	Butene	275 online; 295 to start up in 1982; 45 planned for Sarnia
Dow	Solution	Octene, or C_7-C_{10}	135 online; 180 online by 1982; 200 planned for Benelux
Technology under development			
U.S.I.	Slurry	Butene	Expected to announce a project in November '81; likely to switch part of HDPE capacity.
El Paso/Rexene/Montedison	Slurry	Butene	11.3 under construction at Odessa, Tex.
Phillips	Slurry	Butene	Has 3.2 capacity now under construction.
Arco	High pressure	Butene	
Amoco	Gas phase	Butene	
Cities Service	Gas phase	Butene	
Solvay	Slurry	—	
DSM-Stamicarbon	Solution	—	
Naphtachimie	Gas phase	—	
CDF Chimie	High pressure	—	
BASF	Gas phase	—	
Imhausen Chemie	Gas phase	—	
Sumitomo	Solution	—	60 at Chiba by 1983
Mitsui Petrochemical	Solution	—	60 at Chiba by 1983
Nippon Petrochemicals	—	—	
Showa Denko	Slurry	—	Has modified part of 125 HDPE unit
Chemische Werke Hüls	—	—	
Gulf Oil	—	—	
Licensees			
Mobil	Union Carbide	Butene	140 at Beaumont, Tex., in 1983
Exxon	Union Carbide	Butene	275 at Mt. Belvieu, Tex.
Essochem Europe	Union Carbide	Butene	
Nova Corp./Shell Canada	Union Carbide	Butene	270 at Joffre, Alta.
Nippon Unicar	Union Carbide	Butene	75 at Kawasaki by mid-1983
ICI Australia	Du Pont	C_5 plus	100 in New South Wales
SABIC	Union Carbide	Butene	
Mitsubishi Petrochemical	Union Carbide	Butene	80 by 1983
Idemitsu Petrochemical	DSM-Stamicarbon	—	60 in 1985
Asahi Chemical Industries	Du Pont	C_5 plus	
Esso Canada	Union Carbide	Butene	135 at Sarnia in 1983
ICI	—	—	

Sources: DeWitt & Co. (Houston), *CE* estimates

for manufacturing HDPE by all commercial processes." Those licensees, of course, could easily gain access to the new linear LDPE knowhow.

Some engineers however, note that the hexene comonomer used by Phillips is a good solvent for linear-polymer product, and that unreacted comonomer could gum up the equipment. Phillips refuses to comment on this.

Other U.S. companies developing proprietary linear LDPE technology are U.S. Industrial Chemicals Co. (New York City), Cities Service Co. (Tulsa, Okla.), Amoco Chemicals Corp. (Chicago) and Gulf Oil Chemicals Co. (Houston). USI, which could make an

announcement later this year, probably will be replacing existing HDPE or LDPE capacity.

Gulf reportedly is piloting a reactor, but has not made a firm commitment to linear LDPE.

W. E. Kennel, executive vice-president of Amoco, says: "Based on pilot plant work, we believe we can produce linear LDPE with our gas-phase technology. However, we are not involved commercially, and do not expect to be in the near future."

EUROPEAN DEVELOPMENTS—So far, no Western European producer has a commercial-scale low-pressure plant onstream. However, most are testing modifications of existing HDPE sys-

tems, or are shopping for new HDPE processes that can economically turn out low-density materials. Here's a sampling of recent activity in linear LDPE:

■ In July, the Australian subsidiary of Britain's ICI signed a licensing agreement with Du Pont Canada for its linear LDPE technology. ICI itself last year announced a cross-licensing and technical exchange agreement with the U.S.'s Cities Service for development of a gas-phase HDPE process.

■ Montedison's catalyst system is now going commercial in the U.S. with the Rexene project. In Europe, the firm, which has plenty of HDPE and conventional LDPE capacity, is running a pilot-sized low-density unit at Ferrara, Italy.

■ In West Germany, Chemische Werke Hüls (Marl) says it is working on a bench-scale facility. BASF (Ludwigshafen), which runs a 12,000-m.t./yr semicommercial gas-phase HDPE process at its joint-venture subsidiary with Shell—Rheinisch Olefinewerke GmbH—says it will decide soon on a low-density technology. And Imhausen Chemie (Lahr) is working to adapt its high-pressure, tubular-reactor process to linear LDPE production.

■ Benelux producers Solvay and DSM both claim that they have made low-density polymers with their existing slurry and solution high-density processes.

■ France's Naphtachimie says it has modified its gas-phase system to make low-density materials and is currently negotiating for licensees.

ORIENTAL DELIBERATION—Japan's Ministry of International Trade and Industry (MITI) is carefully monitoring linear LDPE developments to ensure that local producers will not make a massive switch to this material and create a situation of overcompetition. MITI planners believe that the country should not have more than 300,000-375,000 m.t./yr of linear-polymer capacity—i.e., about 20% of the current 1.6-million-m.t./yr LDPE capability.

Here's the cast of characters in Japan's linear LDPE efforts:

■ Nippon Unicar Co. (Tokyo), a joint venture of Union Carbide and Tonen Petrochemical Co. (Tokyo), licensed Carbide's Unipol process last year, and plans to build a 75,000-m.t./yr plant at Kawasaki, scrapping equivalent LDPE capacity.

■ Mitsubishi Petrochemical Co. and Mitsubishi Chemical Industries Ltd. (both Tokyo) have decided to join forces to avoid overcompetition. The former will supply the latter with 20,000 m.t./yr of linear polymer from a $45.4-million, 75,000-m.t./yr plant scheduled for completion in 1983. Mitsubishi Chemical will repay this with raw materials—ethylene and/or naphtha.

(Japan is not the only country to try barter deals in linear LDPE production. In the U.S., Exxon reportedly has received concessions in its royalty arrangement with Carbide in ex-

---

Conversion of HDPE technology to linear LDPE

Plastics producers have known for years that it is possible to produce lower-density polymer in high-density processes, but they now are more interested in such conversions because of rising energy costs. Because HDPE processes operate at low pressures (30-100 atm) and temperatures (80-250°C)—compared with 1,000-3,000 atm and 200-275°C for conventional, high-pressure LDPE systems—they offer capital- and operating-cost savings. SRI International (Menlo Park, Calif.) calculates that fixed capital costs for low-pressure units are less than half those of high-pressure systems. However, the value of the HDPE product is nearly 25% lower.

According to Western European high-density producers, there are some basic technical problems in switching over to making linear low-density material.

One inconvenience, for example, is the varying condensation points of different comonomers, some of which run the risk of condensing and fouling in gas-phase processes. Solution processes, on the other hand, can probably switch monomers with little difficulty, though split-recycling systems would be needed to remove the comonomer from the solvent. Some say that slurry processes might run across complications in removing liquid hexene from the polymer, requiring additional drying or purging steps.

Another frequently cited problem is that the polymerization rate of comonomers is lower than that of ethylene. And because 10% of comonomers must be introduced to induce branching of the linear HDPE chain, this may lower conversion rates.

Finally, the lower melting point and higher viscosity of low-density PE may affect such variables as reaction temperature, catalyst activity and amount of solvent or diluent. These complications support the feelings of many experts that HDPE slurry methods will be the most difficult to modify to linear LDPE. Any changes in temperature could interfere with external heat removal and reduce output.

Low-density polymer made in low-pressure processes (such as Union Carbide's or Dow's) has a molecular structure between that of the linear chains of HDPE and the highly branched pattern of conventional low-density product. The comonomer used to obtain branching in the HDPE molecule implies changes involving capital expenditures for downstream plastics processors. But the linear end-product has higher tensile and impact strengths and better elongation characteristics than conventional LDPE.

---

change for providing some of Carbide's ethylene requirements. Such a pact softens the impact of royalty costs, which, in the case of Carbide's Unipol process, are considered quite high by some linear LDPE experts.)

■ Mitsui Petrochemical Industries Ltd. (Tokyo) wants to have a 60,000-m.t./yr facility in operation at Chiba by March 1983 and is expecting MITI approval.

The Ministry has already delayed (until 1985) plans by Idemitsu Petrochemical Co. (Tokyo) to build another plant of equal size in Chiba.

■ Showa Denko K.K. (Tokyo) is making test quantities of linear LDPE, using a modified HDPE plant based on

the Phillips process. The company hasn't decided whether to build its own plant or buy the resin on the market.

■ Asahi Chemical Industry Co. (Tokyo) considered several processes for licensing before deciding on Du Pont's. Meanwhile, Asahi Chemical's joint venture with Dow—Asahi-Dow Ltd. (Tokyo)—has been importing and evaluating linear-polymer resin made by Dow prior to licensing the U.S. firm's process.

■ Sumitomo Chemical Co. (Tokyo) started operating a linear LDPE pilot plant last spring at its Chiba works. The company says it will decide soon when to build a commercial plant.

Guy E. Weismantel

New catalyst controls LLDPE's particle geometry

Due to be tested on a semicommercial scale, this new slurry process for making linear low-density polyethylene produces a resin that requires no pelletizing. The selective catalyst also prevents undesirable intermediate compounds from forming during the polymerization.

Cipriano Cipriani and Charles A. Trischman, Jr., El Paso Polyolefins Co.

☐ El Paso Polyolefins Co. (Paramus, N.J.) and Montedison, S.p.A. (Milan, Italy) have jointly developed a low-pressure, energy-efficient, linear low-density polyethylene (LLDPE) process based on El Paso's liquid-pool-reactor technology for making polypropylene (see the article on p. 165). Coupled with a new "particle-form-catalyst" technology developed by Montedison,

the process is expected to have the same low operating and capital cost advantages as the established Union Carbide low-pressure vapor-phase Unipol process for making LLDPE (*Chem. Eng.*, Dec. 3, 1979, p. 80). The capital investment for the El Paso process is expected to be 35% lower than that for conventional high-pressure plants of the same size.

At presstime, an already completed 10,000-metric-ton/yr demonstration plant will soon produce film and injection molding grades of LLDPE for selected customers at El Paso's polyolefin complex in Odessa, Tex., where the facility's lowest-capacity polypropylene line was converted for this purpose. Process and product development studies will be conducted to provide design data for future world-scale plants (150,000 m.t. tons/yr).

Development work on the process was done over a two-year period on a 30-50-lb/h pilot plant that made resin grades for films, injection molding, extrusion, and pipe applications.

The rapid growth and market acceptance of LLDPE, spurred on by companies such as Union Carbide, Exxon, Dow and Mobil, have gener-

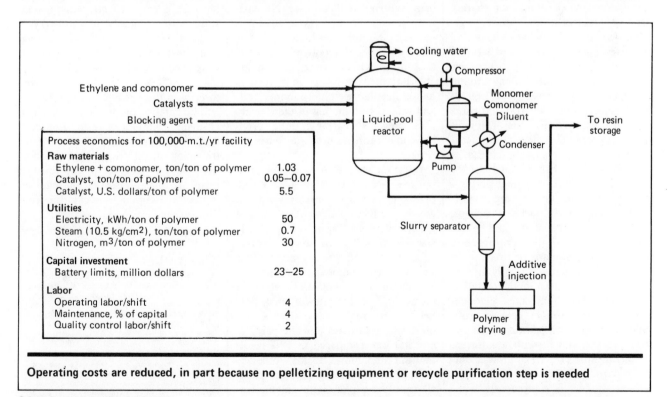

Process economics for 100,000-m.t./yr facility	
Raw materials	
Ethylene + comonomer, ton/ton of polymer	1.03
Catalyst, ton/ton of polymer	0.05—0.07
Catalyst, U.S. dollars/ton of polymer	5.5
Utilities	
Electricity, kWh/ton of polymer	50
Steam (10.5 kg/cm^2), ton/ton of polymer	0.7
Nitrogen, m^3/ton of polymer	30
Capital investment	
Battery limits, million dollars	23—25
Labor	
Operating labor/shift	4
Maintenance, % of capital	4
Quality control labor/shift	2

Operating costs are reduced, in part because no pelletizing equipment or recycle purification step is needed

Originally published May 17, 1982

The spherical particles formed in this process (left, 10× magnification) compared to particles in a powdered product (right, 50× magnification)

ated interest among polyolefin producers to evaluate their opportunities in this segment of the polyethylene market (*Chem. Eng.*, Aug. 24, 1981, p. 47). It is expected that LLDPE will eventually replace the use of approximately 70% of conventional low-density (LDPE), and 20% of high-density (HDPE) polyethylene.

Two of the major reasons for the high degree of interest in LLDPE center around its improved physical properties as compared with conventional LDPE (e.g., higher tensile and impact strength and better elongation characteristics) and less-severe operating conditions in manufacturing, which lead to lower operating costs and reduced capital expenditures. High-pressure processes for making conventional LDPE require pressures from 30,000-50,000 psi, while only 100-300 psi is needed for LLDPE methods. The temperatures are also lower: approximately 100°C for LLDPE, instead of about 300 C° for conventional LDPE.

SLURRY PROCESS—Ethylene and butene-1 (the comonomer) are continuously fed into a jacketed reactor, where they are copolymerized in the presence of a light hydrocarbon diluent (e.g., isobutane or propane), a chain-blocking agent (usually hydrogen), and a catalyst and cocatalyst. The polymerization takes place at operating pressures of between 200-400 psig, and temperatures ranging from 130 to 160°F, with a sufficient residence time (1-2 h) to grow a uniform particle size.

The product is continuously withdrawn from the reactor in the form of a slurry, which contains about 30-35%

resin particles. This slurry stream enters a separating unit (see flowsheet), where the pressure is reduced almost to atmospheric, causing the diluent and the remaining monomer and comonomer to flash off. The LLDPE is removed, and then dried and stabilized.

After passing through a condenser, the recycle stream is separated into liquid and gas components; the liquid (diluent) is pumped back into the reactor, while the gas (monomer and comonomer) is compressed before being returned to the reactor. No purification step for the recycle stream is necessary because the highly selective catalyst allows the polymerization to take place without forming undesirable intermediate compounds, such as low-molecular-weight polymer oils and waxes. Elimination of these materials prevents reactor fouling problems, and improves heat transfer in the reactor.

UNIQUE CATALYST—At the heart of the new process is Montedison's proprietary transition-metal catalyst, which produces LLDPE resin with a highly uniform spherical particle size—from 400-2,000 microns (averaging 1,200 μm); the average particle size for a conventional LLDPE powder is 150 μm (see photos). In addition, the process produces fewer small particles than are found in competitive products; less than 5% of the particles are 500 μm, compared to 12% below 500 μm for Union Carbide's product.

In order to obtain a polymer particle from the alpha-olefin polymerization that would reproduce the catalyst shape, a catalyst meeting the following

criteria was developed for the process:

1. The specific surface of the granular catalyst has to be high (10-100 m²/g or higher).

2. The catalyst granules must have high porosity, with channels penetrating to the innermost region, to provide access for thorough diffusion of the monomer throughout the particles.

3. The catalyst granules must be sufficiently resistant to mechanical stresses to withstand the effects of the various treatments required before polymerization.

The products' high bulk density (26-27 lb/ft³) and flowability (2.3 g/s) permit direct use by customers without the need for pelletizing, thus eliminating the cost of purchasing and operating such equipment. An advantage of the products' narrow particle-size distribution is that conventional bulk-transfer systems can be used with minimal dusting problems.

Based on pilot-plant studies, the new catalyst's high selectivity permits the polymerization of ethylene and the alpha-olefin copolymer into a wide variety of low-density products. Densities from 0.915 to 0.960 g/cm³, and melt indices of 0.5 to 30 g/10-min, have been achieved.

DESIRABLE ECONOMICS—Ethylene conversion of more than 90% per reactor pass is expected on a commercial scale, with essentially 99% recovery as finished product. It is also expected that the recovery of the diluent and unreacted comonomer will average about 98%, since they are recycled directly to the reactor.

The catalyst produces a high yield, ensuring 20,000 g of resin/g of catalyst, or 1 million g of resin/g of transition metal (titanium). In addition, catalyst residues in the finished resin are insignificant, with titanium content at 1-5 ppm, and chlorine from 10-30 ppm.

Leonard J. Kaplan, Editor

The Authors

Cipriano Cipriani, Director of Product Development at El Paso Polyolefins Co. (P.O. Box 665, Paramus, N.J. 07652), holds a Ph.D. in industrial chemistry from the University of Bologna. He is responsible for the development and market introduction of the firm's polyethylene and polypropylene resins for extrusion and molding applications. His background includes R&D work in the areas of polymers, synthetic fibers and plastics.

Charles A. Trischman, Jr., a chemical engineering graduate of Virginia Polytechnic Institute and State University, is currently Director of Process Development for El Paso Polyolefins Co. He has held positions in marketing, product development, customer technical service and plant process engineering with the company.

Section VII
Tar Sands and Shale Oil Processes

Oil recovery is higher in new tar-sands route

Holding the bitumen together instead of dispersing it seems to improve recovery efficiency in a new process that takes place at milder operating temperatures.

☐ A new look at an old process has led to a novel technique for extracting more bitumen from poor grades of tar sands, while using less energy and water than conventional methods. The key is a gentler washing action—done at lower temperatures—that doesn't break up the bitumen as much, but nevertheless attains a higher recovery efficiency (95%) than the traditional Clark process (85%), in use since the 1920s.

Devised by L. M. Cymbalisty, senior research engineer for Syncrude Canada Ltd. Research (Edmonton, Alta.), the two-stage process has been under development for more than a decade, and is now going through final optimization in a 2.5-ton/h (tar-sand feed) pilot plant operated by Syncrude.

Cymbalisty believes that the route has good potential for application in existing tar-sands plants.

Tests at Edmonton show 95% bitumen recovery from medium-quality tar sands (10-11% bitumen content), vs. the usual 85%. The pilot plant operates at 45°C, which is considerably below the 80°C needed to break up the bitumen in conventional processing—an energy-saving feature. And because of recycling, the unit uses only half the water required by other methods.

THE OLD WAY—Conventionally, tar sand, hot process water, steam and small amounts of sodium hydroxide (a separation aid) are fed into a cylindrical rotating tumbler in which the tar sand is disintegrated by mechanical and thermal energy, and the bitumen, locked between the sand grains, is liberated. The resultant oily slurry passes through a screen to remove any oversized material, and is flooded with more hot water to further break up the sand and bitumen. This material next goes to a primary separation vessel where aerated bitumen rises to the surface as a froth and is collected for further treatment. Sand settles and is withdrawn as tailings.

A mixture—consisting primarily of clay and water, and depleted of sand and bitumen—referred to as "middlings" is processed further in flotation cells where, by means of air addition and agitation, a second yield of froth is recovered.

Cymbalisty says the primary froth typically consists of 65% bitumen by weight, while the secondary froth analyzes at around 25%. A drawback of the process, he notes, is that for some very lean (low bitumen content) feeds, as much as 90% of the bitumen is recovered in the flotation cells. Cymbalisty adds that Canadian regulatory agencies have determined that tar sands as low as 6% bitumen are to be processed, and that a smooth and efficient operation may not be possible with conventional techniques.

THE NEW WAY—In Syncrude's two-stage route, the tar sand is slurried and

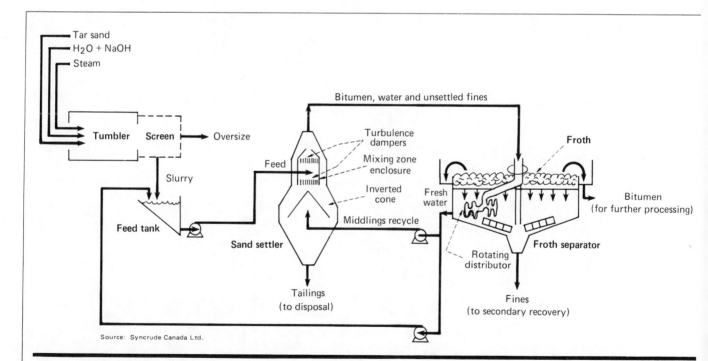

Source: Syncrude Canada Ltd.

Syncrude technology employs two vessels to extract the bitumen

Originally published December 15, 1980

screened in the conventional manner. The mass then goes to a sand settler, where it impinges on an inverted cone that encourages separation of coarse sand. The sand slides down the cone's surface and meets a countercurrent wash by an aqueous phase (middlings) from the froth separator. This wash removes any remaining bitumen.

Water and separated bitumen are pumped from the top of the sand separator to the froth separator. Here the mixture is evenly distributed by means of a submerged rotating distributor. As this unit passes any one point, and the turbulence of feeding subsides, the bitumen rises and any sand and fines (such as clay) settle.

A fresh-water wash (the only use of fresh water after the initial slurrying of tar sand) introduced above the rotating distributor further cleanses the bitumen as it rises, and the downward flow of water helps settle the fines. Settled sand and fines move on to a secondary recovery operation.

Philip M. Kohn

Oil shale commercialization: The risks and the potential

Market forces may favor the development of shale oil over coal-based synfuels. However, retorting techniques must be proven, and logistics and environmental problems have to be addressed.

Robert B. Taylor, Brown & Root, Inc.

☐ Interest in retorting oil from shale in order to produce a competitively priced synthetic crude oil has been intensified in the last few years. Operators of pilot plants have produced plans for commercialization. And many plants that will begin as demon-

This article is based on a paper delivered at the McGraw-Hill Synfuels International Conference held in Frankfurt, West Germany, May 11-13, 1981.

stration modules will eventually be duplicated and run in multiple parallel trains for commercial output (see news story, p. 47).

Market forces will be the catalyst for oil shale development. Besides competing with conventional crude oil and natural gas, shale oil will have to compete favorably with coal-derived fuels for similar markets. The liquid fuels derived from coal will be methanol or the products of indirect or direct liquefaction.

COAL COMPETITORS — The processes used for coal and shale oil synfuels (see Table I) all have similar energy efficiencies — from 58-63% [1] — yet only one of the routes (coal to methanol) has been proven, via coal gasification technology used in the manufacture of ammonia. The product slates produced by indirect liquefaction of coal via the Fischer-Tropsch method are varied, and many products are not worth as high a price per million Btu as mid- and upper-distillates produced from shale oil. In addition, indirect liquefaction based on methanol-to-gasoline technology (*Chem. Eng.*, Apr. 21, 1980, p. 86) still has to be commercially proven.

Direct-coal liquefaction routes represent the least developed technology and therefore generate the most expensive products. In contrast, here in the U.S., shale oil plants are expected to start up before any of the proposed coal-based units.

Projected U.S. demand for liquid fuels favors shale. The mid-distillate market, which would be the prime outlet for shale oil products, is expected to surpass the gasoline market by the year 2000; i.e., mid-distillate demand would be 5.6 million bbl/d and gasoline 5.3 million bbl/d [2, 3].

In contrast to the market for fuels, the market for methanol as a chemical feedstock is small (0.08 million bbl/d in 1979); and even if the market for methanol expands in the future to include transportation and utility fuels [4], demand will only increase to approximately 0.52 million bbl/d in 1990. A major drawback to the use of methanol as a fuel is the lack of methanol production capacity. Refineries cannot be utilized and, therefore, a

Product	Technology status	Market	Commercial plant startup (probable)
Shale oil	Pilot plants up to 2,000 tons/d	Mid-distillates (jet, diesel fuel)	1990
Coal liquids	Direct liquefaction in pilot plants (250 tons/d). Indirect liquefaction proven in SASOL	Varied product slate, light and mid-distillates, petrochemical feedstocks	Mid-1990s
Methanol	Coal gasification proven in 1,000-ton/d NH_3 plants, and in SASOL	Chemical feedstock (currently). Future potential--gas, turbine fuel, gasoline extender, gasoline replacement	1990

Synfuels products and markets — Table I

Originally published September 7, 1981

177

new infrastructure would have to be developed to use methanol in this way.

TECHNICAL RISK—The basic components of an oil shale project are mining, retorting, upgrading, and operation of utilities and general facilities (offsites). Mining includes crushing, shale handling and spent-shale disposal. Retorting is the recovery of the oil. Upgrading involves fractionating, hydrotreating and related processing to lower the pour point of the raw shale oil to make a synthetic crude (low in sulfur, nitrogen and arsenic) suitable as a refinery feedstock.

Looking at capital costs for the four project components (Table II), we see that retorting—which represents the greatest technological risk because it still must be proven commercially—accounts for a little over a third of the total construction costs [5]. Two-thirds of the capital cost includes technology that has been substantially demonstrated or proven.

The table also shows operating costs (not included are water supply, administration, insurance, taxes and overhead); again we note that the majority of the costs are not associated with retorting. This is significant because operating costs can be considerable. Total operation and maintenance can be 15% of the total capital cost; also, operation cost escalates year by year with inflation.

TO RETORT—The basic types of retorting methods tested to date are the Surface (direct or indirect techniques), Modified in-situ (MIS) and True in-situ.

Surface-direct technology is shown in Fig. 1. In this technique, heat is generated within the retort by the combustion of the residual carbon left on the retorted shale before the shale leaves the vessel [6]. During processing, the temperature in the shale bed reaches 1,200-1,300°F. Under these conditions, the residual carbon content of the spent shale is low, and this reduces the chance of organic pollution after shale disposal.

Mechanical operation of the system is relatively simple [7]. And due to the high temperature of retorting, the product oil has a lower naphtha content but higher aromatic content than the oil produced by indirect retorting. The off-gas, due to dilution with nitrogen from the air and carbonate decomposition in the shale, is of low Btu

Typical cost distribution for an oil shale project		Table II
	% total cost	
	Construction	**Operation**
Mining (crushing and spent-shale disposal)	16%	43%
Retorting	37	28
Upgrading	22	29
Utilities and offsites	25	--
Total	100%	100%

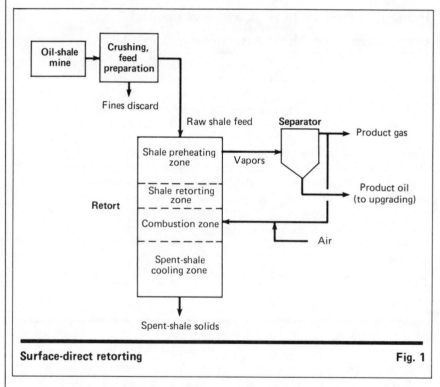

Surface-direct retorting   **Fig. 1**

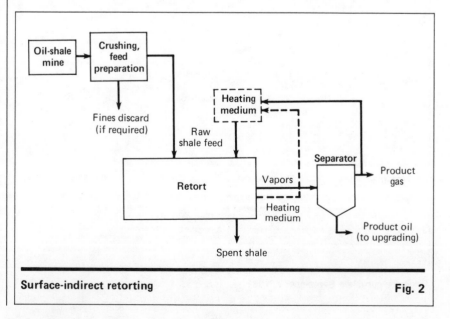

Surface-indirect retorting   **Fig. 2**

(100 Btu/scf) quality—see Table III.

Surface-indirect retorting, shown in Fig. 2, generates the heat required for retorting in a separate unit, by burning either product gas or residual carbon. The raw oil shale is heated either by the hot gas or by means of a heat-carrying medium (solids). The retorting temperature is approximately 900°F, and the process yields a higher naphtha content than does direct retorting. Because of the lack of air in the heating process, the off-gas has a high heating value (950 Btu/scf).

Modified in-situ retorting, shown in Fig. 3, involves partial mining of the oil shale deposit, and blasting (so-called rubblizing) the remainder into the void created by the mining so as to increase the overall permeability of the shale. The underground rubblized shale is then ignited, using an external fuel source.

The flame front moves downward, and the heat of combustion pyrolyzes the lower layers; the oil vapors and gas are collected in a sump and pumped to the surface. The retorting temperatures (1,600°F) are higher than those used in most above-ground processes, which contributes to a low naphtha yield and low-heating-value off-gas. In current plans, MIS retorting will be complemented with aboveground processing of the mined portion of the shale deposit to increase the overall yield of the project [8].

True in-situ processing, shown in Fig. 4, is similar to MIS except that no mining is done except for the drilling of injection and recovery shafts. Fracturing or rubblizing is accomplished to increase permeability, and the underground burning of the shale can be either horizontal or vertical [1].

In Table IV the more prominent retorting processes are organized by type. And Table V shows the extent to which these technologies have been demonstrated.

PLUSES AND PROBLEMS—As previously discussed, the three basic retort types yield a different kind of oil and gas, with the Surface-indirect process yielding a higher-Btu off-gas. Yet, the carbon content of the spent shale in the Surface-indirect retort (Table III) increases the cost of spent-shale disposal, since the leaching pollution potential would increase.

The mechanical operation of the direct retorts developed thus far has been simpler than that of the indirect

	Shale Carbon content (% of spent shale)	Oil Naphtha content (vol. %)	Off-gas (heating value Btu/scf)
Direct	1-1.5	less than 1	100
Indirect	3-5	6-7	700-950
MIS	--	less than 2	less than 100
True in-situ	--	less than 2	less than 100

Products and wastes vary with the retort technology used [13]   Table III

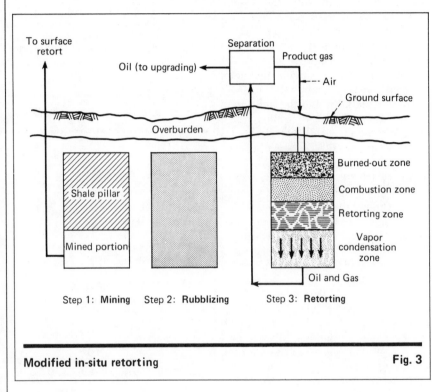

Modified in-situ retorting   **Fig. 3**

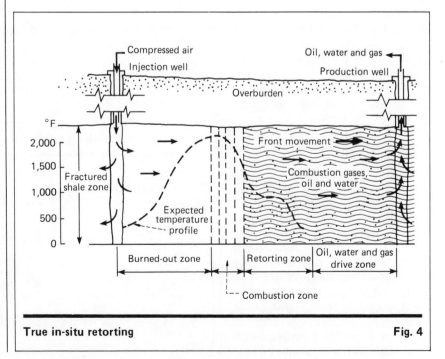

True in-situ retorting   **Fig. 4**

retorts, in which it is necessary to control the critical balance between the recycle gas temperature, flowrate, shale feedrate and retorting conditions. However, in the direct retort, a stable combustion zone must be maintained, and restricting this zone could be a problem [6].

Modified in-situ and True in-situ retorting methods offer the benefits of low mining costs, low secondary processing and waste disposal costs, and the possibility of recovering deep or varied grades of shale not economically mined by conventional methods. The main disadvantages of these techniques are: the possible contamination of underground aquifers by the leaching of spent shale salts, and the difficulty (during retorting) of controlling the burn and the oil yield. (Oil yields of in-situ routes are 60% or less of those associated with surface methods.)

OUT OF THE GROUND—Mining technology has progressed to a point where oil shale mines can be operated with a high degree of confidence. The biggest problem facing these mines is logistics. To produce 50,000 bbl/d of shale oil, 60,000 to 100,000 tons/d of shale will have to be mined. This is a greater amount of material than what is presently mined in the largest operating U.S. coal mine.

Nevertheless, room-and-pillar mining (in which large rooms are mined, leaving 60-ft by 60-ft pillars for support) is recommended by the U.S. Bureau of Mines as being the most suitable in Colorado and has been tested in four mines in the Piceance Basin there. Open-pit mining, although not currently planned by U.S. shale developers, has not been tested with oil shale, but has been developed for copper, iron ore and lignite mines throughout the world.

Open-pit mining is suitable for near-surface deposits of shale and offers 90% recovery, but this technique does have environmental disadvantages such as groundwater disturbance, spent-shale disposal, difficulty in controlling particulate emissions and the need for extensive land reclamation. The room-and-pillar method is suitable for deep shale deposits but offers a lower recovery (about 50-70% of the shale) and is a more costly technique (in $/ton) than the surface open-pit system [12].

For example, in the environmentally sensitive and mountainous Piceance

Basin, room-and-pillar mines will be used. Open-pit mining is being planned for the Rundle shale project in Queensland, Australia, where the overburden is only 40 to 50 ft [14].

In general, shale-oil recovery depends to a large extent on the thickness, location and richness of the seam, which determines the optimum retorting and mining methods. Table VI

Oil-shale retorting technologies   Table IV

Surface		Modified in-situ	True in-situ
Direct	**Indirect**		
Kiviter*	Tosco II	Occidental	Geo-Kinetics
	Lurgi-Ruhrgas		Equity
Union A	Union B		Dow
Superior	Superior		LERC**
Paraho	Paraho		IITRI R/F††
	Petrosix†		
	Galoter*		

* Russian technology
† Developed by Petrbras, Brazilian national oil company
** Laramie Energy Research Center (Laramie, Wyo.)
†† Illinois Institute of Technology Research Institute (Chicago, Ill.)

Oil-shale-retort process demonstrations [9,10,11]   Table V

Retorting process	Maximum operating rate demonstrated by pilot plant	
	Tons/d	Bbl/day
TOSCO II	1,000	700
Lurgi-Ruhrgas	25	NA
Union A	1,200	800
Union B	5	NA
Superior	250	180
Paraho	450	300
Petrosix*	2,200	1,000
Occidental Modified in-situ	—	300
		NA
Kiviter†	1,100	NA
Galoter†	1,100	

NA--Not available
* Located in state of Paraná, Brazil
† Located in Estonia, U.S.S.R.

Mining and retort methods by project   Table VI

Project*	Mining method	Retort method
Cathedral bluffs	MIS	MIS and surface
Colony	Room and pillar	Surface
Union	Room and pillar	Surface
Rio Blanco	MIS	MIS and surface
White river	Room and pillar	Surface
Superior	Room and pillar	Surface
Paraho	Room and pillar	Surface
Rundle	Open pit	Surface
Petrobras	Open pit	Surface

* All projects are in the U.S., with the exception of Rundle in Australia and Petrobras in Brazil.

Overall shale-oil recoveries for mining and retorting (300-ft-thick deposit)   VII

Retorting technology	Mining method	Additional processing	Overall shale oil recovery
Surface	Room and pillar	None	36%
Surface	Open pit	None	90%
MIS (Occidental)	20% shale mined	None	29%
MIS (Occidental)	20% shale mined	Surface	41%

Assumptions:
1. Surface retorting yield 100% of potential oil (Fischer Assay)
2. MIS retorting yields 60% of potential oil
3. 40% of total shale left behind in MIS barrier pillars

gives the mining and retorting techniques that will be used for various projects. And Table VII lists overall oil recoveries associated wth different processing choices.

CRUDE UPGRADING — Much research has been done in the area of shale oil upgrading. Early developers planned to use the raw shale oil directly as a boiler fuel, but most current plans are for the refining of the raw oil to make it a suitable refinery feedstock.

Although the raw shale oil is heavier than most conventional crudes, the upgrading processes will be similar to those used in existing refineries. Thousands of barrels of shale oil have already been refined with these methods, and the end-product has been tested as jet fuel for both the U.S. Navy and Air Force. The configuration of fractionation, coking, hydrotreating (to remove nitrogen, sulfur, oxygen and trace metals), gas treating, sulfur and NH_3 recovery will vary with the particular raw-oil properties, byproducts, and the desired quality of the refinery feedstock [15].

FUTURE PROSPECTS — The commercial potential of oil shale technology depends on: the successful demonstration of commercial-sized plants, the ability to produce an acceptable product at an acceptable price (equal to or cheaper than imported crude oil) and the ability to mitigate environmental effects.

With these factors in mind, the following trends concerning oil shale technology can be identified:

■ Indirect-surface retorting will be more extensively developed than the other retort types, due to the high quality of product oil and gas.

■ Modified in-situ will be utilized for leaner shales that are uneconomical to mine underground and are located in areas where offsite spent-shale disposal is impractical. MIS technology will be developed more slowly than other systems, as the first major projects mine the richest shale. MIS may have potential as a second-stage process to increase oil yield from a completed room-and-pillar mine.

■ Room-and-pillar mining will be utilized in environmentally sensitive areas such as Colorado, and surface mining in less-sensitive areas such as in Australia.

In addition, some observations about the commercialization of the oil shale industry can be made:

■ In the U.S., the Reagan Administration will support privately financed commercial projects. This will result in the operation (by 1990) of a few major commercial plants using only the proven technologies.

■ Limited availability of electrical power in the Rocky Mountain area will restrict expansion of oil shale plants requiring an outside source of power [18].

■ In the U.S., the greatest constraint to shale oil development will not be technological, but environmental problems. The risk of damage to air and water quality will have to be minimized. Permits and approvals from government agencies must be secured. In the arid regions of Utah and Colorado, water availability will be critical, demanding water-conserving methods of operation and spent-shale disposal.

Reginald I. Berry, Editor

References

1. U.S. EPA, *Environmental, Operational and Economic Aspects of Thirteen Selected Energy Technologies*, Vol. EPA-600/7-8-173, Sept. 1980.
2. Shell Oil Co., "National Energy Outlook for 1980-1990," Dec. 1980.
3. Pace Energy & Petrochemical Seminar, Houston, Tex., Oct. 19-31, 1980.
4. Bowden, J. R., *Applications for Methanol from Coal*, presented at Coal Technology '80, Houston, Tex., Nov. 18-20, 1980.
5. Nutter, J. F., and Waitman, C. S., *Oil Shale Economics Update*, presented at the 14th Annual Meeting of AIChE, Anaheim, Calif., Apr. 18, 1978.
6. Baughman, G. L., *Oil Shale Processing Technology*, presented at the 7th Annual Mineral & Metallurgical Division Symposium & Exhibition, Denver, Colo., Nov. 1-3, 1978.
7. Ranney, M. W., "Oil Shale and Tar Sands Technology: Recent Developments," Noyes Data Corp., Park Ridge, N.J., 1979.
8. Occidental Petroleum Corp., *Shale Oil*, 1980.
9. Colorado Energy Research Institute, Report, *Oil Shale in Colorado, the 1980s*, Oct. 1979.
10. Cameron Engineers, Inc., "Synthetic Fuels Data Handbook," 2nd Ed., 1978.
11. Pace Company, *Cameron Synthetic Fuels Quarterly Report*, Vol. 17, No. 4, Dec. 1980, pp. 2-4.
12. Cameron Engineers, U.S. Bureau of Mines, *A Technical and Economic Study of Candidate Underground Mining Systems for Deep, Thick Oil Shale Deposits*, Vol. PB-249 884, July 1975.
13. Atwood, M. T., Above-Ground Oil Shale Retorting: The Status of Available Technology, *Engineering & Mining Journal*, Sept. 1977.
14. 'This Month in Mining," *Engineering & Mining Journal*, Apr. 1980, p.35.
15. Office of Technology Assessment, *An Assessment of Oil Shale Technologies*, Vol. OTA-M-118, Washington, D.C., June 1980.
16. TRW Inc., NTIS for U.S. DOE, *Oil Shale Data Book*, Vol. PB80-125636, June 1979.
17. Jee, C. K., et al. (Booz-Allen and Hamilton) for U.S. DOE, *Review and Analysis of Oil Shale Technologies: Above Ground or Surface Technology*, Vol. FE-2343-6-V4, Aug. 1977.
18. "Colorado Oil Shale Development Scenarios 1981-2000," Colorado Energy Research Institute, 1981.

The Author

Robert B. Taylor is a Senior Project Analyst with Brown & Root, Inc., Houston, in the Synthetic Fuels Dept. His experience includes project engineering management, economic and financial analysis of coal and synthetic-fuels facilities, and market analysis for oil-shale and coal-industry activities.

He graduated with a B.S. degree in civil engineering from Lehigh University in 1969 and an M.B.A. degree from the University of Connecticut in 1973. He is a registered professional engineer in Texas and New York.

Improved routes for making jet fuels from shale oil

Three new processes and a unique catalyst are the results of research work aimed at evaluating the potential of domestic sources of jet fuels.

Leonard J. Kaplan, Assistant Editor

☐ Over the last three years, four companies working under U.S. Air Force contracts—UOP, Inc., Ashland Petroleum Co., Suntech, Inc., and Amoco Oil Co.—have been looking at new ways to convert crude shale oil into military fuels that would be cost-competitive with conventional routes using petroleum crudes. This work is part of the Air Force's overall program for finding alternative sources for military fuels to reduce dependence on imports.

The results of the final phases of this work were recently presented at a technical meeting sponsored by the Air Force Wright Aeronautical Laboratories, Fuels Branch of the Aero Propulsion Laboratory, Dayton, Ohio (and arranged by the Dept. of Chemical Engineering, University of Dayton).

The original Air Force contracts for the shale oil work called for a four-phase program (the first two phases were reviewed at a similar meeting held in May 1981):

- Phase 1—Preliminary process analysis.
- Phase 2—Bench-scale process evaluation.
- Phase 3—Pilot-Plant process evaluation.
- Phase 4—Overall optimized economic evaluation.

Each company proposed a different approach and processing scheme, but the final goal was to be the same—a high-yield, economically-viable refining process with the potential for commercialization.

PROPERTIES—Shale oil (along with various hydrocarbon gases) is the resulting product of heating sedimentary rocks that have a relatively high content of bituminous substance named kerogen, to temperatures of 700-900°F. Unlike conventional petroleum, shale oils have several properties that require them to undergo pretreatment prior to distillation or other upgrading operations such as hydrocracking, fluid catalytic cracking or reforming. Shale oils have high organic-nitrogen contents (as high as 2.2 wt% compared with 0.05 for various U.S. crude petroleums), sizable amounts of arsenic and iron (approximately 20 and 45 ppm, respectively), and a large olefin content.

Nitrogen compounds create problems in refinery operations by poisoning catalysts, and therefore must be removed before upgrading processes begin. Arsenic and iron also contribute to catalyst problems, causing deactivation and formation of deposits. A high percentage of olefins require increased quantities of hydrogen during cracking

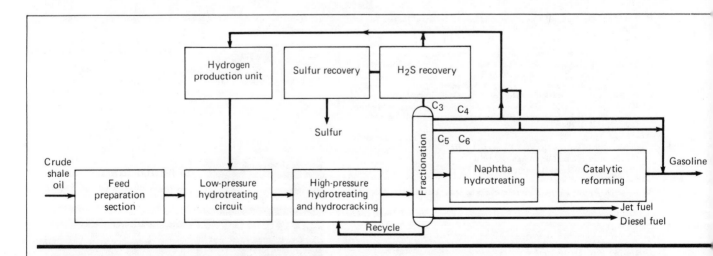

The UOP process features two-step hydrotreating and hydrocracking prior to fractionation   **Fig. 1**

Originally published December 28, 1981

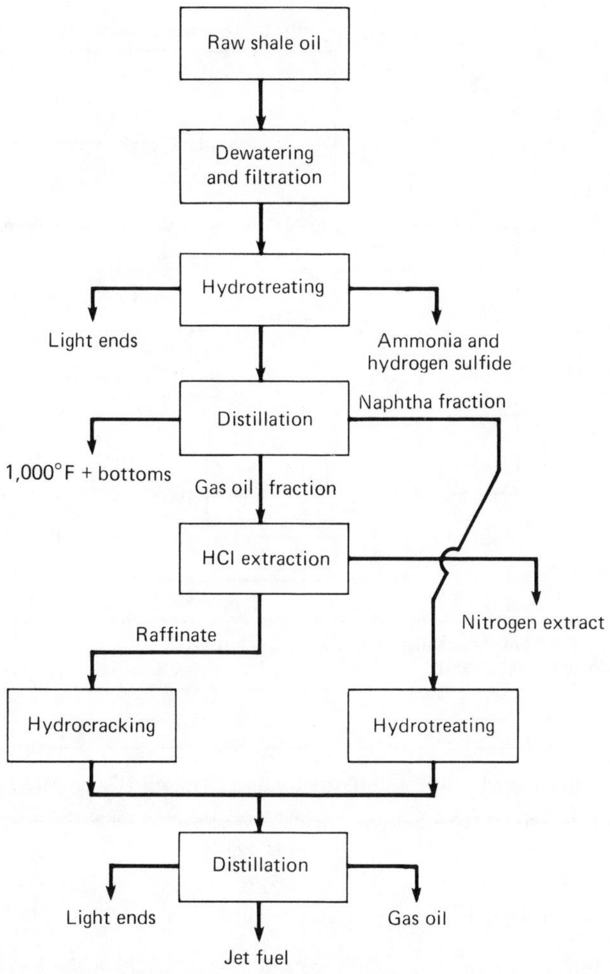
Raw shale oil
Dewatering and filtration
Hydrotreating
Light ends
Ammonia and hydrogen sulfide
Distillation
Naphtha fraction
1,000°F + bottoms
Gas oil fraction
HCl extraction
Nitrogen extract
Raffinate
Hydrocracking
Hydrotreating
Distillation
Light ends
Gas oil
Jet fuel

The Suntech extraction step uses anhydrous HCl **Fig. 2**

operations and produce large amounts of heat on hydrogenation. Because of these problems, a good part of the research projects concentrated on methods of pretreating the crude shale oil.

DIFFERENT APPROACHES—The companies worked with similar shale oils—an Occidental oil from Colorado and a Paraho oil from Utah. The basic processes used by each:

■ UOP, Inc.—Fig. 1 shows the entire processing scheme, but the project was most concerned with: low-pressure hydrotreating for the removal of metals and a degree of olefin saturation; high-pressure hydrotreating to reduce nitrogen content; and what the company calls an advanced hydrocracking flowscheme.

■ Suntech, Inc.—This multiple-step process (Fig. 2) features: hydrotreating for metals removal; distillation; an HCl extraction of the gas oil fraction to lower the nitrogen content; hydrocracking; and hydrotreating of the naphtha fraction.

■ Ashland Petroleum Co.—The Extractacracking process (Fig. 3) combines: hydrotreating for metal removal, olefin saturation, some nitrogen removal, deoxygenation and desulfurization; fluid catalytic cracking to control the quantities of distillate products; a mineral-acid extraction step to remove most of the nitrogen compounds; and reforming to modify the product freeze-point.

■ Amoco Oil Co.—Unlike the previous companies' methods, Amoco's research was limited to developing improved catalysts. According to the company, a novel catalyst, capable of direct conversion of crude shale oil into jet-fuel boiling-range material without separate hydrotreatment to remove the nitrogen compounds, resulted from the work.

Although all of the research produced samples of jet fuels for evaluation, the pilot-plant equipment was usually quite small (catalyst volumes ranged from approximately 100 cc to 2 gal) and the various steps in each system were tested separately—run as individual operations rather than sequentially.

UOP PROCESS—The operating conditions reported by UOP in its work are compared with commercial processes used in petroleum refining (referred to as the base case). For instance, the low-pressure hydrotreating base case in the firm's process involves average operating parameters used in the hydrotreating of coke-oven light oil.

After a crude shale oil is deashed and dewatered, it enters a low-pressure hydrotreating reactor filled with a catalyst, where arsenic and iron are removed to an acceptable level. In pilot tests using an Occidental shale oil, with arsenic and iron contents of 27.5 and 42 ppm, respectively, both metals could be reduced to a maximum of 1 ppm, with the reactor operating at the base pressure and 50°F above the base temperature. UOP's DSA and DRA catalysts were both found acceptable for this application.

With ammonia produced as a byproduct, the next step—high-pressure hydrotreating—uses hydrogen under pressure, to lower the nitrogen content of the oil to a maximum of 1,000 ppm. Process conditions were evaluated in comparison with parameters required for high-pressure hydrotreating of a petroleum gas oil. This process was tested at a pressure of 1,700 psi above the base pressure, and temperatures ranging from −20 below and +50°F above the base case. The company reported that while target nitrogen levels were obtained with both its DSA and DCA catalysts on a low-pressure hydrotreated Occidental oil (1.1 wt% N_2), the latter catalyst had a greater activity level. The result of overall hydrotreatment of an Occidental crude shale oil is shown in the table on p. 184.

In order to upgrade the resulting hydrotreated oil to jet fuel, the material was then hydrocracked, producing the necessary range of boiling points. Several hydrocracking methods were tested at conditions analogous to those used in processing a petroleum vacuum-gas-oil—150 psi below the base case, with temperatures of −90 below and +10°F above the base case. The final hydrocracker process selected was a "parallel flow" system developed for conventional petroleum use.

SUNTECH SYSTEM—This approach features the use of anhydrous HCl to remove nitrogen compounds from the higher-boiling fractions of a mildly hydrotreated shale oil. The processing scheme shown in Fig. 2 applies only to the Occidental shale oil; using Paraho crude requires some modifications because of its different chemical properties. In place of an HCl extraction step for processing Paraho crude, severe hydrotreating is employed.

Crude Occidental oil is dewatered and filtered before undergoing hydrotreating to reduce the total nitrogen content (from approximately 1.5 wt% down to 0.50), lower the arsenic level (from about 33 ppm down to less than 1 ppm), remove sulfur (from 6,200 ppm down to 160 ppm), and hydrogenate olefins and aromatics. Ammonia and hydrogen sulfide are yielded as byproducts.

Preparation for additional processing is accomplished by atmospheric and vacuum distillation. The resulting naphtha fraction undergoes further hydrotreatment to meet product specifications, and a wide-boiling gas oil fraction (450-1,000°F) is treated with anhydrous HCl to lower the nitrogen level to about 700 ppm in the final raffinate. After extraction, the raffinate is water-washed and then hydrocracked in a two-reactor system to maximize the yield of aviation fuel. The company says that the HCl-extraction pilot work was carried out in batch equipment, but feels that continuous operation will be possible while obtaining similar results.

ASHLAND PETROLEUM CO.—The Extractacracking process is said to have advantages including: reduced hydrogen consumption, compatibility with various feedstocks, and elimination of the need for ultra high-pressure equipment. The method differs from the two previously discussed by its use of fluid catalytic cracking to control the boiling-point distribution of the finished products. In addition, Ashland uses an extraction process for removal of the large portion of the basic nitrogen compounds. The reason given for using this approach, which employs an unnamed mineral acid, is to maximize

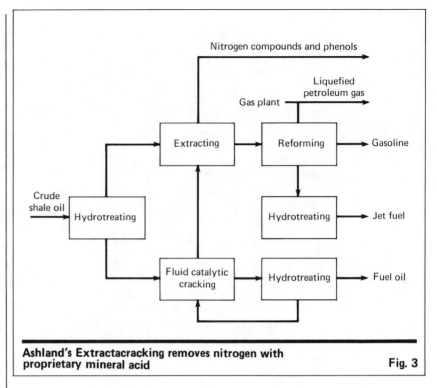

Ashland's Extractacracking removes nitrogen with proprietary mineral acid   **Fig. 3**

Typical properties of crude and hydrotreated shale oil (Occidental)

	Crude shale oil	After hydrotreating
Gravity, °API, 60°F	22.9	34.3
Specific gravity, 60°F	0.9165	0.8654
Distillation		
Initial boiling point, °F	376	255
50 vol. %	712	619
End point, °F	953	1,052
% recovered	87	99
Nitrogen, wt. %	1.51	0.01
Sulfur, wt. %	0.64	<0.03
Arsenic, ppm	27.5	<1
Pour point, °F	+75	+75

nitrogen removal while minimizing hydrocarbon loss without the need for hydrogen addition. Both the hydrotreating and fluid catalytic cracking steps shown in Fig. 3 include stripping units, which produce distillates having a maximum boiling point of approximately 600°F. These two streams both undergo the extraction step.

After the extraction raffinate has been washed and dewatered, it is passed through a hydrotreating step (not shown in the flow diagram) to remove traces of nitrogen and sulfur.

AMOCO OIL CO.—One of the more interesting results of these Air Force contracted research projects was the development of a single-catalyst system capable of direct hydrocracking of a

crude shale oil high in nitrogen. According to Amoco, this new catalyst was demonstrated in a 103-d test run. The company says that although there were several test-unit upsets, perhaps affecting catalyst life, a high yield of JP-4 jet fuel was sustained.

From catalyst-composition studies that were run to determine the optimum metal oxide concentrations and support composition, a material with the following makeup on alumina was produced: 1.5% cobalt oxide, 10% chromium oxide and 15% molybdenum oxide. Further tests were run with different supports, and it was found that a base of 50% molecular sieve/50% alumina was more effective than the alumina alone.

New ways to process oil shale

Several shale-beneficiation techniques now under review aim at cutting down the volume of feed required to obtain oil. Also, use of radio frequency is being studied as an alternative to current retorting processes.

☐ While the first commercial oil-shale facilities are still under construction, and most operating companies are awaiting better market conditions, the settlement of legal problems, or technology improvements (*Chem. Eng.*, Sept. 7, 1981, pp. 47-51), development is under way on a number of processes that may upgrade or substitute for the aboveground or *in situ* retorting methods planned by those companies.

A report submitted to the U.S. Dept. of Energy's Div. of Oil Shale (Germantown, Md.) early last month lists a dozen such techniques. The study—"The Technical and Economic Feasibility of Proposed Oil Shale Beneficiation Techniques for Surface Retorting Processes"—was the result of a survey by Fuel & Mineral Resources, Inc. (Reston, Va.). DOE's purpose is to review the state of the art in beneficiation techniques and determine which processes need and merit further financial support.

At presstime, Douglas B. Uthus, acting director of the Oil Shale Div., says that while he has not yet had time to review the report thoroughly, "This is an area that warrants further research. We need to define what that research will be."

A separate, one-year contract was let last fall by DOE's Office of Energy Research (Germantown) to Massachusetts Institute of Technology's Energy Laboratory (Cambridge) to study the economics of oil-shale beneficiation techniques as part of a production process.

One problem is that the DOE budget has already been cut back, and the funding outlook is uncertain. However, at least some of the process developers are known to be seeking industry support.

FOCUSING ON FEEDS—The processes covered by the report fall into two basic categories: physical separation of high-grade and low-grade crushed shale, and grinding and flotation to extract the oil-bearing kerogen from shale. Both methods are based on the fact that kerogen (or shale rich in kerogen) has a lower specific gravity than shale with little or no kerogen. The general idea is to produce a more concentrated feed and avoid heating a large volume of shale to obtain oil. (Not covered in the report are proposed alternatives to current retorting methods, such as the use of radio frequency to liberate oil.)

"Beneficiation is a step that's been largely ignored by the oil-shale developers, which have spent their research dollars on retorting," says Robert Reeves, of Hazen Research, Inc. (Golden, Colo.). "The situation is very similar to the way coal processing was in the 1950s. Power plants were content to burn high-ash coal, then it dawned on them that they could produce electricity more efficiently if they burned clean coal."

PHYSICAL SEPARATION—Heavy-medium separation, originally developed for coal beneficiation, is one method that is being pursued for upgrading shale feed. Crushed shale is put into an aqueous suspension of finely ground magnetite and atomized ferrosilicon particles.

Shale rich in kerogen floats, while pieces heavier than the medium sink.

The process may be carried out either in a tank or in a cyclone.

"A vessel is used for coarse material up to about 8 in., and the smallest particle it can handle is about $1/4$ in.," says Reeves. "A cyclone is used for finer material; the maximum size it can work with economically is about $1^1/_2$ in." He notes that while a vessel system operates on simple gravity separation, a cyclone also uses centrifugal force to aid separation.

Hazen, a research and process-development company, does assays for oil-shale companies, and has a 5-ton/h heavy-media vessel that it uses to upgrade clients' shale samples for retort tests. Shale is dropped down a chute into the vessel. Heavy material sinks while lighter, high-kerogen-bearing particles float over a weir with the overflow (the heavy medium is made to flow so that the magnetite and ferrosilicon stay in suspension).

Reeves explains that while shale processing is similar to that of coal, sale's specific gravity is higher—about 2.1, vs. 1.6 for coal. Thus, the heavy medium must contain about 75% ferrosilicon and magnetite by weight, compared with about 60% for coal separation.

Although crushing to a fine particle size will result in higher oil recovery, it may not be economical to do this. Reeves explains: "It all depends on the quality of the shale in a particular mine. It has to be assayed so that you can determine the optimum particle size for the most efficient oil recovery. If the assay is the same for fine and coarse particles, it will be more economical to process coarse material."

A leading proponent of this approach is Roberts & Schaefer Resource Service Inc. (Rolling Meadows, Ill.), whose parent company, Roberts & Schaefer Co. (Chicago), builds coal preparation plants and coal handling facilities. Philip C. Reeves, president of R & S Resource Service, says the company has the country's largest

Originally published February 22, 1982

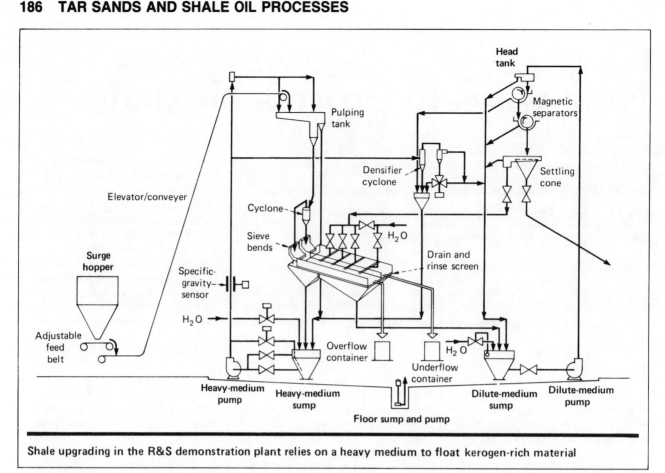

Shale upgrading in the R&S demonstration plant relies on a heavy medium to float kerogen-rich material

demonstration plant (a 25-ton/h cyclone) for heavy-medium separation of shale.

"Most of the major oil companies have expressed interest in having tests run, and so far we have processed shale from three of them," says Reeves. R & S is a licensee of the Dutch State Mines heavy-medium cyclone process. Crushed shale and the heavy medium are fed tangentially into the cone body. Heavy shale material sinks and is discharged, while pieces rich in kerogen float and are collected with the overflow. Shale particle size may vary from 1/4 in. to 2 in., depending on the operating company's retort needs.

Reeves notes that the process may be tailored to meet various requirements. "We have taken material that contains only about 13-15 gal. of oil per ton of shale and upgraded it to 28 gal/ton, with 30% recovery—i.e., 70% of the feed to the cyclone was discarded. We have also produced 40-gal/ton material from 22-gal/ton feed, with 50% recovery."

To illustrate the potential economics of heavy-medium separation, Reeves cites a hypothetical case in which three

retorts might be designed to process 20-gal/ton oil shale. If a heavy-medium plant were installed at the front end to upgrade the shale to 30 gal/ton, two retorts could produce the same amount of oil. The heavy-medium plant would cost only $10,000-$15,000 per ton/h of capacity, compared with about $250,000 per ton/h for a retort, he says. (For more information on the R & S process, see *Chem. Eng.*, Nov. 16, 1981, p. 41).

TWO VARIATIONS—Minerals Separation Corp., a subsidiary of Mountain States Mineral Enterprises Inc. (Tucson) has tested its Dyna Whirlpool Process (DWP) on oil shale, and has made proposals to several operating companies for larger-scale tests, says Martin C. Kuhn, vice-president and general manager. DWP is distinctive in that the vessel used is a straight-walled, inclined cylinder, rather than a cone, and only the finely ground medium is pumped, while the coarser material to be processed is gravity-fed.

A separation system in which either kerogen-rich or kerogen-lean pieces of shale are "labeled" by a reagent, then detected and separated, has been tested on ores from three different sources by

Occidental Research Corp. (Irvine, Calif.). One of those ores was supplied by a sister company, Occidental Oil Shale, Inc. (Los Angeles)—a partner of Tenneco Inc. (Houston) in Cathedral Bluffs Oil Shale Co., which has a 5,094-acre shale lease in Colorado.

Occidental's patented technique is called Oxylore. A carboxylic acid containing a fluorescent dye is sprayed onto the crushed shale, and is selectively adsorbed by the kerogen-rich material. Then a single layer of ore is placed on a moving belt and dropped in front of an array of photoelectric sensors and ultraviolet-light sources. The photocells pick up fluorescence emitted by the coated particles, and activate valves at a lower level, allowing 80-psi water jets to push those pieces out of the falling curtain of rock.

The process has been successfully piloted at rates up to 150 tons/h to separate impurities from limestone in California Portland Cement Co.'s Colton (Calif.) cement plant. That facility has a 20-in.-wide (rock) curtain and 40 water jets. Each jet may be activated as much as 50 times/s. In tests with oil shale, 16-gal/ton feed material has been upgraded to 25 gal/ton by rejecting 10-gal/ton material.

FLOTATION METHODS—Although separation appears to offer economic advantages and would fit in with many retorting operations, one disadvantage is that some oil is discarded with the low-grade shale. An answer to this problem—and a potential alternative to crushing and separation—is to grind all the shale and use flotation to separate kerogen from the rest of the material.

"We have been able to recover more than 95% of the organic carbon content," claims H. Shafick Hanna, of the University of Alabama's Mineral Resources Institute (Tuscaloosa). Hanna is developing a flotation process with funding from DOE and the university's School of Mines and Energy Development.

In Hanna's process, shale is ground to 10-30 microns, then the kerogen-bearing compounds are floated (using water), cleaned and filtered. Hanna declines to give further details because of a pending patent application. So far, work has been done only at the laboratory level (up to 10 lb/h), but Hanna plans to scale it up to about 100 lb/h. He has been working with Devonian shale, which is of lower grade than Western shale, but is found extensively from Texas to New York (an estimated 423 billion barrels could be recovered by surface mining). The flotation process has upgraded this type of shale from about 10 gal/ton to 24 gal/ton, Hanna says. He estimates that the method could cut the net capital cost of a retorting operation by about one-third.

One drawback is that the finely powdered product is unsuitable for conventional retorts. The Fuel and Mineral Resources, Inc. report suggests that R & D work on agglomeration techniques might be undertaken to develop an appropriate retort material (the report also says that flotation's water requirements might be a problem in the West).

Hanna believes he may have an answer to the fines problem. He has made some briquettes by pressure in the laboratory and says they have good grain strength.

SRI International (Menlo Park, Calif.), which is also working on a flotation process, has some proprietary design concepts for retorts that could handle its powdered product without

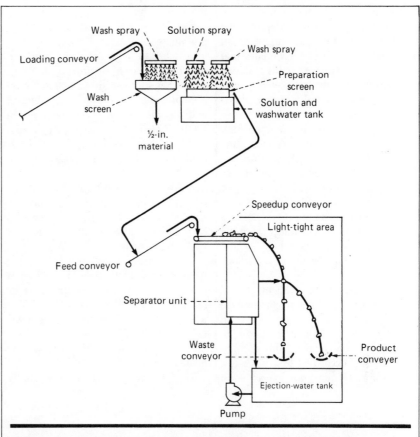

Oxylore process achieves physical separation through use of a dye

agglomeration. "If you go to agglomeration, you are destroying an opportunity you should take," says Robert Murray, project supervisor. SRI has been developing its froth flotation process in the laboratory under contract to DOE's Laramie Energy Technology Center (Wyoming). Shale is ground to 10-20 microns, then floated with water and commercially-available frothing agents.

Results have been better with Eastern shale: SRI has been able to increase the kerogen content from 12% to 48%, compared with 12% to 24% for low-grade Western shale. Murray explains that dolomite found in Western shale tends to float with the organic material, so it is more difficult to obtain a good separation.

He declines to give details on the retort design concepts, except to say that they involve the use of hydrogen or syngas under pressure for more-efficient production of liquids. (Phillips Petroleum Co. is building a demonstration plant for hydrogen retorting of Devonian shale, which is hydrogen-deficient compared to Western shale—see *Chem. Eng.*, Nov. 2, 1981, p. 23).

Murray adds: "In the long range, we would like to concentrate kerogen to a fairly high degree so that it could be directly hydrogenated during the retort process."

TRW Inc. (Redondo Beach, Calif.) is also working on a flotation-type process, but declines to give details.

OTHER PROCESSES—*In situ* radio-frequency (RF) heating of oil-shale deposits to liberate the oil is another technique under development. The general idea is to drill a series of holes for the RF antennas or electrodes that heat up the deposit and break down the kerogen to release oil and vapors. Researchers explain that shale is generally nonpermeable and a poor thermal conductor. But radio waves are able to heat the shale by heating the polar molecules of water (mostly bound water) in the deposits. The big attraction of this approach is the potential for eliminating the need for mining, retorting and disposal of spent shale.

(There are some exceptions to the general nonpermeability of oil shale. Equity Oil Co., Salt Lake City, has been injecting superheated steam at

900°F and 1,400 psig to liberate oil from a 760-1,360-ft-deep deposit near Meeker, Colo. This particular deposit contains water-soluble carbonates that have permitted fracture paths to develop. Equity recently completed a two-year test program and is now evaluating the results.)

An RF process of this type is being developed jointly by Texaco Inc. (White Plains, N.Y.), Raytheon Co.'s Equipment Development Laboratories (Wayland, Mass.), and The Badger Co., a Raytheon subsidiary (Cambridge, Mass.). A pattern of holes is drilled, and an RF heating/pumping unit is lowered into each hole, so that liberated oil may be directly pumped out.

The companies have spent several million dollars field-testing the process over the last two or three years on Texaco-owned oil-shale property in Uintah County (Utah), with encouraging results, according to a Raytheon spokesman. He declines to say what frequency (or frequencies) is used, except to say it is in the range of "a few megahertz."

Another RF process is being pursued by IIT Research Institute (Chicago), which has successfully field-tested it on a small scale near Duchesne (Utah), and has also tested it on tar sands in Utah (*Chem. Eng.*, Sept. 21, 1981, p. 17). Jack Bridges, of IITRI, explains that a commercial system would use 10-14 rows of RF electrodes in a 20-acre lot about 100-200 ft deep. A key feature, he says, is that the electrical field set up by the electrodes is contained within the zone, instead of radiating outward.

The system would probably use a frequency of about 2 MHz, which is more suitable for penetrating a large block of shale than the 13.56 MHz used in the small-scale test, he says. Oil would flow downward for collection in premined drifts below the heated zone.

IITRI's work has been partially funded by DOE, but that support is ending. However, Bridges says, "a major oil company has funded a major engineering company to investigate the economics and engineering feasibility of our process." He estimates that the method could extract three barrels of oil and gas for every barrel of equivalent coal energy used to produce electricity for the process.

Colorado School of Mines Research Institute (Golden) has applied for a patent on microwave retorting of mined and crushed shale. So far, CSMRI has done only laboratory-scale work, but the premise is that this method may be more energy-efficient and yield a better-quality oil than *in situ* heating.

Exxon Research and Engineering Co. (Florham Park, N.J.) also is working on an oil-shale beneficiation process, but declines to give details. The firm is a partner of Tosco Corp. (Los Angeles) in the Colony Oil Shale Project in Colorado.

Gerald Parkinson

Section VIII
Waste Gas Treatment/Recovery Processes

Sulfur Removal
NO$_x$ Removal
Other Processes

Flue-gas desulfurization produces salable gypsum

Offering high reliability, this process is based on lime but avoids the problems of plugging and scaling common to wet scrubbing systems.

D. R. Kirkby, Davy Powergas Inc.

☐ Unlike conventional lime- and limestone-based systems, the Davy S-H flue-gas-desulfurization process uses a solution instead of a slurry to absorb SO_2. Employing a chemistry that produces a soluble intermediate, this process is able to avoid the problems of wet scrubbing, provide reliability, and produce a gypsum that may be used for wallboard. In the U.S., this system is offered by Davy Powergas Inc., which has an exclusive license from Saarberg-Hölter Umwelttechnik GmbH.

The high reliability of this method has been proved at the 40-MW Saarberg Weiher II power plant (Saarbrücken, West Germany) where 80,300 scfm of tail gas has been treated for over 20,000 h. The system has had an onstream factor (availability) of 96%. An additional unit, which will treat 25% of the flue gas for the Saarberg Weiher III power plant (700 MW), is scheduled for startup this month.

Also, this route produces a gypsum byproduct suitable for sale or for use as landfill. Some of this gypsum has been sold in France to wallboard manufacturers and it is being tested for this application in the U.S.

The process can handle the relatively higher SO_2 concentrations found in chemical plants. A unit for Veba Chemie (Gelsenkirchen, West Germany) went onstream in November 1978. This system treats gas from two Claus units and a sulfuric acid plant. Concentrations of SO_2 may be as high as 6,500 ppm (by volume).

If halogens are in the feed gas, they can be treated. Operating experience has included the removal of HCl and HF as well as SO_2 from the flue gas of a waste product incinerator. Approximately 98% of the HCl and HF, and 90% of the SO_2 were removed, even with variations in SO_2 concentration.

TECHNOLOGY FEATURES—The Davy S-H process is divided into three parts: absorption, oxidation and separation. It is in the chemistry of absorption and oxidation that this route differs the most from conventional scrubbing. When SO_2 is absorbed, a soluble calcium salt is formed, and it is this salt that is later oxidized and converted to gypsum, $CaSO_4 \cdot 2H_2O$.

Conventional lime- and limestone-based processes circulate a slurry and form insoluble calcium sulfite hemihydrate, $CaSO_3 \cdot \frac{1}{2} H_2O$, which causes scaling and erosion. The Davy S-H process employs an aqueous solution of $Ca(OH)_2$ to scrub SO_2 and form the water-soluble intermediate, calcium bisulfite, $Ca(HSO_3)_2$.

Formation of this intermediate is ensured by the use of formic acid as calcium formate, $Ca(COOH)_2$, which buffers the solution and keeps the pH within the proper range. Removal of SO_2 from the flue gas is by chemical reaction:

$$2\ SO_2 + Ca(OH)_2 \longrightarrow Ca(HSO_3)_2$$
$$2\ SO_2 + Ca(COOH)_2 + 2\ H_2O \longrightarrow Ca(HSO_3)_2 + 2\ HCOOH$$

The Davy S-H process can tolerate high chloride levels. The absorption of chlorides actually helps to lower the pH to 4, which is the optimum level for oxidation.

With the addition of air, calcium bisulfite formed during absorption is converted to gypsum in the oxidizer:

$$Ca(HSO_3)_2 + O_2 + 2H_2O \longrightarrow CaSO_4 \cdot 2H_2O + H_2SO_4$$

Economic data

Basis:
500-MW coal-fired boiler— 3% sulfur coal (1.1 million scfm)
90% SO_2 removal
U.S. Gulf Coast location, 1978 costs

Raw-materials and utility requirements:

Lime, 95 wt% CaO	22,300 lb/h
Formic acid	17.3 lb/h
Makeup water	180–220 gpm
Electric power (includes oxidation air blower)	5.5 MW

Operating personnel
3 people per shift plus 1 during day shift for supervision.
Personnel for routine maintenance included.

Capital cost
Total installed cost, $20 million.

Gypsum Production

Gypsum (including 20% water)	81,500 lb/h

Originally published January 1, 1979

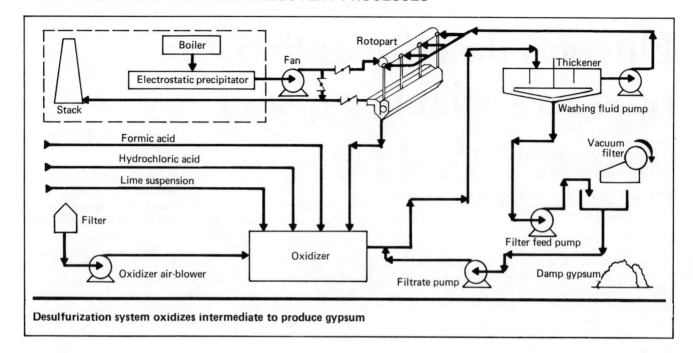

Desulfurization system oxidizes intermediate to produce gypsum

At the outlet section of the oxidizer, calcium hydroxide (as lime and water) is added to replenish the calcium ion in solution. This ion in turn reacts with the sulfuric acid, and neutralizes any HCl present and HCOOH formed during absorption:

$$H_2SO_4 + Ca(OH)_2 \longrightarrow CaSO_4 \cdot 2H_2O$$

$$2HCOOH + Ca(OH)_2 \longrightarrow Ca(COOH)_2 + 2H_2O$$

$$2HCl + Ca(OH)_2 \longrightarrow CaCl_2 + 2H_2O$$

Lime also raises the pH of the solution to the level required for absorption. And a small amount of formic acid is also added. When gypsum—which is suspended in the resulting solution—is removed, some of the calcium formate is lost. The same is true of calcium chloride, and additions of hydrochloric acid may be necessary if the chloride content of the feed gas is low.

SECOND-GENERATION SCRUBBING—In the Davy S-H process, flue gas with temperatures of up to 400°F can be accepted. Although the system can accept high particulate loadings, pretreatment of the gas for particulate removal is recommended.

Flue gas (see flowsheet) enters an absorption device called the Rotopart. This device has no moving parts and is made of carbon steel with a thermally hardened epoxy lining. The gas goes into a header and then into individual 6-ft-dia. venturi-shaped absorbing ducts.

Absorbent recycled from a thickener downstream also enters the ducts and flows cocurrent to the gas flow. Here, with the absorption of SO_2, calcium bisulfite is formed. The treated gas is then removed from the liquid in the separator section of the Rotopart, and discharged through a stack.

Spent absorbent flows by gravity to the oxidizer, where air is blown through the liquid. The sparged air (approximately 15 psi) oxidizes the calcium bisulfite to gypsum.

As gypsum is formed, it goes into suspension. Additions of formic acid, calcium hydroxide and hydrochloric acid are made. These chemicals restore the absorbent by regenerating its active components and adjusting its pH.

The suspension flows by gravity to a thickener where gypsum crystals are separated from the liquid. This liquid is the regenerated absorbent and it recycles to the Rotopart.

A slurry that is 15% gypsum by weight is pumped from the bottom of the thickener to a vacuum filter. Here, a cake is produced that contains approximately 80% solids by weight (95% of the solids content is gypsum, with less than 0.5% being calcium sulfite). This product is conveyed to storage or shipping as required. The clean filtrate is then returned to the thickener, from where it is also recycled as absorbent.

ECONOMICS—Process economics for a typical installation are provided in the table. Battery limits include lime-receiving and storage facilities, slaking, gas handling, SO_2 absorption and gypsum filtration.

However, costs for the process may vary considerably for individual cases. Gas-handling costs may outweigh those of chemical treatment. Power plants typically have large volumes of gas and low SO_2 concentrations. Chemical plants, such as sulfuric acid units, have higher concentrations but lower volumes.

Also, a credit has not been taken for gypsum production, even though such production may be a special advantage. For conventional wet-scrubbing systems, the capital and operating costs of sludge stabilization and disposal have been estimated at a total $3.20 per ton of coal fired. And sulfite sludge, which is a byproduct of these systems, may be classified as hazardous material, due to its chemical oxygen demand.

The gypsum produced by the Davy process is chemically stable and, if not used by the gypsum industry, may be stacked in a relatively small area, minimizing load costs.

The author

D. R. Kirkby is Manager of Process Engineering for Davy Powergas Inc. (Houston, Tex.). He is a Registered Professional Engineer in the state of Florida and holds a B.Ch.E. degree from the University of Delaware.

Dry scrubbing looms large in SO₂ cleanup plans

The latest standards on SO₂ emissions from new coal-fired utilities favor the use of dry scrubbing, which, according to its supporters, boasts lower investment and operating costs and avoids sludge-disposal problems.

☐ The newest fashion on the SO_2-control scene is the emergence of nonregenerable, dry-scrubbing systems for flue-gas desulfurization (FGD). The first U.S. commercial installation based on this technique went onstream last month at Strathmore Paper Co.'s pulpmill in Strathmore, Mass. (*Chem. Eng.*, Aug. 13, p. 89); four more such facilities will start up within the next three years (see table); and several other projects are expected to be announced by the end of this year.

The underlying technology, based on SO_2 removal either by injection of a dry alkaline sorbent or by an alkaline slurry or solution in a spray dryer (see box), has been available for a few years. But potential users, noting a lack of commercial operating experience, and perhaps wary of the extravagant claims made years ago for such cleanup methods as wet scrubbing, have been slow in switching to dry scrubbing—despite what its proponents say are impressive advantages over wet scrubbing (lower investment and operating costs, no sludge-disposal problems).

Nonregenerable dry scrubbing, however, does have one important limitation: it appears to be economic only at low SO_2 concentrations in flue gas. This is because the dry methods (dry injection, spray dryer) use expensive sorbents—lime, sodium carbonate, or such naturally occurring carbonates as trona (a hydrous sodium carbonate) and nahcolite (sodium bicarbonate). Wet scrubbing, the most widespread FGD method, employs cheaper limestone, so it has an operating-cost edge at high SO_2-removal levels.

GOVERNMENTAL BOOST—The U.S. Environmental Protection Agency (EPA) had dry-scrubbing technology very much in mind when it issued tougher standards in June, covering emissions of SO_2, particulates and nitrogen oxides from new coal-fired power plants (350 will be built in the U.S. between now and 1995).

Until now, utilities could lower SO_2 emissions by either burning low-sulfur coal or using scrubbers to remove the emissions given off by high-sulfur coal. But after the June regulations, all new coal-burning facilities (except those using low-sulfur anthracite) will have to install scrubbers.

Observing that it would probably be more economical to employ a wet scrubber when the coal's sulfur content is above 1.5%, the agency devised a sliding-scale emissions-removal requirement aimed at fostering the growth of dry-control methods. For example, new facilities emitting more than 0.6 lb of SO_2/million Btu must install scrubbers able to remove 90% of the emissions. But those that use low-sulfur coals (with emissions under 0.6 lb/million Btu) can employ scrubbers only 70% efficient. This rule is expected to generate additional interest in dry scrubbing among utilities planning new construction.

DRY VS. WET—Among the strongest proponents of dry scrubbing is the Basin Electric Power Cooperative, which plans to install units of the spray-dryer type at plants near Beulah, N.Dak., and Wheatland, Wyo. According to Kent E. Janssen and Robert L. Eriksen, two company spokesmen who presented a paper at a recent EPA symposium on FGD (March 5-8, at Las Vegas, Nev.): "Wet scrubbers are complex and cumbersome, cause a heavy capital burden, are costly to operate, difficult to maintain, and consume a sizable part of the energy generated by the plants they serve. The byproduct, sludge, produces additional environmental problems by taking up otherwise useful land for settling ponds."

As Janssen and Eriksen see it, dry scrubbing comes out ahead in the following aspects of an FGD system:
- Waste handling. Dry control yields a dry waste, so there is no need for sludge-handling equipment.
- Wet/dry interface. Scaling and plugging is common in wet scrubbers at wet/dry interfaces and in scrubber packing materials and demisters. In the spray-drying method of dry scrubbing, only dry powder comes into contact with the scrubber walls.

Guide to nonregenerable flue-gas desulfurization

In contrast with regenerable FGD methods, in which the scrubbing agent is recovered for reuse and the waste is usually converted into a useful product—e.g., elemental sulfur, sulfuric acid—nonregenerable techniques do not recycle the scrubbing medium and must dispose of either a sludge or a dry waste.

Wet scrubbing and dry scrubbing are the two available nonregenerable FGD methods. In the first, an aqueous slurry or solution, frequently of lime or limestone, is brought into direct contact with flue gas. The waste product is usually a sludge of calcium sulfite.

Dry scrubbing can be done either by dry injection, in which a dry sorbent (soda ash, trona, nahcolite) comes into contact with flue gas, or by spray drying, in which the heat from flue gas is used to evaporate the water off a sprayed alkali (lime or soda ash) slurry or solution. The outcome in either case is a dry-powder mixture of fly ash and sulfates, which is collected by filter bags or in an electrostatic precipitator.

Originally published August 27, 1979

Roundup of U.S. commercial-size dry-scrubbing projects

User/location	Developer/licensor	Plant/fuel	System	Status and comments
A. Industrial				
Strathmore Paper Co. Strathmore, Mass.	Mikropul Corp.	Pulp mill Flue-gas flow: 60,000 acfm Existing pulverized coal boiler burns 2.5% sulfur coal	Lime-based Spray dryer (1 unit—15 ft dia.) Two-fluid nozzle Baghouse collector (pulsed air)	Started up July 1979. SO_2 removal efficiency 75%.
Celanese Corp. Cumberland, Md.	Rockwell International and Wheelabrator-Frye in a joint venture	Acetate-fiber plant Flue-gas flow: 65,000 acfm Stoker-fired boiler (from storage) burns 1.5-2% sulfur coal	Lime-based Spray-dryer (1 unit) Baghouse collector (4 compartments)	Slated for completion in January 1980.
B. Utility				
Otter Tail Power Co. leads group of five utilities Coyote Station, near Beulah, N.Dak.	Rockwell International and Wheelabrator-Frye in a joint venture	410 MW (under construction) Flue-gas flow: 1.89 million acfm Fuel: lignite, 0.78% sulfur (avg.)	Sodium-based (soda ash) Spray dryer (4 units) Baghouse collector	Construction to be completed in late 1981. Contract for $36 million awarded in December 1977.
Basin Electric Power Cooperative (Bismarck, N.Dak.) Laramie River Station-Unit 3 Wheatland, Wyo.	Babcock & Wilcox	500 MW (under construction) Flue-gas flow: 2 million acfm Fuel: Wyoming sub-bituminous coal, 0.81% sulfur (max.)	Lime-based Spray dryer—horizontal reactor Electrostatic precipitator	The first of the utility's three units, now under construction, is expected to start up in 1980. Units 1 and 2 are wet-scrubber systems. Identical electrostatic precipitators will be used with both the wet and dry systems.
Basin Electric Power Cooperative (Bismarck, N.Dak.) Antelope Valley, near Beulah, N.Dak.	Joy Mfg. Co. and Niro Atomizer in a joint venture	440 MW (under construction) Flue-gas flow: 2 million acfm Fuel: lignite, 1.22% sulfur (max.)	Lime-based Spray dryer Baghouse collector	Onstream date for the first dry scrubber is targeted for April 1982. Bids for a second dry-scrubbing system will be going out this year.

■ **Materials of construction.** Wet scrubbers require corrosion-resistant alloys or coatings; dry-control systems can use low-carbon steel for vessels.

■ **Maintenance.** In wet systems, maintenance for the slurry-handling equipment is high, because of the need to recirculate corrosive materials at high pressures and volumes. Dry systems operate at low material-volumes and low pressures; liquid/gas ratios are about 0.2-0.3 gal/1,000 acf, compared with about 40-100 gal/1,000 acf for a wet scrubber.

■ **Energy requirements.** Dry systems require about 25-50% of the energy needed by a wet system.

■ **Water consumption.** This is much less for dry systems. For instance, the dry scrubber at Basin Electric's Laramie River Station Unit #3 will use about 50% as much water as is needed for the wet scrubber in Unit #1 or #2. Cooling-tower blow-down or ash water may be used in the spray dryer.

WHO'S ACTIVE—The list of firms doing research on spray-dryer systems includes such names as Carborundum Co. (Niagara Falls, N.Y.), Joy Mfg. Co. (Montgomeryville, Pa.), Niro Atomizer Inc. (Columbia, Md.) and Rockwell International Corp. and Wheelabrator-Frye Inc.'s Air Pollution Control Div., both in Pittsburgh. Among those studying dry-injection systems are Carborundum, and Energy & Pollution Controls, Inc. (Bensonville, Ill.).

The Electric Power Research Institute (EPRI) also is interested in dry scrubbing, mainly in dry-injection systems. According to project manager Navin Shah, dry injection is simpler than spray drying, in terms of both equipment and processing. He notes that, because the latter employs a solution or slurry, the flue gas may need reheating after evaporation of the sorbent liquid. Should flue-gas temperature drop for some reason, clogging may occur in the subsequent collection equipment (a baghouse or electrostatic precipitator).

In an EPRI-sponsored program, KVB Inc. (Tustin, Calif.) tested a sodium-based dry-injection system for one and a half years on a bench-scale basis. Among the parameters studied: sodium/sulfur stoichiometric ratios, injection methods and temperatures, baghouse operation temperatures, incoming SO_2 concentrations, and sorbent particle size.

EPRI is also supporting a Bechtel Corp. economic-comparison study involving a dry-injection system (based on nahcolite and trona) and wet scrubbing. According to Shah, a draft report that may be available by the end of the year indicates that dry-injection economics are highly dependent on reagent costs.

The Tennessee Valley Authority (TVA) too is in the throes of completing an economic assessment of dry scrubbing—under contract with the EPA. The goal is to compare a conventional wet-limestone scrubbing system with lime-based spray-drying.

Another EPA-funded study has been awarded to Radian Corp. (Austin, Tex.), which will review dry-scrubbing technology; a draft report was due at the end of July.

In addition, EPA is funding a few small demonstration-scale projects on dry scrubbing. Envirotech Corp. (Lebanon, Pa.) is about to start a dry-injection study at the Martin Drake Station in Colorado Springs, Colo., using a baghouse collector.

Looking ahead, an industry expert predicts that use of additives to enhance FGD at high SO_2 concentrations will be tried in dry-scrubbing applications, as has been the case in wet-scrubbing systems.

Irene Miller

Fluid-bed gets the nod

It's either that or flue-gas desulfurization for coal-fired plants, according to recent standards for SO₂ emissions, so federal agencies are pouring more money into fluid-bed research, and manufacturers are rushing to make the hardware.

☐ The recent government regulations for new, major fuel-burning installations have made it clear to utilities and chemical manufacturers that want to install a new boiler or process heater: They must either put in a conventional coal-fired unit and clean up the emissions with a flue-gas desulfurization unit, or use a fluid-bed boiler, which burns fuel cleanly.

This situation is helping to speed up developments in fluid-bed combustion (FBC) technology. For example, a number of manufacturers are coming out with new FBC units. And the pace of federal funding of FBC research is accelerating. Consider the following:

■ This spring, the U.S. Dept. of Energy (DOE), which this year budgeted $38 million for R & D in both atmospheric and pressurized FBC technology, said it would fund 20 to 40% of design, construction and startup costs of atmospheric fluid-bed (AFB) boilers with capacities of at least 100,000 lb/h of steam. The agency openly invited industry to build units of this type.

■ In August, DOE signed a $1.8-million contract with three firms—Westinghouse Electric Corp., Babcock & Wilcox Co., and Stone and Webster Inc.—to develop FBC technology for cogeneration projects.

■ Also in August, the Environmental Protection Agency (EPA) completed a study of the SO₂-retention capabilities of pressurized fluid-bed (PFB) combustion. Both EPA and DOE are sponsoring R & D efforts to determine the environmental impact and potential use of solid wastes discharged by FBC equipment.

■ The Electric Power and Research Institute (Palo Alto, Calif.) has said it will spend $5 to $6 million on FBC technology next year. (Under an EPRI contract, Babcock & Wilcox has built an AFB pilot plant.)

■ Overseas, IEA Coal Research (London)—a branch of the International Energy Agency—is presiding over a three-nation (the U.S., the U.K. and West Germany) program to build a PFB boiler in Yorkshire, England. The facility will be finished this year.

INDUSTRIAL ACTIVITY—Meanwhile, lured by a bright sales potential, several equipment makers have set up enterprises to manufacture FBC units.

In February, for instance, Babcock Contractors Inc. (Pittsburgh, Pa.) and Riley Stoker Corp. (Worcester, Mass.) announced that they will jointly manufacture and market AFB boilers capable of producing steam at the rate of 50,000 to 500,000 lb/h (*Chem. Eng.*, May 7, p. 47). Foster Wheeler Boiler Corp. (Livingston, N.J.) and Fluidyne Engineering Corp. (Minneapolis, Minn.) also offer AFB equipment. And Johnston Boiler Co. (Ferrysburg, Mich.), which started offering a range of boilers (capacities: 2,500 to 50,000 lb/h) last Fall (*Chem. Eng.*, Nov. 6, 1978, p. 91), recently made its first sale—to Central Soya (Marion, Ohio).

In the U.K., Babcock Contractors Ltd. and BP Trading Ltd. have formed a joint venture, called Fluidized Combustion Contractors Ltd. (East Grinstead), to engineer and supply FBC systems. And Energy Equipment Co. (Olney, Buckinghamshire), a company that focuses on conversions of conventional equipment, recently sold its first FBC boiler.

Advanced FBC systems also are getting their share of attention, at least in the U.S. In July, Battelle Develop-

---

How expensive are FBC systems?

Where FBC economics are concerned, the only thing experts are sure of is that coal-fired FBC systems are more expensive than oil-fired ones of any kind. However, there is a good probability that FBC types will be competitive with coal-fired conventional boilers.

According to DOE's Freedman, "A new AFB power plant is about 10% less than a conventional plant without scrubbing." Robert Gamble, of Foster Wheeler, feels that operating costs and maintenance also should be less. "The potential for reducing manpower is better for FBC than for conventional boilers," he notes.

Further on the plus side of the ledger, high-sulfur fuels are cheaper, and limestone and dolomite are relatively cheap. "Besides, they are used in amounts that range from only 5 to 10% of the coal charged," says Alan Smith, commercial manager of Fluidised Combustion Contractors Ltd. (East Grinstead, U.K.).

Exxon's René Bertrand believes that even PFB has the potential for being less expensive than conventional systems fitted with scrubbers based on the use of mechanical hot-gas cleanup.

Foster Wheeler has made some preliminary estimates, as follows:

	Atmospheric FBC plant (sized high-sulfur coal)	Conventional plant (pulverized coal with scrubber)
Capital cost, $/kWe	583	661
Cost of electricity, mills/kWh		
Capital	13.1	14.8
Fuel	21.1	22.3
Overhead and maintenance	5.7	8.1
Total	39.9	45.2
Cost ratio	1.0	1.13

Originally published October 8, 1979

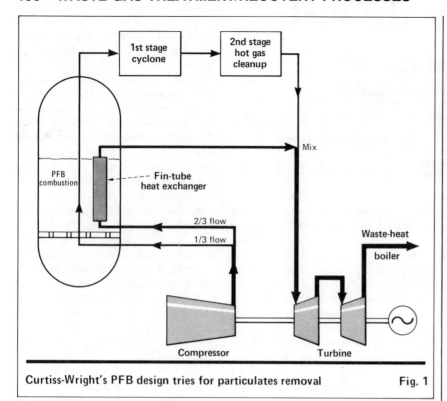

Curtiss-Wright's PFB design tries for particulates removal   Fig. 1

ment Corp. (Columbus, Ohio) signed an agreement that gives Struthers Wells Corp. (New York City) exclusive world rights to an FBC process that combines two kinds of fluid beds. And Lehigh University (Bethlehem, Pa.) researchers, working under DOE sponsorship, are developing a centrifugal fluid-bed unit.

Despite this activity, only two large FBC installations are now operating in the U.S.; both are of the atmospheric type. One went onstream in late 1976 at the Rivesville, W. Va. station of Monongahela Power Co., but has been plagued with operating problems. The other—a demonstration unit at Georgetown University (Washington, D.C.)—started up in late June and has a capacity of 100,000 lb/h of steam.

MANY PLUSES—The advantages of FBC, in which fuel is burned in an air-fluidized bed of limestone, dolomite, sand or other inert material, are well known. (For a detailed review of FBC technology, see *Chem. Eng.*, Aug. 14, 1978, pp. 116-127.)

All sorts of fuels can be burned cleanly with a high combustion efficiency. An FBC boiler occupies less space than a conventional unit of equivalent capacity. And because its combustion temperature is lower, FBC emits lower amounts of nitrogen oxides. Also, it can switch fuels without requiring basic design changes, and it leaves a dry solid waste that could have commercial applications. Further, there are some estimates (see box) showing that FBC systems cost less than coal-fired boilers equipped with scrubbers.

FBC uses more limestone than wet scrubbing—the Ca/S ratio is around 3 for 90% SO_2 cleanup, vs. 1.1 for wet scrubbing—but FBC's dry powdered wastes (CaO, $CaSO_4$, ash), on the other hand, are easier to handle than the wet sludge left over from traditional scrubbing.

Possible uses of FBC waste are being investigated by two DOE-funded programs. In one that involves the U.S. Dept. of Agriculture and West Virginia University, the material is being tested as an agricultural liming compound. The second program, conducted by Valley Forge Laboratories Inc. (Devon, Pa.), will evaluate the waste as a neutralizing agent for municipal wastewater treatment, and as a road aggregate and construction material.

PFB PROS AND CONS—The units now on the market are of the atmospheric type, in which combustion-zone pressure is but a few inches of water. Combustion-zone design is not much different in pressurized (PFB) units,

although the bed is deeper to allow for more heat-transfer area. This is because the combustion rate is higher, since more air happens to be present at the higher pressures.

There are other differences between AFB and PFB designs. In the former, hot flue gases are vented, whereas PFB units use the gases to operate a turbine that supplies power to the feed-air compressor. And instead of limestone, PFB boilers employ mostly dolomite as a sorbent.

Some FBC experts believe that commercial PFB units will be less expensive and more efficient than atmospheric ones. And tests conducted by Exxon Research and Engineering at a PFB pilot plant in Linden, N.J., show lower levels of NO_x and SO_2 in PFB emissions.

The Exxon tests, done for EPA, also suggest that sorbent utilization is higher than in AFB units, probably because of a deeper combustion zone and longer gas-residence times. The company reports a Ca/S ratio of 1.5 to 2 for a sulfur retention of 94 to 98%, vs. a Ca/S ratio of 3.5 to 7 for a sulfur retention of 91 to 98% in a typical AFB boiler.

The PFB advantages, however, may not reach the marketplace for some time. "PFB may not be ready until the 1990s," says Steven I. Freedman, chief of the direct-combustion branch of DOE's Fossil Fuel Utilization Div. According to Robert Gamble, manager of development of Foster Wheeler Energy Corp.'s engineering department, "There are serious technical problems that must be overcome before PFB can be commercialized."

Engineers still haven't found a cost-effective way to clean up hot flue gases before they reach the turbine. "If anything more complicated than cyclones are needed to adequately clean this gas, the system may be too expensive," says René Bertrand, of Exxon Research and Engineering.

Curtiss-Wright Corp. (Wood Ridge, N.J.), which has been commissioned by DOE to build a 13-MW PFB pilot plant, is attempting to remove particulates from flue gas by heating air in the combustion zone and blending it with gas that has been cleaned in cyclones (Fig. 1); the combined stream then goes to a turbine. Air leaving the turbine is sent to a heat-recovery boiler. Steam produced here will drive another turbine for additional power.

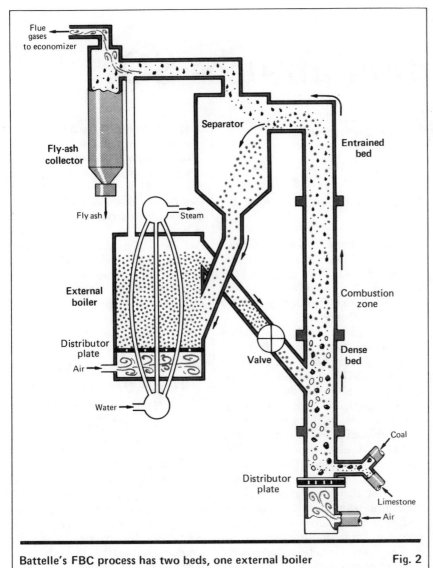

Fly-ash collector

Flue gases to economizer

Separator

Entrained bed

Fly ash

Steam

External boiler

Combustion zone

Distributor plate

Air

Valve

Dense bed

Water

Coal

Distributor plate

Limestone

Air

Battelle's FBC process has two beds, one external boiler   Fig. 2

Wells is promoting for use in secondary oil-recovery steam generators.

The system has a dense bed of silica and an entrained bed of finely ground limestone (see Fig. 2). High-velocity air (about 30 to 40 ft/s) fluidizes the dense bed and fuel in the combustion zone, and the entrained bed is blown out, passing through a collector that recovers sulfated limestone. This hot material then enters an external boiler, in which air velocity is low (1 to 2 ft/s) and little or no combustion occurs. Some solids are recycled to the dense bed.

Battelle claims the following advantages: Because combustion is kept separate from heat exchange, the combustion zone is smaller, and turndown, startup and shutdown are smoother; system response to load variations is better; corrosion has been reduced in the heat-exchanger tubes; and limestone utilization is higher because of long residence times.

Researchers at Lehigh University have been experimenting with a centrifugal fluidized bed (CFB) for the last five years. Although the system is far from commercialization, the University asserts that the CFB's relatively high fluidizing velocities result in much higher rates of reaction per unit volume than can be achieved in conventional beds Also, power output of the CFB can be varied over an extremely wide range by changing both the speed of rotation of the bed and the fluidizing velocity.

The CFB uses a cylindrical distributor that rotates about its vertical axis of symmetry. The circular motion forces the bed material into the annular region of the distributor, and the bed is fluidized against the centrifugal forces by gas that is flowing radially inward through the distributor's porous surface.

Reginald I. Berry

Coal-feeding problems, which have plagued some AFB units (notably the one at Rivesville) may present an even bigger challenge to PFB boilers.

But despite the difficulties, PFB development work continues. In addition to the IEA and Curtiss-Wright projects, American Electric Power Service Corp. (New York City) plans to build a PFB system; the 170-MW plant will be operational by 1985.

ADVANCED FBC SYSTEMS—Among the FBC designs that depart from the typical pressurized or atmospheric configurations, perhaps the most interesting is Battelle's Multi-Solid fluidized-bed-combustion approach (*Chem. Eng.*, Aug. 27, p. 47), which Struthers

Citrate solution absorbs SO_2

This flue-gas-desulfurization technique uses steam

to regenerate a liquid absorbent

and produce a concentrated stream of SO_2

James Farrington and Sune Bengtsson, Flakt, Inc.

☐ The Flakt-Boliden process can remove 90% of the SO_2 from flue gas, using a sodium citrate solution as an absorbent.* This solution can be regenerated via steam stripping to produce a concentrated stream of from 25 to 90% SO_2 that can be used directly in a sulfuric acid plant; reduced to form elemental sulfur (e.g., in a Claus plant); or liquefied and stored to be sold or used later.

The process is based on work begun in the early 1970s by the Norwegian

*For U.S. Bureau of Mines citrate process see the following article.

technical institute SINTEF, Boliden AB (Sweden) and AB Svenska Flaktfabriken (Sweden) and its U.S. subsidiary, Flakt, Inc. Now it will be demonstrated at the Tennessee Valley Authority's Colbert steam plant. Process equipment is in place and startup is scheduled for August. Here, the system will treat 4,000 acfm (equivalent to a 1-MW power plant) of flue gas from a coal-fired boiler.

The Electric Power Research Institute (Palo Alto, Calif.), which is sponsoring this trial, is now planning a 100-MW demonstration in order to develop a commercial flue-gas-desulfurization technology that will offer utilities an alternative to conventional lime/limestone-based scrubbing. These so-called throwaway systems generate large volumes of waste sludge. Several utilities, particularly those in eastern urban areas, cannot operate these processes because there is a shortage of landfill area required for disposing of the waste.

In contrast, the Flakt-Boliden technology avoids many of these problems. No sludge is produced; the absorbent is a nontoxic clear solution that prevents the abrasion caused by scrubbing with slurries; makeup reagent demand is low, since the absorbent is regenerated; and most of the material-handling operations are eliminated.

Also, the system can accept a wide range of SO_2 concentrations—from 3,000 ppm produced by a coal-fired boiler to over 3% discharged from a nonferrous smelter.

In a pilot plant at the Boliden works in Sweden, copper and lead smelter flue gases having SO_2 concentrations ranging from 0.5% to 5% were treated. (Here, the concentrated SO_2 stream

produced is liquefied, stored, and then later used in acid plants.)

DISSOLVED SO₂—Absorption of SO_2 in aqueous solutions is determined by pH—as pH increases, absorption increases. But dissolution of SO_2 forms the bisulfite ion (HSO_3^-); this results in a decrease in pH, a phenomenon that limits the absorption of sulfur dioxide. The Flakt-Boliden system incorporates a buffering agent (citrate) to inhibit the pH decrease; therefore, high SO_2 loadings and, as a result, high SO_2 removal can be achieved. The process can be characterized by the following equations:

$$SO_2\ (g) \rightleftarrows SO_2\ (aq)$$
$$SO_2\ (aq) + H_2O \rightleftarrows H^+ + HSO_3^-$$
$$H^+ + Ci^= \rightleftarrows HCi^-$$
$$H^+ + HCi^- \rightleftarrows H_2Ci^-$$
$$H^+ + H_2Ci^- \rightleftarrows H_3Ci$$

Citrate is denoted by Ci. The three citrate dissociation equilibria provide excellent buffering to keep the pH in the optimum range for absorption (3-5). For a specific application, the exact pH needed (determined by the SO_2 concentration in the feed) is maintained by adding sodium hydroxide or soda ash to form the sodium citrate absorbent solution.

Since flue gas contains oxygen, the SO_2 absorbed by the system can oxidize to form sodium sulfate. But the process does minimize oxidation, and pilot-plant tests have reported oxidation rates less than 1% of the total absorbed SO_2.

According to theory, the low oxidation rates are the result of the formation of chelate complexes between transition-metal ions and citrate ions. The metallic ions, normally present in flue gas, enhance oxidation, but in the complex they are effectively blocked.

PACKED COLUMNS—The flowsheet illustrates the Flakt-Boliden route. Flue gas containing SO_2 passes through a prescrubber, where it is washed with water to remove particulates and chlorides. The gas then

Desulfurization economics

Basis
2.1 million acfm flue gas (500-MW coal-fired boiler—4%-sulfur coal); 90% SO_2 removal; Midwest location, 1979 costs

Capital costs	$, million
Flue-gas handling	6.5
Absorber stripper	18.5
Steam processing	7.0
Citrate recovery	2.0
Total cost	34.0

Raw materials	
Soda ash	1,735 lb/h
Citric acid	52 lb/h

Utilities	
Steam (50 psig)	22,000 lb/h
Steam (350 psig)	370,000 lb/h
Process water	560 gpm
Cooling water	22,000 gpm
Electric power (includes flue-gas fan)	9,100 kW

Byproducts	
Sodium sulfate (decahydrate)	3,800 lb/h
Elemental sulfur (with an SO_2-reduction process)	24,000 lb/h

Originally published June 16, 1980

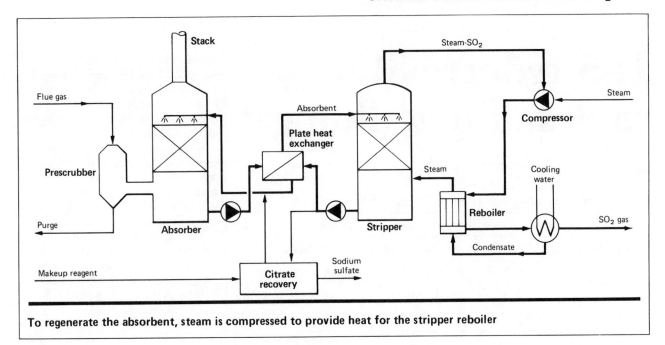

To regenerate the absorbent, steam is compressed to provide heat for the stripper reboiler

enters the bottom of the absorption tower, which is a packed column. For the TVA unit, this tower, constructed of carbon steel and lined with FRP, is 3 ft. 6 in. dia. and 40 ft. high.

In the tower, the flue gas flows countercurrently to the absorbent. Absorption of SO₂ takes place at atmospheric pressure between 100-130°F. The clean gas leaves via a stack.

Before entering the stripper, absorbent now rich with SO₂ is heated in a plate heat exchanger with regenerated absorbent leaving the stripper.

The stripper is also a packed column of FRP. Here, the SO₂-rich absorbent enters the top of the tower and flows countercurrently to low-pressure steam that accepts the sulfur dioxide from the citrate solution, reversing the chemical reactions given above. The regenerated absorbent flows through the plate exchanger and passes to the absorption column.

A small slip stream, about 1% of the clean absorbent, is processed by the citrate recovery unit. Sodium citrate and sodium sulfate are selectively crystalized to recover the citrate absorbent and purge the sodium sulfate salts.

VAPOR RECOMPRESSION — In order to minimize costs and steam requirements, steam that is used for stripping also is used to provide heat in the stripper reboiler.

The steam-SO₂ mixture leaving the stripper is mixed with makeup steam. The combined gas is compressed slightly and sent to the reboiler. Here,

it flows through the tube side, providing heat while the steam is partially condensed.

This stream enters the condenser, where more water comes out of the vapor phase, and SO₂ gas is recovered. (SO₂ concentration will depend on the temperature of the cooling water.)

The condensate, which contains little sulfur dioxide, leaves the condenser and flows to the shell side of the reboiler where it is vaporized. The steam produced is used in the stripper to regenerate the citrate absorbent.

Steam is the major input to the Flakt-Boliden process, and the vapor recompression scheme not only minimizes steam consumption but also makes it attractive to integrate the process with an existing power plant or non-ferrous smelter.

The stripper can operate either at atmospheric pressure or under vacuum. For a specific application, e.g. a power plant, the source of steam can be the exhausted or extracted steam from a low-pressure turbine, and the flue-gas-desulfurization route can function to reduce the duty of the power plant's condenser.

Various vapor-recompression operations can use intermediate-pressure steam (100-400 psig) to reduce steam consumption further. Waste-heat boilers and sulfuric-acid plant waste heat can be tapped as a steam source in nonferrous-metal smelting facilities.

Economic data for the Flakt-Boliden process are presented in the table.

Capital cost estimates represent a turn-key installation that will yield a 30% SO₂ product. Each process unit includes foundations, support steel, electrical and instrumentation systems, process piping, field installation, project engineering and management, and initial startup. Costs for producing a sulfur byproduct are not included; however, these have been estimated at about $15/kW for an SO₂ reduction process.

All raw-material and utility requirements are based on 90% SO₂ removal at the maximum capacity rating of the boiler. The steam usage corresponds to the vapor recompression scheme shown on the flowsheet. This can be further reduced depending on the site-specific conditions.

Also, if warranted by local market conditions, it is possible to add a drying step to the sodium sulfate recovery unit so as to obtain an anhydrous sodium sulfate. However, since the formation rates of sodium sulfate are very low, the quantity produced may not justify the investment.

Reginald Berry, Editor

The Authors

James Farrington is the Product Manager of Flue Gas Desulfurization Systems for Flakt, Inc. (Old Greenwich, Conn.). He is a member of AIChE and holds a B.Ch.E. degree from the University of Notre Dame.

Sune Bengtsson is a Project Manager for FGD research and development at Flakt's Industrial Div. laboratories (Vaxjo, Sweden). He has a D.Sc. in chemical engineering from the Technical University of Munich, West Germany.

H₂S reduces SO₂ to desulfurize flue gas

This citrate-based system produces pure elemental sulfur; and it can treat off-gases from refinery Claus sulfur-recovery units and fluidized catalytic crackers as easily as it absorbs the stack gas from a powerplant.

R. S. Madenburg and T. A. Seesee, Morrison-Knudsen Co.

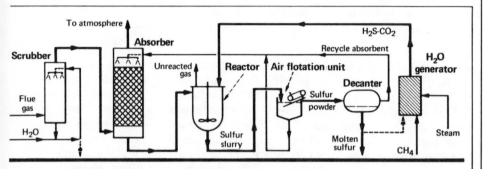

Simplified scheme for a powerplant application   Fig. 1

☐ The U.S. Federal Bureau of Mines (BuMines) has developed and now commercialized a flue-gas-desulfurization system that uses a citric acid buffered solution to absorb SO₂. This absorbed sulfur dioxide is then reacted with hydrogen sulfide in order to generate elemental sulfur.

BuMines technology is particularly suited to the abatement of SO₂ from Claus sulfur recovery units, process heaters, fluid catalytic cracking units, and industrial boilers burning high-sulfur petroleum coke or residual fuels. And the availability of hydrogen sulfide can make the economics of refinery applications very attractive.

The process offers a number of other advantages:

■ The elemental sulfur produced is more than 99.5% pure, and suitable for acid plant feedstock.

■ For refinery applications, sulfur recovery efficiencies can be in excess of 99%.

■ Since the absorbing solution is clear, the process is free of scaling or plugging problems.

■ The citric acid buffer is a non-toxic, biodegradable and environmentally acceptable material.

■ Under process conditions, the citric acid is chemically stable, and very little decarboxylation/decomposition occurs.

■ The sulfur product and the Glauber's salt (Na₂SO₄•10H₂O) byproduct do not contain vanadium or other heavy metals.

IN OPERATION—The BuMines process, which was devised in 1968 for the nonferrous smelting industry, was first tested in a pilot program completed in 1976.

Recently, in October 1979, construction was completed on the first commercial unit, through a joint effort between the U.S. Environmental Protection Agency, BuMines and industry. The citrate-based facility was designed to treat 180,000 scfm of tail gas from a powerplant burning 2.5-4.5%-sulfur coal. This powerplant, located in western Pennsylvania, is owned and operated by St. Joe Minerals Co. (New York).

By midsummer, the powerplant is expected to be at or near design loading. Then operational testing and documentation of the BuMines process will be accomplished through a formal one-year demonstration program.

During this demonstration, materials and equipment selected for a 25-year plant life will be tested. Accordingly, the unit is arranged as a single train with inline redundancy for all major components.

All vessels, process reactors and tanks are fabricated of carbon steel and lined with flaked-glass vinylester. Process piping systems are constructed of polypropylene and Teflon-lined pipe. Trim, such as agitator blades, pump impellers, internal linkage and wetted structural supports, is all of corrosion-resistant materials such as rubber-lined carbon steel, Hastelloy and Inconel. (Special-alloy materials have been used in the demonstration unit because chlorides—derived from the combustion of coal—are present in the citrate solution. In applications where chlorides would not be present, less-exotic materials can be utilized.)

TWO MODES—The BuMines route has two slightly different schemes: one for powerplant applications (where H₂S must be generated) and one for refinery applications (where H₂S is available). Nevertheless, the process follows four basic steps: the flue gas is scrubbed; SO₂ is absorbed; SO₂ is reduced to elemental sulfur; and the sulfur is separated from the regenerated absorbent.

For a powerplant application (see flowsheet in Fig. 1), flue gas enters a venturi scrubber, where the gas is cooled, and particulates, sulfuric acid mist and chlorides are removed.

The stream then enters the bottom of the absorber (a packed column) and flows upward countercurrent to the citrate absorbent. Here in the aqueous solution, absorption is pH-dependent, but as SO₂ enters solution the pH can decrease. The citrate acts as a buffer,

Originally published July 14, 1980

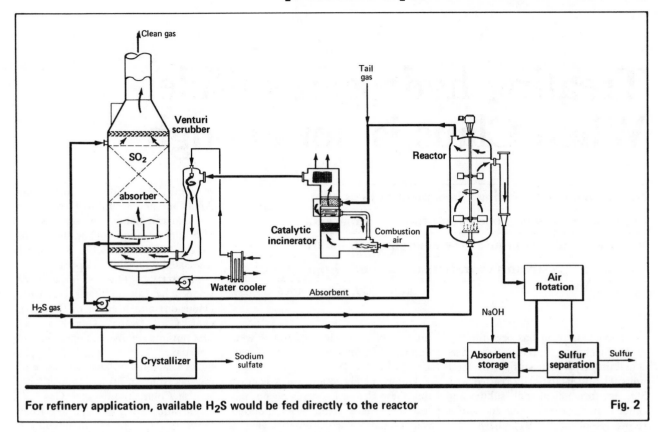

For refinery application, available H₂S would be fed directly to the reactor   Fig. 2

maintaining pH in the desired range for absorption.

After absorption, the clean gas goes to a stack. Absorbent now loaded with sulfur dioxide passes to a closed reactor that is agitated and operates at 125-130°F and about 15 psig. H₂S from the hydrogen sulfide generator is introduced, and the following occurs:

$$SO_2 + 2H_2S \longrightarrow 3S^0 + 2H_2O$$

This reaction regenerates the absorbent. The sulfur formed precipitates, and the unreacted H₂S is recycled to the powerplant for combustion.

The slurry of sulfur and regenerated absorbent that leaves the reactor is separated into two streams by air flotation: one stream is a concentrated sulfur (10%) slurry, the other is the absorbent. Sulfur is separated from the concentrated slurry by melting the regenerated sulfur and decanting the remaining absorbent.

This absorbent, which will be recycled, will contain a small amount of sodium sulfate. (During absorption, some of the SO₂—1.5% of the amount absorbed—is oxidized, forming sulfate ions. NaOH is added to the absorbent to neutralize any sulfuric acid generated by forming sodium sulfate. Before the absorbent is recycled to the absorp-

tion tower, a small stream is sent to a crystallizer, where the sulfate is removed as Glauber's salt.)

To generate the needed hydrogen sulfide, superheated product sulfur is catalytically reacted with steam and methane. The reaction is:

$$4S + 2H_2O + CH_4 \longrightarrow 4H_2S + CO_2$$

The hydrogen sulfide formed is 75-80% pure, with the balance being carbon dioxide and trace sulfur compounds, such as COS and CS₂. A patent is pending on the H₂S generator designed by Morrison-Knudsen Co. (Boise, Idaho) and Home Oil Co. (Calgary, Alta.) for the demonstration unit.

For refinery applications where high-purity H₂S is available from the amine strippers, the BuMines process scheme would be slightly different (see Fig. 2). Here, the H₂S from the refinery would enter the sulfur precipitation reactor. The unreacted gas would then be combined with Claus-plant tail gas. This stream would pass through a catalytic (or thermal) incinerator, where H₂S would be oxidized to SO₂. The stream would then enter a venturi scrubber and follow the basic process steps.

Sulfur production from a citrate unit would be slightly more than three times the equivalent sulfur contained in the Claus tail gas. The additional sulfur would come from the consumption of hydrogen sulfide.

TREATING COSTS—For a BuMines citrate facility associated with a 100-ton/d refinery Claus sulfur-recovery plant, the direct annual operation cost would be approximately $25/ton of recovered sulfur product. Initial capital investment for such a plant would be approximately $3 million (however, this should be considered on an individual basis, since the quality of H₂S, siting restraints, and performance criteria can vary.)

For a powerplant application, the cost would be approximately $140/kW based on a 500-MW facility and 3.5%-sulfur coal.

Reginald Berry, Editor

The Authors

R. S. Madenburg is technical director for the Morrison-Knudsen Power Group. He has a B.S. in mechanical engineering from the U.S. Merchant Marine Academy, is a registered Professional Engineer in California, and a member of AIChE, ASME, ANS, and Air Pollution Control Assn.

T. A. Seesee is a senior chemical engineer with the Morrison-Knudsen Power Group. He has B.S. and M.S. degrees in chemical engineering from the University of Idaho, is a registered Professional Engineer in the state of Idaho, and is a member of AIChE and ACS.

Treating hydrogen sulfide: When Claus is not enough

New systems for cleaning Claus plant tail gas offer

improved economics: one even produces a marketable

fertilizer. And techniques now emerging can handle

the dilute H$_2$S streams characteristic of synfuel operations.

Reginald I. Berry, Assistant Editor

☐ The Claus plant has been the traditional technology used to treat gases containing hydrogen sulfide. In these units, H$_2$S is catalytically reacted with SO$_2$ to form elemental sulfur and water. This process is efficient: sulfur recoveries of up to 97% have been achieved by using multiple stages.

Nevertheless, the current air-pollution regulations, promulgated in 1978, require a better than 99% recovery. This has forced industry to add on systems to clean Claus tail gas. A number of these cleanup routes were developed in the early 70s (see *Chem. Eng.*, May 15, 1972, p. 66).

Unfortunately, many users confirm that these cleanup systems do not pay for themselves. And even popular routes—particularly those having a hydrogenation step that requires fuel—are becoming increasingly expensive to operate, states one expert.

Now, some processes are being commercialized that, if they do not yield a valuable product, offer better economics than older cleanup routes. And some new techniques are emerging that can handle gas streams with H$_2$S concentrations below the normal range of Claus units but particularly associated with synfuel operations.

FERTILIZER—The ATS process, developed by Coastal States Gas Corp. (Houston) and licensed by The Pritchard Corp. (Kansas City, Mo.), was first put into commercial operation this February. The system, which produces a salable fertilizer, is treating the tail gas from an 80-long-ton/d Claus unit at Colorado Interstate Gas Co.'s (Colorado Springs) Table Rock, Wyo., facility. The ATS plant will produce 18,140 metric tons/yr of ammonium thiosulfate, (NH$_4$)$_2$S$_2$O$_3$, fertilizer solution.

The new process (see Fig. 1) first incinerates Claus tail gas, generating a stream containing SO$_2$. This gas is

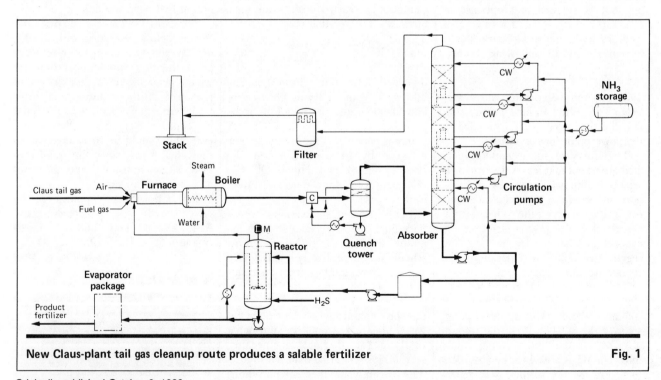

New Claus-plant tail gas cleanup route produces a salable fertilizer   Fig. 1

Originally published October 6, 1980

cooled and sent to a packed tower where the SO_2 is absorbed in an aqueous ammonia solution. Ammonium sulfite, $(NH_4)_2SO_3$, and ammonium bisulfite, NH_4HSO_3, are formed in the column, which is staged to maintain temperatures and NH_3 concentrations. Clean gas exits through the stack. The ammonium sulfite/bisulfite solution from the bottom of the tower enters a reactor that is sparged with H_2S. Here the fertilizer is formed:

$$H_2S + NH_4HSO_3 + (NH_4)_2SO_3 \rightarrow$$
$$3/2 \, (NH_4)_2S_2O_3 + 3/2 \, H_2O$$

After the reaction, the product is concentrated in a falling-film evaporator to produce a solution that contains 12% nitrogen and 26% sulfur.

The advantage of the route is said to be the fact that even considering the cost of the ammonia consumed, it makes a salable product that has a higher value (about $80/ton of solution) than sulfur itself. However, the product may only be marketable in areas, like the West and Southwest, where the soil is deficient in sulfur.

BUFFERING—Stauffer Chemical Co. (Westport, Conn.) is now licensing its phosphate process that first burns Claus tail gas and then absorbs the SO_2 formed in a sodium phosphate-buffered aqueous solution. The SO_2-laden absorbent is reacted with H_2S to form pure (99.9%) elemental sulfur.

The phosphate process is similar to citrate-based technology such as the Flakt-Boliden and the Bureau of Mines routes (see *Chem. Eng.*, June 16, p. 88, and July 14, p. 88, respectively). Compared to amine absorption/stripping systems for tail gas treatment, Stauffer's method is said to have equal capital costs, but its operating costs are given as about 75% of the amine units', since its utility requirements are lower.

SYNFUEL PROBLEMS—Coal conversion projects will produce gas streams that are relatively dilute in H_2S (25% or less) and high in CO_2 content. Now processes are being developed to handle these applications.

Union Carbide Corp.'s (New York) UCAP process can be added onto or integrated with Claus, and it is said to handle gases with a low H_2S content and still meet air pollution regulations. The integrated route, which can eliminate the need for a multiple-stage Claus unit, was first commercialized in September 1978 at Gulf Oil's Venice,

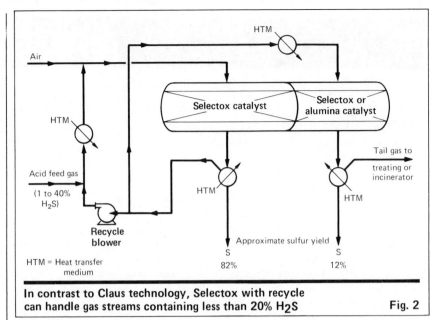

In contrast to Claus technology, Selectox with recycle can handle gas streams containing less than 20% H₂S   Fig. 2

La., refinery; this was accomplished through a joint effort between Union Carbide and Ford, Bacon & Davis Texas, Inc. (Dallas).

In the integrated process, acid gas is not burned as in the usual Claus plant. The feed is instead mixed with recycled SO_2. The combined gas is heated and then sent to a Claus reactor.

After reaction, sulfur is condensed, and the gas is oxidized to convert all remaining sulfur components to SO_2. The stream is cooled and dehumidified in a packed column. Then the SO_2 is absorbed in a trayed tower with an aqueous solution of triethanolamine sulfite. CO_2 is not absorbed, and exits with the stack gases. The rich absorbent is thermally stripped under vacuum to yield SO_2 which is recycled to mix with the feed. Heat-stable salts (sulfates and thiosulfates) formed during absorption and stripping are removed by ion exchange.

UCAP does require the use of acid-resistant materials (such as 304 and 316 stainless steels) in the portion of the system that comes in contact with SO_2. But the SO_2 produced by the process is said to be of high purity, so it can be sold as a liquid or used in a sulfuric acid plant, if UCAP is not integrated with a Claus facility.

A REFINEMENT—The Science and Technology Div. of Union Oil Co. of California (Brea, Calif.) and The Ralph M. Parsons Co. (Pasadena, Calif.) have developed a new process from existing technology. The new route, called Selectox with recycle, is

based on Union Oil's Selectox catalyst, and it covers the "no-man's land" between 1% and about 20-25% H_2S where a standard Claus system is either ineffective or marginally effective, according to David K. Beavon, vice-president, technical, for Parsons.

The new process is similar to Claus in that it uses a series of catalytic stages (up to three). But, in contrast, there is no thermal reaction stage at the front end. Instead, H_2S is oxidized directly at a relatively low temperature.

The acid gas (see Fig. 2) is fed directly to the Selectox catalyst after being mixed with a stoichiometric amount of air. The feed temperature is about 350°F, against 450°F for a standard Claus catalyst. The reaction heats the gases, which are then cooled to condense elemental sulfur.

Part of the cooled gas is recycled to the feed in order to control the reaction temperature. An H_2S concentration of 1% causes the temperature to rise 70°F during reaction. (Above 5% H_2S, the reaction must be stopped at about 700°F to avoid corrosion problems.) The recycled gas dilutes the H_2S content to limit the temperature rise. Gas that is not recycled continues to a second and third stage, which may use Selectox or less-expensive conventional alumina catalysts.

The system has been tested on gases containing up to 20% H_2S in a pilot plant at Union Oil. Beavon expects that a plant to process gas with a 25% H_2S content would cost about one-quarter less than a Claus plant.

LETTERS

Sulfur-recovery knowhow

Sir: This is in regard to the article, 'Treating hydrogen sulfide: When Claus is not enough," on pp. 92-93 of your Oct. 6, 1980, issue, concerning its discussion of the Selectox process of The Ralph M. Parsons Co. I wish to point out that this type of process has been used for many years.

The process was mentioned in the literature in 1953. Our company conducted development work and evaluated applications at an Amoco plant. One of the early commercial applications was reviewed in the literature in the Apr. 1, 1963, issue of *Chemical Engineering.* The flowsheet in the 1963 *Chemical Engineering* reference is quite similar to the one in the 1980 reference. Preheated acid gas along with air goes to the first catalytic reactor; the effluent gas is cooled to remove sulfur, then is reheated before going to the second catalytic reactor to form additional sulfur, which is removed by condensation. Various methods have been used to control the temperature rise in the first catalytic stage.

Other publications concerning the process for direct oxidation of hydrogen sulfide in a catalytic reactor include: *Oil & Gas Journal*, July 20, 1959; and Gas Conditioning Conference, U. of Oklahoma, Apr. 4-5, 1967. I am surprised that in your

article you referred to this method as a "new process."

ROBERT L. REED
Amoco Production Co.
Tulsa, Okla.

Comment from Parsons:

No one at Parsons or at Union Oil [also involved in the Selectox process development] claims to have invented the broad concept of direct catalytic oxidation of hydrogen sulfide to sulfur; the work of Amoco along this line is recognized and respected.

We do think that we have made some developments that significantly aid the industrial application of direct oxidation. Among these are Union's development of a catalyst that works at temperatures lower than any mentioned in the 1963 article, or any others known to us; the catalyst is remarkably selective in "ignoring" compounds other than hydrogen sulfide. Also, we know of no other Claus-type operation in which the temperature is moderated with gas recycle.

D. K. BEAVON
The Ralph M. Parsons Co.
Pasadena, Calif.
Respondent Beavon won this magazine's Personal Achievement Award in 1980, partly in recognition of his contributions to sulfur-removal technology and control of sulfur emissions; for details, see our Dec. 15, 1980, issue, pp. 54-58. — ED

Originally published February 23, 1981

NO$_x$ controls: Many new systems undergo trials

The roster includes methods to nip nitrogen oxides at the combustion end and in flue gases, even though some experts feel that combustion controls alone can meet increasingly stricter U.S. emissions standards.

☐ Last year, when Joy Industrial Equipment Co.'s Western Precipitation Div. (Los Angeles) licensed a Japanese process for removing nitrogen oxides (NO$_x$) from flue gas, it thought it was one step ahead of the competition for this specialized market. But Joy and its rivals (e.g., Exxon, Babcock & Wilcox, several Japanese firms) may be barking up the wrong tree after all, because flue-gas treatment, which is a complement—not an alternative—to combustion controls (e.g., low-NO$_x$ burners) in NO$_x$ removal, may not be needed to meet emissions standards set by the U.S. Environmental Protection Agency.

Indeed, combustion controls alone may be able to do the job, says G. Blair Martin, a program manager at EPA's Industrial Environmental Research Laboratory (Research Triangle Park, N.C.). These controls already meet new-source performance standards (NSPS) for coal-fired boilers—0.4-0.6 lb/million Btu. And while flue-gas treatment can cut this by about 80%, it is expensive, costing around $40/kW for a new, coal-fired, utility-size boiler (about 4-5% of total plant cost). The cost is even higher for smaller boilers.

EPA is pushing for more-stringent emissions standards—0.2 lb/million Btu, or about 150 ppm in the flue gas—for coal-fired boilers in the next two or three years. But it may be possible to meet even those levels with combustion controls alone, since the

devices, notes Martin, have yet to reach their NO$_x$-removal potential.

The agency's $12-million NO$_x$ research and development budget for 1981 reflects this credo. About 95% of the amount will go to develop and demonstrate combustion modification technology and low-NO$_x$ burners; only

two flue-gas treatment demonstration projects are included.

TALK OF STANDARDS—The situation is not the same everywhere. Plenty of work on flue-gas NO$_x$ removal takes place in Japan, where the government imposed the world's most stringent NO$_x$ standards in six prefectures in 1979. The controls include 60 ppm for new gas-fired boilers; 130 ppm for large oil-fired boilers; and 100 ppm for furnaces in ammonia plants and ethylene crackers.

Currently, the EPA's NSPS cover only large boilers of 250 million Btu/h or more. The standards are: 0.2 lb/million Btu for natural gas; 0.3 lb for oil,

---

NO$_x$ sources and how to deal with them

There are two sources of NO$_x$ in the combustion process. One is the formation of oxides of nitrogen by the heat of combustion, which may be controlled by lowering the flame temperature. Control methods include reduction of excess air; recirculation of flue-gas; and the use of special burners and combustion modifications to burn a fuel-rich mix initially, then the addition of more air later to complete fuel combustion (so-called "staged combustion").

The other source of NO$_x$ is fuel nitrogen, found particularly in coal and some oils, that is released by combustion. This process is not temperature-sensitive, but may be controlled by limiting the availability of oxygen.

Post-combustion or flue-gas treatment systems have been developed and commercialized by a number of Japanese companies because of Japan's special need. Generally, they involve selective catalytic reduction of NO$_x$ to nitrogen and water by injection of ammonia in the presence of a catalyst. A variation on this theme is the Thermal DeNox system developed by Exxon Research and Enginering Co. (Florham Park, N.J.): it uses ammonia, but no catalyst (*Chem. Eng.*, June 19, 1978, p. 85).

One U.S. concern about these systems is that they have been used mainly for oil and gas, whereas EPA's emphasis is on coal-fired boilers, which represent a significant and growing segment of U.S. boilers. Another worry is the solid ammonium sulfate and liquid ammonium bisulfate that form inside the systems; the salts can cause corrosion and plugging problems in equipment, plus contamination of catalysts.

Originally published March 9, 1981

and 0.6 lb for pulverized bituminous coal. These numbers translate into about 150 ppm (in the flue gas) for gas, 225 ppm for oil and 450 ppm for coal (in the cases of oil and coal, the ppm content depends on the particular fuel).

EPA plans to extend these standards to industrial boilers in the 50-250-million-Btu/h range. An EPA spokesman says they are not likely to go into effect before the end of 1981 and will be achievable with existing combustion-modification systems. The agency also is working on a NO_x standard for cement plants.

Only the California Air Resources Board (CARB) has so far imposed NO_x controls stricter than federal requirements. It has ruled that oil-fired power plants in the Los Angeles area must reduce NO_x emissions by 80% from present levels. This will require installation of catalytic flue-gas treatment systems, which a CARB spokesman says will reduce emissions to 0.06 lb/million Btu. Southern California Edison Co. and the Los Angeles Department of Water and Power are contesting the rule in court, on the ground that the technology is not proven and the cost too high (*Chem. Eng.*, Dec. 29, 1980, p. 10).

CARB also plans statewide NO_x controls of 0.8 lb/million Btu on refinery boilers and heaters. Alan Goodley, chief of CARB's Energy Strategy Development Branch, says this would be about a 50% reduction from present levels and could be met by reducing excess air, and some combustion modification. CARB has also approved a model rule that calls for a 50-60% reduction of NO_x emissions from glass furnaces, whose uncontrolled emissions average about 1.5 lb/million Btu. The board is recommending that the South Coast Air Quality Management District (Los Angeles basin) adopt the rule (CARB does not normally enforce rules, but develops models for adoption by local air pollution control authorities).

DOING IT WITH BURNERS—An important part of EPA's burner program involves a "distributed mixing burner" developed by Energy and Environmental Research Corp. (Irvine, Calif.). The work is divided into two projects for purposes of field evaluation. The industrial-boiler project (35-175 thermal MW) is being conducted by EER, with Foster Wheeler Energy

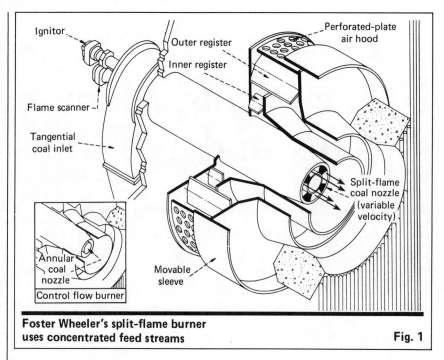

Foster Wheeler's split-flame burner uses concentrated feed streams   **Fig. 1**

Corp. (Livingston, N.J.) as a major subcontractor. The utility-boiler project (up to 300 MW electrical) is being handled by Babcock & Wilcox Co. (Barberton, Ohio), with a subcontract to EER. Both projects are scheduled for completion in late 1983.

EER is testing a full-scale single burner, at its El Toro test site in Santa Ana (Calif.). The dual-register unit, for wall-fired boilers, has three concentric feed systems: pulverized coal, with transport air, is fed through the central tube, and air is fed through the two outer registers. The burner is operated in a highly-staged manner at about 70% stoichiometry. Then tertiary air is added through separate ports to obtain about 20% excess air and complete carbon burnout.

"We hope to achieve emissions of 0.2 lb per million Btu, or about 150-155 ppm, of NO_x," says William H. Nurick, director of marketing for EER. "The results in our research furnace suggest we will meet our goal."

Four of the burners, each rated at 85 million Btu/h, will be installed in a 285,000-lb/h steam boiler at Western Illinois Power Cooperative's Pearl (Ill.) station in April for a two-year, full-scale commercial test. This is being done in cooperation with Foster Wheeler. A 125-million-Btu/h burner of the same design will be tested by Babcock and Wilcox in a 300-MW B & W boiler at an electrical utility;

the site has not yet been selected. A third test of the same burner design is also planned by EPA on a 600-MW boiler, with Foster Wheeler as the prime contractor.

Foster Wheeler and B & W are also working independently to develop low-NO_x burners for wall-fired, coal-burning boilers to meet EPA's future requirements. These are improvements on existing boilers the companies now market to meet current NO_x standards.

Foster Wheeler's approach is a split-flame burner. Coal is injected through an annular nozzle in concentrated streams (typically four), each of which forms an individual flame. This minimizes mixing between the coal and primary air, permitting 60-70% stoichiometry up to 6 to 10 ft from the burner. After this, a swirling secondary air stream from the outer annulus completes carbon burnout. Foster Wheeler has already sold a few of these systems and says it has lowered NO_x emissions to 0.45 lb/million Btu, without overfire air or flue-gas recirculation.

B & W markets a coal burner that meets current NO_x standards by delaying mixing of fuel and air. Coal is fed through a central pipe and primary and secondary air through two outer annuli. Using this same burner, the company claims NO_x emissions of 0.15 lb/million Btu in a 35-million-Btu experimental combustion system

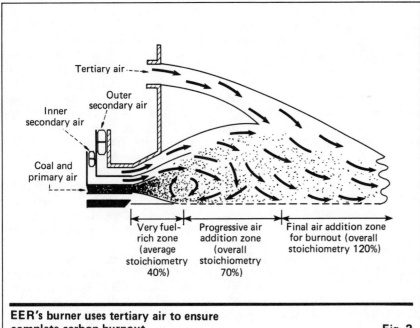

Tertiary air

Outer secondary air

Inner secondary air

Coal and primary air

| Very fuel-rich zone (average stoichiometry 40%) | Progressive air addition zone (overall stoichiometry 70%) | Final air addition zone for burnout (overall stoichiometry 120%) |

EER's burner uses tertiary air to ensure complete carbon burnout

Fig. 2

at its research center in Alliance, Ohio. A spokesman says this is done by interaction between several rows of burners: the lower ones are fired fuel-rich and the upper ones fuel-lean. A commercial test is planned for May at an undisclosed location.

Combustion Engineering Inc. (Windsor, Conn.) has an EPA contract to demonstrate a low-NOₓ concentric firing system based on CE's long-commercial tangential-firing burner. Separate streams of fuel and air are injected from the four corners of a furnace into an imaginary circle in the middle. The delayed mixing of fuel and air reduces NOₓ formation. In the new method, the air is directed away from the coal streams, to delay mixing even longer. CE hopes to cut NOₓ to 0.3-0.35 lb/million Btu in tests planned on a 400-MW utility boiler at the end of 1981.

CE also has a license agreement with Mitsubishi Heavy Industries (Tokyo) for a low-NOₓ burner that divides the coal feed into two streams, one fuel-rich and the other fuel-lean, which are separated from the combustion air by recirculated flue gas. MHI claims NOₓ emissions of 0.13-0.2 lb/million Btu in laboratory tests on high-grade coal. The company is testing U.S. coals under a contract with the Electric Power Research Institute (EPRI), of Palo Alto, Calif., and plans the first commercial test of the burner late this year.

EPRI is also working with CE and B & W on their programs and has a contract with Foster Wheeler on the engineering feasibility of applying the old technique of arch-fired furnace design to the design of new furnaces. The latter work follows an earlier study by KVB of some old furnaces, still in use, in which powdered coal is fed through the top and air is blown from a side wall. Tests revealed that these old furnaces had NOₓ emissions of only about 200 ppm.

Riley Stoker Corp. (Worcester, Mass.) is testing an improved version of its directional-flame burner, in which air is injected through a central slot, and coal, with some air, through two outer slots. The company is also working on a concentric, distributed-mixing type of burner, but declines to give further details.

In the smaller-boiler range, EER has an EPA contract to test a distributed mixing burner on a package boiler, using heavy oil. The burner is also being evaluated for use on steam generators used to recover heavy oil in California's Kern County, where oil producers must obtain NOₓ offsets in order to expand production (*Chem. Eng.*, April 7, 1980, p. 54).

North American Manufacturing Co. (Cleveland, Ohio) is already doing a good business in oilfield steam generators, using lances to introduce secondary air downstream (the burners are standard). KVB is developing a similar concept for process heaters, adding pipes for the secondary air (*Chem. Eng.*, Jan. 26, 1981, p. 17).

John Zink Co. (Tulsa, Okla.), which makes low-NOₓ burners for various sizes of boilers, claims emissions of 50 ppm from heavy oil in developmental work.

FLUE-GAS TREATMENT—EPA is funding two demonstration projects on coal-fired boilers: at Georgia Power Co. (Albany, Ga.), where the Hitachi Zosen Co. process is being evaluated,

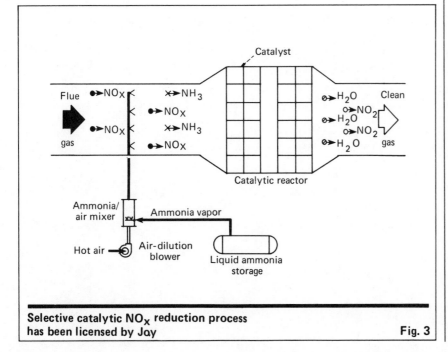

Catalyst

Flue gas → NOₓ, NOₓ — NH₃, NOₓ, NH₃, NOₓ — Catalytic reactor — H₂O, NO₂, H₂O, NO₂, H₂O → Clean gas

Ammonia/air mixer

Ammonia vapor

Hot air — Air-dilution blower — Liquid ammonia storage

Selective catalytic NOₓ reduction process has been licensed by Joy

Fig. 3

and Tampa Electric Co. (North Ruskin, Fla.), the test site for the Shell/UOP method (UOP is the U.S. licensor) for removal of both NO_x and SO_x. The Kawasaki process is being tested on a coal-fired unit by EPRI at Public Service Co. of Colorado's Arapahoe Station (Denver) and by Southern California Edison on an oil-fired unit at its Huntington Beach Station (scheduled startup: late 1981). Los Angeles' Department of Water and Power is scheduled to start testing Exxon's Thermal DeNox system at Long Beach about mid-year.

Both the Hitachi Zosen and Kawasaki processes are catalytic systems that use ammonia as a reducing agent. (Chemico Air Pollution Control Corp., New York, has the North American license for the Hitachi Zosen process). The Shell process uses copper oxide as the sorbent for SO_2, and copper sulfate as the catalyst for reduction of NO_x with ammonia. The copper sulfate is regenerated (with hydrogen) to copper oxide.

Exxon's system, which uses ammonia, but no catalyst, operates within a fairly narrow temperature range at around 1,700-1,800°F, although the reaction temperature may be varied down to around 1,300-1,400°F by introducing hydrogen. If the temperature is too high, ammonia and oxygen react to form nitric oxide. If the temperature is too low, no reaction occurs and ammonia is emitted.

Catalytic systems operate at lower temperatures and within a wide range—around 550-850°F. They can remove 80-90% of the NO_x remaining after commonly used combustion controls, vs. 40-60% for the Exxon system, and they use less ammonia. However, the capital cost is at least triple that of the Exxon system, according to a spokesman for KVB, which installs the Exxon system and is also running EPRI's tests in Denver. Also, the typical catalyst life is currently only one-to-two years.

Marketers of flue-gas treatment systems expect to find business in areas where industry must obtain NO_x offsets in order to build new facilities. KVB has already installed Exxon's system in more than 20 units in two California refineries for this purpose. USA Petrochem Corp. operates a proprietary catalytic system to meet tough NO_x rules at its Ventura (Calif.) refinery. Combustion Engineering, which has a license from Mitsubishi Heavy Industries, and B & W, which shortly expects to conclude a license agreement with Japan's Babcock-Hitachi, are also seeking this type of business.

One way to avoid the hassle with NO_x control is to use a fuel that has low NO_x emissions. Southern California Edison successfully tested methanol in a 26-MW gas turbine and got NO_x emissions of 45-50 ppm, against 200 ppm for distillate fuel in an identical unit. Now, Edison is testing methanol in a 45-MW gas-or-oil-fired boiler. One catch is where all the methanol would come from to meet the potential demand.

Gerald Parkinson

Catalytic burning tries for NO$_x$ control jobs

This technique is said to reduce formation of nitrogen oxides by promoting combustion at lower-than-normal temperatures. One aspect that could stand improvement is the duration of the catalysts currently used.

☐ Tests scheduled to get underway at the end of June on natural-gas-fired package boilers retrofitted for catalytic combustion are expected to show reductions of 90% in emissions of nitrogen oxides (NO$_x$). The tests will be done by Acurex Corp. (Mountain View, Calif.), which is developing catalytic combustion systems under a contract with the U.S. Environmental Protection Agency's Industrial Environmental Research Laboratory (Research Triangle Park, N.C.).

Elsewhere, under two separate programs, catalytic methods of reducing NO$_x$ emissions from gas turbines are being investigated. Westinghouse Electric Corp.'s Combustion Turbine Systems Div. (Concordville, Pa.) is working on one concept under contract to the Electric Power Research Institute (Palo Alto, Calif.). In a much more modest effort, General Electric Co.'s Gas Turbine Div. (Schenectady, N.Y.) is working on a preliminary design of a catalytic combustor for the GE Series 7001 turbine under contract with Southern California Edison Co. (Rosemead, Calif.).

AN EVOLVING TECHNIQUE—Catalytic combustion is still some years from commercialization, but proponents feel it offers a relatively simple and inexpensive way to cut NO$_x$ emissions to very low levels. EPA is sponsoring the Acurex work as part of its program of developing control technology to pave the way for stricter emissions rules. The California Air Resources Board (CARB) has already recommended that catalytic combustion be required in gas turbines installed in southern California on or after Jan. 1, 1989, provided that a review of the technology proves its technical and economic feasibility. Future regulations aside, improved control methods may help industry

Originally published June 15, 1981

obtain pollution offsets to permit construction of new facilities in areas that do not come up to existing environmental standards.

The principle behind catalytic combustion is simple. The catalyst promotes combustion at a temperature lower than normal, reducing the formation of NO$_x$. Its main drawback is the relatively short life of catalysts—a problem that is receiving much attention in the development programs. Also, while catalytic combustion seems promising for such "clean" fuels as natural gas and light distillates, it is not readily adaptable to heavier oils, which are not easily vaporized and tend to foul the catalyst.

Some industry spokesmen, although cautious about the method's prospects,

hope that it may provide a relatively low-cost alternative to the possible imposition of more costly methods, such as selective catalytic reduction (SCR) systems. In these, ammonia is injected into fluegas to reduce NO$_x$ to nitrogen and water. Southern California Edison and the Los Angeles Department of Water and Power are starting court action against CARB's ruling that they must install SCR systems—they contend that the technology is not proved and the cost too high (*Chem. Eng.*, Dec. 29, 1980, p. 10).

"Catalytic combustion would certainly be less expensive than post-combustion treatment," says Leonard Angello, a project manager with EPRI. "Here [in a gas turbine], we are talking about catalyst sections of several inches thick by maybe 12 in. diameter. "With an SCR system, you are looking at catalysts as big as boilers—it's a whole chemical plant at the back of your plant."

Current EPA standards can be met by combustion modifications (*Chem. Eng.*, Mar. 9, p. 39). The rule for natural-gas-fired, 250-million-Btu/h boilers is 0.2 lb/million Btu, which is about 170 ppm at 3% excess oxygen. A similar standard has been under consideration for smaller industrial units. (The standards can be met by operating with low excess air and staged combustion.) There is also a

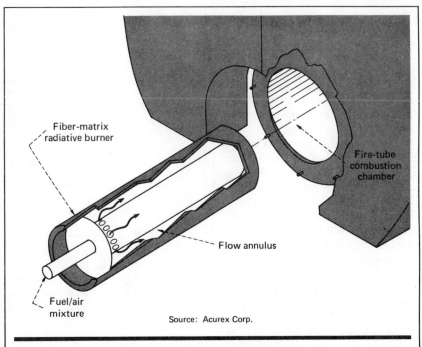

Fiber-matrix radiative burner

Fire-tube combustion chamber

Flow annulus

Fuel/air mixture

Source: Acurex Corp.

Fire-tube burner fits into combustion chamber of boiler　　　　　**Fig. 1**

regulation for coal syngas of 0.5 lb/million Btu (about 425 ppm).

The standards for gas turbines are 75 ppm for engines of 10,000 hp or more (achievable by water or steam injection to reduce the temperature), and 150 ppm for 1,000-10,000 hp turbines (obtainable by combustion modifications).

EPA has proposed to rescind the 75-ppm standard for industrial turbines only; the agency is not expected to make a decision on this before the end of July.

California's toughest standards, which would limit utility turbines in the Los Angeles area to 25 ppm (natural gas) and 40 ppm (distillate fuel), have been proposed by CARB for adoption by the South Coast Air Quality Management District. The possible future standard, involving catalytic combustion, would set limits of 12 ppm and 20 ppm.

BLUEPRINT FOR TESTS—If Acurex's package boiler tests go as planned, NO_x emissions will be reduced from 150 to 15 ppm. "The fire-tube boiler looks very promising," says Wayne Krill, a project engineer with Acurex. "We have many results from smaller-scale burners of the same type and we think the scaleup will be fairly straightforward." Subscale tests of the water-tube boiler configuration also indicate a 90% NO_x reduction, he adds.

The fire-tube catalytic burner will be fitted into the combustion chamber of a 25-hp boiler (about 850,000 Btu/h). The burner is a tube made of packed alumina-silica fibers, closed at one end. It is about 3 ft long, 12 in. dia., and ranges in thickness from $1/4$ to $1/2$ in.

The gas-and-air fuel mix is injected into the open end of the cylinder and diffuses through the walls to the outer surface, which is the reaction zone. Catalytic material may be sprayed onto the tube or woven in, says Krill. The catalysts will be various types of alumina-silica, which has a low level of activity, but is considered adequate for the demonstration.

When the burner is in use, "it appears to be just a radiating surface," says Krill. "You use that hot surface to provide radiation transfer to heat the boiler fluids." Tests are scheduled to start at the end of the month and will be completed in September. Water-tube boiler tests will be done on an intermediate scale (about 1 million Btu/h) in June and July, followed by 2- to 3-million-Btu/h-scale tests in the fall.

RIGHT-ANGLE INJECTION—The water-tube boiler consists of parallel ceramic tubes (silicon carbide) coated with catalytic material that has an external coating. Acurex is using chromium- or nickel-based catalysts. Water tubes are located concentrically inside the ceramic tubes. Fuel and air are injected at right angles, and heat from the resulting reaction is transferred to the water tubes. A one-million-Btu/h boiler requires 25 catalytic water-tube assemblies, says Krill. He notes that the configuration has one drawback—the tubes are set in a vertical position, completely different from current commercial boilers.

Catalytic combustion occurs at 1,900-2,100°F in fire-tube boilers, and at 2,400-2,500°F in water-tube boilers, Krill says, compared with 3,000-3,600°F for flame combustion.

"You get the same amount of heat release and the same amount of energy per unit of fuel, but it is transferred differently," he explains. "In flame combustion, you burn all your fuel in the gas phase and liberate the heat in one area."

SUITABLE FOR TURBINES—A key aspect of catalytic combustion, discovered by Engelhard Corp. (Iselin, N.J.) in the early 1970s, is that when the combustion temperature goes above 1,600°F, less catalyst is needed. This is because of the thermal gas reaction that occurs after the initial catalytic reaction.

"Homogeneous gas reactions occur at a very rapid rate, because they are not limited by mass transfer, so you need a much smaller catalyst," explains Kenneth Burns, the firm's manager of combustion technology. He notes that, prior to Engelhard's work, which involved the development of higher-temperature catalysts, calculations of catalyst size were based on

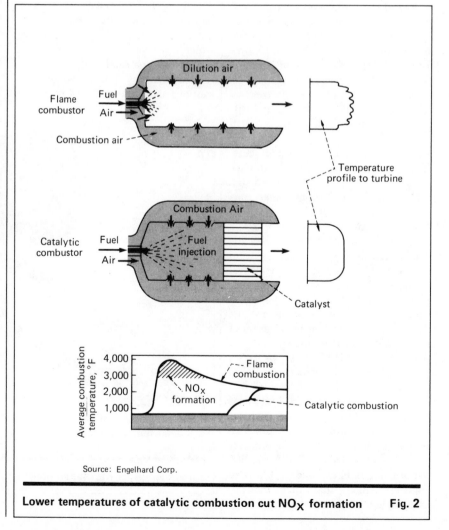

Lower temperatures of catalytic combustion cut NO_X formation   **Fig. 2**

Source: Engelhard Corp.

the mass transfer that occurs at lower-temperature catalytic combustion. Engelhard prefers the term "catalytically supported thermal combustion" when talking about higher-temperature processes, which involve both catalytic and thermal reactions.

Actual catalyst size is determined by such factors as pressure, temperature and the type of fuel, he adds, but as a rough measure, a process operated at a 2,000°F outlet temperature requires only about one-tenth as much catalyst as is needed for a mass-transfer reaction.

Burns feels that catalytic combustion is ideally suited for gas turbines because they run lean on fuel and because it promises better operating efficiency. He notes that conventional turbine operation requires that a lot of air be added to cool the hot combustion gases in order to protect the turbine blades. Efficiency is also lost by the injection of water or steam into the combustion zone to cut NO$_x$ emissions. Because catalytic combustion operates at a low-temperature and can handle a lean mix, dilution air could be added before combustion, and the NO$_x$ problem can be handled at the same time.

Acurex has done some small-scale gas-turbine combustor tests using catalytic combustion, and has achieved combustion efficiencies of 99.98%—as good as a conventional combustor, but with lower NO$_x$ emissions, says John Kesselring, associate manager of combustion technology. He adds that, while a large-scale demonstration is needed, the important thing is that

catalytic combustion can be scaled—unlike a conventional combustor, because "you can't really scale flames."

EPRI and Westinghouse hope to demonstrate a full-scale model of a catalytic combustor in a 74-MW Westinghouse turbine by 1985, using No. 2 distillate fuel. Westinghouse's concept is of a two-stage burner consisting of a conventional pilot burner to bring air to the reaction temperature, followed by a catalyst to complete the reaction. Tests of a full-scale fuel-air preparation system are expected to start at the end of the year.

EPRI's Angello points out that the normal way to control inlet temperature is to reduce the flame temperature by adding about 300% excess air, which must be injected in three zones to avoid quenching the flame. Because of the fuel/air ratio limitations, the flame cannot operate below about 3,000 to 4,000°F in the primary zone. Catalytic combustion permits a leaner mix, and combustion at 2,300°F. Angello notes that the pilot burner allows the turbine to operate at half load when the inlet temperature is too low for catalyst operation.

SOME PROBLEMS—Catalyst life is generally considered one of the chief problems to be dealt with. Another difficulty, in the case of gas turbines, is the substrate—a ceramic honeycomb structure that is made thin in order to maximize surface area. These structures tend to be fragile—like eggshells, says one critic.

Also, heavy liquid fuels, which are not easily vaporized, tend to foul the

catalyst. So far, very little work has been done to solve this problem, although Acurex has had limited success in tests in which waste heat from the process was used to vaporize most of the fuel, leaving only 20% to vaporize in the catalyst bed.

United Technologies Research Center (East Hartford, Conn.) claims to have developed a process in the laboratory that allows liquid fuels to contact the catalyst without any need for prevaporization. A spokesman says that more testing is needed before the work can continue.

As for catalyst materials, platinum-group metals are the most promising, says Engelhard's Burns, because they promote reactions at lower temperatures than do other materials.

Engelhard has tested proprietary palladium catalysts at 2,400°F for 2,000 h under a contract with National Aeronautics and Space Administration's Lewis Research Center (Cleveland). There was some loss of activity, says a NASA spokesman, but the catalyst still functioned. However, critics say the tests were inconclusive, especially as they did not involve full thermal cycling to test the response of both catalyst and substrate.

There is general agreement that more development work and tests are needed. Krill says that commercial boiler systems would probably require catalyst lifetimes of 3,000-5,000 h, and turbines 5,000-10,000 h. Acurex will be running catalyst tests of 1,000 h or more in its boilers.

Gerald Parkinson

Breakthroughs ahead for cleaning hot gases?

Gasification and pressurized fluidized-bed combustion of coal pose some tough air pollution control problems. But testing will soon begin on novel cleanup techniques.

☐ Techniques exist to remove pollutants from coal-combustion or coal-gasification exit-gas streams. But to work effectively (see box), they require that the pressure or temperature be lowered, resulting in reduced energy efficiency.

Developers are aiming to get around these shortcomings, however. Testing begins this fall, in fact, on several prototype systems or devices that can cleanse from high-temperature, high-pressure gases pollutants that are damaging to downstream turbines or other equipment, or are harmful to air and water quality. Yet other systems, not quite as advanced as those to be tested, will be brought to the point of conceptual design, for further evaluation.

Much work is progressing under a Turbine Cleanup Project established by the U.S. Dept. of Energy (DOE) in April 1979. Under a Program Research and Development Announcement (PRDA), DOE has awarded nine contracts worth a total of $9.5 million for the development of cleanup equipment, in two categories:

■ Category I: Less-mature technologies that need concept definition, and laboratory and bench-scale testing.

■ Category II: More-mature techniques scheduled for process-design testing at sub-pilot scale.

In addition to the PRDA-funded activity, a number of firms and organizations are developing other cleanup systems.

THE ALKALI CONNECTION—Gaseous alkali metals (from sodium chloride or potassium chloride) in the hot combustion offgas pose problems for such downstream equipment as turbines. Upon reduction of either temperature or pressure, these gaseous metals react with sulfur in the gas stream to form sulfates that deposit on turbine blades, causing corrosion.

Argonne National Laboratory (Argonne, Ill.) is investigating the use of either diatomaceous earth or activated bauxite as an alkali-metal scavenging medium. Although not part of PRDA, but funded by DOE, the Argonne work is important to some of the DOE contracts because those methods would use the Argonne scavengers. Irving Johnson, Argonne's program manager, says that diatomaceous earth removes alkali by chemical reaction (to form silicates that are affixed to the scavenger), while activated bauxite achieves removal by physical adsorption. Spent bauxite can be regenerated

---

The heart of the problem

DOE's program is geared to gas temperatures of 1,500 to 1,700°F and pressures of around 10 atm, and focuses on removal of alkali metals—corrosive to turbine blades—and particulates. Systems developed will have to clean gases sufficiently to meet the U.S. Environmental Protection Agency (EPA) standard of 0.03 lb of particulates per million Btu of fuel, the equivalent of about 0.02 grains/std ft³ of offgas. Additionally, alkali removal will have to be such that downstream turbines are unaffected.

Kenneth Markel, project manager for the Turbine Cleanup Project at DOE's Morgantown (W.Va.) Energy Technology Center, explains the shortcomings of conventional cleanup equipment:

■ *Cyclones*—Inexpensive to design and operate, their small-particle collection efficiency falls off when cleaning high-pressure, high-temperature gases, especially at the high gas volumes typical of coal-combustion systems. Increasing their size decreases collection efficiency for a given gas flow, because gas velocity is reduced.

■ *Electrostatic precipitators* (ESPs)—Generally quite effective, these units run into problems at severe conditions (even "hotside" ESPs only operate at temperatures to around 800°F): insulating materials become conductive, abetting short-circuiting that adversely affects collection efficiency. In addition, the hot, corrosive environment can damage the electrodes and collection plates, thereby also diminishing efficiency.

■ *Granular-bed filters*—Widely used for particulate removal in the chemical process industries, these devices are unable to handle the sticky ash generated by coal-combustion systems, says Markel.

■ *Baghouses*—Conventional bag materials just cannot withstand the high temperatures that coal-combustion systems generate.

Originally published February 23, 1981

simply by leaching in water, says Johnson, but he notes that bauxite is from two to five times as costly as the diatomaceous earth.

Laboratory tests have shown that 99.9% of the alkali can be captured by using the two materials, says Argonne, and the laboratory is currently accumulating more data on them; the materials will be tested later this year with a bench-scale combustor. Such companies as Exxon Research and Engineering Co. and Air Pollution Technology Inc. are planning to incorporate the scavengers into their own particulate-collection designs, and are eagerly awaiting the results of Argonne's work.

DOE WORK: TESTS NEAR— DOE has awarded four contracts for development work that is to culminate in sub-pilot-scale tests, with some to begin as early as this fall:

■ Westinghouse Electric Corp. (Pittsburgh, Pa.): *Granular fixed-bed filtration.*

This system consists of a pressure vessel that contains a number of modules, each comprising as many as a dozen 1- to 1½-in.-deep sand beds. Hot, dirty gas flows downward through the beds, with particulates settling out onto the sand. When the pressure drop through the unit increases, the dirty gas flow is stopped and cleaned gas is blown back through the bed, fluidizing it and backflushing the collected particles into a holding chamber upstream. Thomas Lippert, Westinghouse's principal engineer on the project, says that because the collected material tends to agglomerate, there is little reentrainment.

Lippert indicates that the system has a collection efficiency greater than 97% to 99% (by weight) for particles up to 10 μ in dia. Westinghouse believes that its filter units can be made relatively inexpensively, being based on standard technology and not incorporating any complicated auxiliary material-transport systems.

The company is currently setting up a 600-acfm hot-gas-cleanup test apparatus at Madison, Pa., and hopes to begin testing the unit sometime in the fall.

■ General Electric Co. (Schenectady, N.Y.): *Electrocyclone.*

In its system, GE will electrically charge the hot, dirty combustion gas that enters a large cyclone having an axial, high-voltage electrode. The radial electric gradient (in combination with the charge on the particle) enhances the centrifugal force induced by the cyclone, to capture smaller as well as larger dust particles. Ralph Boericke, GE's manager of gas cleanup programs, says that cold-testing of an 18-in.-dia. model indicates that electrostatic enhancement can give a 12-ft-dia. cyclone the same (good) capture efficiency as 10-in. apparatus.

GE will later this year test its 18-in. electrocyclone on a Curtiss-Wright pressurized fluidized-bed combustor (PFBC operating at 7 atm and 1,550°F. In addition, the firm is building a

Clean bed material →

Valve —

Bed-material admixture:
● =Ferromagnetic
○ =Alkali scavenger

Magnetic coil

Gas to turbine

Louvers

Hot, dirty gas →

Blowout recovery

Spent bed material →

Source: Exxon Research and Engineering Co.

Schematic of Exxon's magnetically stabilized bed device   **Fig. 1**

36-in. cold-test unit, to determine scaleup characteristics.

■ Cottrell Environmental Sciences (Somerville, N.J.): *High-temperature, high-pressure electrostatic precipitation.*

Cottrell, a subsidiary of Research-Cottrell Inc., will design, fabricate, erect and test (in 1982), on Curtiss-Wright's PFBC, an ESP that will be contained within a pressure vessel and will use tubes in lieu of plates as the collectors. Sam Kumar, Cottrell's project manager, says that tubes allow more collection area per unit volume than do plates. In addition, he states that the tubes will be suspended (not attached at the bottom) to allow for thermal expansion not typically experienced in ESPs.

Kumar also says that the device will be redesigned to keep insulators away from the gas stream, and that external cooling might be used to keep insulators under 1,000°F, so they do not become conductive.

Cottrell's unit is now in the design phase.

■ Acurex Corp. (Mountain View, Calif.): *Ceramic-bag filtration.*

Acurex's system features bags made from 95%-alumina matted (felted) ceramic fibers that can withstand high temperatures. In the technique, hot, dirty gas is first rough-cleaned by a cyclone, then enters the bag. The firm says that small-scale tests have demonstrated collection efficiencies of 95-99%. And accelerated pulse-cleaning tests have shown that the bags can survive the mechanical flexing of 50,000 cleaning pulses (about a year's service) at 1,500°F.

Acurex is currently assessing the costs of its baghouse system on a GE computer model of a PFBC installation. The firm will then design, build and demonstrate its apparatus in early 1982 on the Curtiss-Wright PFBC unit.

DOE WORK: FEASIBILITY—Five firms have won DOE contracts, most to be completed in 1982, to show that the following techniques are feasible:

■ Westinghouse Electric Corp.: *Ceramic cross-flow filtration.*

The firm's Research and Development Center is developing a ceramic-block filter that can handle 2,000°F temperatures. Dirty gas enters alternate rows of tubes in the block, passes through the permeable ceramic (where the particulates are left behind), and exits via adjacent tubes as clean gas. Clogging is prevented by backpulsing with cleaned gas.

David Ciliberti, project manager, says the unit has passed virtually no particulates in tests where limestone dust below 10 μ was fed. He also notes that the unit's design yields 10 to 20 times the filtration area of a bag filter of the same volume.

■ Southeastern Center for Electrical Engineering Education (Orlando, Fla.): *Acoustic agglomeration.*

A university research service organization, SCEEE has subcontracted work on this collection-enhancement technique to SUNY, the State University of New York (Buffalo).

In this route, a 900-1,000-Hz sound wave causes small particles to agglomerate into larger ones (around 10 μ in dia.). Particles 0.3μ or smaller remain unagglomerated, but a spokesman says these represent only a very small mass. SUNY is currently fine-tuning the system to obtain maximum agglomeration with minimum vibration in the equipment. The technique would ultimately find use in combination with other particulate-collection devices.

■ General Electric Co.: *Electrostatically enhanced granular-bed filtration.*

This system uses a standard-design moving-bed granular filter. GE will electrically charge the hot, dirty gas; collection efficiency will be enhanced by attraction of the charged particles to the uncharged moving bed, says the firm.

GE's Boericke notes that cold-flow and lab-scale tests show a collection efficiency of 99.0 to 99.9%, depending on original loading and final needs. (PFBC cleanup devices typically must cut gas loading from 20,000 ppm to 20 ppm to meet EPA limits, says Boericke.)

■ Exxon Research and Engineering Co. (Linden, N.J.): *Magnetically stabilized-bed filtration.*

In this route, rough-cleaned (by cyclone) hot gas enters a cross-flow

panel bed (see Fig. 1) that is suspended by a magnetic field. The bed consists of ferromagnetic granules and alkali-metal scavengers. Particulates are captured by impaction. Bed material is continuously removed and sent to a separate system for removal of captured particulates and alkalies; cleaned bed granules are returned to the filtration unit.

Exxon's Edward S. Matulevicius

Media-return legs

Clean media

Gas inlet

Clean-gas outlet

inlet duct

Outer sleeve

Media-flow annuli

Inner sleeve

Dirty media

Source: Combustion Power Co.

Moving granular-bed filter   **Fig. 2**

says that a magnetically stabilized bed accommodates gas velocities four times greater than can be used with an unstabilized bed.

■ Air Pollution Technology Inc. (San Diego, Calif.): *Dry-plate scrubbing.*

APT's device uses a shallow (1- to 2-cm deep), moving bed of granules semifluidized by the hot gas to collect particulates by impingement. Three

or four stages may be needed, says APT, some of which could consist of alkali scavengers.

Richard Parker, APT's research manager, says the firm is accumulating data on its system, looking at such questions as how rapidly the bed granules must be removed, and whether alkali scavengers are hard enough to achieve particulate removal without crumbling.

OTHER WORK—In other work funded by DOE (part of the Turbine Cleanup Project, but not the PRDA program), Combustion Power Co. (Menlo Park, Calif.) has been developing its Granular Bed Filter. According to the firm, 1,000 h of testing show that the unit's overall particle-capture efficiency remained near 99%, with no sign of irreversible plugging on actual coal-combustion offgas at 1,540°F and 17 psia. The vertical unit captures particulates by impaction: dirty gas is forced through a continuously downward-moving packed bed of granules (see Fig. 2). CPC is currently modifying slightly the design of its unit, and expects to test the revised device on an atmospheric fluidized-bed combustor later this year.

Late last year, the Institute of Gas Technology (Chicago) announced a particulate-removal system that combines sonic agglomeration with ceramic cross-flow filtration (*Chem. Eng.*, Dec. 1, 1980, p. 19). IGT has applied for a patent on both the concept and the equipment, and is now discussing funding for further work on the device with both DOE and the Electric Power Research Institute (Palo Alto, Calif.).

Finally, the Buell Emission Control Div. of Envirotech Corp. (Lebanon, Pa.) is developing ceramic bags that are woven (as opposed to Acurex's matted ones). Buell's bags are made from 50%-alumina/50%-silica fibers. Both the Acurex and Buell systems have been tested by EPRI at 1,650°F and 250 psia (about 17 atm), with both performing well, says the Institute. EPRI adds that in light of DOE's funding of further Acurex work, it will continue to support more work on the Buell system.

Philip M. Kohn

Cost-saving process recovers CO_2 from power-plant fluegas

This technology features a new alkanolamine-based solvent that is said to be less corrosive at higher concentrations and have greater resistance to oxygen than other CO_2 stripping solvents. The high-purity CO_2 will be used for enhanced oil recovery in Texas.

Leonard J. Kaplan, Assistant Editor

☐ Dow Chemical Co. (Midland, Mich.) and Procon International Inc. (Houston) have teamed up to develop a new stripping process that overcomes some of the problems that have made the recovery of CO_2 from fluegas uneconomical in the past (*Chem. Eng.*, Oct. 4, p. 17). The new technology combines the improved properties of the Dow proprietary alkanolamine-based solvent, called Gas/Spec FS-1L, with an energy-conserving system design developed by Procon. Both companies are marketing their technology under separate licenses.

A facility now under construction by Carbon Dioxide Technology Corp. (Houston), slated for startup in January, will use the new technology to recover CO_2 from the fluegas of a gas-fired power plant belonging to Lubbock Power & Light (Lubbock, Tex.). The purified, compressed CO_2

(2,000 psig) will be piped approximately 35 miles for use in enhanced oil recovery at the Garza oilfield south of Lubbock. According to Carbon Dioxide Technology, this plant will be among the world's largest for separating CO_2 from fluegas.

Both the solvent and equipment design were tested in a 1.5-ton-CO_2/d pilot installation operated for $1\frac{1}{2}$ years at the Lubbock power plant. The fluegas had a concentration of O_2 and CO_2 of 4-7 mol% and 8-10 mol% (dry basis), respectively.

Dow reports that the solvent has also been tested independently of the Procon equipment system, recovering CO_2 from the exhaust gas from an ammonia reformer at a plant in Carlsbad, N.M., operated by N-ReN Corp. of Cincinnati. In this application, CO_2 at a concentration of about 8-10 mol% was removed from the gas by the Dow solvent in a retrofitted amine stripping system.

CONVENTIONAL SYSTEMS—Solvents such as monoethanolamine (MEA) and diethanolamine (DEA) are used for a variety of gas-cleaning operations, including the sweetening of natural gas, and refinery-gas purification. Conventional MEA gas-scrubbing systems have had extensive use in CO_2 removal from shift-gas streams in ammonia plants, with the typical concentration of the solvent about 15%. Concentrations as high as 30% can be used if inhibitors are present to reduce the severe corrosive properties of the amine solutions. Robert L. Bixler of

Dow notes that in older CO_2 recovery systems, the amine concentrations ran about 12% because anything higher caused corrosion problems.

According to the developers of this new process, recovery of CO_2 from fluegas is complicated by three factors: the low CO_2 concentration of the gas; the low pressure of the gas; the corrosion of absorption equipment having high solvent concentrations, especially in the presence of oxygen.

The low CO_2 concentration in a typical fluegas (about 8 vol%), is a result of using air to supply the oxygen requirements for the combustion process. Greater circulation rates of solvent are necessary to absorb sufficient amounts of CO_2, increasing operating and capital costs. Higher concentrations of the new solvent are possible (about 20% at the Lubbock plant), reducing the total liquid circulated.

In addition to the greater solvent concentration used, the system employs a patented enrichment boiler that not only supplies the steam needed for the plant but also increases the CO_2 concentration of the gas.

Low fluegas pressure also affects process economics. The CO_2 partial pressure in the fluegas is insufficient to produce the desired solubility in solvents such as MEA at low concentrations—kept at low concentrations to avoid solvent degradation and equipment corrosion. Energy costs for increasing the fluegas pressure, to improve absorption, are said to be prohibitive.

Dow says the new solvent offers higher CO_2 loading than other products (approximately 8 ft³ of CO_2/gal of solvent). This higher CO_2 loading also reduces the required solvent circulation rate in the absorber, lessening equipment size and utility use.

CORROSION CONTROLLED—The firms say that the corrosion inhibitors available up till now could not provide

190-ft absorbers at the Lubbock plant

Originally published November 29, 1982

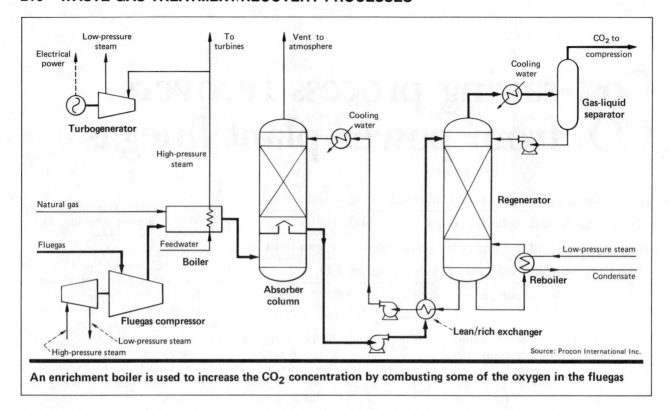

An enrichment boiler is used to increase the CO₂ concentration by combusting some of the oxygen in the fluegas

adequate protection when oxygen was present in the gas stream. The presence of free oxygen increases the rate of degradation of the solvent, increasing the amount of solvent used. Dow reports that its new product is stable in the presence of oxygen and provides corrosion protection sufficient to permit the use of mild steel for most of the equipment at the Lubbock plant.

ENERGY-SAVING DESIGN—Among the features of the system's design said to contribute to its more efficient operation are a patented boiler where the CO_2 concentration is increased by about 25%, and a packed absorption column that is claimed to operate with a more uniform temperature profile than did previous units.

Fluegas, with a typical CO_2 concentration of 8 mol%, enters the recovery system (see flowsheet) at slightly above ambient pressure and a temperature generally in excess of 250°F. First, in order to accommodate the pressure drop through the system, the gas is compressed to somewhat less than 5 psig. It then enters the enrichment steam boiler, where residual oxygen in the fluegas is combusted with natural gas to provide additional CO_2, and generate high-pressure steam (500-800 psi). The steam will be used to produce electricity, and also low-pressure steam for the stripping column,

after being reduced to 25-100 psi. The gas stream now enters an absorber column where it is stripped of about 90% of the CO_2 with the Dow solvent. The residual fluegas is released to the atmosphere, while CO_2-rich solvent (at about 110-150°F) goes through a preheater where its temperature is increased to 200-250°F before it enters the stripping column.

It is in the stripping column or regenerator that enough energy is supplied to the solvent to release the chemically bound CO_2. The CO_2 gas stream then leaves the regenerator, to be cooled, compressed and dehydrated; while the hot lean solvent flows through the preheater and a cooler

before it reenters the absorbing column. A filtration system maintains the required purity levels of the solvent.

ECONOMICS—According to both companies, the new technology makes obtaining CO_2 from fluegas more economical than was possible before with conventional solvents and equipment designs. The table, supplied by Dow, compares the performance of its new solvent—used with both conventional and Procon system-design—with MEA at two different working concentrations. Along with an expected reduction of over 50% in energy requirement compared with a 12% MEA system, the new technology is said to lose only half as much solvent.

Comparison of both conventional and Procon systems using monoethanolamine (MEA) and Gas/Spec FS-1L

Basis: 1,000 tons/d of CO_2 from fluegas

	MEA with conven.	MEA with conven.	Gas/Spec FS-1L with conven.	Procon
Solvent concentration	12%	20%	20%+	20%+
Heat duty required, million Btu/ton of CO_2	8.1	5.4	4.8	3.5
Electric power, kWh/ton of CO_2	48	20	13	15
Solvent losses, lb/ton of CO_2	8	8	<4	<4
Estimated capital cost, including dehydration and compression to 2,000 psia, $ million	90	60	35	35

Source: Dow Chemical Co.

Section IX
Wastewater Treatment/Recovery Processes

Waste treatment boosted by bacterial additions

A number of benefits are claimed for the use of specialized bacterial cultures in biological facilities. Firms that sell such products say that business is growing rapidly.

Getting the "bugs" out of biological treatment of industrial wastes may involve putting in specialized bugs—i.e., bacterial cultures formulated for a given set of conditions. According to firms that market such cultures, a regular schedule of bacterial additions can improve the operation of activated-sludge, trickling-filter, or lagoon-treatment plants—providing faster system response to such problems as startups, plant upsets due to variable and shock loads, and cold-weather operation.

These claims are not easily verifiable, and experts are divided in their opinions on the cultures' effectiveness. For example, Richard L. Raymond, an environmental consultant with Suntech, Inc. (Marcus Hook, Pa.), says that "most systems are well inoculated; they just need a good nutritional balance for each waste being degraded. There is little evidence—published and documented—that special cultures are needed." While basically agreeing with Raymond, Ed Barth, chief of biological treatment at the U.S. Environmental Protection Agency's Cincinnati Research Lab, concedes that "there may be a place for special bacterial additions in small, specialized systems."

A more-positive opinion is that of Edward Stigall, director of the Office of Quality Review, EPA's Effluent Guidelines Div.: "Bacterial additives will speed up the process of one culture dominating the system, and they aid plant recovery after an upset. As such, they are a very useful opera-tions and maintenance tool for help on infrequent bad days."

Although no data exist on the number of waste treatment plants that have tried cultures, use of the specialized bugs seems to be increasing. A marketing study by Business Communications Co. (Stamford, Conn.) predicts that the biological-waste-treatment market for the cultures will jump from only $5 million in 1975 to about $50 million by 1987.

TAILORED PRODUCTS—Bacterial cultures are sold either as a liquid preparation, or as an air-dried or freeze-dried product. The formula may contain bacteria specific to a particular waste material, or a number of bacteria in a "shotgun approach," for treating a general class of wastes—e.g., organic/hydrocarbon.

General Environmental Science Corp. (Cleveland, Ohio), one of many firms in the business, offers a patented liquid product called LLMO (for liquid live microorganisms). The live bacterial concoctions, grown to maturity, come in a dormant state, suspended in sodium sulfite, which acts as an inhibitor. There are seven different strains, including both aerobic and anaerobic bacteria for treating most organic wastes. Shelf life is guaranteed for two years; reactivation is by dilution.

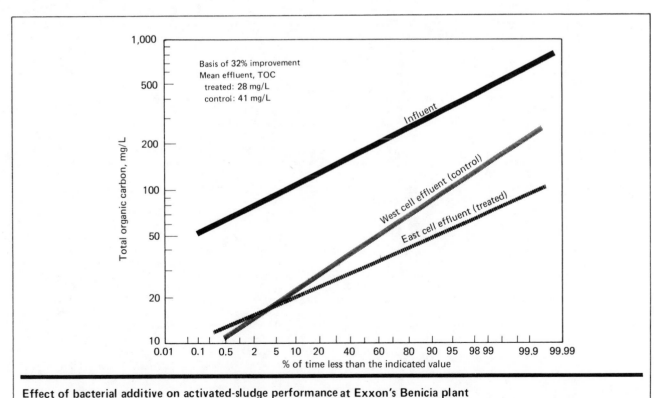

Effect of bacterial additive on activated-sludge performance at Exxon's Benicia plant

Originally published April 23, 1979

Polybac Corp. (New York City) selects bacteria on the basis of their ability to survive in environments containing uncommon or unnatural organic compounds. The bacteria are then exposed to increasing concentrations of those compounds to increase the tolerance of subsequent strains. Use of radiation helps to obtain strains with even more enhanced properties— i.e., bacteria with very specialized waste-disposal capabilities.

Because the ability of the cultures to survive when introduced into a wastewater-treatment system may be marginal, companies that market the bacterial brews suggest a regular reinoculation schedule. This ensures that the bacteria will be present even if the contaminants they feed on—e.g., phenol, ammonia, lignin—are in low concentrations or entirely absent.

Sellers of bacterial mixes acknowledge that their products need a proper nutritional balance. General Environmental Science, for example, insists that wastewater analyses show the presence of at least 5 ppm of nitrogen (as ammonia) and 1 ppm of orthophosphate for every 100 ppm of BOD.

CONCRETE RESULTS—At least one plant—Exxon's 1-million-gal/d activated-sludge system at the Benicia, Calif. oil refinery—has conducted controlled tests of bacterial addition. At the plant, which has parallel trains, specialized bacteria were added (in freeze-dried form) to the east aeration basin, while the left basin was allowed to undergo conventional treatment.

The company found that, in normal operation, there was a 32% improvement in the performance of the activated-sludge system—on the basis of organic matter removed (total organic carbon content was used as the key indicator). Phenol and ammonia effluent-levels were lower than in the control unit. Upon seeding the west aeration basin, Exxon obtained equivalent results.

Other observed benefits were faster unit startups, a more stable operation with variable loads, and a reduction in foaming. The saving in antifoaming chemicals exceeded the cost of the bacterial additives. Exxon concluded that regular maintenance doses of bacteria are not necessary for oil refineries with constant crude processing.

J. T. Baker Co.'s 3-million-gal/d secondary treatment plant at Phillipsburg, N.J., started using bacterial cultures last fall. Two different broad-spectrum dried additives were chosen: one for ammonia removal, the other for hydrocarbon degradation. Daily maintenance doses are 2.5 lb of the former product and 5 lb of the latter.

After weighing the correct amounts, the operator activates the bacteria with water; the mix is kept as a slurry at 100°F for two hours before it is dumped into the aeration tank at a point downstream of the recycled sludge stream.

The company believes that the cultures have improved removal efficiency. For example, BOD levels, which had hovered at about 18 ppm—perilously close to the discharge-permit limit of 20 ppm—now average about 6 ppm; and ammonia levels, which had been in the 40 to 45-ppm range (indicating little treatment of the 50-ppm influent), are now well below 20 ppm.

J. T. Baker feels that without the use of additives, its facility would have exceeded the violation level of 30 ppm/d of ammonia at least nine days in the past three months.

David J. Deutsch

Biological phosphorus removal

Without using chemicals, this wastewater treatment system reduces BOD, removes ammonia, phosphorus and nitrates, and produces a sludge that can be sold as a fertilizer.

John V. Galdieri, Air Products and Chemicals, Inc.

☐ During the late 1960s, nitrates and phosphorus were recognized as the nutrients responsible for eutrophication of confined bodies of water, most noticeably the Great Lakes of North America. In order to control phosphorus in the environment, bans were imposed that limited the amount of phosphorus in detergents, and the phosphorus that did appear in wastewater was treated by chemical means. However, the cost of those methods slowed their implementation.

Now, a new activated-sludge process, called the A/O system, can biologically remove phosphorus and, if desired, nitrates—in addition to simultaneously reducing biochemical oxygen demand (BOD) and removing ammonia from wastewater. This process was developed and patented by, and is now commercially available from, Air Products and Chemicals, Inc. (Allentown, Pa.).

With the A/O system, phosphorus levels can be reduced to less than 1 ppm. Residence times are low, since the rate of reaction is high. And no methanol is needed for denitrification. Moreover, the process uses equipment common to all types of activated-sludge-treatment plants.

Also, since no chemicals are used for phosphorus removal, no chemical sludge is generated. The biological sludge produced is similar to conventional secondary sludge, with a higher density due to the presence of fixed phosphate.

This biological sludge does not bulk, since filamentous organisms responsible for bulking are eliminated by the phosphorus-removing organisms. In addition, the sludge can be an economic benefit because the high phosphorus content upgrades it to a slow-release organic fertilizer.

The city of Largo, Fla., put an A/O system onstream in August. This unit, which includes nitrification/denitrification capability, treats 3 million gal/d of wastewater. The sludge produced here is dried and pelletized for sale as a fertilizer (LarGrow).

DIVIDED REACTOR—In the A/O process, a concrete tank is used as the reactor (see figure). This tank is divided into two sections: In one, wastewater is agitated and oxygen is excluded (anaerobic); in the other, the liquid is vigorously mixed and air or pure oxygen is introduced (aerobic). The anaerobic section is equipped with mechanical mixers; the aerobic section uses turbine aerators. The two sections are subdivided even further into stages to provide plug flow through the reactor, which is covered and sealed with prestressed concrete slabs and steel plates. A conventional secondary clarifier follows the reactor.

A feed of raw degritted wastewater is mixed with sludge recycled from the clarifier. This combined feed enters the anaerobic section (where dissolved-oxygen content is less than 0.7 ppm). The anaerobic section is divided into two parts—a purely anaerobic zone, and what is known as an anoxic zone. In the latter, wastewater is mixed with liquid recycled from the aerobic section.

After the anoxic zone, there is a conventional aerobic section in which the dissolved-oxygen level is maintained in the 2-4 mg/L range. The mixed-liquor volatile suspended solids (MLVSS) are maintained in the 3,000-5,000 mg/L range.

The liquid passes to a clarifier that separates sludge from the effluent; sludge concentration is 2-4% by weight solids.

To treat wastewater, a residence time of 60 to 90 min in the anaerobic section (this includes 30-45 min in the anoxic zone) and 90 to 180 min in the aerobic section is required for 90%-plus BOD removal, 60 to 80% nitrogen removal and a phosphorus content of 1 mg/L in the effluent.

STRESSED ORGANISMS—By putting the biology in the wastewater under anaerobic/aerobic conditions, the A/O system encourages the proliferation of a type of microorganism that can take advantage of the alternating stress/growth conditions. These organisms have the ability to store energy in the form of polyphosphate chemical linkages.

In the anaerobic section, BOD concentration is high. In the absence of oxygen, organisms that have stored energy in the form of polyphosphate employ that energy to actively transport BOD through their cell walls while they decompose stored polyphosphate to orthophosphate. The BOD of the liquid decreases while the phosphate concentration increases.

In the anoxic zone, the absence of dissolved oxygen and the presence of nitrates recycled from the aerobic zone favor organisms capable of biological denitrification. Bacteria that normally use dissolved oxygen for oxidation of BOD employ the oxygen chemically combined with nitrogen in the nitrate (NO_3^-) ion. Carbon from the BOD is oxidized to carbon dioxide, and nitrogen gas is formed and removed from the wastewater stream.

Originally published December 31, 1979

221

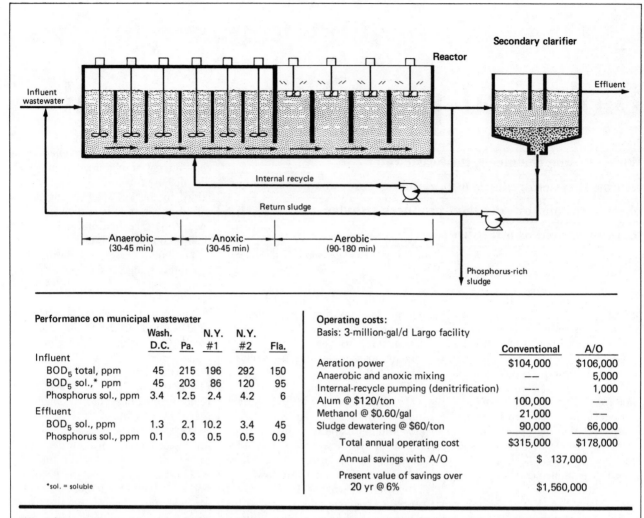

The A/O system produces less sludge than conventional wastewater processes based on chemical methods

Performance on municipal wastewater

	Wash. D.C.	Pa.	N.Y. #1	N.Y. #2	Fla.
Influent					
BOD_5 total, ppm	45	215	196	292	150
BOD_5 sol.,* ppm	45	203	86	120	95
Phosphorus sol., ppm	3.4	12.5	2.4	4.2	6
Effluent					
BOD_5 sol., ppm	1.3	2.1	10.2	3.4	45
Phosphorus sol., ppm	0.1	0.3	0.5	0.5	0.9

*sol. = soluble

Operating costs:
Basis: 3-million-gal/d Largo facility

	Conventional	A/O
Aeration power	$104,000	$106,000
Anaerobic and anoxic mixing	––	5,000
Internal-recycle pumping (denitrification)	–––	1,000
Alum @ $120/ton	100,000	––
Methanol @ $0.60/gal	21,000	––
Sludge dewatering @ $60/ton	90,000	66,000
Total annual operating cost	$315,000	$178,000
Annual savings with A/O		$ 137,000
Present value of savings over 20 yr @ 6%		$1,560,000

In the aerobic section, the organisms use the oxygen that is added to convert stored BOD to CO_2, water, and additional cell mass. The excess energy from this reaction is directed to recreating the cellular polyphosphate pool from the phosphate present in solution. Due to the growth of new cells and a net change of the phosphorus within existing cells, the amount of polyphosphate taken in by the biology is greater than the amount that was given up in the anaerobic section, thus providing net phosphate removal.

The amount of phosphorus in the sludge will depend on the concentration of BOD and phosphorus in the feed and on the sludge yield. The sludge has contained as much as 10% phosphorus on a dry basis, and concentrations between 3 and 4% are average. There may also be a nitrogen concentration of 7 to 9%. (Sludge from a conventional system may have nitrogen and phosphorus values of 4% and 2%, respectively.)

SAVING ON COSTS— The high nitrogen and phosphorus concentrations increase the sludge's value as a fertilizer. On the bulk market, nitrogen fertilizer is worth approximately $8/ton for each percentage point, and phosphorus $3 to $9/ton. If the sludge is sold as a fertilizer, the cost of plant operation can be reduced.

Using the Largo plant as a basis, the operating costs of a conventional system and an A/O unit are compared in the table. A conventional approach for the Largo facility might be air-activated sludge for BOD reduction and nitrification, followed by alum addition for phosphorus removal. Denitrification would be accomplished in sand filters, using methanol as the carbon source. The secondary clarifier would be the same in the conventional system as in the A/O unit.

Since the A/O process has a high reaction rate (residence times are short), by making use of existing aeration basins, older activated-sludge plants can be converted to it. For example, an existing five-pass basin at Largo was converted and only three of the five passes were needed for the same throughput of wastewater.

In contrast to an A/O system, the conventional chemical-addition method will produce two different sludges with a greater total poundage, because of the addition of alum, thereby requiring additional dewatering equipment, with consequent increased cost.

Reginald Berry, Editor

The Author

John V. Galdieri is Manager, Advanced Wastewater Treatment Systems, Environmental Products Dept., of Air Products and Chemicals, Inc. He holds bachelor's and master's degree in mechanical engineering from Lehigh University and is a professional engineer in the state of Pennsylvania.

COAL-BASED SYNFUELS

Biological, mechanical methods compete for wastewater-cleanup job

Designers of coal-conversion facilities are unsure of whether to use slower, but more-extensively tested biological treatments instead of high-throughput, high-energy-consuming mechanical techniques.

☐ According to engineers working to solve the environmental problems associated with coal-conversion technology, air-pollution control and solid-waste disposal probably will not be too difficult or costly. Wastewater and process-water treatment, however, is a different story—one that combines negative as well as positive aspects.

On the positive side: (1) Both coal gasification and liquefaction leave waste streams of similar characteristics, even though the actual amount of wastes varies with type of coal and conversion process; (2) adding either air, oxygen, hydrogen or a solvent during coal combustion reduces the level of toxic byproducts; and (3) no new technologies are needed for wastewater treatment. Methods developed by steelmakers to cope with coke-oven effluent can serve as the starting point for coal-conversion waste treatment.

The bad news, however, is that there are many variables—e.g., choice of coal and combustion process—involved in scaling up the steelmakers' techniques to deal with the enormous volumes of liquid-waste effluent from coal conversion. Another uncertainty is the extent of control that is needed (the U.S. government, for instance, has not decided whether demonstration plants should be subjected to rules mandating zero-discharge levels; many designs for commercial plants, however, do anticipate zero discharge).

RACKING UP COSTS—Experts are saying that these factors will push up the cost of liquid-waste treatment—in some cases to as much as 60% of total pollution-control costs for coal-conversion projects.

Originally published June 16, 1980

Michael J. Mujadin, manager of process engineering and operations planning for the Great Plains gasification complex, says that 10% of the estimated $1.2-billion project cost is earmarked for water-treatment systems. Of the current design, he notes that "the wastewater treatment part has changed more than any other area. Tests were needed to resolve certain unknowns, and each test uncovered more unknowns. It has taken us about six years to work things out."

Officials of the International Coal Refining Co., which designed the new $300-million SRC demonstration plant now operating in Kentucky, estimate that 12.5% of total capital expenditures are for pollution control; 59% of this will be for water treatment.

Noting these outlays, many experts question the wisdom of the zero-discharge concept. "It's not whether it can be done or not, but whether it is economical," says Preston Junkin, a biologist with TRW Energy Systems Planning Div., McLean, Va. The firm is part of a group that will soon begin a $3.5-million study of the costs and technologies of coal conversion.

TECHNICAL ANSWERS—Typically, process water in coal-conversion facilities may contain up to 8,000 ppm of phenols, 14,000 ppm of ammonia or nitrogen, and up to 30,000 ppm of BOD and COD—not to mention a long list of metals, cyanides and other substances. Engineers are using two different approaches (or combinations thereof) to cope with varying feeds and processes: biological, and mechanical/chemical. The first is considered conservative because it has undergone

much testing in coal-conversion schemes, as opposed to mechanical/chemical treatment. Those in favor of a conservative concept admit that biological processing can be inefficient and time-consuming, yet doubt that such mechanical alternatives as cooling towers and evaporators will be able to perform in a cost-effective manner.

Waste treatment for the H-Coal project, now being tested in Kentucky in a 600-ton/d plant, is an example of the biological approach. The technology was developed by Hydrocarbon Research, Inc.—a division of Dynalectron Corp.—and the $296-million project is being funded by the U.S. Dept. of Energy, oil companies, utility groups, and the state of Kentucky.

According to John Gray, supervisor of environmental protection, Ashland Synthetic Fuels, Inc., following ammonia stripping, inorganics are flocculated and separated via air-flotation skimming; they are then dewatered, and the filter cake is used as landfill. Organics are removed via activated-sludge treatment, which takes place in aeration tanks and an aerobic digestor. The effluent is discharged into a local river.

The Cogas gas-and-liquids plant proposed for southern Illinois by five pipeline-gas utilities organized into the Illinois Coal Gasification Group will have a half-and-half combination in its wastewater treatment facility. Designed by Dravo Engineers and Constructors' Chemical Plants Div. (Pittsburgh), it will employ evaporators for taking out inorganics, and biological treatment for removing organics prior

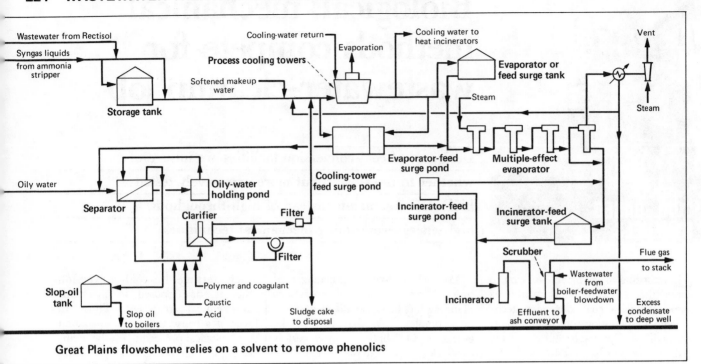

Great Plains flowscheme relies on a solvent to remove phenolics

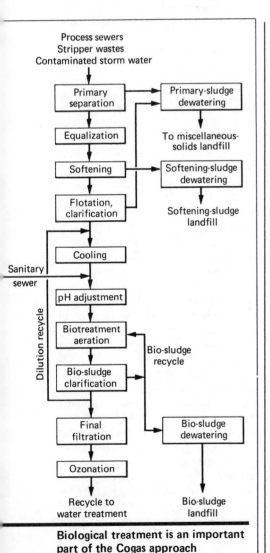

Biological treatment is an important part of the Cogas approach

to recycling the water through a waste-heat cooling tower. Although water from the Mississippi River can be used, a zero-discharge system was selected because of local regulations.

Water demand for the project, designed to convert 2,300 tons/d of coal into 2,100 bbl/d of naphtha and fuel oil plus 24 million ft³/d of syngas, is about 900 gal/min; half of this is recovered and recycled. The highly phenolic waste stream is diluted after ammonia stripping to ease biological oxidation.

A DIFFERENT IDEA—The Great Plains project bypasses biological methods in favor of an entirely mechanical approach. As the first step, the wastewater is concentrated in a cooling tower; then it is sent through multiple-effect evaporators for further concentration prior to solids incineration. Initially, American Natural Resources Co. (Detroit), which will operate the complex through a subsidiary, and the engineering contractor (C-E Lummus Co., Bloomfield, N.J.) had proposed a typical biotreating system. However, test runs conducted in a SASOL pilot plant in South Africa indicated that the phenols "were difficult to biotreat," says Mujadin.

Mujadin contends that the Great Plains approach, which is based on actual designs in operation at U.S. oil refineries and steel mills, is anything but dangerous pioneering. "We're not

the first to use it, even though we're the only ones in this country proposing it for coal gasification," he adds.

Tar and oils are removed from the 2,100 gal/min of process water via a proprietary, three-stage gravity-separation method. After sand filtration, solvent extraction with an isopropyl ether removes the phenolic compounds before ammonia stripping. Treatment in the cooling tower and evaporation train reduces the waste volume and the plant's water requirements. Inorganics are removed by zeolite softening, reverse osmosis and demineralization.

PRO AND CON—TRW's Junkin believes that use of evaporation translates into "a heavy energy penalty, high capital costs, and increased toxic emissions" because of the carcinogenic nature of several byproducts of the coal-conversion process. Polynuclear aromatic hydrocarbons, he notes, "appear to be present in significant concentrations." Junkin contends that "advanced control techniques [beyond those now available] may be necessary for the toxics."

However, David J. Goldstein, of Water Purification Associates, Inc. (Cambridge, Mass.)—a consultant to DOE and industry—says that solvent extraction of the phenolics, in combination with cooling-tower treatment, could be more cost-effective than biological methods. In fact, Goldstein advocates using the cooling tower for partial biological oxidation of the organics. He asserts that BOD pollutants at concentrations of up to 1,000 ppm—about three times the conventional treatment level—could be handled by a cooling tower. "People are worried about using a tower because they are scared," he notes.

Still, the lack of data from the various coal-conversion processes and feedstocks leaves many knowledgeable observers fearful that a commercial-size, first-of-its-kind project—i.e., the Great Plains complex—will try such an innovative design. One such observer is Donald K. Fleming, director of process evaluation at the Institute of Gas Technology (Chicago), who nevertheless concedes that the water-treatment problem is not considered as serious a challenge now as it was several years ago.

"The Great Plains project includes relatively minimal treatment," he says. "I'm somewhat amazed at the design and would be tempted to put additional equipment into it, just in case." Fleming would prefer to hold back the cooling tower/evaporator approach for plants to be built later.

Professors Richard G. Luthy and James R. Campbell, of the Dept. of Civil Engineering, Carnegie-Mellon University (Pittsburgh), agree. They advise that "the lack of information to predict minimum makeup-water quality criteria with respect to acceptable loadings of organics and trace contaminants . . . make it prudent for the first demonstration plants to be designed with a conservative approach including solvent extraction, ammonia recovery, biological oxidation, and other treatments."

DISSATISFIED WITH BOTH—Many engineers weighing the alternatives are not happy with either the biological concept or the mechanical approach. For example, one source at Exxon's synthetic fuels department, which is considering the economic potential of a medium-Btu coal-gasification plant feeding on Texas lignite, notes that "biotreatment of the whole stream has not been demonstrated, and we don't think that bio-oxidation will work on straight gas liquor. But the cooling tower/evaporator idea [the preferred method] leaves me uneasy."

If the Great Plains project's finances are worked out by DOE and the participating firms this summer, the question of what approach is best may be answered by early 1983. Engineers versed in the design of wastewater treatment facilities say that the outcome will probably reinforce the concept of a combined approach.

According to an official of Fluor Corp., which has extensive experience in the construction management of SASOL projects, "The key is the way the technologies are linked. Many plants are using conventional designs, and some other approaches are being ignored."

Larry Marion
World News (Chicago)

Activated carbon removes pesticides from wastewater

Up to five different organic compounds are being removed simultaneously from an industrial wastewater stream by a once-through adsorption system. Removal efficiencies are better than 99%, and treatment costs are significantly lower than for disposal.

Allan D. Holiday and *David P. Hardin*, Farmland Industries, Inc.

☐ Although activated carbon has been used for the adsorption of organic compounds, design data for the simultaneous adsorption of a variety of such compounds has not been available. Now, a system that uses activated-carbon adsorbent is treating an industrial wastewater stream containing a number of pesticide agents, including phenoxies, organophosphates and chlorinated organic compounds.

The Farmland Industries plant at St. Joseph, Mo., synthesizes and formulates insecticides and herbicides for the agricultural sector. In the course of plant operations, a water stream is generated that contains a number of pesticide compounds in relatively low concentrations. This results primarily from cleaning process equipment when changing the formulations being processed through the plant. Recirculation or reuse of this water does not appear feasible, considering the contamination potential. And since the cleaning operation is intermittent, and different insecticides and herbicides are involved, the composition and concentration of these substances vary substantially in the washwater.

Prior to the installation of the present treatment unit, the washwater left after cleaning was transported from various locations within the plant to an approved disposal site at substantial expense. Concern about rapidly increasing disposal costs led to the investigation of onsite treatment methods that would be environmentally sound as well as cost-effective.

An evaluation was made of various treatment processes that might be applicable, including biological degradation, natural evaporation (lagoon), thermal/chemical degradation, incineration, forced evaporation, and activated-carbon adsorption. All were evaluated considering the usual factors of capital cost, operating cost, technical feasibility, etc. Special consideration was given to environmental factors and the availability of reliable design data.

NO DATA—Activated-carbon adsorption was selected as the most promising approach. A literature survey provided design data for the continuous-adsorption columns and some idea of expected capacities and removal efficiency for some of the insecticides and herbicides involved. References were located on the removal of a single compound from water, but nothing could be found on the simultaneous adsorption of multiple compounds, including chlorinated hydrocarbons, organophosphates and carbamates in varying concentrations. Also, no information could be located on the impact of major changes in feedwater contaminants and variations in concentration.

Nevertheless, in the absence of the desired specific data and in the interest of expediency, the full-scale adsorption system was designed and constructed without process development work. The time and cost of a full-scale program appeared unjustified.

The range of carbon-column design parameters for treating industrial wastewater, according to literature references, was 1.0-3.0 gpm/ft² feedrate, 60 to 120-min superficial residence time, and 10-ft minimum carbon-column height. A conservative design (see Table I) for the system was prepared, reflecting reference parameters and the constraints of the available, used equipment that was rebuilt to serve as columns.

In addition to adsorber considerations, the overall system provides feedwater conditioning (or pretreatment) and a guard against the possible

Design and cost data for activated-carbon adsorption units — Table I

Basis: 300,000 gal/yr wastewater containing one to five pesticides

Actual Design

Adsorbent	activated carbon
Adsorber	3.5 ft dia. x 10 ft carbon depth
Carbon charge	3,150 lb/adsorber
Wastewater flowrate	10 gpm (1.04 gpm/ft²)
Water residence	72 min (superficial)
Feed pesticides	50-1,000 ppm
Pesticides in effluent	1.0 ppm total
Pesticide loading on carbon	1.0-5.0 wt %

Actual operating costs/gal of water treated

Carbon purchased	$0.070
Operating labor	0.074
Maintenance	0.010
Chemicals (caustic and emulsion breaker)	0.021
Electrical power	0.001
Disposal costs (carbon, sludge, hydrocarbon)	0.054
Total	$0.230

Originally published March 23, 1981

Adsorption removes 99% of the pesticides in industrial wastewater stream   Table II

Pesticide	Feed concentration, ppm Range[2]	(Avg.)	After pretreatment, ppm Range[2]	(Avg.)	% removal by pretreatment[3]	Effluent concentration, ppm Range[2]	(Avg.)	% overall removal
Phenoxies								
2,4-D esters	0.5– 65.1	(13.1)	0.1– 8.9	(3.8)	81	0.05–1.4	(0.2)	99.0
Organophosphates								
Terbufos	0.5–595.0	(71.7)	0.5–178.1	(46.5)	3	0.05–0.29	(0.08)	99.8
Phorate	0.5–356.0	(148.0)	0.5– 12.1	(3.6)	>97	0.05–0.2	(0.09)	99.9
Malathion	0.5– 29.2	(24.1)	0.5– 29.2	(24.1)	0	0.05–0.5	(0.4)	98.3
Miscellaneous								
Chlordimeform	0.5– 38.4	(12.7)	0.4– 29.7	(7.6)	0	0.05		99.3
Carbaryl	0.1–169.7	(64.0)	0.1–169.7	(64.0)	0	0.05–0.5	(0.09)	99.9
Atrazine	0.5–430.0	(144.7)	0.5–336.0	(107.9)	>29	0.05–1.5	(0.2)	99.9
Metolachlor	0.5– 41.1	(16.4)	0.1– 41.1	(10.0)	0	0.05–0.07	(0.05)	99.5

1— Results based on analysis of 59 batches.
2— Only values above the analytical detection limit are reported.
3— Based on mass balances, not average concentrations.
4— Analysis by Ron Emberton, Farmland Industries Laboratory.

release of water containing too much contaminant. The adsorption columns, although in series, are also piped for parallel flow to process at double the series flowrate. The system is now processing water at an annual rate of around 300,000 gal.

Overall treatment of the washwater is in two stages: the first is a batch pretreatment; the second, adsorption on activated carbon. Water from the process-plant tanks is transferred to the feed tank by truck. A typical batch is around 2,000 gal. The water, which appears to be an emulsion, is adjusted for pH to around 9.0, using 17 wt % NaOH. The amount of caustic required varies significantly from batch to batch, depending on washwater pH and on how much hydrolysis occurs in the feed tank. Also, 1,000-5,000 ppm of a cationic-polymer emulsion breaker is added to enhance water clarification.

After mixing, settling and decanting, the clarified water is pumped through the carbon adsorbers at a controlled rate. The treated water from the adsorbers is collected, mixed and analyzed prior to being released.

After several batches are processed, a small amount of hydrocarbon phase is withdrawn from a decant tank connected to the feed vessel. A flocculant sludgelike material is also withdrawn and dewatered. The water is returned to the feed vessel and the dewatered sludge and hydrocarbon phase are disposed of at an approved site.

EFFICIENT REMOVAL—To date,

110,000 gal of washwater have been processed through the unit. The compounds in the washwater appear in Table II, along with their concentrations before and after pretreatment and after the carbon adsorption treatment. In the case of some compounds, it appears that the pretreatment results in significant removal of contaminants. By analysis, overall removal efficiency has averaged 99% or better for all compounds involved except malathion, which averaged 98.3%.

The calculated combined loading of all pesticides on the carbon has averaged 1.0 wt % of dry carbon charge. (This is on the low end of the assumed 1-5 wt % that was used in the original design and evaluation.) The water treated per pound of dry carbon at saturation has averaged 12.5 gal.

Although the process does eliminate large disposal problems, disposal is still required for the spent carbon, the small amount of hydrocarbon phase withdrawn, and the dewatered sludge. Since disposal cost is weight-related (transportation and disposal-site fees), the overall weight reduction experienced was considered to be a system performance criterion. Based on operations to date, the weight of materials to be disposed of (wet spent carbon, hydrocarbon phase and dewatered sludge) is approximately 4% of the weight of the wastewater treated.

Treatment costs have been compared to the alternative of wastewater disposal at a properly permitted disposal facility. Currently, the cost is

$0.043/lb for transportation, and $0.037/lb for disposal-facility fees. This amounts to $0.664/gal, a rate that is expected to escalate rapidly in the years to come. The actual costs for the activated-carbon system, based on an annual treatment rate of 300,000 gal, is $0.23/gal (see Table I).

The cost reduction for activated-carbon treatment is $0.434 per gallon of wastewater. The largest costs of activated-carbon treatment are those for carbon purchase and disposal. If the spent carbon is found acceptable for regeneration, a further cost reduction will result. Consideration also is being given to increasing the tankage size, which will have the effect of reducing the batch pretreatment time and thus the operating labor cost. The system as designed incorporated used tankage that was already on hand and was recognized to be of less than optimum capacity. These tanks were used in the interest of capital cost reduction and expediency.

Reginald I. Berry, Editor

The authors

Allan D. Holiday is Administrative Engineer for the Fertilizer/Agricultural Chemicals Div. of Farmland Industries, Inc. (P. O. Box 7305, Kansas City, MO 64116). He holds eleven U. S. patents on chemical and hydrocarbon processes, reactor designs and separation operations, and has also authored several papers on energy conservation and pollution control. He has a B.S.Ch.E. from the University of Missouri at Rolla, and is a member of AIChE.
David P. Hardin is Superintendent—Quality Control and Laboratory Services at the Farmland Industries, Inc., chemical plant at St. Joseph, Mo. His background is in process and analytical chemistry relating to the synthesis and formulation of pesticide chemicals. Holder of a patent on herbicidal compounds, he has a B.S. in chemistry from the University of Missouri at Rolla.

Big waste-treatment job for water hyacinths

Several system designers in the U.S. sunbelt have become enamored of this sewage-loving weed that holds promise for lower-cost waste treatment. But will it work in colder, cloudier areas?

☐ A pesky weed that clogs waterways and reservoirs in the southern U.S. is being viewed by the U.S. Environmental Protection Agency (EPA) as perhaps the best prospect for a cheaper way to process waste in sewage-treatment facilities. The plant—the water hyacinth—will be evaluated in a one-million-gal/d demonstration plant to be built in San Diego, Calif. One big goal: production of potable water.

The $3-million project, the largest to date for water hyacinths, will use the plant for both primary and secondary treatment of city waste and also in combination with other primary treatment processes. EPA is funding 75% of the cost, the state of California 12½%, and San Diego the balance. The city plans to advertise the project nationwide in the near future, then screen the responses and invite three to five companies to bid on the work. The facility is scheduled to go onstream in late 1982.

Treatment of industrial wastes is also in the cards for the aquatic plant. This month, Exxon Co. U.S.A. will start planting water hyacinths in 25 acres of ponds at its Baytown, Tex., refining and chemical complex. The large-scale experiment will study the cleanup of wastes from various unit operations. And Seabrook Seafoods, Inc. (Kemah, Tex.) started testing water hyacinths last year to clean wastewater from its shrimp processing plant. The firm has had good results with a series of eight ponds, says plant manager Arthur Hults. The first four use algae; the others contain water hyacinths.

MANY PLUSES—The attraction of the water hyacinth is that it is one of the world's fastest-growing plants that thrives on sewage. Since it floats on the surface, its roots extend into the sewage medium for efficient absorption of nutrients, including toxic wastes and heavy metals (see illustration). After harvesting, hyacinths may be used for animal feed (if they meet U.S. Food and Drug Administration standards), or for compost, or may be fed into a digester for production of methane to fuel the facility and/or to generate electricity. Estimates of the annual dry tonnage (including roots) that can be harvested vary widely, ranging from 25-70 tons/(acre)(yr). This compares with about 10 tons/(acre)(yr) for alfalfa, a high-yield field crop.

Economics is EPA's main interest, says William R. Duffer, an aquatic biologist with the agency's Robert S. Kerr Environmental Research Laboratory (Ada, Okla.). "We want to develop alternative water treatment processes that cost less or consume less energy," says Duffer, who is a member of a technical advisory committee overseeing the project.

A facility that grows and uses water hyacinths for secondary treatment may cost 50% less than an activated-sludge plant and save about 40%-50% in annual operating costs, says George Tchobanoglous, another committee member. Tchobanoglous, a professor of environmental engineering at the University of California (Davis), says that a 1-million-gal/d activated-sludge plant, plus chlorination, would cost about $1.6 million to build, compared with $830,000 for a water hyacinth installation, plus chlorination. In both cases, primary treatment consists of clarification and land spreading. The biological system would require only about two-thirds the labor, parts and supplies needed annually for an acti-

vated-sludge plant, Tchobanoglous says, and annual primary and secondary energy consumption would be only about 3.95 billion Btu, against 7.48 billion Btu. The energy consumption figures do not take into account the potential production of methane from the crop.

Land requirements for the 1-million-gal/d biological system are calculated at 21½ acres, against only four acres for a comparable activated-sludge system. Tchobanoglous says, however, that the land cost does not really affect the economics when amortized over 20 years, and notes that activated sludge is typically used for large systems in heavily populated areas.

Treatment criteria for the above systems are the current EPA standards: 30 mg/L for BOD and 30 mg/L for total suspended solids.

ADVANCED TREATMENT—When tertiary treatment is required the economics are even more favorable for water hyacinths.

"A physical-chemical treatment plant to achieve advanced treatment standards is very expensive to construct and operate," explains Charles Padera, research manager for Coral Ridge Properties, Inc., a land development company that manages water and waste treatment facilities for part of an area it developed in Coral Springs, Fla. Two years ago, the company added a 100,000-gal/d water hyacinth system to its existing activated-sludge plant to meet local requirements for tertiary treatment. The cost, says Padera, was $60,000 (60¢/gal), compared with around $4/gal for a conventional system. The facility has five ponds in series; they are harvested to maintain a 75% coverage level of plants.

Coral Ridge properties is now considering expanding the water hyacinth facility to treat more than 3 million gal/d. Padera says this would require about 40 acres of ponds, compared with the 1.25 acres currently used.

The present system is designed to meet a maximum content of 5 mg/L each for BOD and suspended solids, 3 mg/L for nitrogen, and 1 mg/L for phosphorus; dissolved oxygen content is at least 4 mg/L. Padera says the only standard that has been a problem is that for phosphorus, and this is because the starting phosphorus content is too high for the hyacinth's

Originally published May 4, 1981

228

biological intake ratio of four parts nitrogen to one part phosphorus. The company is considering adding nitrogen to the ponds to balance the ratio.

THE GOAL: POTABLE WATER—The potential for water recycling is the main reason for the interest in water hyacinths by cities like San Diego. In fact, that city hopes that hyacinth aquaculture ultimately will permit tertiary treatment of all its sewage (current volume is 128 million gal/d) for recovery of potable water.

"The main problem now with conventional treatment such as activated sludge and aerated lagoons is the electrical cost," says Ray Dinges, a specialist with the Texas Dept. of Health's Div. of Wastewater Technology and Surveillance (Austin). "We increased time-space efficiency by using cheap energy, but now the equation has changed," he says.

Dinges calculates that for comparable wastes a hyacinth pond requires about a 5-d retention time, compared with 10-20 d for a standard lagoon, 16-24 h for a lagoon with extended aeration, and 6 h for an activated-sludge plant. Dinges ran a 350,000-gal/d experimental system that used hyacinths to remove algae from stabilization pond effluent for the city of Austin, and now the city plans to build three acres of greenhouse-covered hyacinth ponds to do this on a permanent basis. Algae and suspended solids are filtered by the hyacinth roots and also eaten by bacteria and micro-organisms that live among the roots.

TACKLING TOXICS—Although various aquatic plants are suitable for waste treatment, water hyacinths are usually preferred, since they have been already tested in a number of smaller projects (mostly in the southern U.S.).

The ultimate goal of Bill Wolverton, a research scientist with the National Aeronautics and Space Administration's National Space Technology Laboratories (Bay St. Louis, Miss.), and a pioneer in the field of aquatic plants for waste treatment, is to develop a space station system that would recycle water from sewage and also produce most of the oxygen needed. In the meantime, for the past six years he has operated open ponds for the treatment of sewage and chemical waste at NSTL. Low-level chemical waste (a few ppm), including silver, cyanide and bromides from the photographic laboratory, and some pesticides, is treated in an open canal 800 ft long, 20 ft wide and 2½ ft deep. The hyacinths do this work unharmed, he says.

Similarly, in a one-year test program with toxic wastes, Solar Aqua-Systems, Inc. (Encinitas, Calif.) processed through its pilot plant five times the level of toxins normally found in conventional sewage, according to Steven Serfling, president.

"We put in three different heavy metals and three different chlorinated hydrocarbons and found that all were removed down to safe levels, to the point where the byproducts we harvested—shrimp and fish—met FDA requirements," he says.

Solar AquaSystems designed a 350,000-gal/d plant for the city of Hercules, Calif. The plant has been operating for about a year and is still being tested. It has an anaerobic pond for primary treatment, then an aerated pond, and finally a hyacinth pond. Serfling states that the main work is done by bacteria, rather than by the hyacinths themselves. Bacteria and other organisms live in the roots of the

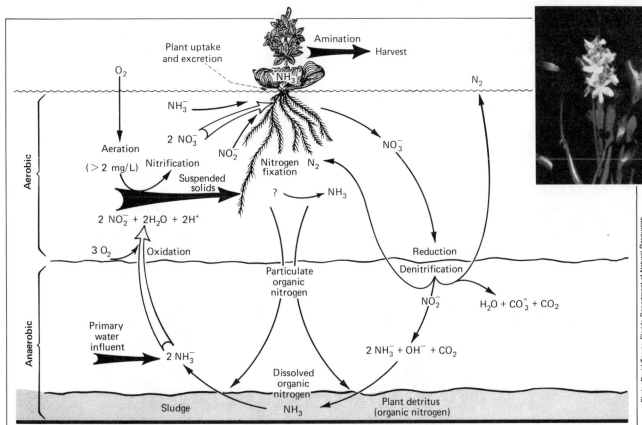

In a typical hyacinth pond, plant roots pick up nutrients from waste; unused materials are discarded as sludge

Photo: David Tarver, Florida Department of Natural Resources
Drawing: Lockheed Ocean Science Laboratories

plants, but, in addition, Solar Aqua-Systems suspends in its ponds substantial amounts of polyethylene film ("bioweb substrate") that contain bacteria. The hyacinth pond is also aerated to promote growth.

However, the Hercules, Calif., plant has been delayed by operating problems. Ralph W. Snyder, Hercules city manager, says that in an effort to cut costs the primary tank was sized below optimum; a larger primary tank is planned. Otherwise, he maintains, "the plant works, and with the change, we are confident we can put as much as 500,000 gallons per day through the hyacinths."

BIOLOGY/ENGINEERING LINK — The design and operation of aquaculture facilities for waste treatment "isn't strictly an engineering problem," says Alan B. Thum, a senior scientist with Lockheed Ocean Science Laboratories (Carlsbad, Calif.). "There are engineering solutions, but there is also management of a crop in water [to be considered]. It takes biologists and engineers working together."

Those involved in aquaculture seem to agree that relatively narrow lagoons are preferable for pond management, particularly harvesting, which should be done at a pace that promotes the desired growth rate. A narrow pond is also more convenient for a greenhouse structure, which is necessary in cooler climates and/or to maintain the ideal 30°C needed for year-round growth.

The need for greenhouses outside the water hyacinth's natural habitat in the U.S. South is one of the plant's disadvantages. Evaluation of greenhouses and the potential for their use in cold climates, in fact, will be one aspect of the San Diego project, where half the ponds will be covered. Solar AquaSystems uses double-wall polyethylene for its facilities and, although the material is rugged, "there's a lot of debate" on whether it would stand up to a cold, rough winter, Serfling says.

A cold climate would also require supplemental greenhouse heating, although a more severe limiting factor would be a very cloudy climate, since the plants must have adequate light. A more general limitation is that hyacinths have a relatively low tolerance to salinity, particularly chlorides, which are found in some chemical wastes, but not normally in domestic sewage.

Gerald Parkinson

New ways to destroy PCBs

Chemical routes to break down these hazardous compounds
are being tested. And exotic thermal methods other than
incineration also are making a bid for PCB disposal.
Both kinds of processes claim advantages over incineration.

Incineration is not the only way to burn PCBs

At the 64th annual conference of the Chemical Institute of Canada, held
on May 31 - June 3 in Halifax, N.S., researchers from the Royal
Military College of Canada (Kingston, Ont.) told of a PCB-destruction
scheme employing a plasma arc that can develop a temperature of
50,000°C. Destruction efficiency is higher than 99%; the College is
building a 1-gal/min pilot plant to further test the concept.

The process is said to be able to handle liquid and solid wastes.
Because it is a pyrolysis (using less than 1 ft³/min of air), its effluent
stream is smaller than for incineration—which reduces capital costs and
simplifies monitoring and scrubbing. A portable, 1-gal/min unit is
estimated to cost about $400,000 (Canadian). Electrical consumption, as
estimated from test results, is about 1.2 kWh/kg of feed. The system
"competes with incineration in cost and efficiency and beats it hands
down," says researcher T. G. Barton. "Its energy cost is about one-tenth
that of incineration."

At the same meeting, D&D Group Inc. (Smithsville, Ont.)—a
consortium of Canadian firms in waste management—described its work
with a modified diesel engine that has run a generator on a blend of
hydrocarbons and PCB-contaminated oils. The engine destroyed 99.998%
of the PCB content.

EOI (Washington, D.C.), a U.S. licensee of D&D, is trying to obtain
a permit to test the engine in the U.S. According to the firm, the diesel
engine is portable, and can run with blends of up to 20% PCBs.

Co-firing of PCB-contaminated material with coal or oil is getting more
attention because heat recovery is easy, and special equipment is not
needed. After running some tests in May, Duke Power Co. (Charlotte,
N.C.) has received EPA approval to fire PCB-contaminated oil along with
coal at its steam plant. And General Motors Corp. (Detroit), which has
burned PCB-contaminated oil in an oil-fired boiler, is seeking a permit
from the Michigan Dept. of Natural Resources to use this disposal
method.

In tests, GM mixed fuel with the waste (containing 50 ppm PCBs);
destruction efficiency was 99.9% with a boiler-flame temperature of
2,500°F.

Originally published August 10, 1981

☐ The pressing need to eliminate
polychlorinated biphenyls (PCBs) still
in use or in storage throughout the
U.S. has spawned a group of new
chemical treatment methods, as well as
some offbeat thermal, non-incineration
techniques (see box at left). One chemical route has already been approved
by the U.S. Environmental Protection
Agency, and two others will be tested
this month for EPA approval.

Incineration at extremely high temperatures (above 2,000°F) is still the
method recommended by EPA for disposing of liquids containing 500 ppm
or more PCBs; the agency recently
approved two sites for this type of
operation (*Chem. Eng.*, Mar. 23,
p. 73). But the new techniques are
said to be cheaper and easier to perform (see box on p. 233). Most are
portable, which means that PCB-
contaminated liquids need not be transported for treatment. And some can
even salvage useful products—e.g.,
transformer oils.

The paths to PCB destruction are
similar in all but one of the chemical
processes (which uses ozonolysis), and
involve employing a reagent containing sodium or calcium to strip chlorine
from the PCBs, which are broken down
into harmless polyphenyls.

A WASTE PROBLEM—PCBs, used as a
dielectric in transformers, or as heat-
transfer fluids and hydraulic fluids,
have not been produced in the U.S.
since 1977. But, according to EPA,
there are approximately 750 million lb
of the compounds still in use or storage
in various locations.

Considered suspect carcinogens,
PCBs have been associated with skin,
nerve and liver damage. Says Edwin
H. Clark, EPA's acting assistant
administrator for pesticides and toxic
substances: "The presence of PCBs in
the environment is potentially one of

the most serious public health issues we face."

The U.S. doesn't have enough capability for PCB disposal; in addition to the two approved incineration sites, only eight landfills have been allowed to handle the compounds. And talk of approving new sites almost always provokes public opposition, based on fears about carcinogens being released from incinerators or into the soil. "There has been a public reaction to every fixed-site operation," notes William Gunter, who heads EPA's PCB-regulation effort.

TWO-MEDIUM CLEANUP—This month, The Franklin Institute (Philadelphia) will stage for EPA a demonstration of its process, said to remove PCBs from oils and destroy PCBs spilled on land or dumped in landfills. One test will take place in Philadelphia, at a storage facility of the Philadelphia Electric Co. Another is being readied in Coventry, R.I., where PCBs in an illegal dumpsite will be destroyed.

Developed with partial funding from EPA and Philadelphia Electric, the Institute's process first makes a reagent by combining molten sodium with polyethylene glycol. The result is a sodium glycolate dubbed NaPEG, which acts as a super base that reacts with chlorine atoms in PCBs.

As NaPEG, insoluble in contaminated transformer oil, strips chlorine from the PCBs, these pollutants also become insoluble in the oil and are extracted by the reagent. NaPEG containing PCBs goes to a separate vessel and is exposed to air; oxygen completes the PCB breakdown. Whatever chlorine remains is replaced by oxygenated compounds, which form oxygenated biphenyls. Sodium chloride is the byproduct. The reaction can take place at room temperature, but higher temperatures decrease reaction time.

For treating spills or landfills, the reagent is simply sprayed on the contaminated area. According to EPA's Charles Rogers, a physical scientist at the agency's branch in Cincinnati, there is no comparable treatment for contaminated soils. "Incineration would work," he says, "but you would have to incinerate thousands of pounds to treat one or two pounds of PCBs. Transportation of the soil, and fuel costs, would make this alternative prohibitively expensive."

This is not the case with NaPEG treatment. According to the Institute's

Reactors to treat PCBs in the PCBX route are mounted inside vehicle

SunOhio Co.

Louis L. Pytlewski, "It appears that the cost will be less than the current price of transformer oil, which is about two dollars per gallon."

UNKNOWN REAGENT—Acurex Waste Technologies Inc. (Mountain View, Calif.), another firm planning to stage a demonstration for EPA this month, also uses a sodium-based reactant to remove chlorine. But Leo Weitzman, the firm's manager of industrial service operations, will not reveal its exact nature, saying only that it is made onsite, and that it decomposes PCBs exothermically to form NaCl and polyphenyls. There is a pretreatment step that removes water, aldehydes and acids from transformer oils.

Acurex claims that the route can handle feed-PCB concentrations as high as 10%. After treatment, the transformer oils have a fuel value of about $1/gal, and can be further purified into commercial-grade product. As for overall treatment costs, Weitzman claims that those are lower than that of incineration (including the cost of shipping and insuring the wastes).

The firm is now building a portable, 250-gal/h demonstration unit, to be mounted on a 35-ft-long flat-bed trailer. Weitzman says that if the EPA

demonstration goes well the process can be approved for use sometime this fall.

CLEANUP ON WHEELS—SunOhio Co. (Canton, Ohio), a company that specializes in the reclamation of transformer oil (via filtration and sludge removal), is also secretive about the nature of the reagent it uses in its PCBX

Royal Military College of Canada

Reactor houses a plasma arc that burns PCBs at high efficiency

process (*Chem. Eng.*, Sept. 22, 1980, p. 35). Others in the business say that it is a reductant containing sodium or calcium. SunOhio chairman Norman E. Jackson will reveal only that the reagent removes chlorine, that the remaining biphenyls form a nontoxic polymer suitable for landfill, and that the reaction takes place at "very modest temperatures and pressures."

According to Jackson, during tests of the PCBX route, the PCB level of transformer oils has been reduced from 1,000 ppm to 1 ppm, and from as high as 10,000 ppm to 50—all in one pass. However, SunOhio says it expects most of its business to be in oils having PCB concentrations of less than 500 ppm. Cost of using PCBX is about $3/gal for a "reasonably sized job," says Jackson; this includes reclamation of the transformer oils.

PCBX units can be mounted on 40-ft tractor-trailers and moved to waste sites. SunOhio already has built one mobile plant, which will tackle the company's first commercial PCB-disposal job—for Georgia Power Co.—in the next few weeks. The firm says it has already trained the operating crew for the first mobile unit; this personnel is now busy training operators for two other portable PCBX installations, whose delivery is expected before the end of August.

PIONEER EFFORT—The first company to develop a chemical PCB-destruction method is Goodyear Tire & Rubber Co., which last fall (*Chem. Eng.*, Sept. 8, 1980, p. 18) announced a process originally invented for use in-house—to remove PCBs from heat-transfer fluids.

Although the firm has applied for patents, it has declared its intention to make the technology freely available to all interested parties.

Allen R. Kovalchik, manager of projects and materials coordination for Goodyear, says that the process reduces PCB levels to less than 10 ppm from a high of "slightly above 500 ppm."

Goodyear makes its reagent (sodium naphthalene) by mixing metallic sodium with a dispersing oil at about 150-170°C. After cooling, the oil, containing fine sodium particles, is combined with tetrahydrofuran (a solvent) and naphthalene.

The reagent is mixed with PCB-contaminated fluids in a reactor blanketed with nitrogen; the minimum ratio

Let's hear it for chemistry—some vocal support

Although some chemical routes have not been sufficiently tested, and cannot handle PCBs in sludges or soils (most cannot deal with PCB concentrations over 10%)—all of which incineration can do—they have their share of supporters, not all of them with axes to grind.

One is H. Nugent Myrick, president of The Process Co. (Houston), a firm specializing in environmental matters. "With chemical methods," he notes, "you are treating and often recovering oil that may have an economic value, in comparison with incineration, which is simply a disposal strategy. Also, chemical treatment is done at the jobsite. You are not trucking the material hundreds or thousands of miles [which entails spending for transportation and risking the possibility of accidents that might release PCBs to the environment]. If you want to burn PCBs, there are a limited number of sites in the U.S."

EPA's William Gunter, who heads the agency's PCB-regulation effort, points out that chemical systems to not release pollutants into water or air, whereas incinerators—even those approved by EPA—are capable of emitting flue gas that contains carcinogens if the combustion is not being conducted under the right conditions.

As for cost, the developers of chemical methods say their systems are considerably cheaper. According to Allen R. Kovalchik, manager of projects and material coordination for Goodyear Tire & Rubber Co., incineration right now costs about $8/gal or slightly less. Rollins Environmental Services (Deer Park, Tex.)—one of the incinerators approved by EPA—is charging $9-$11/gal for waste shipped in drums; the price drops to below $5/gal for bulk quantities of 20,000 lb or more.

Transportation costs must be added to these figures. In comparison, developers of chemical routes are quoting treatment costs averaging about $3/gal or less.

of reagent to chlorine content is 50-100:1.

In an exothermic reduction-reaction, chlorine is stripped from the PCBs to form sodium chloride and nonhalogenated polyphenyls.

After quenching with water, the products of reaction go through vacuum distillation to recover tetrahydrofuran and naphthalene. The bottoms are sent to another vacuum-distillation unit that recovers clean fluid (transformer oil or heat-transfer fluid) overhead.

Goodyear says it has thought of making the technology portable—a task considered impractical by others in the field because of the difficulties of mounting two large distillation units on wheels. But Kovalchik says that this is feasible.

Raw-materials cost for the Goodyear method is about $3/100 lb of oil treated to reduce PCB levels from 130 ppm—a degree of contamination typical of transformer oil—to less than 10. Capital cost would depend on plant size, of course, but Kovalchik claims

that "there is a considerable saving over incineration."

OXIDATION OPTION—A process that uses ozone to oxidize PCBs was described at the Chemical Institute of Canada's 64th annual conference, held May 31 - June 3 in Halifax, N.S.

The technique, under development at the Royal Military College of Canada (Kingston, Ont.), is still in the laboratory stage. However, studies show a 90-95% destruction of PCBs when a six- to eight-fold excess of ozone (about a 1:1 weight ratio with PCBs) is used. Assuming an industrial-size facility, about 1,000 kWh of electricity would be needed to generate 100 lb of ozone to treat 100 lb of PCBs.

D. Diaper, one of the researchers, concedes that electricity outlays would make such a system more expensive than other destruction options. However, he says that the ozone method should prove useful in the destruction of "environmental PCBs in air, water and soil."

Reginald I. Berry

Rapid oxidation destroys organics in wastewater

Ultraviolet irradiation, cavitational shock and hydrogen peroxide team with air to oxidize high-strength biological and chemical wastes in a continuous process.

Mark Lipowicz, Assistant Editor

□ Enercol, Inc. (Red Bank, N.J.) has developed a new process that cleans up organic-laden wastewaters. The Enercol Oxidation Process (EOP) can handle a variety of organics—dissolved or suspended—at concentrations of 10,000 to 20,000 mg/L*, and higher with recycle. According to its developers, EOP is flexible in that inlet concentrations can vary widely, extent of cleanup is controllable, and shock loads do not ruin performance.

Commercial EOP systems already ordered include one for an organic-chemicals plant, and one for a bottling plant where treatment via EOP is replacing evaporation/incineration. Nicholas Zaleiko, Enercol's founder and president, expects EOP installations will be favored over biological-treatment systems because landfill for biological sludge is becoming prohibitively expensive and difficult to obtain at any price.

ONE STEP—Like other direct-oxidation schemes (e.g., incineration, ozonation, wet combustion), EOP enjoys an advantage over conventional treatment methods such as biological digestion or activated-carbon adsorption: Organics are destroyed in one step, not merely removed from the wastewater for later destruction or disposal. However, the EOP process requires neither high temperatures (as in incineration) nor excess concentrations of expensive oxidants (as in ozonation) to deliver rapid and thorough oxidation. EOP operates below 205°F and uses air as its primary oxygen source.

To gain these pluses, Enercol combined several innovative technologies in its EOP package. The flowsheet shows where the process gets its oxidative punch: Free radicals—largely the potent hydroxyl radical (OH·)—are generated by cavitational shock, by catalytic splitting of hydrogen peroxide (HOOH), and by ultraviolet (UV) irradiation. The free radicals split organic molecules, yielding new free radicals that can split other molecules—a chain reaction. In the presence of oxygen, the organics are ultimately broken down—carbon to CO_2, sulfur to SO_4, etc.

Tests in a commercial-scale EOP module have shown reductions of about 98% in organics for a variety of 5,000-15,000-mg/L (total solids) wastes: high-sugar bottling-plant cleaning water; high-sulfide tannery dehairing waste; weak ammonia liquor from a coker; activated sludge from a sewage-treatment plant. The EOP module has a nominal capacity of 6 gpm, but its performance and capacity are flexible because each system is designed and operated for a specific type of waste, concentration, and desired effluent quality. Thus EOP can achieve cleanup to 1 mg/L organics or

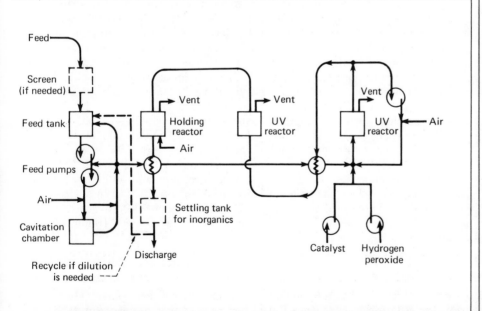

The Free Radical Oxidation process module may have one or two UV reactors, depending on the waste to be treated.

Originally published November 2, 1981

* Organic wastewaters are characterized by mg/L (ppm) of (1) specific compounds present; (2) total organic carbon (TOC); (3) chemical oxygen demand (COD), a measure of chromate-oxidizable matter; and (4) biological oxygen demand (BOD), a measure of biologically digestible matter.

less, says Enercol, and can handle wastes more concentrated than the 20,000-mg/L guideline limit.

HOW IT WORKS—The first flowsheet step is screening, necessary if coarse solids are present. If the wastewater is too concentrated, it can be diluted with effluent water from the discharge end—fresh water is unnecessary, except during a 15-min to 1-h startup. Of course, dilution cuts the net capacity of the EOP module, and the need for dilution can also limit the types of wastes that can be treated. For example, highly concentrated (i.e., 30,000-100,000 mg/L) aromatics that do not mix with water might not be treatable.

The next step is the cavitation chamber, with recycle. Cavitation is usually avoided by process designers, as it can pit and erode many materials of construction. But Enercol induces it on purpose, with a critical-flow nozzle inside a cylindrical chamber. As water speeds up in the nozzle, its pressure falls and vacuum bubbles are formed. These bubbles collapse when the water leaves the nozzle, creating tremendous pressures at the bubble walls that break up solids (to increase surface area for reaction), and actually split molecular bonds to form free radicals. The collapse is cushioned somewhat by the air injected downstream.

Wastewaters high in suspended solids, such as activated sludge rich in microorganisms, require several cycles through the cavitation nozzle. Solids-free streams may not require cavitation at all. The cost of this step is the fluid energy lost in the nozzle: about a 20-psi pressure drop, at a flowrate that depends on the number of cavitation cycles.

Hydrogen peroxide and catalyst are metered into the flow after the cavitation system. Since air is the main oxygen source, the level of HOOH needed is only about 0.05 mg/L per mg/L of TOC (if HOOH were the only source, at least 5.7 mg/L would be needed). Rate of reaction, and thus effluent quality, depends on the amount of HOOH added.

The catalyst—originally 2-10 mg/L of copper sulfate, but now 0.1 mg/L of an undisclosed organic—splits some of the HOOH molecules into OH· radicals in a redox reaction. The stream passes into the first UV reactor, where more peroxide is split photolytically, but first the stream is

Capital cost and operating inputs for two EOP applications

Capital	Bottling waste	Mixed primary and waste-activated sludge
Input analysis	7,000 mg/L COD	13,000 mg/L COD
	4,000 mg/L BOD	9,000 mg/L BOD
Output analysis	200 mg/L BOD	200 mg/L BOD
Capacity (8-h day)	160,000 gal/d	80,000 gal/d
Capital cost	$600,000	$690,000

Operating inputs, per 1,000 gal processed		
Labor	0.05 man-h	0.1 man-h
Power	22.5 kWh	90 kWh
Catalyst	0.01 lb	0.70 lb
Peroxide (50% sol.)	0.67 gal	0.84 gal
Lamp replacement	$0.36	$0.71
Direct maintenance	$0.36	$0.72

Source: Enercol, Inc.

heated to about 180°F in a counterflow, plate-type heat exchanger that uses the discharge from the holding reactor (about 200°F) as a heat source.

The UV reactor itself is an aluminum-lined stainless-steel cabinet with ten quartz tubes (8 ft long, 1½ in. dia.) arranged around an 8-ft-long tubular UV lamp. Each tube is fitted with a cavitation nozzle. The lamp is a custom built medium-to-high-pressure mercury-vapor lamp with a broad spectrum and a high UV intensity, not the low-pressure, narrow-spectrum (peak 253.7 nm wavelength) lamp used for sterilization, because EOP seeks to excite a variety of molecules that absorb UV light at different wavelengths, while sterilization focuses on DNA and other proteins that absorb strongly near 253.7 nm. The UV lamp draws about 15 kW at full power, but can be run at half power for materials (such as sugars) that need less UV irradiation.

Exposure to UV light is generally ¼-2 min, but depends on the nature of the waste. For example: Keratin, a heavy protein, needs 2 min at full power, while sugar needs 0.5 min at half power. Exposure time is varied in several ways: by varying the number of UV reactors, by installing piping to get two passes per reactor, or by varying the flowrate.

Of the 15-30 min total processing time, about 90% is spent in the holding reactor, where there is no energy input. This is an insulated gas-absorp-

tion tower in which the remaining organics, mixed by injected air, continue to react with oxygen in self-sustaining chain reactions. Effluent can be polished here by holding it longer.

If the wastewater contains inorganics, they either are oxidized to a precipitable form—$Ca(OH)_2$, for example—or leave in the effluent stream. A clarification step may be required if the amount of precipitated inorganics is high. Oxidation products vented with excess air include CO_2 and H_2O vapor, but not CO, says Enercol.

WHAT IT COSTS—The table lists estimated capital costs and operating inputs for two EOP applications—pretreatment of mixed primary and waste-activated sludge and of bottling-plant waste to 200 mg/L BOD. Because the sludge is rich in protein and suspended solids, it needs 1.5-2 min of UV exposure (bottling waste needs 0.5-1 min), and the flowrate per module is half that of the bottling-waste system. Capital costs shown are for complete skidmounted packages, including instrumentation, power transformers and operator training.

Assuming 20% per annum capital recovery, $0.06/kWh for power, $0.60/lb for catalyst, $3/gal for 50% HOOH solution, and a 40-h operating week, the estimated costs per 1,000 gal treated are about $7.50 for the bottling-plant waste and $17.50 for the sludge. The cost for sludge translates to about $275 per ton dry solids, which is somewhat higher than the cost of dewatering and incineration.

Waste treatment with hydrogen peroxide

A versatile, environmentally acceptable agent, hydrogen peroxide can oxidize a wide variety of aqueous and vaporous pollutants. It is cost-effective because it can be handled with inexpensive equipment.

George W. Ayling and *Harry M. Castrantas, FMC Corp.*

☐ Hydrogen peroxide is used in a wide variety of industries to treat a host of effluents (Table I). It also can help reduce biochemical oxygen demand (BOD) and chemical oxygen demand (COD), and control filamentous bulking in secondary clarifiers.

Unlike classic inorganic oxidizing agents (such as chlorine and hypochlorite), hydrogen peroxide yields no noxious or polluting byproducts. Because its only byproducts are water and oxygen, it provides a secondary benefit—dissolved oxygen, which stimulates the activity of the aerobic organisms that break down many organic wastes, and attacks the anaerobic organisms that produce sulfides, and the filamentous organisms that cause bulking and poor settling in activated-sludge systems.

Hydrogen peroxide is clear, colorless, nonflammable, and has a characteristic pungent odor. Miscible with water in all proportions, it is sold in percentage-by-weight of a water solution. It contains a small amount of stabilizer, the amount and nature depending on the solution grade.

Economics of using hydrogen peroxide

A major advantage of hydrogen peroxide in waste treatment over other oxidants is the low capital investment required. Equipment costs range from a few dollars for a drum drip-feed system to $18,000–24,000 for a 7,000-gal aluminum storage tank (such as that shown above) with aluminum transfer lines, heavy-duty pump and injection fittings.

Although hydrogen peroxide's unit cost may be higher than other oxidants, less of it is needed. Hydrogen peroxide costs from 28¢ to 58¢ (depending on grade and container) per pound of 50% solution. Approximate costs in Table II show that hydrogen peroxide can be substantially more cost-effective than potassium permanganate or chlorine dioxide.

Factors other than unit cost must be taken into account, too. For example, chlorine is not practical for treating phenols because of the chance of forming chlorinated phenolics. (Although hydrogen peroxide is often more environmentally acceptable for treating sulfides, a plant already using chlorine for other purposes may find it more expedient to expand use of the latter.) Ozone's high capital and operating costs and limited water solubility put it at a disadvantage in wastewater treatment. And although air or oxygen may be cheaper than hydrogen peroxide on a unit cost basis, their limited oxidizing powers and long reaction times, as well as high capital costs, restrict their application.

Labor costs for storage and handling hydrogen peroxide are low and vary with the type of storage system

Originally published November 30, 1981

(larger systems being less labor-intensive). Only periodic checks need be made of feedrate and vessel level.

Contamination can be hazardous

Although safer than most oxidants if normal safety precautions are observed, hydrogen peroxide is, however, subject to vigorous decomposition by various organic compounds, heavy metals, dirt, heat and alkali. Storage and handling systems are, of course, designed to minimize such contamination.

Contact with hydrogen peroxide can cause mild burns and eye injury. Therefore, goggles and gloves should be worn when handling it. Other oxidants, however, pose even greater handling hazards. Chlorine, which is confined under pressure in cylinders, can cause severe skin lesions and irritation of the mucous membranes, even at low concentrations. Ozone is corrosive and toxic. Chlorine dioxide is explosive at levels above 11% in air, is also toxic, and can irritate the respiratory tract. Potassium permanganate must be dissolved before use, making skin contact and resulting burns more likely.

Special applications and benefits

Pollutants in gaseous effluents can be scrubbed with hydrogen peroxide in a conventional scrubber that provides the necessary contact time between the liquid and gas phase. Packed-bed scrubbers are probably the best because their characteristics are known and operating parameters can be calculated. Venturi and spray-chamber scrubbers usually do not provide sufficient contact time.

When biological treatment systems are overloaded and mechanical aerators cannot satisfy the oxygen demand, hydrogen peroxide is often used temporarily to add dissolved oxygen. In such cases, H_2O_2 can be introduced rapidly. Once in the system, it decomposes and releases its oxygen.

In certain wastewaters, hydrogen peroxide contributes to BOD enhancement—that is, a greater amount of BOD reduction occurs than can be accounted for solely by the oxygen content of the H_2O_2.

Nitrogen gas carries sludge particles to the surface of secondary clarifiers. This can be prevented by adding 1–2 ppm of hydrogen peroxide to the clarifiers. Slowing the denitrification process reduces the amount of suspended solids in the effluent.

In filamentous bulking, long bacterial strands trap sludge particles and form mats, which cause settling problems in secondary clarifiers. Adding 100-ppm doses of hydrogen peroxide continuously to the return sludge line for five to seven days will usually bring bulking under control.

Usually, a laboratory evaluation of hydrogen peroxide in treating a waste will result in the early optimization of H_2O_2 application at minimum cost. For example, even though the control of phenol with hydrogen peroxide is well established, the presence of catalytic decomposition agents and other reactive substances in the waste stream can influence the amount of hydrogen peroxide needed. When the significant variables—temperature, pH, catalyst concentration and peroxide-to-pollutant ratio—are poorly defined (as with mixed organics), these should be analyzed in the laboratory.

Plant tests normally should be conducted after a laboratory evaluation. However, such an evaluation is not always useful. In BOD reduction and filamentous bulking control, the most practical approach often is to proceed with a full plant trial, because of the difficulty of simulating biological conditions in the laboratory. Pilot tests should be run when a full-scale unproven test would be costly or the equipment for a full-scale test is not available. For field tests, a drum, tubing and pump setup is enough.

Injection methods

Rapid dispersion of hydrogen peroxide into the waste stream is critical, because catalytic decomposition agents are present in almost every waste stream and these compete with the pollutant in the reaction with hydrogen peroxide.

The reaction rate between hydrogen peroxide and the pollutant will usually dictate the point where the H_2O_2 should be added. An addition point 15 to 30 minutes upstream of the control point should be chosen. If this is not practical, because of poor mixing conditions or other physical limitations, a compromise addition point should be found and the variables (such as peroxide injection rate, catalyst and temperature) adjusted. In some cases, the addition of a holding tank might even become necessary to increase the reaction residence time.

For wastewater applications, a 50% hydrogen peroxide solution is normally used and is recommended. Lower concentrations, such as 35%, offer no significant safety advantage and carry a cost penalty in freight charges for additional water. Higher strengths, such as 70%, should be avoided because this grade can form detonable mixtures with a wide range of organics.

When choosing an injection method, consider the amount of hydrogen peroxide to be used, the precision needed, equipment costs and setup time.

The simplest, cheapest and fastest method is the gravity-feed system. To set it up, attach a plastic spigot to a drum, lay the drum on its side (preferably on a drum rack) and feed by gravity. This system is recommended when flowrate precision is not critical. A rotameter will improve flow control.

Pump-feed systems are most precise and require less attention. A low-cost system for permanent service or field trials consists of a drum, polyethylene tubing, polyvinyl chloride fittings and a plastic pump. A more durable system requiring less maintenance consists of drums or an aluminum storage tank, aluminum or stainless-steel transfer and injection lines, and a high-quality pump.

Most industrial applications call for 35 to 50% concentration of hydrogen peroxide, in which form it is shipped in drums and tank wagons. To most bulk users, however, it is shipped as a 70%-by-weight solution and diluted at the storage point.

Storage safety

Stabilizers are added to hydrogen peroxide to inhibit catalytic decomposition caused by metals and other impurities that might contaminate it during shipment, storage or handling. However, this inhibition is limited. Gross contamination will result in decomposition despite the stabilizers.

Hydrogen peroxide's stability is also sensitive to pH. Alkaline solutions are generally less stable than acid ones. For this reason, the chemical is usually shipped and stored as a slightly acid solution.

Temperature is also a factor in stability. The decomposition rate increases approximately 2.2 times for each $10°C$ rise from 20 to $100°C$ (1.5 times for each $10°F$ rise from 68 to $212°F$). Low temperature has little effect unless substantially below $0°C$.

Decomposition produces heat, as well as water and oxygen. In dilute solutions, the heat can be readily absorbed by the water. In concentrated solutions, the heat raises the temperature of the solution and accelerates decomposition.

Because freezing points are low and boiling points are high for hydrogen peroxide solutions ($-52°C$ and $114°C$, respectively, for a 50% concentration), storage tanks can usually be located outdoors in both cold and hot climates.

Storage must be in accordance with National Fire Prevention Code 43A. Under this code, a 70% solution is listed as a Class 3 oxidizer, and solutions between 27.5 and 52% as Class 2 oxidizers.

Fires are unlikely from spills of less than 35% solutions, but can follow if such a spill is allowed to dry on combustible material.

To ensure hydrogen peroxide purity, storage must be restricted to original containers, or tanks of compatible materials, properly designed and thoroughly passivated. Once removed from the original container, hydrogen peroxide should not be returned to it.

Tanks must be vented and located away from sources of direct heat. Shipping drums must always be stored head up, preferably on a concrete floor in a cool, clean ventilated fireproof area having a source of water for washing away spills.

Materials of construction

A high-purity aluminum alloy—specifically, AL 5254, a 99.6% alloy—is recommended for bulk storage systems. Aluminum alloys such as 1060 are used for transfer piping, and Type 316 stainless steel for transfer

pumps. Iron, steel, copper, brass, nickel and chromium are not recommended for handling concentrated solutions of H_2O_2.

Aluminum pipe should be Van Stone flanged, and welded. Screwed fittings should be kept to a minimum. Schedule 80 pipe should be used if threads are to be cut. Teflon tape is suggested for sealing screwed fittings.

Valves are generally of aluminum alloy 356 but may be of 300 series stainless steel or porcelain under certain conditions. Preferred valves are ball-type with internal relief seals. However, valves of other designs, made of compatible materials, may be used under certain conditions. If diaphragm valves are used, Kel F or Teflon diaphragms are preferred.

As a rule, any valve that could trap hydrogen peroxide must be vented in some manner to prevent pressure buildup in the valve body while the valve is closed.

Centrifugal pumps of aluminum alloy 356 or 300 series stainless steel are recommended for most installations. Mechanical seals may be of glass-filled Teflon,

A variety of industrial wastes treated with hydrogen peroxide   Table I

Industry	Chlorine	Cyanide	Formaldehyde	Hydrogen sulfide	Hydroquinone	Mercaptans, thiols	Phenolics	Sulfite	Thiosulfate	Other sulfur compounds
Electroplating		●								
Textiles	●		●	●			●			●
Rubber processing					●	●				●
Paint and ink						●				
Organic chemicals	●	●	●	●		●	●	●	●	●
Carbon manufacturing								●		
Leather tanning				●						
Pulp and paper				●		●				●
Inorganic chemicals	●	●		●			●	●	●	●
Plastics and synthetics		●		●						●
Petroleum refining				●		●	●	●		●
Iron and steel		●		●			●			●
Ore mining and dressing							●			●
Food processing										●
Miscellaneous chemicals							●			●

Comparative costs of common oxidants in waste treatment*   Table II

Oxidant	Oxidant cost, $/lb	Hydrogen sulfide		Hydrogen cyanide		Phenol	
		Weight ratio	Cost, $/lb	Weight ratio	Cost, $/lb	Weight ratio	Cost, $/lb
Hydrogen peroxide (50% solution)	0.25 - 0.30	2:1	0.54	2.2:1	0.59	4:1	1.08
		8:1	2.16				
Chlorine	0.08 - 0.14	8.9:1	1.07	2.7:1	0.32	Not practical	
Potassium permanganate	0.50 - 0.70	3.9:1	2.34	12:1	7.20	15.7:1	9.42
Chlorine dioxide	2.00 - 20.00	8:1	20.00	—	—	1.5:1	3.75
Ozone	0.35 - 0.70	14:1	0.70	1.5:1	0.75	2.5:1	1.25

*Excludes capital costs for equipment

ceramic, or 300 series stainless steel. Other types of packing should be Teflon, never graphite, bronze, copper, lead or common packing materials.

Gaskets of Koroseal 700, Kel F and Teflon are satisfactory.

Flexible steel hoses for unloading or transferring should be of 300 series stainless steel.

All equipment for storing and handling hydrogen peroxide solutions must be thoroughly cleaned and passivated before being placed in service. Contaminants left behind or embedded in the surface of the storage container could cause decomposition. Gross contamination could result in the rupturing of storage tanks and other equipment.

Aluminum and stainless steels should be passivated as follows: (1) preliminary detergent wash to remove all oil, grease and loose contaminants; (2) trichloroethylene rinse; (3) thorough rinse with clean water, followed by draining to remove the contaminated washing medium; (4) exposure of all parts to an acid solution (preferably nitric); and (5) another thorough rinsing with clean water, followed by draining to remove all acid and water-soluble contaminants.

The size and type of facility will generally determine the specific passivating procedure. In some cases, tanks, pipes and other equipment may be passivated separately before they are installed. In other cases, it may be necessary to passivate after installation.

Transportation modes

Hydrogen peroxide is shipped in drums, tank wagons and tank cars:

Standard-size drum containers come in 15, 30 and 55-gal sizes. These are polyethylene, polyethylene-lined fiber, or steel overpack drums. Drum rockers, bung wrenches, valves and pouring spouts are available.

Tank wagons with a shipping capacity of 4,000 gal are used for delivery to customer-owned bulk storage tanks. These meet MC 312 specifications, and can carry 70% solutions for dilution on delivery, or 50% and 35% solutions. Most tank wagons are equipped with pump, hoses and fittings for attachment to storage systems.

Shipment to large consumers can be by dedicated aluminum tank cars, which range in capacity from 4,000 to 20,000 gal. They conform to AAR specifications for delivering up to 70% solutions. Most cars have an expansion dome with an inspection manhole, a combination safety vent with filtered compressed-air connection, and an unloading connection. Unloading and dilution are usually performed by the customer.

Safety considerations

Although hydrogen peroxide and its decomposition products are not systemic poisons, contact can cause irritation.

Concentrated vapors irritate the mucous membranes and the eyes. Eye contact with hydrogen peroxide is particularly dangerous, because corneal burns occur very rapidly. If the eyes should be contacted, they should be flushed thoroughly with water. Also consult a physician immediately. Safety glasses or, preferably, goggles should always be worn when handling concentrated solutions.

In addition to eye protection, rubber gloves and suitable protective clothing (such as aprons or coveralls made of polyester acrylic fiber, polyvinyl chloride, polyethylene or neoprene) should be worn. Protective clothing that is not fire-resistant must be washed thoroughly with water after contact. If the hydrogen peroxide is allowed to dry on the fabric, a fire could result, particularly if the clothing is soiled.

Moderate concentrations will cause whitening of the skin and a stinging sensation. This whitening is due to the formation of gas bubbles in the epidermal layer. In most cases, the stinging subsides quickly after thorough washing.

High concentrations can cause blistering if left on the skin for any length of time. Such blistering should be treated as if it were a burn.

Inhaled vapors can cause irritation and inflammation of the respiratory tract. According to the American Conference of Government Industrial Hygienists, a threshold limit value of 1 ppm ($1.4 \ mg/m^3$) of hydrogen peroxide vapor in air has been determined as a maximum exposure limit for any 8-h work day during a normal 40-h work week.

If vapors are inhaled, fresh air should be sought at once; if inhalation has been prolonged, a physician should be consulted immediately.

A mild disinfectant, hydrogen peroxide is useful in counteracting various microorganisms. Because of this antiseptic action, dilute hydrogen peroxide solutions are frequently used to treat open wounds and as a gargle or mouthwash. However, the contact of concentrated solutions (above 3%) with the membranes of the mouth should be avoided.

Under no circumstances should hydrogen peroxide be taken internally. If it is swallowed, water should be drunk immediately to dilute it, and a physician contacted. Do not attempt to cause vomiting.

Jay J. Matley, Editor

The authors

George W. Ayling is a technical service specialist with FMC Corp., specializing in research concerning the application of hydrogen peroxide in municipal, industrial and geothermal pollution control, and in-situ mining of uranium. He holds a B.A. degree in biology/chemistry from Mansfield State College (Pa.).

Harry M. Castrantas is a technical service specialist in municipal, industrial and geothermal pollution control with FMC Corp. (Industrial Chemical Group, P.O. Box 8, Princeton, NJ 08540; telephone 609-452-2300). He has chiefly worked in product, process and applications research involving hydrogen peroxide. A graduate of the University of Buffalo with a B.A. in chemistry, he is a member of the American Chemical Soc.

Process gives new life to contaminated sulfuric acid

Dilute waste acid, containing volatile and nonvolatile organics, is reconcentrated and purified sufficiently for recycling. The system features corrosion-resistant equipment.

Hans Rudolph Kueng and Peter Reimann, Bertrams AG

☐ A process developed jointly by Bertrams AG (Muttenz/Basel, Switzerland) and Bayer AG (Leverkusen, West Germany) for the concentration and cleaning of waste H_2SO_4 from 20 to 96-97% (a) requires less energy, particularly in concentrating from 20 to 78%, (b) uses a forced-circulation concentrator for concentrating from 78 to 96%, and (c) contains a purification reactor that removes nonvolatile organic contaminants and minimizes off-gas pollution problems (*Chem. Eng.,*

May 4, 1981, p. 18). As part of an overall program to eliminate the need to dispose of large quantities of waste acid generated at Bayer's Leverkusen plant, a concentrator and two purification units were installed in mid-1980 to recover 43 metric tons/d of 100% H_2SO_4.

Recently, concerns over pollution have led to restrictions on the disposal of waste acids generated in many chemical processes, making it necessary to develop methods for dealing

with the problem of purifying them. There are many different types of waste acid, some of which have high levels of nonvolatile organics and must be treated by thermally decomposing a preconcentrated acid (approximately 70%) to sulfur dioxide (SO_2) and water, and then oxidizing the SO_2 to SO_3 for use in making H_2SO_4. Because this method has high capital and energy costs, reconcentration and purification (by oxidation of organic contaminants) are preferable if the impurity level is low enough so that this method can be used.

Slightly contaminated waste acid has been reconcentrated in Pauling pots (mechanically agitated cast-iron vessels) at the Leverkusen plant.

OLD METHODS—Alternative systems for the recovery of waste sulfuric acid having low concentrations of organics have been used:

- Simonson-Mantius concentrator—Acid is concentrated in brick-lined vessels under vacuum, with heat supplied by tantalum steam heaters. It is used for concentrating H_2SO_4 from approximately 40 to 93%.
- Chemico drum concentrator— Waste acid is evaporated in brick-lined vessels by direct contact with hot furnace gases. The process has a range of 40 to 94% H_2SO_4.
- Pauling concentrator—Acid is heated directly in cast-iron vessels, with the escaping gases contacting feed acid countercurrently in a packed tower. The acid can be concentrated to 96% and purified to some extent.

All of these processes were found to have certain limitations, including: formation of deposits on the heating surfaces, which impaired heating efficiency; equipment constructed of materials that limited the processing temperatures; and final concentrations that did not meet the minimum of 96% H_2SO_4.

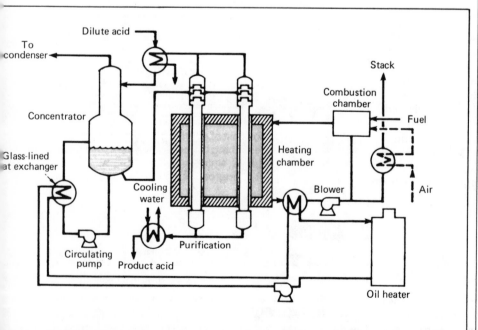

A 43-m.t./d H_2SO_4 (100%) installation at Bayer AG concentrates waste acid to 97% and removes most of the organic contamination   **Fig.1**

Originally published April 19, 1982

In addition, organic contaminants were not removed to acceptable levels, and undesirable offgases (SO_2 and NO_x) were created.

The key components of the new system (see Fig. 1) include a forced-circulation concentrator with a separate heat exchanger—both made of glass-lined steel for corrosion resistance, and a purification unit constructed of a silica tube encased in a metal housing. Waste acid undergoes preconcentration from 20 to 78% in a multieffect arrangement of falling-film evaporators and a forced-circulation concentrator, with the vapors used for evaporating heat. The final step of this operation uses the new Bayer/Bertrams concentrator because temperatures can be as high as 220°C, while concentrating from 78 to 96% H_2SO_4. If it is necessary to remove nonvolatile organics, the acid then passes through the purification unit, where it undergoes high-temperature (over 300°C) oxidation.

NEW CONCENTRATOR—The Bayer/Bertrams concentrator, made entirely of glass-lined steel, operates on a forced-circulation principle to avoid deposit formation on the heating surfaces. The waste acid is pumped from the evaporator through a glass-lined exchanger, where heat is supplied by heat-transfer oil that moves countercurrently to the acid flow. This serves to reduce any local overheating and the formation of deposits. Concentrations of at least 96% can be obtained at a vacuum of 60 mm Hg. Concentrating from 60 to 97% consumes about 185 m^3/h of natural gas.

A unit of this type has been used since 1979 to concentrate sulfuric acid obtained from the drying of chlorine at Cellulose Attisholtz (Lutterbach, Switzerland). It was observed that although the acid contained large amounts of iron sulfate (from previous equipment corrosion), the heat-exchanger surface remained clean after months of operation. Three other concentration plants are under construction for chlorine drying facilities, with capacities of 1 to 20 metric tons/d of 100% H_2SO_4.

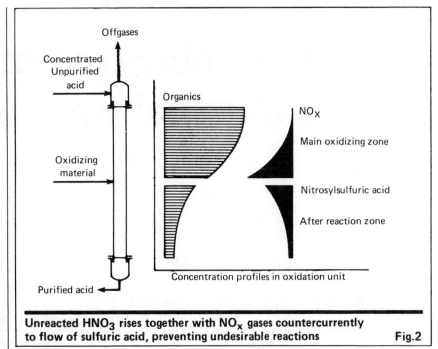

Unreacted HNO_3 rises together with NO_x gases countercurrently to flow of sulfuric acid, preventing undesirable reactions   Fig.2

ACID PURIFICATION—It was found that if conventional reaction vessels were used to purify waste acid with nonvolatile organics, the following undesirable effects occurred:

■ SO_2 was formed from the reaction of sulfuric acid and the organic materials, creating a pollution problem.

■ Unreduced NO_x escaped with the offgases, also adding to pollution.

■ Nitrosylsulfuric acid ($HNOSO_4$) was formed, impairing the quality of the purified acid.

To reduce these problems to an acceptable level, a vertical, tubular reactor that operates without any backmixing was developed. The unit consists of a silica cylinder, surrounded by a red-hot (900°C) metal tube heated by hot gases or electrically. To avoid damage to the silica cylinder, heat is transferred to the concentrated acid by radiation.

As the acid flows slowly down through the reactor at atmospheric pressure, it is gradually heated to the required oxidation temperature (300°C), while the oxidizing agent, in most cases 65% nitric acid, is added at a suitable location below the top of the unit. The main part of the oxidation

takes place close to the point of entry of the HNO_3 (see Fig. 2). Part of the unreacted HNO_3, however, rises together with NO_x in countercurrent flow to the treated H_2SO_4. As this happens, the NO_x reacts further, preventing the undesirable reaction between the organics and the H_2SO_4.

In addition, the exhaust gases contain little NO_x because most of the HNO_3 reacts with the organics:

$$5C + 4HNO_3 \longrightarrow 5CO_2 + 2N_2 + 2H_2O$$

In the lower section of the reactor, nitrosylsulfuric acid reacts with the residual organics:

$$2HNOSO_4 + C \longrightarrow CO_2 + 2H_2SO_4 + N_2$$

Both materials are reduced to acceptable levels in this manner.

Leonard J. Kaplan, Editor

The authors

Hans Rudolf Kueng, head of the Chemical Plant Div., Bertrams AG (Muttenz, Switzerland), graduated with a B.S. in mechanical engineering from the Winterthur College of Technology. He studied chemical engineering at the Swiss Federal Institute of Technology.

Peter Reimann manages the process development section at Bertrams AG. He holds a B.S. in chemistry from Winterthur College of Technology.

Anaerobic treatment of industrial wastewaters

Anaerobic wastewater treatment uses less energy and generates less sludge than does aerobic treatment. Now that energy and sludge disposal are so costly, anaerobic treatment is gaining favor.

Meint Olthof and *Jan Oleszkiewicz*, Duncan, Lagnese & Associates, Inc.

☐ Biological treatment of industrial wastewaters is now generally carried out in an aerobic reactor, one that converts organic matter to water and carbon dioxide in the presence of air. Anaerobic treatment, in contrast, converts wastes to methane and carbon dioxide in the absence of air. Its key advantages are low energy consumption, and production of a useful fuel gas—methane—rather than a large volume of useless sludge.

Despite these advantages, anaerobic treatment is not yet widely used in the U.S. for chemical-process-industries (CPI) wastes. It is still associated most with municipal-sludge stabilization and with treatment of easily biodegradable wastes such as food-industry effluents.

Besides low energy consumption and low sludge production, other benefits of anaerobic treatment include: ability to break down complex organic compounds at high concentrations; absence of odor; and suitability for seasonal operation. Also, established fears about toxicity and instability problems with anaerobic systems are not supported by recent experience. When properly designed and started up, anaerobic wastewater-treatment systems can: work at low temperatures (10–30°C); resist slugs of toxic compounds and eventually degrade them; provide steady performance under varying load conditions; and offer treatment efficiency as high as, or higher than, aerobic processes—using comparable reactor volumes and with lower operating cost.

Table I aims to correct some of the misconceptions about anaerobic treatment, and to show that it is a good alternative to aerobic treatment for many CPI wastewaters. We will now detail how anaerobic processes degrade waste, what the particular process options are, and what types of wastes each approach is suited for.

How anaerobic treatment works

In an anaerobic reactor, bacteria convert most of the organic matter in wastewater to methane gas and carbon dioxide in the absence of air. If the wastewater is concentrated enough, the energy value of the gas is greater than the energy needed to operate the reactor at

Common misconceptions about anaerobic treatment	Table I
Misconceptions	**Facts**
Anaerobic treatment processes are . . .	In fact, anaerobic treatment . . .
applicable only to concentrated wastes, slurries and sludges.	has been applied to streams with COD as low as 1,000 mg/L.
not applicable to streams containing difficult-to-degrade organics.	may be acclimated to degrade organic compounds, even some that aerobic treatment cannot degrade.
not applicable to streams having no suspended solids.	processes soluble wastes more quickly.
slow, requiring 8-10 day retention times and, therefore, high reactor volume.	requires hydraulic retention times comparable to those in aerobic treatment.
energy-inefficient, as the reactors must be heated	generates surplus energy when treating streams with more than 3,000 mg/L COD.
requires costly chemicals for process control.	requires only 10-20% of the nutrients that aerobic treatment does, and controls alkalinity where needed by recycle.

its optimal temperature (e.g., 33–35°C for one class of anaerobic bacteria).

Fig. 1 illustrates such a system, and contrasts it with an aerobic system treating the same waste. The aerobic approach requires a large amount of air, and results in conversion of at least 40–60% of the organic matter (measured as COD, or chemical oxygen demand) into excess sludge. In other words, the aerobic bacteria devote most of their energy to reproduction, creating a large mass of cells (sludge), while anaerobic bacteria devote most of their energy to producing methane.

The performance of an anaerobic treatment system depends primarily on two parameters: the rate of organics removal (degradation) and the rate of biomass growth (yield). Gas-generation rate is also important, but it is a function of the removal and growth rates.

Originally published November 15, 1982

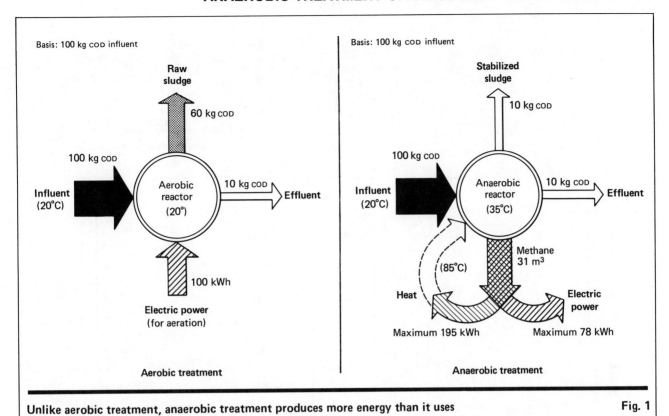

Basis: 100 kg COD influent

Raw sludge — 60 kg COD

Influent (20°C) — 100 kg COD → Aerobic reactor (20°) → 10 kg COD → Effluent

Electric power (for aeration) — 100 kWh

Aerobic treatment

Basis: 100 kg COD influent

Stabilized sludge — 10 kg COD

Influent (20°C) — 100 kg COD → Anaerobic reactor (35°C) → 10 kg COD → Effluent

Methane 31 m³

(85°C)

Heat — Maximum 195 kWh   Electric power — Maximum 78 kWh

Anaerobic treatment

Unlike aerobic treatment, anaerobic treatment produces more energy than it uses   Fig. 1

Compared with aerobic treatment, anaerobic treatment typically has a lower rate of organics removal and a lower rate of biomass growth. This means that, for the same removal efficiency (defined as % drop in COD), the anaerobic system:

1. Requires a longer solids residence time (SRT). This does not necessarily mean a larger reactor, however, because hydraulic residence time (HRT) need not be greater.

2. Produces less sludge. This is a key advantage of the anaerobic approach because disposal costs for sludge are typically at least $70–150 per 1,000 kg dry matter.

3. Recovers more slowly after a toxic shock, because the bacteria reproduce more slowly. But anaerobic bacteria attached to a substrate can protect themselves from toxics by becoming dormant when exposed. In some systems, the bacteria become active again once the toxic stress is removed.

The growth and removal rates are affected by various wastewater characteristics. To understand how, it is important to know the mechanisms of anaerobic degradation. Roughly, there are three stages: hydrolysis of suspended solids; acetogenesis, or conversion of soluble organics to volatile fatty acids (mostly acetic acid); and methanogenesis, or conversion of the volatile fatty acids to methane. Acetogenesis and methanogenesis are carried out by different anaerobic bacteria.

Wastewater characteristics

Let us look at several wastewater characteristics to see how they affect the removal and growth rates:

Transient toxicity. A sudden influx of a toxic material not usually present in the wastewater can kill some of the bacteria. In general, the methanogenic bacteria are the most vulnerable, so one would see a drastic drop in gas generation and a rise in volatile-fatty-acid concentration almost immediately. Where transient toxicity is expected, an alarm system and an easily-flushed reactor (e.g., fixed-film) will limit the damaging effects.

Chronic toxicity. A steady influx of a toxic chemical can be handled either by diluting it (e.g., in a recycle reactor) below its threshold level or by acclimating the bacteria to it—exposing them to the compound over a long period of time so that they become able to degrade it. There is very little correlation between a compound's toxicity to unacclimated bacteria and to acclimated bacteria. In general, acclimation enables a culture to withstand 5–15 times the unacclimated threshold level (see Speece in [1]). For example, formaldehyde and phenol are toxic to unacclimated bacteria at concentrations of 100–200 mg/L (ppm), but acclimated bacteria can withstand 1,000 mg/L or greater. In general, halogenation, double bonds, benzene rings and aldehyde structures in a chemical compound increase its toxicity and retard its degradation by anaerobic bacteria.

Solids concentration. Suspended-solids content affects the time needed for hydrolysis; thus hydrolysis may be the rate-limiting step for degradation of a concentrated high-suspended-solids wastewater. A high level of dissolved inorganics (e.g., above 30,000 mg/L) inhibits the metabolism of anaerobic bacteria, so such a wastewater may have to be diluted. In terms of dissolved organics, expressed as COD, anaerobic systems typically handle wastes having more than 3,000–5,000 mg/L, while aer-

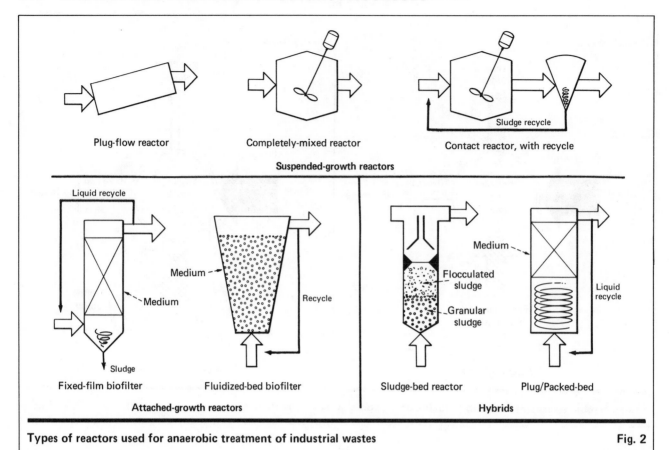

Types of reactors used for anaerobic treatment of industrial wastes   **Fig. 2**

obic systems are limited (by oxygen mass-transfer) to wastes below 5,000 mg/L.

Temperature. A steady temperature in the correct range is essential to efficient operation. Anaerobic treatment uses one of three types of bacteria, each of which has an optimal temperature: psychrophilic (15°C); mesophilic (35°C); and thermophilic (55°C). As one would expect, the higher-temperature bacteria have higher growth and organics-removal rates. Since temperature control is so important, anaerobic systems generally provide for heating the wastewater, often using the generated methane as a fuel. The anaerobic reactions themselves generate too little heat to be considered in the energy balance (see Rozzi in [*1*]).

Alkalinity. Ammonium-yielding compounds (e.g., proteins) and metal cations can neutralize excess volatile acids, aiding methane generation. Controlling the pH may be necessary if too little alkalinity is produced during treatment.

Sulfates. High concentrations of sulfate ion (i.e., exceeding 1,000–2,000 mg/L SO_4^{-2}) may require special measures to prevent reduction of sulfates to toxic sulfides or corrosive hydrogen sulfide (H_2S). When there is H_2S in the product gas, scrubbing is required before the gas can be combusted to provide heat, mechanical power or electricity.

Anaerobic reactors

Anaerobic treatment is typically carried out in continuous reactors. Compared with aerobic reactors, the anaerobic ones require greater solids residence time—

due to slower degradation—but do not need a greater reactor volume because they produce less solids and handle greater solids concentrations. Fig. 2 shows the various types currently in full-scale or pilot-plant use, and Table II compares their features. Let us now look at how these anaerobic reactors work and what types of wastes they are applied to.

Flow-through reactors without solids recycle may be either plug-flow or completely-mixed. The plug-flow reactor, which operates at medium (35°C) or high (55°C) temperature, can handle the most concentrated wastes and semisolid materials—i.e., over 6% volatile solids. For less-concentrated wastewaters and municipal sludges (2–6% volatile solids), the completely-mixed flow-through reactor is more efficient.

Contact reactors with solids recycle offer long solids residence times (about 20 days) and short hydraulic residence times (0.5–5 days), resulting in high efficiency despite small reactor volumes. These reactors are best used for wastes that are warm (25–50°C) and relatively dilute (0.5–1% volatile solids), like many wastes generated by the CPI.

Attached-growth reactors (also called biofilters) are recommended for relatively cold (15–25°C) and dilute (0.5% or less volatile solids) wastewaters because of their high SRT/HRT ratios (over 50). Unlike flow-through and contact reactors, where the bacteria are simply suspended in the fluid, biofilters provide a substrate for bacteria to grow on. Biofilters offer stability, resistance to toxics and freedom from sludge-separation and -recycle problems. Their key disadvantage is a tendency

Comparison of anaerobic reactors (the more stars, the better)   Table II

| | Suspended-growth reactors | | Attached-growth reactors | | |
	No recycle	Solids recycle	Fixed-film biofilter	Fluid-bed biofilter	Sludge-bed reactor
Ease of startup	*	*****	****	***	**
Ease of operation	**	**	*****	***	***
Controllability	*	*****	***	***	***
Resistance to shocks:					
Temperature	*	***	*****	*****	*****
Toxic	*	***	*****	*****	****
Organics-load	*	****	*****	*****	*****
Solids-load	****	***	*	**	**
Ability to process sludges from pretreatment and polishing steps	*****	****	*	*	*

to plug with influent solids. There are two principal types of anaerobic biofilters:

Stationary-medium use plastic packing to enhance contact and retain solids. While random (dumped) packing has typically been used, oriented packing reduces plugging problems. Upflow is more typical, but downflow allows for pushing higher-solids streams through the reactor.

Fluidized-bed biofilters increase the surface area available for growth, allowing higher loading and providing greater resistance to toxics. Their drawback is greater energy use, since recycle ratios of 0.5 to 10 or higher may be needed to achieve fluidization. Media used may be either inert (e.g., sand) or reactive (e.g., carbon), with sizes from 0.3 mm to 50 mm.

Sludge-bed reactors are hybrids of attached and suspended growth that feature a bed of granular and floc-culated sludge suspended by upflow. They cost less than biofilters because they need no medium, but this is also a disadvantage, in that part of the sludge bed may be lost in case of a hydraulic surge or toxic upset.

Other reactors used for anaerobic treatment are: hybrids of attached and suspended growth (as in Fig. 2, for example); bacteria encapsulated in ceramic media; rotating biological contactors; and baffled sludge-bed reactors.

Today's reactors are largely the product of efforts to get high SRT in a minimum volume. While current research is still trying to increase solids residence time, greater effort is being devoted to improving reaction rate:

Operation in the thermophilic (55°C) temperature range rather than the mesophilic (35°C). This increases the reaction rate by about 50%, and is feasible for concen-

Examples of full-scale anaerobic-treatment plants processing industrial wastes in the U.S.   Table III

| Wastes | Reactor type | Influent | | Reactor | | Removal, % | Reference |
		Flow, m^3/d	Load, kg/d	Volume, m^3	Load, $kg/m^3/d$		
Meat packing	Contact	5,330	7,360 BOD	2,660	2.5 BOD	91 BOD	[5], 1961
Wheat starch, gluten	Fixed-film biofilter	454	3,200 BOD	1,140	3.9 COD	75 COD 49 BOD	[6], 1972
Guar gum	Fixed-film biofilter	N.A.	N.A.	N.A.	7.9 COD	60 COD	[7], 1979
Acetic acid; MEK; methyl acrylate and formate	Fixed-film biofilter	2,460	34,000 COD	5,670	6.0 COD	80 COD	Mfr. claim, 1981
Methanol; formaline; polyols	Fixed-film biofilter plus aerobic lagoon	4,160	53,100 COD	5,700	9.3 COD	88 COD	Mfr. claim, 1981
Beet sugar	Contact plus activated sludge	25,000	30,000 COD	25,000	1.2 COD	95 BOD	Mfr. claim, 1981
Beet sugar	Contact plus activated sludge	28,000	56,000 COD	28,000	2.0 COD	97.5 BOD	Mfr. claim, 1981
Starch	Sludge-bed	912	20,000 COD	1,800	11 COD	85 COD	Mfr. claim, 1981
Brewery	Sludge-bed	23,000	11,000-30,000 COD	4,700	8-16 COD	80 COD	Mfr. claim, 1982

N.A. Not available   BOD Biochemical oxygen demand   COD Chemical oxygen demand

trated effluents from plants having enough waste heat to maintain a 55°C reactor. This is used for municipal sludges.

Separation of the process into stages: that is, an acidic reactor where organics are broken down to volatile fatty acids, followed by a methanogenic reactor where the acids (and perhaps the CO_2) are converted to methane. This is feasible for waste streams in which conversion of acids to methane is the rate-limiting step (e.g., petrochemical wastes that contain secondary and tertiary branched organics).

Applications experience

The best applications for anaerobic treatment are in handling warm, concentrated organic wastewaters. Most of the present full-scale installations are in the grain-milling, sugar-refining, food-processing, fermentation and organic-chemicals industries, but recent applications include pharmaceutical, textile, tanning, pulp and paper, petrochemical, coal-processing and synfuels effluents. The current trend toward concentrating waste streams through in-plant recycle means that anaerobic treatment will be more widely used, because aerobic systems handle mainly dilute streams that are largely free of suspended solids.

Table III provides details of nine full-scale applications in the U.S.; there are at least a dozen other CPI installations, plus hundreds of municipal-sludge digesters. Among the plants listed, the greatest loadings are handled by sludge-bed reactors treating easily-degraded food-processing wastes.

Recent research has turned up some new applications. For example:

Organics considered inhibitory or toxic to anaerobic bacteria can be degraded efficiently if the system provides adequate retention of acclimated bacteria (high SRT). Such compounds include: pesticides, phenols, hydroquinone, nitrobenzene, vinyl acetate, nitrosamines, phthalic acids and phthalate esters, trihalomethanes, lignin, and wood fiber [3,4].

Anaerobic treatment can break down or change the structure of compounds that aerobic treatment cannot degrade. For example, anaerobic treatment followed by aerobic polishing may be used in tanning, textile or pharmaceutical applications where color frequently remains if aerobic treatment alone is used.

Anaerobic pretreatment of effluents, before discharge to a municipal sewer system, can cut COD levels by 80–90% at relatively low cost when compared with aerobic treatment. For phenolic or petroleum chemicals, loadings of 2–5 kg COD/m³-d would be reasonable; loadings of 5–8 kg COD/m³-d and above could be considered for more degradable (e.g., sugar-refining or corn-milling) wastewaters.

Economics

Whether anaerobic treatment is economical depends on whether or not the process generates enough methane to supply the necessary heat. If it does, then the power-cost and sludge-disposal-cost advantages make anaerobic treatment preferable over aerobic treatment. Capital cost may be a factor but, in general, anaerobic reactors are about the same size as aerobic ones and cost

Example: Economics of treating a concentrated waste		Table IV

Basis: Removing 1,000 kg COD from a wastewater having 7,500 mg COD/L.

Cost component	Aerobic process (activated sludge)	Anaerobic process (biofilter)
Power for mixing and aeration, kWh	1,000	75
Excess sludge produced, kg	500	50
Sludge-conditioning chemicals, kg	3	0
Phosphorus, kg	10	1
Nitrogen, kg	50	5
Cost*	$139	$12
Methane production, m³	0	320
Net cost (income)*	$139	($40)

*Assuming $4.58/1,000 ft³ for natural gas; 8¢/kWh for electric power; $80/m.t. solids for disposal of sludge; $250/m.t. for phosphate and ammonia nutrients.

only about 10% more for a given waste stream. Let us now look at the key economic factors:

Organics concentration affects the amount of methane produced. At 0.350 m³ CH_4/kg COD (theoretical), methane production must be enough to keep the reactor at 35°C year-round (for the usual mesophilic system). This depends on climate, input temperature and other factors, but in general a stream having less than 2,000–3,000 mg/L COD will not generate enough methane. In such cases, one can run the reactor at lower temperature, and compensate for lower removal rate by higher solids residence time and greater mass of solids in the reactor. Or one can heat the anaerobic reactor for part of the year with steam or with a waste-heat source from the plant.

Incoming-waste temperature affects the heat requirements. Of course, many CPI waste streams are warm or even hot compared with other wastes handled anaerobically (e.g., municipal sludge).

Scale (kg COD/d) affects the economics of gas recovery. For very small systems, it may not be economical to use the methane produced, especially if cleanup is required before use.

Nutrient (nitrogen and phosphorus) requirements are linked to sludge generation, so anaerobic systems typically need only 10–20% as much nutrients as do aerobic systems. Food-type wastes usually have nutrients available, but chemical-synthesis wastes may require addition of nutrients.

Table IV compares operating costs for aerobic (activated-sludge) and anaerobic (biofilter) treatment of a relatively concentrated (7,500 mg/L COD) wastewater. On the basis of 1,000-kg/d COD removal, the anaerobic system in this case costs about $180/d less to operate when the value of the methane is considered.

A cost comparison performed for an organic-chemicals plant discharging 1,900 m³/d of 7,500-mg/L-COD wastewater showed that an anaerobic reactor would be smaller in volume and no greater in cost than a conven-

Example: Total cost of treating easily biodegradable wastewaters		Table V

Basis: 10°C wastewater containing 20,000 mg COD/L

Type of system	Relative total cost* $0.04/kWh	Relative total cost* $0.12/kWh
Aerobic (coagulation; activated sludge; denitirifer; sludge treated chemically)	100	127
Anaerobic/aerobic (flow-through anaerobic digester; activated sludge; denitrifier; sludge recycled)	115	111
Anaerobic (anaerobic contact reactor; anaerobic biofilter; trickling filter; denitrifier; sludge recycled)	77	73
Anaerobic (clarifier; anaerobic biofilter; trickling filter; denitrifier; sludge processed in separate reactor)	69	39

*Relative cost includes electric power at given price, 10-yr capital amortization at 8% interest, operation and maintenance, and methane and heat recovery.

tional activated-sludge aerobic reactor. Power cost for the aerobic system was figured at $260,000/yr (at 8¢/kWh), while for the anaerobic reactor it was only $20,000. The anaerobic reactor would also produce $370,000 worth of methane per year (at $4.58/1,000 ft³, which is 70% of the 1981 price of natural gas in Philadelphia).

Design concerns

The theoretical maximum yield of methane is 0.350 m³/kg COD; methane typically comprises about 75–80% of the total gas produced. If H_2S is present, this gas must be scrubbed before it can be stored or used as fuel. (There are several ways to remove the H_2S, one of which is to pass the product gas through iron "sponge" so that the H_2S precipitates as ferrous sulfide.)

In Europe, anaerobic-treatment plants frequently burn the gas in "total-energy modules," which convert 24% of the gas's energy to electricity and 60% to hot water. The gas may also be used as steam-boiler fuel, or sold to an outside user if the supply is large and reliable (the City of Los Angeles does this). In a smaller plant, the economics of collecting, compressing and desulfurizing the gas are less favorable, and such a plant may even require a backup supply of natural gas from a utility company.

Of course, economic comparisons such as in Table IV

do not mean that anaerobic treatment by itself is the best approach to treating concentrated wastes. Rather, some design creativity is needed, and perhaps a combination of aerobic and anaerobic treatment may be more economical. For example: In a plant discharging wastes to a municipal sewer system, aerobic polishing of anaerobically-pretreated waste may pay off if sewer-system charges are high.

Table V offers some perspective on system design by showing four ways to treat the same concentrated wastewater. Here, the systems include all of the steps needed in a biological treatment process (e.g., denitrification), and the relative costs include the cost of capital. The least expensive treatment for this particular case happens to be anaerobic treatment of the wastewater in a biofilter, with sludge treatment in a separate flow-through anaerobic reactor. Note that the methane value is linked to the power cost here.

In view of the many systems being marketed today, it is important to use bench-scale treatability studies and pilot-plant tests before choosing any anaerobic system. This allows one to tailor the process to the wastewater characteristics, rather than tailoring the wastewater (through solids-removal, for example) to a preselected process. The preferred testing approach is: toxicity studies; tests at bench scale in attached-growth and flow-through reactors installed in parallel; and finally pilot-scale tests on the selected process to determine scaleup factors. The final system may be ready-made or custom-designed.

Mark Lipowicz, Editor

References

1. "Proceedings of the International Seminar on Anaerobic Wastewater Treatment and Energy Recovery," Nov. 3–4, 1981, Duncan, Lagnese & Associates, Inc., Pittsburgh.
2. Cooper, P. F., and Atkinson, B., eds., "Biological Fluidized-Bed Treatment of Water and Wastewater," Ellis Horwood Ltd., Chichester, U.K., 1981.
3. Chou, W. L., Speece, R. D., Siddiqui, R. H. L., Acclimation and Degradation of Petrochemical Wastewater Components by Methane Fermentation, in "Biotechnology and Bioengineering Symposium No. 8," 1978, pp. 391–414.
4. Kobayashi, H., Rittmann, B. E., Microbial Removal of Hazardous Organic Compounds, *Env. Sci. and Tech.*, Vol. 16, No. 3, 1982, pp. 170A–183A.
5. Steffen, A. J., Bedker, M., Operation of Full Scale Anaerobic Contact Treatment Plant for Meat Packing Wastes, in "Proceedings of the 16th Industrial Wastes Conference," Purdue University, Lafayette, Ind., 1961, pp. 423–437.
6. Taylor, D.W., Full-Scale Anaerobic Filter Evaluation, in "Proceedings of the 3rd National Symposium on Food Process Wastes," U.S. Environmental Protection Agency Report R2-72-018, 1972.
7. Witt, E. R., Humphrey, W. J., Roberts, T. E., Full-Scale Anaerobic Filter Treats High Strength Industrial Wastewater, in "Proceedings of the 34th Industrial Waste Conference," Purdue University, Lafayette, Ind., 1979, pp. 229–234.

The authors

Meint Olthof is the manager of the Industrial Waste Section of Duncan, Lagnese & Associates, Inc., 3185 Babcock Blvd., Pittsburgh, PA 15237, where he has worked for a number of years on industrial-waste projects in the chemical, tanning, metal-finishing and iron-and-steel industries. He holds a Ph.D. degree in sanitary engineering from Vanderbilt University, and is a registered professional engineer in Pennsylvania, Ohio and West Virginia. He belongs to the Water Pollution Control Federation (WPCF), the American Soc. of Civil Engineers (ASCE), the International Assn. on Water Pollution Research (IAWPR), and several trade organizations.

Jan A. Oleszkiewicz is project manager in the Industrial Waste Section of Duncan, Lagnese & Associates, Inc., where he is responsible for projects involving both anaerobic and aerobic waste-treatment processes. Prior to joining the firm, he was principal investigator on two U.S. Environmental Protection Agency anaerobic-treatment projects overseas. He holds a Ph.D. degree in sanitary engineering from Vanderbilt University, and belongs to WPCF, ASCE, IAWPR and several other organizations.

Section X
Solid Waste Disposal

Waste-sludge treatment in the CPI

Expedient sludge processing can reduce many industrial waste-disposal problems. This study focuses on thickening and dewatering equipment, presenting design and operational parameters, and plant and laboratory performance data.

Industries reviewed:

☐ Organic chemicals
☐ Petrochemicals
☐ Inorganic chemicals
☐ Fine and pharmaceutical chemicals
☐ Pulp and paper
☐ Mineral processing
☐ Secondary-metals processing

Processes covered:

☐ Thickening . . . gravity and flotation
☐ Dewatering . . . centrifugation, filtration and heat treatment
☐ Secondary treatment . . . aerobic and anaerobic digestion
☐ Chemical fixation
☐ Byproduct recovery

R. W. Okey, D. DiGregorio and E. G. Kominek, Envirotech Corp.

☐ The chemical process industries (CPI) produce a great volume and variety of waste solids, which are separated as sludge from process waters and aqueous wastes. The sludge must be treated and prepared for environmentally sound disposal. Here is a CPI-wide review of operational principles and design approaches for sludge-processing equipment and techniques. Volume reduction is the key to economical disposal and is the theme of this article.

A typical waste-processing flowsheet describing solids capture and concentration is shown in Fig. 1. Raw wastewater is treated in clarifiers specifically designed to provide a clear overflow, regardless of underflow concentration. Primary sludge is defined as the liquid waste containing suspended material that remains after primary clarification. The clear overflow is sent to wet-line bioconversion and secondary clarification units, which produce secondary sludge. The primary sludge is the main focus of this article because it possesses a unique character that differs from industry to industry. The secondary sludge is basically waste-biological cell tissue and is generally similar for all industries.

Primary sludges vary widely in specific gravity, chemical and biological stability, solubility, toxicity and particle size. However, some general classifications can be made. Table I lists primary sludges and waste solids generated by the CPI.

Table II lists thickening and dewatering performance data for sludges typically generated by certain industries. Though available cost data are insufficient to bracket the various industries, thickening and dewatering usually cost about $10-$50/ton of dry solids.

Originally published January 29, 1979

Pretreatment

Most sludges require chemical conditioning prior to thickening and dewatering. Conditioners can be divided into two broad categories: inorganic coagulants, such as lime, hydrated aluminum sulfate (alum) and ferric salts; and organic flocculants, which have high molecular weight and are cationic, anionic or nonionic.

Inorganic coagulants have widespread use, and in most cases effectively capture colloids. They have the

Solid-wastes generated in the CPI		Table I
Origin	**Physical and chemical characteristics**	**Comments**
Coke manufacture	Coke and coal fines	Generally contained in scrubber wastes
Dyes and pigments	Reaction or raw-material sludge, highly variable	—
Fine-chemical and pharmaceutical	Raw-material solids, biological wastes	—
Inorganic chemicals	Insoluble salts, tailings, slimes	—
Metal processing (secondary)	Ash, scrubber wastes, metal hydroxide sludges	—
Petrochemical	Oily, greasy, asphaltic	Usually float
Plastic and rubber	Latex or plastic crumbs, often alum coagulated	—
Pulp and paper	Fibrous; some fine filter, often with lime or alum	—

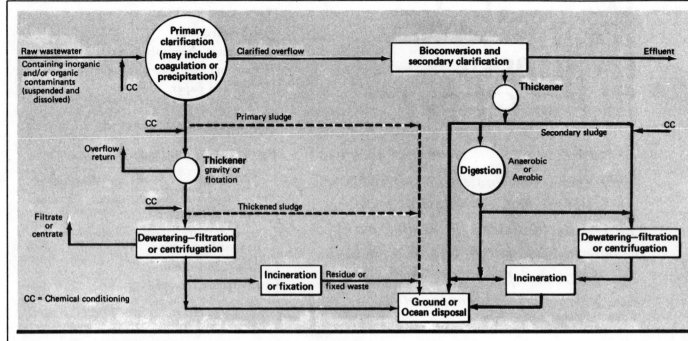

Typical waste-sludge processing flowsheet employed in the chemical process industries   **Fig. 1**

disadvantage of contributing substantially to the sludge volume, which frequently increases the overall dewatering problem.

Organic flocculants are specific in use and effective at low dosages. However, they do not scavenge colloids as well as the coagulants, and under conditions of very high or low pH undergo size degradation, with consequent loss of effectiveness.

Bench-scale tests are used to determine the optimum conditioning agent and procedure.

Heat treatment also is effective in preparing a sludge for thickening and dewatering, because it destroys the gelatinous components, which bind water. This method is discussed in the dewatering section.

Thickening

The purpose of thickening is to reduce sludge volume as much as possible prior to dewatering and/or disposal. Sludge-dewatering characteristics improve as solids concentration increases. Maximizing thickening results in cost and equipment performance benefits during dewatering operations.

Consider the dewatering of a calcium carbonate sludge by filtration. This type of sludge cannot be filtered effectively at feed concentrations of less than 6 or 7 weight %. At a concentration of approximately 10%, a filtration rate of 10–12 lb dry solids/h/ft^2 of filter area can be obtained. However, if the feed sludge is thick-

Industry	General characteristics of sludge	Concentration after thickening, % solids	Concentration after dewatering, % solids	Comments
Coke manufacture	Scrubber fines	20-25	40-50*	Vacuum filtration
Dye and pigment	Primary	7-10	20-30*	Vacuum filtration
Fine chemical and pharmaceutical	Biological	2-5*	15-25*	—
Inorganic chemical	Slimes, fines, tailings	10-30*	30-80*	—
Metals (secondary)	Metal fines, metal hydroxide	20-50	40-80	—
Petrochemical	Oily, greasy asphaltic	2-5	10-40	Thicken by flotation
Plastic and rubber	Latex or plastic crumbs, often alum-coagulated	7-8	15-25	—
Pulp and paper	Fibrous, fillers, often with lime or alum	2-50 2-5	20-55 12-20	Combined primary and waste-activated, 30-50% with pressure filtration

*Estimated

CPI waste-sludge thickening and dewatering performance data (summary of Table III, V, VI, and VII)   **Table II**

Industrial performance data for thickening of waste sludge							Table III
Industry	Origin or nature of sludge	Feed-solids concentration, %	Loading lb/d /ft^2	Underflow solids, %	Polymer or coagulant type, concentration	Comments	Reference
Coke	Scrubber blowdown (coke and coal dust)	Unknown	Unknown	20-25	Betz 1115 0.25-0.50 mg/L	—	34
Dye and pigment	Primary sludges	Unknown	Unknown	7-10	Nalco 683 4-5 mg/L	Solids contact clarifier	34
Glass etching	Calcium fluoirde	Unknown	Unknown	15	Lime to pH 11.3	Solids contact clarifier	34
Latex production	Alum coagulated primary	Unknown	Unknown	Unknown	Alum	4-6 days settling in lagoon	6
Lead smelter	Scrubber wastewater	0.1-0.2	7.8	40-50	Unknown	—	34
Petrochemical	Oily general	Unknown	Unknown	2-5	Unknown	Flotation thickener	13
Pharmaceutical	General	Unknown	6	Unknown	Unknown	—	12
Pulp and paper	Mixed primary and secondary	0.3-0.35	3.1	3.3	Magnaflox	Primary to secondary > 20:1	34
Pulp and paper	Ground-wood or waste-activated	Unknown	Unknown	4.5 (float) 7-12	Unknown	Flotation disk-centrifuge	4
Pulp and paper	Deinking Glassine Waste-activated	1 0.5 1.6	2.2 2.0 13.3	3-5 3-4 3-4	Unknown Unknown Unknown	— — —	34
Pulp and paper integrated kraft	Biological waste Primary Mixed primary and secondary	Unknown Unknown Unknown	34.1 13.5 34.0	2.5 3.0 3.0	Unknown Unknown Unknown	— — 1:1 and 2:1 ratios	10
PVC production	Chemical or primary	2.5	26.6	8	Unknown	Design values from bench scale	5
Rubber	Primary	Unknown	Unknown	15.4 (overflow)	Unknown	Flotation	7
Rubber, synthetic latex	Primary Secondary	144 mg/L 129 mg/L	5.5 gal/min/ft^2 4.2 gal/min/ft^2	0.62-10.63 7.80 Avg 1.24-18.08 5.75 Avg	Alum 20-62 mg/L Cation 0.54 mg/L Anion 0.61-0.85 mg/L	Recycle rate 22-30% Air to solids 0.1 lb/lb	8
Steel (specialty)	Rollingmill and furnace cooling water, pickling rinse and cooling, tower blowdown, metal hydroxides	8-15	Unknown	25	Lime	—	11
Zinc electrolytic refinery	Reactor clarifier, gypsum from H_2SO_4 neutralization, metal hydroxides	Unknown	Unknown	Unknown	Unknown	Disposal to pond	9

ened to 20–22% dry solids, filtration rates can be as high as 50–60 lb dry solids/h/ft^2. At higher filtration rates, the size and cost of equipment is smaller.

Thickeners can concentrate slurries as dilute as a few hundred ppm to sludges of 5–15 weight % dry solids (or higher). Thickening rates depend on properties of the feed, feed solids concentration, retention time, feed concentration (underflow) required for dewatering, and overflow clarity required for disposal or recycle. Generally, thickening is accomplished by gravity sedimentation or flotation.

Sedimentation

A typical gravity thickener, Fig. 2, consists of a feedwell that baffles the slurry prior to entry into the sedimentation zone, and a rotating floor scraper that transports the thickened slurry to the centrally located underflow discharge-hopper. The feedwell serves as a flocculation zone when chemical conditioning is used.

The most common configuration for a gravity thickener is a circular tank. Side-water depths vary from 10 to 12 ft. Tank diameters vary from as small as 10–15 ft to as large as 600 ft. Floor slopes for thickeners are generally 10–15 deg.

Gravity-thickener design criteria are almost always established from experience or bench-scale tests. Traditionally, these tests are conducted using 2-L cylinders equipped with slow-turning rakes, and are interpreted by the Coe-Clevenger or the modified Kynch (Talmadge and Fitch) method. Each method has certain requirements that must be met, which involve selecting an operating point on a settling curve to calculate thickener sizing. These tests are also useful in predicting overflow clarity, underflow solids content, and the efficacy of chemical flocculants.

Gravity thickening of organic sludges (particularly waste-activated sludge) is complicated by anaerobic action. If the temperature is warm, bacteria in the sludge will decompose organic matter, releasing gases. This causes flotation problems, hinders compaction and creates noxious odors. Sedimentation-thickener loadings for organic sludges range from 4 to 12 lb/d/ft^2, while flotation thickening can handle loading rates of 36 to 48 lb/d/ft^2.

Flotation

Flotation thickening is used for slurries containing solids that float rather than settle, or have slow settling rates or poor compaction. Flotation keeps the system aerobic. A typical dissolved-air-flotation (DAF) thickener is shown in Fig. 3. Most flotation thickeners use recycle pressurization. Part of the flotator subnatant is pressurized. The incoming feed does not pass through the pressurization system but rather is mixed with pressurized recycle in the inlet-diffuser.

Flotation thickeners are rectangular or circular. Standard rectangular flotators vary in size from 100 ft^2 to as large as 1,800 ft^2. Rectangular flotators are usually used for small applications and where space considerations are paramount. Both types are equipped with surface skimmers and floor rakes. The surface skimmers remove float from the thickening tank to a float sump.

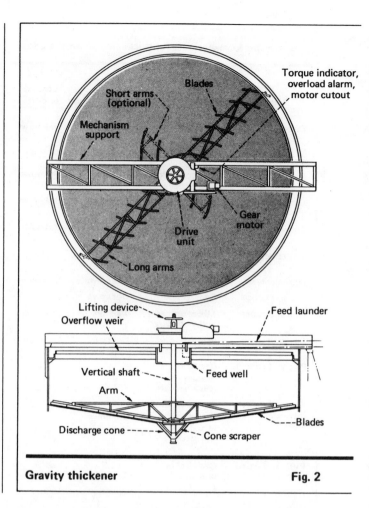

Gravity thickener   **Fig. 2**

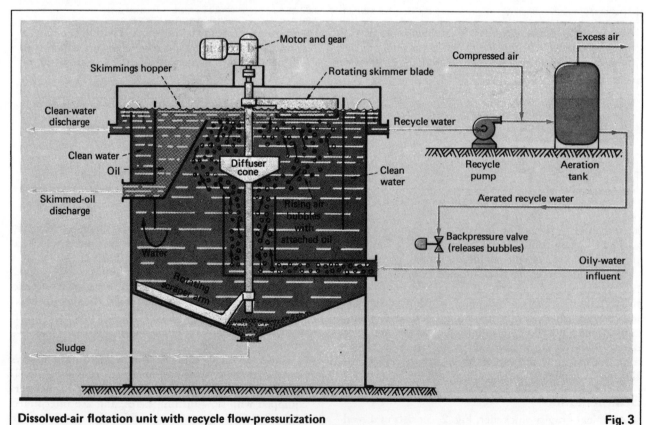

Dissolved-air flotation unit with recycle flow-pressurization   **Fig. 3**

Variables affecting centrifuge performance	Table IV

Machine variables	Process variables
Bowl diameter	Type of sludge
Bowl length	Loading rate
Differential speed	Chemical conditioning
Centrifugal force	
Beach length	
Beach angle	
Pool depth	

These skimmers should be of variable speed, 2–25 ft/min, to provide maximum flexibility. The floor rakes remove heavier solids that cannot be floated.

Flotation thickening may be evaluated by bench- or pilot-scale tests. Bench-scale apparatus is normally used to investigate flotation on a batch basis, predicting thickening performance in terms of float solids content and solids capture at various air/solids ratios. Bench-scale tests are also useful in evaluating the effect of chemical flotation aids on float solids and solids capture. Pilot-scale tests are useful for sizing as well as for predicting flotator performance. Bench and pilot devices are available from thickener manufacturers.

Table III lists industrial performance data for several sludge thickening operations.

Dewatering

The purpose of dewatering is to remove sufficient liquid from the thickened sludge to produce a cake that has optimal handling properties and solids content for subsequent processing or disposal. Dewatering is accomplished by mechanical methods, the most common being centrifugation and filtration, which include: vacuum, pressure, horizontal belt, belt press and precoat.

Flocculants are almost always used in dewatering industrial sludge to improve solids capture rate and cake discharge characteristics. Even though the cake may form well, the fine fraction usually contains an appreciable amount of material that tends to seal the cake. Flocculation coagulates the fine particles and helps prevent this. In a few cases, such as the filtration of water-softening sludge, no flocculation is required because the sludge does not contain a significant amount of fines. Determining the most economical chemical addition for a specific sludge should be based on bench-scale testing or pilot-plant studies.

Heat treatment of sludge prior to dewatering stabilizes and improves the dewatering characteristics of the sludge. It also destroys organics. It can be oxidative or nonoxidative, depending on the specific process involved. Waste sludge is treated at temperatures of 350–450°F and maintained at 150–250 psig for 20–30 min. Part of the sludge is solubilized, producing liquor having a COD, chemical oxygen demand, that can be as high as 4,000 mg/L.

Heat treatment reduces the specific resistance to filtration of a sludge by destroying the gelatinous components, which bind water. Dewatering improvement is similar for oxidative and nonoxidative systems. Sludges that gravity thicken to 2–3% total solids can be thickened to 10–15% total solids after heat treatment. Organic sludges that filter at 2–3 lb/h/ft^2 with chemical conditioners can be filtered without conditioners at 5–8 lb/h/ft^2 after heat treatment. Cake solids content is also improved by heat treatment.

Centrifugation

The solid-bowl centrifuge (Fig. 4) consists of a rotating bowl that concentrates and dewaters a slurry by separating the mixture into a cake and a dilute stream (centrate). The cake is transported within the bowl and discharged from it by a screw conveyor, which rotates within the bowl at a speed slightly less than the bowl

Industrial performance data for centrifugation of waste sludge							Table V
Industry	Origin or nature of sludge	Feed-solids concentration, %	Loading gal/min	Cake-solids concentration, %	Polymer or coagulant type, concentration	Comments	Reference
Chemicals	Primary	Unknown	Unknown	35	Polymer type unknown	Carried out in 40 x 60-in. machines, solids landfilled	21
Pulp and paper sulfite & NSSC*	Biological waste	1-2	Unknown	8-10	None	—	16
Pulp and paper						Solids capture	4
	Fine paper	3.5-12.0	56-120	16-35	Unknown	80-95%	
	Kraft	1.0- 4.5	95-320	15-40	Unknown	80-95%	
	Tissue	1.0- 4.0	20-38	21-42	Unknown	65-90%	
	Hardboard	2.0	43	30-35	Unknown	85%	
	Book	5.0-12.0	150	40-45	Unknown	85-92%	
PVC production	Chemical or primary	8	28	15-25	Unknown	—	5

*Neutral sulfite semi-chemical

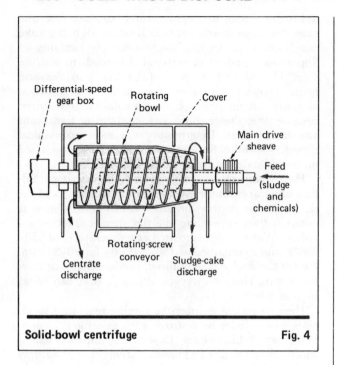

Solid-bowl centrifuge   **Fig. 4**

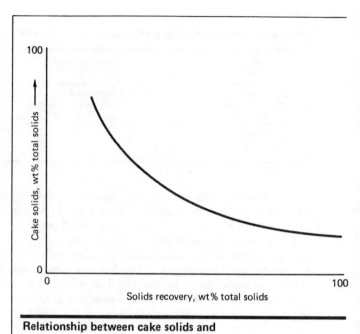

Relationship between cake solids and solids recovery for centrifugation   **Fig. 5**

speed. The centrate overflows weirs on the opposite end of the bowl.

There are numerous machine and process variables that affect centrifuge performance. Table IV lists the most important ones.

The capacity of a centrifuge is related to its size (bowl diameter and length). During operation, a pool is formed in the inner periphery of the bowl. The pool volume is determined by the bowl diameter and length, and pool depth. For a specific loading, solids capture generally improves as pool volume increases.

If all variables remain constant, an increase in the bowl diameter will increase pool volume and retention

time. Increased retention time permits a higher capture of the smaller and/or lower-specific-gravity particles. However, these fines decrease cake drainage characteristics, increasing the liquid content of the cake.

Increasing bowl length increases pool volume and retention time. Generally, long-bowl machines (length:dia. = 2–3:1) are more effective and have greater capacity than short-bowl ones (length:dia. $\leq$ 2:1).

The centrifugal force required to achieve a specific level of performance for a given size of machine is related to the characteristics of the sludge being dewatered. Most commercial machines are operated at

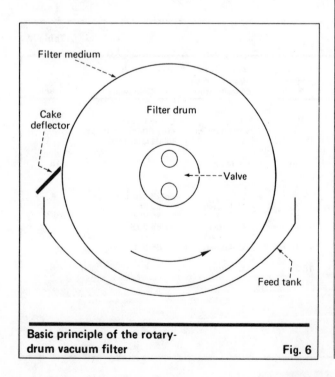

Basic principle of the rotary-drum vacuum filter   **Fig. 6**

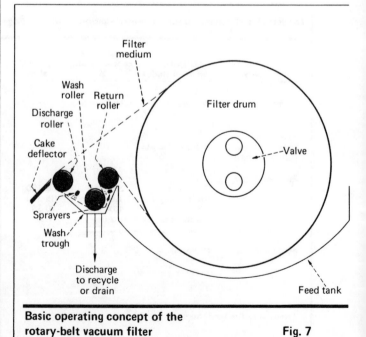

Basic operating concept of the rotary-belt vacuum filter   **Fig. 7**

Industrial performance data for vacuum filtration of waste sludge   **Table VI**

Industry	Origin or nature of sludge	Feed-solids concentration, %	Loading lb/d/ft^2	Cake-solids concentration, %	Polymer or coagulant type, concentration	Comments	Reference
Aluminum extrusion	Clarifier, aluminum hydroxide	4-6	Unknown	15	Unknown	Roll discharge employed	34
Chemicals	Anhydrous sodium metasilicate manufacture	0.86-1.10	0.1-0.5 (gal/min/ft^2)	16.0-27.2	Unknown	Precoat filter	17
Coke manufacture	Scrubber blowdown (coke and coal dust)	20-25	150	Unknown	Unknown	Horizontal-belt filter	34
Ground wood pulping	Primary	Unknown	Unknown	24	2:1 Chipper fines added	—	15
Lead smelter	Thickened scrubber wastewater	40-50	18.7	80	Unknown	Agitated disk filter	34
Pharmaceutical	General	Unknown	2-biological 4-mixed primary and biological	Unknown	Up to 20 lb/ton	—	12
Pigmented paper	Primary	2.5-3.0	Unknown	30	Unknown	—	19
Pulp and paper	White water (W.W.)	1.33-4.70	1.7-13.4	23.3-33.0	Unknown	—	4
	Decoating and W.W.	5.85-10.02	2.1-11.0	34.6-42.9	Unknown	—	
	Boardmill	0.87-2.36	1.2-5.8	26.1-30.7	Unknown	—	
	Deinking and W.W.	5.87-7.15	3.1-10.0	31.4-36.4	Unknown	—	
	Felt mill	5.20-5.27	3.7-5.9	21.4-25.8	Unknown	—	
Pulp and paper	Combined primary and secondary	3.3	1-2	15.3	Unknown	—	34
Pulp and paper integrated kraft	Primary	2.6	Unknown	18-22	Unknown	—	14
Pulp and paper integrated kraft	Combined primary and secondary	1.38-2.70	2-13	12-18	FeCl$_3$ 5%	Ratio 1:1 and 2:1, formtime varied from 30-90 s, coil filter	10
Pulp and paper integrated kraft	Combined primary and secondary	1.63-1.90	2.4-5.5	12-13	None	Ratio 1:1 and 2:1, formtime varied from 30-90s	10
Pulp and paper integrated kraft	Combined primary and secondary	1.57-1.98	3.5-10	19-22	1:1 Bark fines added	Ratio 1:1 and 2:1 Rates do not include bark fines	10
Pulp and paper, kraft	Lime slurry from color removal process	16-26	17-51	44-53	None	Precoat and belt filters	20
Pulp and paper, sulfite and NSSC*	Waste activated	Unknown	Unknown	25-40	Unknown	Precoat filter	l6
Steel	Pickling liquor and limestone	Unknown	22-68	51-67	Unknown	Lime Pilot operation	18
Steel	Limestone treated mill scale	Unknown	56	70	50% Excess of limestone	—	18
Steel (specialty)	Thickened rolling mill and furnace cooling wastes, cooling tower blowdown, metal hydroxides	25	17	40-60	Unknown	—	11

*Neutral sulfite semi-chemical

speeds sufficient to develop forces 1,000–3,000 times that of gravity. Increasing centrifugal force usually increases solids capture and cake solids content. However, cake solids content may remain constant or even decrease, depending on the quantity and characteristics of the additional solids captured.

The solids forced to the inner wall of the bowl are conveyed up the cone of the bowl, the beach, for discharge from the machine. The length and angle of the beach have a strong influence on centrifuge performance, particularly with sludges containing light, gelatinous solids (waste-activated or metal hydroxide sludges). As the sludge is conveyed along the beach (especially where contact with the pool ceases), additional moisture drains off prior to discharge. For most sludges, increasing dry-beach residence time will increase cake solids content.

As the cake is conveyed along the beach, it is subjected to centrifugal force acting perpendicular to the bowl and to a force acting parallel to the beach. The parallel force tends to push the solids down the beach into the pool. This "slippage" force is related to the centrifugal force and the beach angle.

For sludges consisting of light, gelatinous solids, the slippage force can be sufficient to cause conveyed solids to flow under the conveyer blades back into the pool. Solids capture deteriorates in this situation and centrifuge capacity is severely limited. Machines having small beach angles, or pools raised to slightly below the cake discharge point, can be used to overcome this problem.

The difference in rotational speed between the screw conveyor and the bowl is termed the differential speed. For a specific loading, it is advantageous to maintain the minimum differential speed necessary to convey all cake from the machine. Increasing differential speed above this level increases pool turbulence and speeds the cake conveying rate in the pool and on the beach. These factors generally reduce solids capture, especially in applications involving waste-activated, metal hydroxide and other difficult sludges.

The major process variables affecting performance are the type of sludge, the use of chemical flocculants, and the loading rate to the machine. For any application, loading rate and chemical conditioning can be varied to control centrifuge performance in terms of cake solids content and solids recovery. Fig. 5 shows the normal relationship between cake solids content and solids recovery. This relationship can be controlled by changing process conditions (loading rate) and machine conditions (pool volume and differential speed).

Centrifuges are best sized from pilot- or prototype-scale units operating at the plantsite under actual conditions. When this is not possible, performance may be estimated using batch laboratory centrifuges. Procedures for conducting laboratory tests are described in the literature [35]. Since these tests are batch and do not include provisions for continuous removal of settled solids from the pool, indirect methods must be used to judge whether solids can be successfully discharged from full-scale machines. This is essential. Generally, granular solids can be readily conveyed out of a full-scale centrifuge, whereas flocculated and/or compressible solids are significantly more difficult.

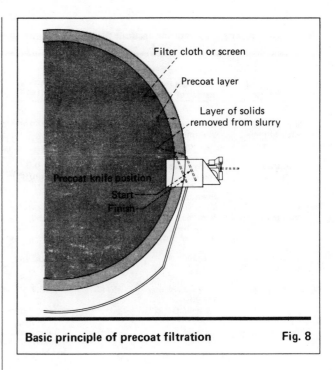

Basic principle of precoat filtration   **Fig. 8**

Table V lists performance data for a few industrial centrifugation operations.

Vacuum-drum filtration

The typical continuous vacuum filter consists of a rotating drum partially submerged in a vat containing slurry. The drum is divided into sections that can be separately evacuated by means of an automatic valve. The filter medium, which can be made of various natural and synthetic materials, overlies the face of the drum and supports the dewatering sludge.

The two most frequently used vacuum filters are the rotary drum (Fig. 6) and rotary belt (Fig. 7). Drum diameters vary from 3 to 12 ft; lengths vary from 3 to 40 ft.

Since the development of continuous filtration, the conventional vacuum-drum filter has been used in more sludge applications than any other filter type. This is mainly due to its operational flexibility, and ability to handle a wide range of slurries. The basic geometry of the drum filter permits the operator to vary the cycle time for cake formation, washing and dewatering. This minimizes inactive time, increases productivity/unit area and makes it easier to handle off-quality feeds. The conventional drum filter also permits discharge of thinner cake than is possible with disk, pan, horizontal and other filters.

The major disadvantage of vacuum-drum filters is progressive blinding of the filter medium, particularly with sludges containing extreme fines. Blinding generally results from: (1) plugging of the interstices within the filter medium by suspended solids in the feed, (2) chemical precipitation within the filter medium or (3) sealing of the surface of the medium caused by the shape of the feed solids.

Slow blinding occurs in the majority of existing applications, but several days to weeks may elapse before the filtration rate is severely reduced. To achieve a

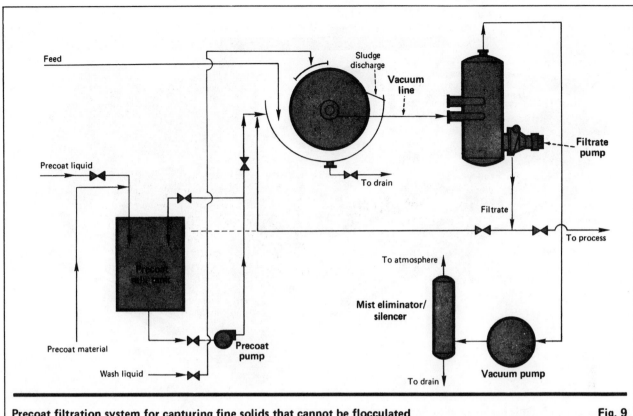

Precoat filtration system for capturing fine solids that cannot be flocculated

Fig. 9

balance between cloth life, operating time and capital cost, the design filtration rate must be 70–80% of the clean-medium rate. When blinding reaches a critical point, the medium must be replaced or rejuvenated with an acid or alkaline wash.

Another problem is poor cake-discharge. If a filter cake of minimum thickness and sufficient dryness is produced, the cake discharge will be practically complete. Generally, $\frac{1}{4}$-in. minimum thickness is desirable, although cakes of $\frac{1}{8}$-in. can be discharged. However, if thin, slimy or moist cakes are encountered, cake discharge can be severely impaired. In addition, reduction in filter capacity, accelerated rate of blinding and difficulty in cake removal can occur if multifilament weaves of cotton and other natural fibers are used in the filter medium, because short fibers imbed in the cake.

Use of a rotary-belt-type drum filter eliminates or greatly reduces problems of blinding and cake thickness. The design (Fig. 7) consists of a sectionalized drum, in which the periphery of each section contains a soft rubber or synthetic strip that is slightly raised from the drum surface. The filter medium lies over these strips, sealing the vacuum side from the atmosphere. Since a typical filter operates at a 20–25 in. Hg vacuum, there is approximately 10–12 lb/in.² of pressure providing sufficient sealing force. The filter medium is an endless belt that travels off the drum to a cake discharge roll and then to a wash chamber, where fluid is applied through high-pressure sprays to both sides of the cloth. Occasionally, an acid wash may be required to reduce blinding.

Cake discharge is accomplished by passing the filter medium over a small-diameter roller, which very abruptly changes the radius of curvature of the medium relative to the cake, causing the cake to break free from the cloth. Generally, a scraper blade is not required to loosen the cake from the medium but rather to deflect the cake to the discharge point of the filter. Cakes as thin as $\frac{1}{16}$ in. can be satisfactorily discharged.

Particulate sludges containing less than 50% of 200-mesh (74 microns) solids can be successfully handled by belt-type units, which can also handle most plating, steel mill, water softening, foundry, organic and pharmaceutical sludges, and even wet-air oxidation sludges.

Filtrate containing suspended-solids levels as low as 500 mg/L can be achieved with belt-type units, from feed concentrations as high as 10–20% solids. With other filters, such as precoat units, filtrate solids of only 25 mg/L or less are possible.

Vacuum-drum filtration may be evaluated from bench- or pilot-scale tests. Bench-scale apparatus procedures have been well developed and are described in the literature [36], [37]. The filter-leaf method is useful for predicting filter performance under a variety of operating conditions and for establishing medium type, bridging and submergence. Pilot-scale vacuum-drum filter tests are particularly useful in identifying and solving cake pickup and discharge problems.

Table VI lists performance data for several industrial vacuum-filtration operations.

Precoat filtration

Precoat vacuum filtration is used primarily for clarification purposes or for difficult filtering applications in

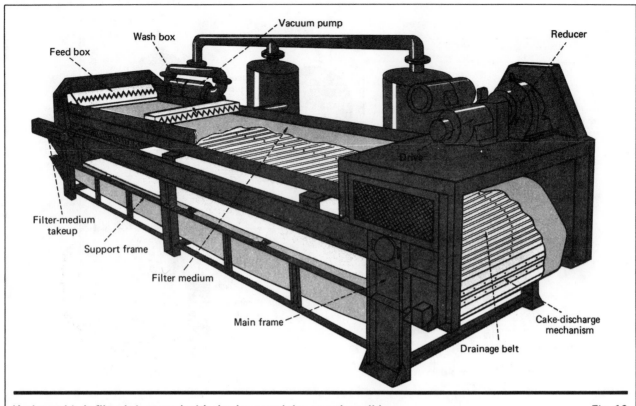

Feed box

Wash box

Vacuum pump

Reducer

Filter-medium takeup

Support frame

Filter medium

Main frame

Drive

Drainage belt

Cake-discharge mechanism

Horizontal-belt filter is best used with slurries containing granular solids   **Fig. 10**

which severe cake-discharge problems are expected. It is also used when feed characteristics are highly variable.

The precoat filter (Fig. 8) is similar to the rotary-drum vacuum filter. A cake of precoat material, such as diatomaceous earth or expanded perlite, is formed on the filter medium prior to sludge contact. Filtration proceeds continuously by shaving a portion of the precoat from the filter along with the filter cake. Shaving is accomplished by a sharp knife that removes 0.003–0.005 in. of precoat per drum revolution. As filtration continues, the knife advances toward the drum surface. The cake of precoat may last for several hours or several days, depending on operating conditions. A system employing a precoat unit is shown in Fig. 9.

For fine colloidal particles that will not settle or cannot be flocculated, the continuous precoat filter is almost the only solution. Precoat filters have been used successfully for clarifying slop oil, removing TNT fines from wastewater at ordnance plants, and dewatering oily or metal hydroxide sludges. Generally, the feed has a low solids concentration, and filtration rates can run as high as 40–50 gal/h/ft.[2] Precoat filter operating costs are usually higher than those for conventional drum or belt-type filters because of the precoat, which on the average is consumed at a rate of 10–15 lb of precoat/1,000 gal of filtrate.

Precoat-filter design criteria may be established from either bench- or pilot-scale tests. Bench-scale tests are similar to those employed for conventional vacuum filtration. A precoat material is initially formed on the filter leaf, and the slurry is filtered using the cake of precoat as the filter medium. Tests are useful in establishing filtration rate, solids capture and filter-cake solids content. Criteria such as precoat type and consumption (which serve to verify cake solids content) are best obtained from pilot-scale tests.

Horizontal-belt filtration

Horizontal-belt filters (Fig. 10) are best applied to slurries containing granular solids that form cake rapidly and have high dewatering rates. They are also used when cake-discharge problems are anticipated. A typical horizontal-belt filter includes a slurry flocculation unit that serves to distribute feed material across the width of the filter. A filter medium overlies the horizontal grids. Dewatering is achieved as the slurry is transported on the medium along the length of the filter. Vacuum, at controlled levels, may be applied at different zones along the filter.

These units vary in width from $5\frac{1}{2}$ to 18 ft and in length from 16 to 110 ft. Filtration areas range from 10 to 1,200 ft.[2]

The horizontal, top-feed belt filter allows extensive washing and countercurrent staging for removal of objectionable solubles (such as mother liquor) from the cake. Continuous belt washing minimizes blinding. The horizontal-belt filter has been used primarily in industrial waste applications where an extremely wide range of particle sizes must be removed or where cake washing is a necessity.

Design criteria are determined by bench-top and pilot-scale testing in similar fashion to vacuum-drum filtration.

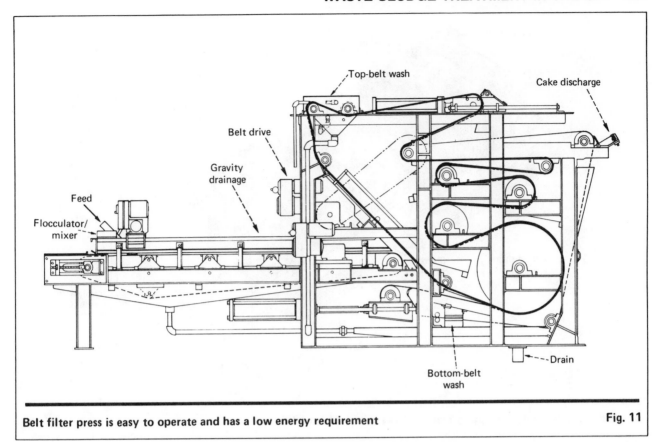

Belt filter press is easy to operate and has a low energy requirement   **Fig. 11**

Belt-press filtration

The belt press (Fig. 11) is widely used in Europe but has only recently been applied to sludge dewatering in the U.S. A typical belt press includes a slurry flocculator that assists in distributing feed across the width of the unit. An adjacent drainage zone removes liquid from the flocculated slurry by gravity or vacuum. Moisture is removed from the drained cake by mechanical and shear forces that are exerted as the cake is sandwiched between two endless belts and transported around several rollers of various diameters. Pressure is applied to the cake by a combination of increasing belt tension and decreasing roller diameters. Scrapers are used to continuously discharge the cake.

The advantages of the belt press over other mechanical dewatering methods include low energy requirements and operating simplicity. The belt press overcomes cake pickup problems experienced in vacuum filtration of sludges that are difficult to dewater. It is ideally suited for operation with polymer conditioning. This eliminates handling problems associated with ferric chloride and/or lime chemical conditioners, which may be required for vacuum or pressure filtration of difficult sludges.

There are numerous belt-press configurations. Units are available with belts up to 10 ft wide. Lengths vary depending on the configuration.

The most widely accepted method of sizing belt presses and predicting their performance is to run pilot-scale tests. Manufacturers of these devices have units available for determining capacity, solids capture, cake solids and polymer conditioning. Until more oper-

ating experience is available on belt presses, pilot tests should be performed prior to specifying and sizing.

Pressure filtration

Recessed-plate or plate-and-frame types of filter presses consist of a series of rectangular plates supported face-to-face in a vertical position. Filter cloth is fitted over the face of each plate. Hydraulic rams or powered screws are used to hold the plates together during dewatering. A typical pressure filter is shown in Fig. 12.

Filter pressing is a batch operation in which chemically conditioned sludge is pumped into the space between the plates. Pressures of 100–250 psi are applied for periods of one to several hours. The solids in the slurry are retained by the filter cloth and gradually fill the space inside the plate while liquid is forced through the filter cloth. At the termination of the dewatering period, the plates are separated and the sludge cake removed.

Pressure filtration can produce cake of higher solids content, containing 5–20 percentage points less moisture, than other filtration processes. It is frequently employed where low cake moisture is required. However, it is a batch process and may require cake breaking and storage facilities.

In applications where very hydrous, cellulose pulps are encountered (pulp-and-paper and municipal sludges), pressure filtration is frequently used. Sludges containing water and oil are also good feeds. The pressures are usually sufficient to prevent blinding of the filter cloth by the oil.

Design criteria for sizing pressure filters are generally

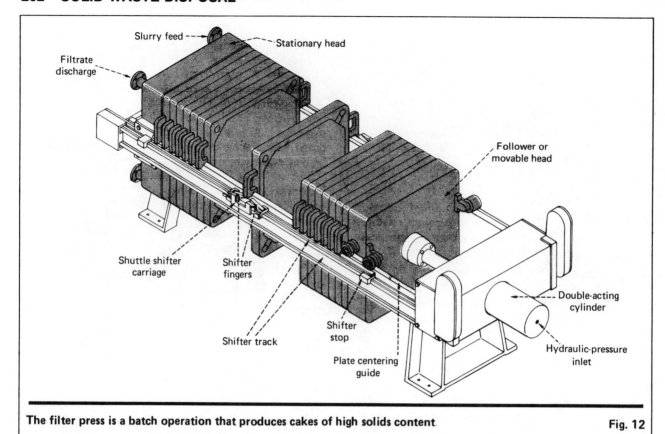

Slurry feed

Stationary head

Filtrate discharge

Follower or movable head

Shuttle shifter carriage

Shifter fingers

Shifter track

Shifter stop

Plate centering guide

Double-acting cylinder

Hydraulic-pressure inlet

The filter press is a batch operation that produces cakes of high solids content.   Fig. 12

obtained from pilot-scale tests. Attempts to use specific resistance to filtration, as measured by laboratory-scale apparatus, have generally been unsuccessful. Pilot-scale apparatus varies in size from several square inches to small prototype units of about 100 in². The results obtained from these devices correlate well with full-scale yields. These units are also very useful in identifying proper filter media and in measuring the effects on performance of cake thickness, time, pressure and chemical conditioning.

Table VII lists performance data for several industrial pressure-filtration operations.

Reverse osmosis and ultrafiltration

The pressure-driven membrane processes—reverse osmosis and ultrafiltration—have found only limited use in dewatering sludges because they cannot tolerate large concentrations of suspended material without serious reduction in flux. For example, ultrafiltration systems can concentrate solids to only 3–5%. However, these processes are useful when it is necessary to separate and concentrate a colloidal or dissolved species from a slurry. Toxic materials present in a liquid stream at very low concentration can be concentrated to 1–3% solids in pressure-driven membrane processes without any severe penalty other than an increase in required system pressure, resulting from the increasing osmotic pressure. Usually, the molecular weight of the species to be separated and concentrated is high enough that the pressure increase is insignificant. The process chosen depends on the size of the species to be concentrated. For ionic species, reverse-osmosis is used, and for or-

ganic species ultrafiltration is used. The concentrated liquid or viscous material can then be coincinerated with sludge.

Secondary sludge treatment

Waste biological sludge is generated in the secondary stages of waste treatment (Fig. 1). Organic chemical, petrochemical and pulp and paper wastes usually require biological treatment for BOD, biochemical oxygen demand, reduction in compliance with EPA's BPTCA (Best Possible Technology Currently Available) standards. This is accomplished in aerated lagoons, contact-media units and activated-sludge-treatment units. This discussion will consider only the latter.

The sludge production of activated-sludge processes is shown in Table VIII. The data are typical and are based on feeds consisting of less than 0.2 lb inerts/lb BOD. The amount of inerts in the feed is critical. The sludge produced has a concentration of 0.5% to 1.5% solids after discharge from a secondary clarifier. Thickening is essential to reduce this volume prior to dewatering operations.

If primary treatment produces a sludge volume that is much larger than that from secondary treatment, the sludges can be mixed and dewatered together. However, they are often kept separate for better control in dewatering and disposal.

Biological sludge stabilization, which is often practiced prior to dewatering or disposal, can be aerobic or anaerobic. Anaerobic digestion has the advantages of lower power requirements and potential recovery of methane.

Aerobic-digestion basins have sludge retention times of 10–20 days, with feeds of 5–10 lb of solids/100 ft³. Up to 40% of the volatile solids are destroyed.

Anaerobic digestion reduces organic sludges to methane, carbon dioxide, ammonia and hydrogen sulfide. The methane can be recovered and used for power generation or other purposes. Digestion is temperature dependent. In the mesophilic temperature range, 50–95°F, the mixed cultures of bacteria require 20–55 days for digestion. In the thermophilic range, 100–140°F, digestion time is reduced to 15–20 days [38].

Chemical fixation

Chemical fixation is a process for binding hazardous waste in a chemical matrix that is impervious to water penetration. This minimizes the hazard of chemicals being leached into groundwater.

One acceptable procedure involves fixing the toxic substance in a silicate matrix [25]. Presumably, this process causes the formation of metal silicates, or the direct occlusion of metals in SiO_2, depending on the pH. Fixation is acceptable for many heavy metals and certain inorganic anions such as arsenates. To date, fixation has not been very effective with toxic organics. Table IX lists sludge types that have been chemically fixated.

Taconite tailings derived from the beneficiation of iron ore have been successfully fixed using a carbonate bonding process [1]. The waste is mixed with water and lime hydrate, and then reacted with a gas rich in CO_2 to form a calcite crystal matrix. This procedure may be broadly applicable to fine inorganic solids that cannot be contained when released to the environment.

The moisture content of sludge has a substantial impact on the cost of fixation. Increasing solids content from 30% to 50% reduced cost in one application from $30 to $10/ton of dry solids.

Tests are required to demonstrate feasibility and cost effectiveness of fixation because of the unpredictable variation in sludge compositions.

Byproduct recovery

Byproduct recovery can be a significant factor in sludge management [2]. Waste tars, spent catalysts and

Industry	Origin or nature of sludge	Feed-solids concentration, %	Loading lb/h /ft²	Cake-solids concentration, %	Polymer or coagulant type, concentration	Comments	Reference
Chemical	Biological and chemical	1.7	Unknown	37-38	5% FeCl₃	Max pressure 120 psi 75-90 min cycle time	22
Chemical	Primary and waste activated	4.5	147,000 lb/d on 5,800 ft²	32	Lime and FeCl₃ CaO-20% FeCl₃-2.5%	66.6% Biological 33.3% Chemical conditioners	34
Glass etching	Reactor-clarifier, calcium fluoride	15	15	40	Unknown	Belt filter	34
Petrochemical	General Oily	2-5%	Unknown	10-40	$2-$10/ton Types unknown	>99% solids capture, gravity-belt press	13
Pulp and paper	Board-mill	Unknown	—	27.5-50.3	Unknown	100-900 psi 1-10 min	4
	Deinking	Unknown	—	38.5-64-3	Unknown		
Pulp and paper, bleached kraft	Primary	Unknown	50 tons/d	35	2.5 lb/ton	Belt press twin wire	23
Pulp and paper, integrated kraft	Primary	18-22	Unknown	37-40	Unknown	V-press,dried to 70-85%	14
Pulp and paper, integrated kraft	Combined primary and secondary	19-22	Unknown	33.5	1:1 Bark fines added	Batch press	10
Pulp and paper, sulfite and NSSC*	Secondary	1-2%	0.4 to 3.6	30-50 filter press 20 belt press 7 ultrafilter	Unknown 10-20 lb/ton None	>99% solids capture	16
Pulp and paper, sulfite and NSSC*	Primary	Unknown	10-25 gal/min/ft²	30-40	0-5 lb/ton	—	16
Steel	Oily chemical	Unknown	Unknown	50-80	Unknown	—	22

Industrial performance data for pressure filtration of waste sludge **Table VII**

*Neutral sulfite semi-chemical

Production from activated-sludge processes   Table VIII

Process	Sludge retention time, d	Waste sludge lb/lb BOD removed
High rate	2-4	0.5 -0.7
Conventional	4-8	0.4 -0.6
Extended aeration	30	0.15-0.13

*Based on a feed containing less than 0.2 lb inerts/lb BOD.

Types of sludges that have been chemically fixated   Table IX

Oily wastes and tars
Mining tailings
Paper fines
Strong acids or alkalis
Dissolved metals
Nonvolatile organics
Resins
Phosphate wastes
Chemical emulsions

other materials from petrochemical processing are often sold for metal and other valuable component recovery. Carbide lime and acetylene process byproducts are used as sludge conditioners. Waste aluminum and iron salts are used as flocculants.

In many cases, particularly in the secondary-metals industry, thickeners and filters may be employed for product recovery and waste management. Consider the recovery of zinc from sludge produced during rayon manufacturing [3]. Starting with a mixed sludge containing zinc hydroxide and cellulose at about 3% solids, the following steps can be taken to obtain a reusable product:

1. Heat-treat the sludge at 280–300°F to increase filterability.
2. Filter the sludge in a press to about 35% solids.
3. Digest the sludge cake with H_2SO_4 (to produce a 25–30% $ZnSO_4$ solution).

Ultimate disposal techniques other than incineration for waste sludge   Table X

Industry	Origin and nature of sludge	Treatment prior to disposal	Rate	Solids concentration, %	Disposal technique	Comments	Reference
Chemical	Aerobic digestor biological waste	Aerobic digestion	1 million gal/d	Unknown	Ground disposal	15-day digestion	24
Chemical	General chemical, primary	Concentration* after separation of toxics	Unknown	35	Landfill	—	21
Chemical	Waste activated	Bioconversion	Unknown	Unknown	Deep-well disposal	—	21
Latex production	Alum coagulated primary	Thickening	Unknown	Unknown	Landfill	—	6
Metal processing (secondary)	Hydroxide formation	Concentration	Unknown	Unknown	Fixation	Costs $50-$100/ton, dependent on solids content	25
Organic chemical	Waste activated and flyash	None	Unknown	3(activated waste)	Lagoon	95% Flyash 5% Waste activated	28
Organic chemical	Aerobic industrial fermentation	None	8,000 to 152,000 gal/d	0.7 to 5.5	Ground disposal	100 acres in 1-acre plots	26
Paint	Solvent recovery, latex washout	None	Unknown	Variable	Landfill	—	1
Paint	Oil and water-based paint	Sedimentation and concentration	Unknown	Unknown	Lagoon	Conditioned with polymer	31
Pesticides	All types	Concentration	Variable	Unknown	Contained burial	—	32
Pharmaceutical	General	Concentration and bioconversion	Unknown	Unknown	Landfill	—	12
Plastics	All sources All wastes	None	Unknown	Variable	Landfill	—	7
PVC production	Chemical or primary	Concentration	23 tons/d	15-25	Sanitary landfill	—	5
PVC and rubber intermediate	Ferric coagulated primary	Clarification	Unknown	Unknown	Pit disposal	—	27
Rubber	All sources	None	Unknown	Variable	Lagoon or landfill	—	7
Synthetic fibers	Polyester manufacture, primary and waste activated	Clarification and bioconversion	Unknown	Unknown	Lagoon	—	29
Uranium processing	Sands and slimes	Leach extraction filtration	Plant flow	Variable	Tailings pond	—	30

*Thickening and/or dewatering

4. Refilter the sludge in a filter press (to remove $CaSO_4$ and organics).

5. Oxidize ferrous iron with H_2O_2; adjust pH to 4.5.

6. Remove ferric iron on a rotary precoat filter.

This is only one example of the many waste-recovery processes that are now or will soon be in use. With the cost of disposal on the rise, recovery and reuse is a potential profit area and an engineering and process design challenge.

Disposal

Though methods of sludge disposal are beyond the scope of this article, Table X lists examples of industrial disposal techniques other than incineration.

References

1. Carbonate Bonding of Taconite Tailings, USEPA, Office of Research and Development, EPA-670/2-74-001, Jan., 1974.

2. Burd, R. S., A Study of Sludge Handling and Disposal, U.S. Dept. of the Interior, Federal Water Pollution Control Administration, Office of Research and Development, WP-20-4, May, 1968.

3. Iammartino, N. R., Wastewater Clean-up Processes Tackle Inorganic Pollutants, *Chem. Eng.,* Sept. 13, 1976, p. 118.

4. State-of-the-Art Review of Pulp and Paper Waste Treatment, USEPA, Office of Research and Monitoring, EPA-R2-73-184, 1973.

5. Wastewater Treatment Facilities for a Polyvinyl Chloride Production Plant, USEPA, Water Pollution Control Research Series, 12020 DJI, June, 1971.

6. Putting the Closed Loop into Practice, *Environ. Sci. Technol.,* Vol. 6, No. 13, Dec. 1972, p. 1,072.

7. Fluidized-Bed Incineration of Selected Carbonaceous Industrial Wastes, USEPA, Water Pollution Control Research Series, 12120 FYF, March 1972.

8. Air Flotation—Biological Oxidation of Synthetic Rubber and Latex Wastewater, USEPA, Environmental Protection Technology Series, EPA-660/2-73-018, Nov., 1973.

9. Environmental Considerations and the Modern Electrolytic Zinc Refinery, *Min. Eng.,* Vol. 29, No. 11, Nov. 1977, p. 31.

10. Kehrberger, G. J., Mulligan, T. J., South, W. D., and Djordjevic, B., Thickening and Dewatering Characteristics of Kraft Mill Sludges from a High-Purity Oxygen Treatment System, *Tappi,* Vol. 57, 1974, p. 119.

11. Wills, Jr., Robert H., Crucible Waste Treatment, *Clear Waters,* J. of N.Y. WPCA, Vol. 6, No. 4, Dec. 1976, p. 15.

12. Interim Final Effluent Guidelines for The Pharmaceutical Manufacturing Industry, USEPA, 440/1—75/060 Group II, 1976.

13. Grove, George W., Use Gravity Belt Filtration for Sludge Disposal, API Special Report, *Hydrocarbon Process.,* May 1975, p. 82.

14. Methods of Pulp and Paper Mill Sludge Utilization and Disposal, USEPA, Office of Research and Monitoring, EPA-R2-73-232, 1973.

15. Bishop, Fred W., and Drew, A. E., Disposal of Hydrous Sludges From a Paper Mill, *Tappi,* Vol. 60, No. 11, Nov. 1971, p. 1830.

16. Miner, R. A., Marshall, D. W., and Gellman, I. L., A Pilot Investigation of Secondary Sludge Dewatering Alternatives, National Council of the Paper Industry for Air and Stream Improvements Inc., USEPA, EPA-600/2-78-014, Feb., 1978.

17. Turkki, E. V., Hildebrand, A. S., and Nemerow, N. L., Removal of Fine Solids from Chemical Waste Stream Through Precoat Rotary Vacuum Filtration, *Proc.,* 30th Purdue, Ind., Waste Conf., 1977, p. 122.

18. Limestone Treatment of Rinse Waters from Hydrochloric Acid Pickling of Steel, USEPA, Water Quality Office, 12010 DUL, Feb. 1971.

19. Sludge Material Recovery System for Manufacturers of Pigmented Papers, USEPA, Waste Pollution Control Research Series, 12040 FES, July 1971.

20. Color Removal from Kraft Pulp Mill Effluents by Massive Lime Treatment, USEPA, Office of Research and Monitoring, EPA-R2-73-086, Feb. 1973.

21. Gossett, J. W., How Dow Chemical Deals with Diverse Wastes, *Chem. Process.,* Vol. 31, No. 1, Jan. 1968, p. 20.

22. Kellogg, Stephen R., and Weston, Roy F., Treatment of Various Industrial Sludges by Pressure Filtration, The Eleventh Mid-Atlantic Regional Meeting of ACS, University of Delaware, E-6158, April 1977.

23. Keener, P. M., and Metzger, L. R., Startup and Operating Experience with a Twin-Wire Moving Belt-Press for Primary Sludge, *Tappi,* Vol. 60, No. 9, Sept. 1977, p. 120.

24. Howard, J. W., Poduska, R. A., and Walls, W. V., Upgrading of Industrial Wastewater Treatment Facilities at Tennessee Eastman Co., 1977 Water & Wastewater Equipment Mfrs. Assn. Industrial Conference, Atlanta, Ga., 1977.

25. Landreth, R. E., and Mahlock, J. L., Chemical Fixation of Wastes, *Ind. Water Eng.,* Vol. 14, No. 4, July/Aug. 1977, p. 16.

26. Woodley, Richard A., "Spray Irrigation of Organic Chemical Wastes," Commercial Solvents Corp., Terre Haute, Ind., 1977.

27. Kemp, C. E., BFG Chemical Upgrade Treatment, *Water and Sewage Works,* Vol. 119, Nov. 1972, p. 94.

28. Kumke, G. W., Hall, J. F., and Oeben, R. W., Conversion to Activated Sludge at Union Carbide's Institute Plant, *JWPCF,* Vol. 40, No. 8, Part I, Aug. 1968, p. 1,408.

29. Reuse of Chemical Fiber Plant Wastewater and Cooling Water Blowdown, USEPA, Water Pollution Control Research Series, 12090 EUX, Oct. 1970.

30. State-of-the-Art Uranium Mining, Milling, and Refining Industry, USEPA, Office of Research and Development, EPA-670/2-74-030, March 1974.

31. Waterborne Wastes of the Paint and Inorganic Pigments Industries, USEPA, Office of Research and Development, EPA-670/2-74-030, March, 1974.

32. The Pesticide Manufacturing Industry—Current Waste Treatment and Disposal Practices, USEPA, Water Pollution Control Research Series, 12020 FYE, Jan. 1972.

33. "Ultimate Disposal of Liquid Wastes by Chemical Fixation," Chem Fix Div., Environmental Sciences Inc., 1977.

34. Data from Envirotech Corp., Eimco Machinery Products Div., Salt Lake City, Utah.

35. Vesiland, P. A., "Treatment and Disposal of Wastewater Sludges," Ann Arbor Science, Mich., 1974.

36. Perry, R. H., et al., "Chemical Engineers' Handbook," 5th ed., McGraw-Hill, New York, 1973, p. 19-60.

37. Nelson, P. A., and Dahlstrom, D. A., Moisture Content Correlation of Rotary Vacuum Filter Cake, *Chem. Eng. Progress,* Vol. 53, No. 7, July 1957, p. 320.

38. "Waste Water Treatment Plant Design," Manual of Practice No. 8, WPCF and ASCE, Lancaster Press, Pa., 1977.

The authors

OKEY   KOMINEK   DIGREGORIO

Robert Okey is Director of Technical Services for Envirotech Corp., Eimco Process Machinery Div., 669 West Second South, P.O. Box 300, Salt Lake City, UT 84110. (801) 521-2000. Previously, he was the water-and-waste staff technologist for an engineering and construction firm, a consultant, and Eastern Washington State District Engineer. He has taught at Seattle University and the University of Southern California. He has a B.S. in Agricultural Engineering from Iowa State College and a B.S. in Civil Engineering and M.S. in Sanitary Engineering from the University of Washington, Seattle. Mr. Okey has authored many articles on wastewater treatment and contributed to the Water Pollution Control Federation Manuals of Practice. He is a member of WPCF, AIChE and ASCE, and is a registered Professional Engineer in eleven states.

Edward G. Kominek is the Manager of Carbon System Operations for Envirotech Corp., Eimco Process Machinery Div. During his 35 years in water- and waste-treatment activities, his major experience has been in process design and research and development. He has received B.S. and M.B.A. degrees from the University of Chicago and is a registered professional engineer in Ohio, Illinois and Arizona. Mr. Kominek has authored several papers on water and waste treatment. He is a member of AIChE, WPCF, AWWA, API and others.

David DiGregorio is Director of Sanitary Engineering Research and Development for Envirotech Corp., Eimco Process Machinery Div. His responsibilities include the development and application of unit processes for wastewater treatment. Before joining Eimco in 1971, he worked in research and development at Dorr-Oliver. Mr. DiGregorio has a B.S. in Civil Engineering from the University of Massachusetts and an M.S. from Cornell University, Ithaca, New York. He has authored papers on waste treatment and solids processing, worked on two EPA-sponsored research contract studies and contributed to the WPCF Manuals of Practice. He is a member of WPCF and a registered Professional Engineer in Utah.

Waste disposal with an energy bonus

A biological and chemical system

produces extra energy, while turning strong organic

wastes into fertilizer and clean water.

Peter R. Savage, London Regional Editor

☐ In the Anox system, anaerobic digestion has been combined with catalytic oxidation to convert strong organic waste into fertilizer. The process, developed by D. Evers & Associates Ltd. (Worcester, U.K.), also produces enough methane so that it can supply more than its own power needs.

Anox is said to be suitable for treating a wide variety of organic materials, including waste from sugar refineries, distilleries, cereal-processing units, cattle farms and poultry farms. According to its developers, this route produces only usable or salable byproducts and can result in a payback time of little more than a year in some cases.

The first commercial Anox plant will be commissioned this month at a pig-breeding unit owned by Etablissement L. Ferard, near Rennes, France. The system will process 120 m³/d of manure (7% solids content), and from this produce 3.2 tons/d of organic-fertilizer base material, 60 m³/d treated water and, from methane-rich gas produced in the digester, 2,400 kWh/d of electrical power (of which only 500 to 1,300 kWh/d is needed for the process).

Experience with the Anox system has been gained through a 3,600-gal digester pilot unit operated for twelve years by the Rowett Research Institution, a government-sponsored body in the U.K. The catalytic oxidation section of the process has been in commercial use for four years.

ANAEROBIC DIGESTION—In the process, manure is collected as a slurry in a large holding sump and pumped through a coarse filter screen to remove inerts such as wood or metal. This raw organic material, which has a biochemical oxygen demand (BOD) of 20,000 mg/L, is fed to the digester (see flowsheet), where it will have a residence time of 10 d. During this time the BOD of the slurry will fall to 5 to 10% of the original figure.

At Rennes, the 1,200-m³ digester is a closed vessel with a glass-coated mild-steel construction and a polyurethane cladding for heat insulation. This unit is fitted with heating coils and is maintained at a temperature of 35 to 40°C to encourage the activity of bacteria and to aid gasification.

A gas rich in methane (typically 70% CH_4, with the balance being largely carbon dioxide) will be evolved in a ratio of 25 m³ for every 1 m³ of slurry. This gas, which has a calorific value of 600 Btu/ft³ (22.8 megajoules/m³), is transferred to a gas holder feeding two 60-kWh electric generators that provide power for the electricity-consuming units at the site. The exhaust gas from the generators is passed through a waste heat exchanger producing hot water that is used to heat the digester.

The slurry from the digester is pumped to a clarifier, where liquids (mainly water) and solids are separated. Solids are transferred to a main sludge storage-tank, while the liquids (with a BOD of approximately 800 mg/L) are sent to a flocculating tank.

In this tank, coagulants and polyelectrolytes (aluminum and ferric salts) are added to the liquid at levels of around 2 to 5 ppm to encourage coagulation and separation of the remaining solids.

In the flotation cells, these solids are removed by a sludge-thickening process—electroflotation—which uses fine bubbles of hydrogen and oxygen produced by electrolysis of water (carried out by electrode panels in the bottom of the cells) to achieve flotation.

Economic data

Annual expenses	($, thousands/yr)
Chemicals	2.0
Operating labor	6.0
Maintenance	2.0
Depreciation (10-yr basis)	60.0
Interest on capital invested	54.5
Total	124.5
Credits	
Electricity, (100 kWh/h)	41.6
Fertilizer base	19.6
Water recycle	15.46
Disposal of slurry	115.95
Total	192.2
Total savings	67.7

Originally published May 21, 1979

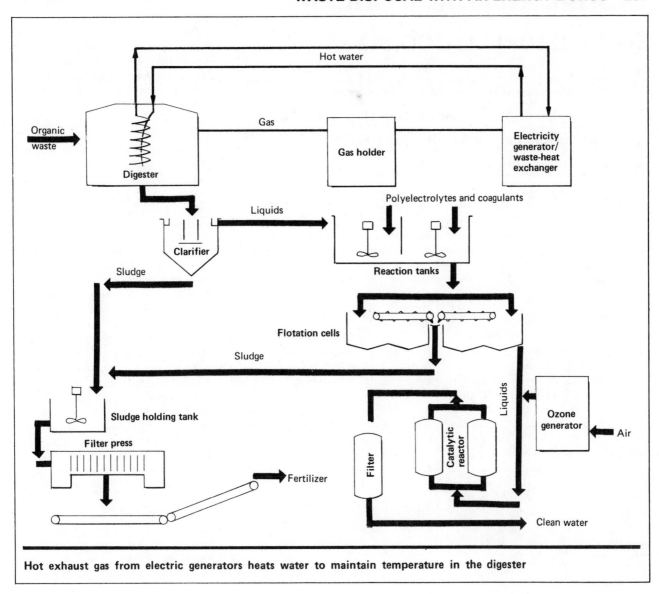

Hot water

Organic waste

Gas

Gas holder

Electricity generator/ waste-heat exchanger

Digester

Liquids

Polyelectrolytes and coagulants

Clarifier

Reaction tanks

Sludge

Flotation cells

Sludge

Sludge holding tank

Liquids

Filter press

Ozone generator

Air

Filter

Catalytic reactor

Fertilizer

Clean water

Hot exhaust gas from electric generators heats water to maintain temperature in the digester

The separated sludge that is carried to the surface by this operation is periodically drawn off and transferred to the main sludge storage-tank.

With a BOD of approximately 200 mg/L, the relatively clear water from the flotation unit is passed to a catalytic oxidation unit, called Catox. An ozone generator, with a capacity of 175 normal m³/h of air, creates an ozone concentration of around 1%. This weakly ozonized air is mixed with the liquid, and the mixture is passed through two fixed-bed reactors that are loaded with a transition-metal catalyst.

USEFUL PRODUCTS—After a final filtration, with a mixed-media filter bed, the water produced has a BOD of less than 10 mg/L, and is said to be suitable for drinking or washing. At Rennes, this water will be eventually passed to a trout farm.

The highly mineralized sludge from the main tank is also a useful byproduct. This sludge is mechanically filtered to give a dry, low-odor organic fertilizer base of about 40% solids content. Its fertilizer assay is 6.8% nitrogen, 6% phosphorus and 1.5% potassium.

Fertilizer from the Rennes plant will be used locally. Since the product is said to be high in protein and low in harmful ingredients such as Salmonella bacteria, the developers say that it can also be used as feed in fish farming or recycled into animal feedstuff bases.

Anox systems have been designed to be largely automatic in operation, requiring only brief daily inspection and periodic topping off of chemical tanks. For the Rennes plant (see table), operating and chemical costs are minimal. Capital costs of the diges-

tion plant and electric generators at Rennes amount to $300,000, with the Catox unit (which includes reactors and ozone generator) costing a further $200,000. This brings the total to $500,000 on a late-1978 basis. Construction costs are around $100,000.

OTHER WASTES—With different feed materials, the process has a similar configuration, though capital cost varies according to throughput. For example, an Anox plant for a cattle farm with 1,000 cattle, producing 72.7 m³/d (19,200 gal/d) of waste and yielding 15,000 m³ of gas from the digester, would require capital costs estimated at $140,000, plus $130,000 for an associated Catox unit.

Currently, D. Evers & Associates are handling marketing of the Anox process directly, but plan to appoint agents soon for the U.S. and western Europe.

Sludge drying—one way to waste-reduction

Though controversial because of its heavy energy demands, sludge drying is nonetheless getting attention from municipalities and industries concerned that they can't meet local land-disposal restrictions.

☐ Amid all the uncertainty over public and governmental attitudes toward the final disposal of waste material, industrial and municipal producers of waste sludge are turning more and more often to drastic measures—including drying.

That's strong medicine under most circumstances. Not only does drying (from typically 15% to 90%-by-weight solids) take a lot of energy, but once it is completed, processors are often left with a major marketing problem—selling the stuff to skeptical fertilizer consumers. But to many a sludge producer, the energy/marketing uncertainty is outweighed by the environmental outlook: More-drastic sludge treatment is coming into favor at regulatory agencies, since it cuts down on waste-disposal headaches. In particular, it reduces the need for landfill.

Drying is one of four overall strategies for sludge treatment. In order of processing severity these are:

Pasteurization. One such method—composting—involves holding wet sludge in a tank or pond, allowing biological action enough time to eat up any pathogens present. The sludge can then be spread over cultivated fields. In some cases, pasteurization is accomplished by a mild heat treatment or by exposure of sludge to ionizing radiation.

Dewatering. In this method, water is removed by vacuum filters, centrifuges or filter presses. Solids content of the treated sludge ranges between 25 and 60%, depending on the severity of treatment. Often, chemical or thermal conditioning is applied to improve mechanical dewatering methods.

Drying. The main subject of this article, drying can take place in rotary, fluid-bed, multi-effect and milling-type equipment. Heat may be applied directly—usually in the form of exhaust gases from a combustion source—or indirectly (typically as steam). Solids content reaches about 90%. However, some processes employ simple air drying—a method that cannot achieve anything like 90% solids, but is drying nonetheless.

Incineration. The ultimate method of disposal, incineration is often used in conjunction with dewatering and drying to produce energy for those methods. Some processors are moving to pyrolysis as an alternative to incineration.

Whatever the disposal process, or combination of processes, the aim is to produce a waste material suitable for landfill, fertilizer or recycling. The ultimate disposition has a deciding effect on the type of process chosen. Drying can produce a fertilizer, provided there are no toxic metals present in the waste, and that odor can be controlled. If fertilizer production is not possible, or simply not worth the marketing trouble, then another method may be chosen, though many of the alternatives employ drying in part.

Besides its environmental advantage, sludge drying adds a lot of flexibility to the waste-disposer's bag of tricks. It's part and parcel of process schemes leading up to incineration or pyrolysis. It reduces volume and weight. And because nitrogen content isn't eaten up, as it is in composting, drying provides a salable product (if no toxic materials are present).

But drying is far from a perfect solution to the waste problem. For one thing, companies or municipalities that commit themselves to this alternative have to be prepared to handle an incessant inventory buildup, in case the market for fertilizer goes flat. The

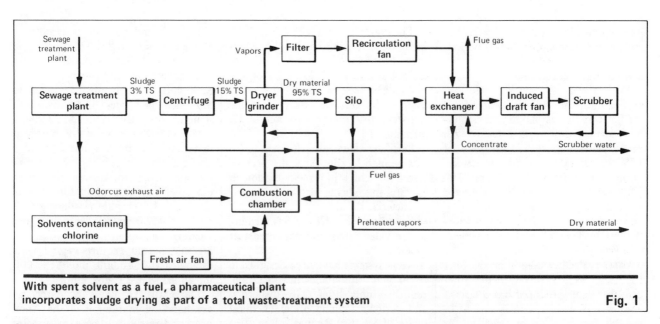

With spent solvent as a fuel, a pharmaceutical plant incorporates sludge drying as part of a total waste-treatment system

Fig. 1

Originally published June 4, 1979

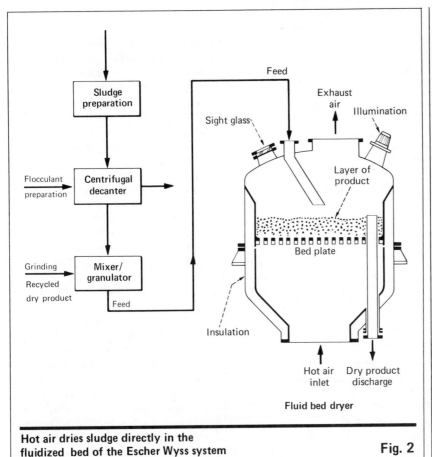

Fluid bed dryer

Feed

Exhaust air

Illumination

Sight glass

Layer of product

Bed plate

Insulation

Hot air inlet

Dry product discharge

Hot air dries sludge directly in the fluidized bed of the Escher Wyss system

Fig. 2

---

supply of sludge would always be present, no matter what the economic conditions.

Certain processing problems also may crop up. Dried sludge may pose a difficult odor-control problem. It could ignite in direct-heated dryers. Its nitrogen content may be too low anyway (at least 5% nitrogen is necessary to make effective fertilizer).

VENERABLE PRACTICE—Sludge drying goes back at least fifty years in some localities. Milwaukee first tried it in the 1920s, and Chicago began using it in the 1940s. Both cities take waste heat from nearby power generators. However, both now are shifting away from this practice, since waste heat is more valuable in other applications. Chicago plans to stop drying sludge altogether by 1985, and Milwaukee will soon cut back its capacity from 200 to 130 tons/d.

Most new processes take heat produced either by incinerating dried sludge or by burning fuel. The major differences between processes involve the methods by which heat is applied—either directly or indirectly, at what temperature, and in what combination with other treatment methods.

DIRECT HEAT—Escher Wyss Ltd. (Zurich) exposes ground, dewatered sludge to direct heat in a fluid-bed dryer. Dewatering is accomplished in a centrifugal decanter, with the aid of added flocculant. In a subsequent mixer/granulator, the sludge is mixed with some recycle dry product before being sent to the dryer. Heat for the dryer comes either from incinerated dry sludge or from purchased fuel. The bed operates at about 600°C.

This system is most economical for treating sewage from a local population of 20,000 to 50,000, Escher Wyss says. Designs are available for producing 100 to 1,000 kg/h of dry product (particle size up to 5 mm).

Chicago's CE Raymond, a unit of Combustion Engineering, Inc., offers two basic systems: a flash dryer that treats sludge in a cage mill, sending it on to a cyclone from which some dry product is recycled to the mill; and a rotary dryer, which also operates with partial recycling. Raymond designs its systems for any of three processing modes—complete incineration of prod-

uct, partial incineration plus partial output of soil conditioner, or conditioner production only. The technology has been in operation in cities such as Chicago and Houston for many years.

Ecological Services Products, Inc. (Dunedin, Fla.) also offers a rotary-dryer design, with prior filter-press dewatering. The complete system, from filter to screw conveyor and into the rotary dryer, followed by cyclone separation of dust and offgases, and scrubbing of the exhaust, aims at yielding a bagged product suitable as a soil conditioner. Systems have been installed in Florida and Georgia.

In Italy, Item S.p.A. (Milan) is now in the pilot-plant stage of development with a "drying tunnel." Its essential aim is to dry and pulverize at the same time (as does Raymond's cage mill and some other techniques) by forcing sludge and direct air through a long chamber at velocities up to 50 m/s.

Organic Recycling, Inc. (West Chester, Pa.), a unit of UOP Inc., has tested a "toroidal dryer" at a municipal waste-treatment plant in Washington, D.C., albeit without great success. While not yet available commercially (the firm intends to produce an improved design in 1980 or later), the system takes dewatered sludge (20% solids), admitting it with some recycled product along the outer circumference of the 16-ft device. Air preheated from 500 to 1,200°F takes the moisture content down to 6%, according to original design objectives. The process's chief advantage, claims Organic, is high thermal efficiency. However, the firm had some difficulty with internal ignition during the method's early development.

In another system that hasn't yet made it to commercialization, Resources Conservation Co. (Renton, Wash.) combines a novel water-extraction step with more-prosaic rotary drying. Dewatered sludge (about 15% solids) is mixed with chilled triethylamine (TEA)—a solvent miscible with water at temperatures below about 20°C, but staying in a distinct layer at higher temperatures. The mixture is then fed to a mechanical dewatering unit where much of the solvent, with bound water and oil, is removed. Then the wet cake is dried at 120-150°C. Because TEA evaporation requires only 133 Btu/lb, drying is faster and less expensive.

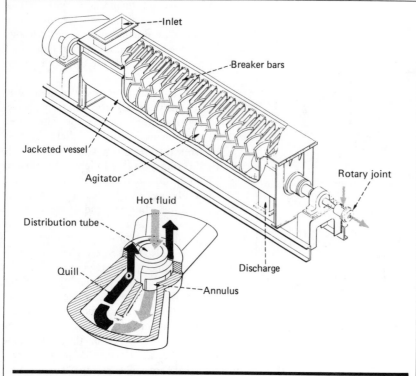

In the Porcupine processor, hot fluid passes through the agitator to supply indirect heat for drying **Fig. 3**

Labels on figure: Inlet, Breaker bars, Jacketed vessel, Agitator, Hot fluid, Distribution tube, Quill, Rotary joint, Discharge, Annulus

Calling its process "B.E.S.T." (for Basic Extractive Sludge Treatment), Resources Conservation has failed so far to sell it to its prime prospect, the nearby city of Portland, Ore. Portland opted instead for a landfill alternative, avoiding altogether the question of sludge drying. The firm does not consider its route useful in industrial sludge treatment, because wastes from these sources are too variable. So for now, the process still awaits commercialization at some municipality.

INDIRECT HEATING—Indirect heating via a heat-exchange fluid—steam, water, hot oil, Dowtherm or the like—negates any worry about ignition hazards, but is not as thermally efficient. The solids content of sludge from such systems never gets much above 60 or 70%, mostly because the heat transfer coefficient decreases rapidly as the material dries.

The Environmental Engineering Div. of Von Roll Ltd. (Zurich) offers both direct and indirect designs. Its indirect system employs a thin-film dryer built by Luwa Corp. (its U.S. subsidiary located in Charlotte, N.C.). In this vessel, sludge is spread thinly over the internal cylindrical surface while heat is applied from the outside. Dry material is scraped off by rotating blades, and then falls to the bottom.

Von Roll's direct system, which gets solids up to 90%, is a dryer/grinder in which sludge and hot gases flow in parallel fashion. A system built for Bayer, at its Eberfeld pharmaceutical plant near Düsseldorf, West Germany, puts centrifuged sludge (15% solids) through the grinder, which is heated from combustion gases derived from several sources: sewer gas, chlorinated solvents, and oil underflow from an exhaust-gas condenser.

Another system for drying sludge is the Porcupine, offered by Bethlehem Corp. (Easton, Pa.). Resembling a grinder more than anything else, the unit mixes and grinds sludge by means of rotating blades that are heated internally with steam. The blades also propel the sludge along the unit's length, discharging the dry product at the far end. The major advantage Bethlehem claims for this system, compared to a thin-film dryer, for example, is higher surface area per unit mass of sludge processed at any one time.

EVAPORATORS—Another alternative is to employ multi-effect evaporators to remove water.

At the Gibraltar, Pa., dyestuff plant of Crompton & Knowles Corp. (Fair Lawn, N.J.), a multi-stage evaporator is treating wastes containing 10% organic salts. Vapor is passed through an activated-carbon adsorption system. The sludge is suitable as landfill.

Foster Wheeler Corp. (Livingston, N.J.) and Dehydro-Tech Corp. (East Hanover, N.J.) offer the Carver-Greenfield process of multiple-effect evaporation. This system, designed chiefly for food-processing applications (e.g., meat rendering), is under study by ITT Rayonier for treating pulpmill wastes at the firm's Port Angeles, Wash., facility. Dried waste would be used as boiler fuel, or, if approved by the U.S. Food and Drug Administration, it may find use as animal feed.

Another firm that may make animal feed via this method is Adolph Coors, at its Golden, Colo., brewery. Eli Lilly Co. uses the process on pharmaceutical sludge now being pyrolyzed.

ULTIMATE DISPOSITION—The pyrolysis option for getting rid of sludge altogether can replace incineration. Though it seems easier, incineration may force costly scrubbing measures to prevent air pollution. Furthermore, heavy metals may show up in the ash and cause land pollution problems anyway—particularly in the case of chromium ions that become oxidized to the harmful +6 valence level.

Mindful of this problem, Japan's NGK Insulators, Ltd. (Nagoya) is developing a system of indirect drying followed by sludge pyrolysis. The dryer closely resembles a Porcupine unit, if in fact it isn't one. The first commercial installation of NGK's process, rated at 60 metric tons/d, will be completed later this year.

In any case, there will probably always be a place for incineration. Under a one-year contract with the U.S. Dept. of Energy, Battelle Memorial Institute (Columbus, Ohio) is investigating a dual fluidizing-bed system of drying and incineration. In the first bed, hot sand fluidizes incoming sludge and imparts heat to the system, generating steam that is piped to another part of the waste-treatment facility. The dried material and sand are then transported to the incineration bed, from which hot sand is recycled back to the first one. Some garbage is incinerated along with the sludge. Ideally, refuse and sludge provide all the energy for the process.

Reginald I. Berry

The cost of shredding municipal solid waste

Sharp hikes in landfill and energy costs have spurred interest in recovering resources from municipal solid wastes. Here are correlations for estimating facility capital and operating costs.

Cheng-Shen Fang, University of Southwestern Louisiana

☐ The first step in all municipal-solid-waste recovery processes is shredding. Such municipal waste varies considerably in size, usually from less than 1 in. to 2 ft. Via shredding, it is reduced to more-manageable sizes, normally 3 to 6 in.

Before a shredding plant is constructed, its capital requirement and operating cost customarily must be estimated. Correlations based on reported data for making preliminary cost estimates are developed and presented in this article.

Installed cost of shredding equipment

The most popular equipment for shredding municipal solid waste is the hammermill. The estimated purchased cost of a hammermill is presented in Fig. 1 [1,2]. Additional capital is needed to cover installation labor and materials, such as electrical parts, piping and instruments. The installed equipment cost—which is the sum of the purchased cost, installation labor cost and material cost—can be estimated from the purchased cost, using the Lang factor [1].

The installed cost of this equipment that has been erected recently in different locations in the U.S. has been reported by Drobny [3] and Boettcher [4]. To obtain the Lang factor for municipal-solid-waste shredding equipment, these data are adjusted to the first quarter of 1979, using the Marshall & Swift Equipment Cost Index (see Fig. 1). The Index shows 577.0 for this quarter. Of the dozen data points reported by Drobny and the four by Boettcher, all except two by Drobny (or 87.5% of the total) are plotted in Fig. 1.

A comparison of the 14 installed-cost data points with the purchased cost of municipal-solid-waste shredding equipment results in a Lang factor of 6.0. A band of ±25% of 6.0 × purchased cost covers most of the reported data points, as shown in Fig. 1.

If the purchased cost of two belt-type conveyors is added to the purchased cost of a hammermill, and the sum is used as the base, the Lang factor for the installed

Originally published April 21, 1980

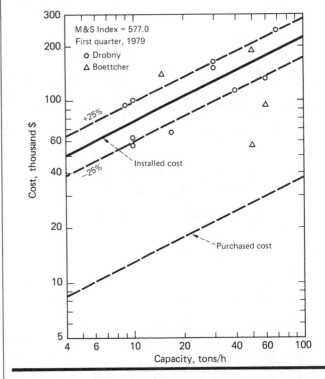

Purchased and installed cost of shredding equipment Fig. 1

shredding equipment is 3.5 ±0.4. This agrees closely with the Lang factor for solid-processing facilities, which ranges from 3.1 to 3.9 [1,5].

From the slope of the line in Fig. 1, the capacity exponent of municipal-solid-waste shredding equipment is 0.46, i.e.:

$$\frac{\text{Cost at Capacity 1}}{\text{Cost at Capacity 2}} = \left(\frac{\text{Capacity 1}}{\text{Capacity 2}}\right)^{0.46} \quad (1)$$

The total capital investment for a complete shredding plant—which includes the costs of a front-end loader, scale, service buildings, land improvement, hammermill and conveyors—varies greatly from one installation to another. Although the ratio of capital requirement for a complete shredding plant to the installed cost of shredding equipment for the Gondard (Netherlands) and Tollemache (Wisconsin) milling plants is, respectively, 2.63 and 2.76 [9], there have been instances of higher ratios.

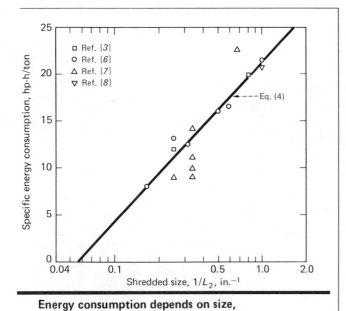

Energy consumption depends on size, moisture and composition   **Fig. 2**

For example, the ratio of the Iberia, La., shredding plant (now in the planning stage) is approximately 4.

Estimating energy consumption

The energy required to reduce the size of a mass from L to L-dL is given by Kick's law [5]:

$$dE/dL = -C/L \qquad (2)$$

Here, E is the specific energy consumption, hp-h/ton, and C is a constant.

If the average size of municipal solid waste before and after shredding is, respectively, L_1 and L_2 inch, Eq. (2) gives, after integration:

$$E = C \ln L_1 + C \ln (1/L_2) \qquad (3)$$

Energy consumptions of more than a dozen municipal-solid-waste shredding facilities at different locations were reported by Drobny [3], the National Center for Resource Recovery [7] and others [6,8]. From the plot of these data, C and L_1 in Eq. (3) are determined to be 7.46 and 17.9 in., respectively. Therefore:

$$E = 7.46 \ln (17.9/L_2) \qquad (4)$$

Energy consumption depends not only on the size of the shredded product, as indicated by Eq. (4), but also on the moisture content and composition of the waste. Moisture content varies from less than 20% to as high as 35%, depending on the climate at the location and the method by which the waste is collected. The composition of the waste also varies greatly, depending on the economic and social level of the community. The data points in Fig. 2 reflect these variations. Nevertheless, a band of ±15% of Eq. (4) includes most of the reported data points.

Unit costs for shredding operations

To calculate the unit costs ($/ton) of equipment, electricity, labor and maintenance, the following base is assumed:

1. The shredding equipment is operated 16 h/d, 5 d/wk, and is online 80% of the time.
2. Two operators at $9.50/h each are needed for 20-40-ton/d units, four operators for a 100-ton/d unit.
3. The waste is shredded to 3 in.
4. The cost of electricity is 4.1¢/kWh.

Calculated costs for equipment, electricity and labor are listed in the table. Experience in solid-waste-shredding facilities indicates that hammermills require frequent repairs and replacement of parts. An average of 22¢/ton for hammermill maintenance in 1970 was reported by Wilson [9]. Adjusting this figure with the Marshall & Swift Equipment Cost Index yields an approximate cost of 42¢/ton in 1979.

Jay Matley, Editor

References

1. Peters, M. S., and Timmerhaus, K. D., "Plant Design and Economics for Chemical Engineers," 2nd ed., McGraw-Hill, 1968.
2. Richardson Engineering Service, Inc., "The Richardson Rapid System— Process Plant Construction Estimating Standards," Vol. 4, 1977-78, Richardson Engineering Service, Inc., Solana Beach, CA 92075.
3. Drobny, N. L., Hull, H. E., and Testin, R. F., "Recovery and Utilization of Municipal Solid Waste," Reports SW-10C for Solid Waste Management Office, U. S. Environmental Protection Agency, prepared by Battelle Memorial Institute, Ohio, 1971, U.S. Govt. Printing Office, Washington, DC 20402.
4. Boettcher, R. A., "Air Classification of Solid Wastes," Report SW-30C for Federal Solid Waste Management Program, U.S. Environmental Protection Agency, prepared by Stanford Research Institute, Calif., 1972, U.S. Govt. Printing Office, Washington, DC 20402.
5. Coulson, J. M., and Richardson, J. F., "Chemical Engineering," Vol. 2, 2nd ed., Pergamon Press, Elmsford, N.Y., 1968, pp. 639-642.
6. Natl. Aeronautics and Space Admin.—American Soc. of Eng. Educ. 1974 Systems Design Institute, "Energy Recovery from Solid Waste," Vol. 2, U. of Houston, College of Engineering, Houston, TX 77004 (order from Dr. C. J. Huang, U. of Houston), p. 43.
7. National Center for Resource Recovery, Inc., "Incineration," Lexington Books, D. C. Heath and Co., Lexington, Mass., 1974, p. 53.
8. Allis-Chalmers Bull. No. 33B5266, 1000 W. College Ave., Appleton, WI 54911.
9. Wilson, D. G., "Handbook of Solid Waste Management," Van Nostrand Reinhold, 1977, pp. 156-163.

The author

Cheng-Shen Fang is an associate professor in the Dept. of Chemical Engineering at the University of Southwestern Louisiana (Lafayette, LA 70504). In addition to teaching, he has participated in a number of research projects involved with energy conservation and developing alternative-energy sources. He holds a B.S. degree from National Taiwan University, and M.S. and Ph.D. degrees from the University of Houston, all in chemical engineering, and is a member of AIChE.

Unit costs of shredding municipal solid waste

Component costs, $/ton	Capacity, tons/h		
	20	40	100
Installed equipment, Fig. 1	0.159	0.109	0.066
Electricity, Eq. (4)	0.407	0.407	0.407
Labor	0.950	0.475	0.380
Maintenance	0.42	0.42	0.42
Total unit cost, $/ton	1.94	1.41	1.27

Petrochemicals from waste: Recycling PET bottles

Important industrial chemicals can be economically recovered from plastic beverage bottles by chemical reduction techniques. And a solid-waste problem can be solved.

Bruce A. Barna, David R. Johnsrud* and Richard L. Lamparter,
Michigan Technological University

☐ As oil prices rise, it is becoming increasingly attractive to recycle petrochemical products. Michigan Technological University (Houghton, Mich.) has developed technology that can economically recover valuable chemicals such as terephthalic acid (TPA) and ethylene glycol from used beverage bottles made of polyethylene terephthalate (PET).

Legislation in several states has mandated the collection of these bottles. And the potential amount of material that can be recovered is great. Plastic bottles are a large market for PET: in 1979, 300 million lb of PET was used for bottles, doubling 1978 production; for 1980, consumption is estimated to range from 375 to 500 million lb.

Due to the great amount of material involved, several companies—including Goodyear, Du Pont, and Eastman Kodak—are actively attempting to recycle PET wastes, including bottles that are now landfilled or incinerated. For this purpose, three different types of technology have been investigated: direct reuse, remelting, and chemical reduction.

PET can be directly reused either by blending it with a virgin stream and reforming, or by grinding it into small pieces for use as a filler material. Another method is remelting, which includes high-pressure molding, extru-

* Now with Exxon, Inc.

Originally published December 1, 1980

sion and depolymerization. However, direct reuse and remelting systems simply recover PET. And, due to impurities such as pigments, paper and aluminum (found in waste bottles), applications for the reclaimed PET are limited to non-food uses or low-purity molded products.

BACK TO BASICS—Chemical reduction systems, however, can produce valuable pure products that have wide markets. Saponification, essentially a reversal of the polymerization process, is the basis of these routes: methyl alcohol will saponify PET to produce dimethyl terephthalate and ethylene glycol. Alkaline saponification to obtain TPA and ethylene glycol has been pursued by Barber-Colman Co. (Rockford, Ill.), an instrumentation manufacturer, and by Du Pont.

Barber-Colman's technology was developed to reclaim silver and TPA from waste polyester photographic and X-ray film. This material was dissolved with the aid of a base to yield a salt of terephthalic acid, as well as ethylene glycol and silver. Water was used as the reaction solvent because the reaction products are soluble in water, while metallic silver, and impurities such as pigment and catalysts, are insoluble. TPA was then recovered from the filtered reaction product by acidification followed by filtration.

In contrast, Du Pont has patented a process that uses an ethylene glycol

solvent for the saponification reaction. This is a substance common to the plastic and it is produced in the reduction process. Also, since ethylene glycol has a high boiling point, the reactor can be operated at atmospheric pressure. Nevertheless, the salt of TPA that is formed during the reaction is not soluble in ethylene glycol, and therefore it must be separated from the other insolubles.

WATER SOLVENT—Patents on the Barber-Colman technique are now owned by Michigan Technological University, which has adapted the method for waste PET bottles. Now the new route, which has been piloted, is available for license.

In the University's system, the entire waste bottle is ground, slurried with water recycled from the distillation column (see flowsheet), and then pumped into the PET reactor. Ammonia, as ammonium hydroxide, recovered in the ammonia stripper is also recycled to the reactor. Additional ammonium hydroxide makeup is added to the feed as required.

The PET reactor, an agitated vessel, is heated to 400°F by the direct injection of high-pressure steam. Here, through reaction with the ammonium hydroxide base, PET is changed to ethylene glycol and the diammonium salt of TPA—both of which are soluble in the water present. The reactor effluent is cooled and then sent to a rotary-drum filter, where insoluble impurities (pigment, polyethylene, etc.) are removed.

Liquid filtrate enters the acidification reactor, where sulfuric acid is added to the process stream. As the TPA salt reacts with the acid, TPA precipitates. This product is recovered by filtration and prepared for market-

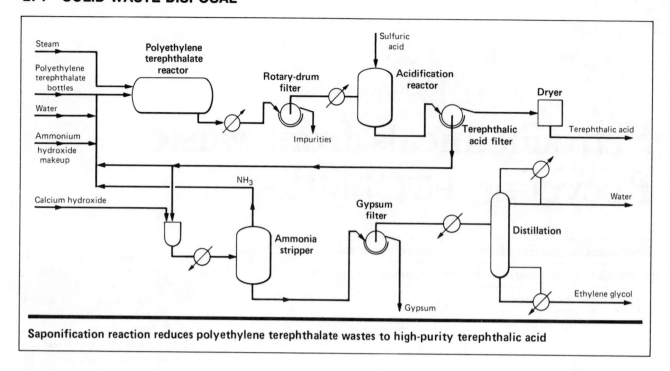

Saponification reaction reduces polyethylene terephthalate wastes to high-purity terephthalic acid

Michigan Technological University's route promises an early payback

Basis: 6 million lb PET processed per year
    M&S cost index: 620.8, 4th quarter, 1979
    300 onstream-d/yr

Products

TPA	713 lb/h
Ethylene glycol	246 lb/h
Gypsum	1,500 lb/h

Utilities and chemicals

H_2SO_4	845 lb/h
CaO:H	635 lb/h
Steam	5,034 lb/h
Electricity	23 kW
Cooling water	160 gpm

	$, thousands
Investment	
Fixed capital	1,436
Working capital	253
Total capital investment	1,689
Income	
TPA (30¢/lb)	1,540
Ethylene glycol (30¢/lb)	516
Total income	2,056
Expenses	
Total direct expenses (includes utilities, chemicals, maintenance, labor)	1,066
Total indirect expenses (insurance and depreciation)	245
Total expenses	1,311

ing by washing, drying and packaging. Purity of the TPA recovered can be better than 99%, the major impurity being ammonium sulfate.

About 40% of the liquid stream leaving the TPA filter is recycled to the PET reactor; the rest enters the ammonia stripper. In the stripper, lime reacts with ammonium sulfate present to form gypsum. As heat is added to the stripper, ammonia is driven off to be recycled to the PET reactor. The gypsum is removed by filtration.

The remaining liquid, ethylene glycol and water, is distilled. Ethylene glycol is recovered in the bottoms and water in the overheads—most of the water is recycled to the PET reactor (the rest is purged).

Two waste-streams are generated by the process—a dilute ethylene glycol (0.4%) stream purged from the overheads of the distillation column, and gypsum cake. The ethylene-glycol/water stream has a low chemical-oxygen-demand and should require only normal activated-sludge digestion. The gypsum cake can be dried to produce essentially pure gypsum for the gypsum industry.

COST OF RECOVERY—The table provides estimated expenses and income for a facility that would process 6 million lb/yr of PET bottles. The purchased equipment for such a plant would cost $348,000 (based on the 4th-quarter 1979 Marshall & Swift equipment cost index). Steam costs were found to be $217,000/yr, using a figure of $6/million Btu. Total expenses and total income are given in the table; however, credit was not taken for the gypsum, pigment, aluminum or polyethylene recovered.

Assuming a 10-yr project and tax life, 48% tax rate and 10% investment tax credit, the discounted-cash-flow rate of return would be 33%. Payback would require 2.7 yr.

The system's economics are not very sensitive to energy costs. A rise in the cost of utilities will likely be coupled with a rise in the price of TPA. If the price for steam doubles, the selling price for the TPA would only have to rise 14%. And as the price of petroleum rises, the economics should be enhanced.

As large markets for TPA and ethylene glycol exist, a significant nationwide recycling effort could take place without hurting process economics. In addition, the system can recover silver from photographic film, oxides from recording tapes, and TPA from old clothes and other polyester products.

Reginald I. Berry, Editor

The Authors

Bruce A. Barna is an instructor in the Chemical Engineering Department and a research engineer with the Center for Waste Management Programs at Michigan Technological University. He holds B.S. and M.S. degrees from Michigan Tech in chemical engineering.

David R. Johnsrud received his B.S. and M.S. degrees in chemical engineering from Michigan Technological University. He is currently employed by Exxon Chemical Company (Linden, New Jersey).

Richard A. Lamparter is technical director of the Center for Waste Management Programs at Michigan Technological University and an adjunct lecturer in chemical engineering. His research efforts have focused on evaluation of new process technology in a wide range of fields.

Firms avidly seek new hazardous-waste treatment routes

Despite some public opposition to siting of treatment facilities, and user reluctance to accept new methods, development and commercialization of treatment techniques for hazardous wastes continue apace.

☐ Even with the impetus for development of new hazardous-waste treatment technology given by such legislation as the (U.S.) Resource Conservation and Recovery Act (RCRA), companies seeking to commercialize such methods face an obstacle course of difficulties.

First, waste-management firms that would ultimately use the techniques tend to be quite conservative. "We're using technology that is 10 to 40 years old," says Robert Pojasek, vice-president of Roy F. Weston, Inc. (Burlington, Mass.), an environmental engineering firm, because "no one wants to take the risk of using unproven technology."

Second, the environmental regulations themselves have added to uncertainties that users may have had about new technologies, having been loosened, tightened and loosened once again in the last year alone. For example, while final regulations for landfills, incinerators, storage ponds and other facilities were promulgated in late July—some seven years after RCRA was passed, there is no assurance that they will not change again, since Congress has just begun hearings on the renewal of RCRA itself.

Still, a raft of chemical-process-industries companies and waste-handling concerns are busy developing new treatment technologies, or dusting off old ones and finding new applications for them. And the firms are doing the work pretty much on their own

hook: The U.S. Environmental Protection Agency (EPA), except for its work on mobile and other kinds of incinerators, has taken a definitely backseat approach to bringing new technology along.

The approaches to ridding the environment of hazardous wastes are many and varied, and so far, no one technique outshines the rest. In fact, waste managers in CPI companies and waste-handling firms seem resigned that there never will be one destruction method that will handle most wastes. "Everybody wants a 'black-box' method where you don't have to worry about what is going in, or how the waste was made [i.e., generated]. But there is no panacea," concludes Jergen Exner, research and development manager at IT Enviroscience, Inc. (Knoxville, Tenn.).

There also seems to be diminishing interest in one oft-stated and long-sought goal of waste managers: the siting of large, centralized waste-treatment centers that would handle a multitude of wastes via a number of techniques (the so-called European model). This is primarily because of the economic and tactical difficulties connected with siting, constructing and operating such facilities.

INCINERATION STATUS—Some 300-odd incinerators in the U.S. are now burning about 400,000 metric tons/yr (wet basis) of wastes, according to surveys done for EPA by consultant Booz, Allen & Hamilton, Inc. (Bethesda, Md.). Although a popular method

of treatment, the cost is high (see table), and just-issued EPA regulations promise to make it even higher. (The regulations mandate: a destruction/removal efficiency of 99.99%, a level often referred to as "four nines"; removal of 99% of gaseous hydrogen chloride; and an elaborate testing procedure before certain wastes can be burned.) But the guidelines have not met with much criticism, and waste managers do not see any intractable problems in dealing with them.

Heat recovery or minimized supplementary fuel use is seen as the best way to get incinerator costs down. Union Chemical Co. (Union, Me.), for one, emphasizes the advantages of its fluidized-bed incinerator (*Chem. Eng.*, May 17, p. 19), which has a silica-sand bed that stores heat while the incinerator is in operation. Raymond Esposito, Union's president, says the unit—which processes still bottoms from the solvent-recovery operation—is completely self-sustaining once it is running. The firm is now looking at installing heat-recovery equipment to supply process steam to the rest of the plant.

Heat recovery is also on the mind of IT Corp., where John Schofield, a vice-president, labels it "probably the

Originally published September 6, 1982

Hazardous-waste disposal costs, 1981

Type of waste management	Type or form of waste	Price, $/metric ton 1981
Landfill	Drum	168-240
	Bulk	55-83
Land treatment	All	5-24
Incineration	Relatively clean liquids, high Btu value	(13)[1]-53
	Liquids	53-237
	Solids, high toxic liquids	395-791
Chemical treatment	Acids/Alkalies	21-92
	Cyanides, heavy metals, highly toxic wastes	66-791
Resource recovery	All	66-264
Deep-well injection	Oily wastewaters	16-40
	Toxic rinse waters	132-264

[1]Some cement kilns and light-aggregate manufacturers are now paying for wastes.
Source: Adapted from Booz, Allen & Hamilton Inc.

biggest trend developing in hazardous-waste control." He says his firm is currently examining the systems of several European manufacturers for use in the U.S.

But Richard Moon, technical director for the Chemical Div. of SCA Chemical Services, Inc. (Boston, Mass.), notes that one common use of excess incinerator energy—the production of electricity—probably will not catch on in the U.S. "We did a comparison of steam and electricity for the Newark, N.J., area, where we have a facility, and found that we could sell steam to nearby industrial customers for about $10/1,000 lb. By converting that same steam to electricity, we could only sell it for $3/1,000 lb," he notes.

In Moon's view, economics would also seem to limit the chances of another waste-incineration process that has received a lot of attention lately—ocean incineration (*Chem. Eng.*, March 22, p. 49). "Ocean incineration makes no sense, longterm," he says, because there is no opportunity for energy recovery, and because the terminal where wastes would be loaded onto oceangoing vessels faces the same siting difficulties as any other waste facility. Still, another firm, Chemical Waste Management, Inc. (Oak Brook, Ill.), is pressing ahead with construction of a second incinerator ship (it already has one), and At-Sea Incineration, Inc. (Greenwich, Conn.) is involved in the financing and eventual operation of two such vessels (*Chem. Eng.*, May 17, p. 18).

MOBILE INCINERATION—Yet another permutation of incineration technology is the mobile unit that has been developed over the past several years by EPA (*Chem. Eng.*, May 17, p. 18). That rotary-kiln system, mounted on four truckbeds, is intended for one-shot cleanups of abandoned wastesites, or possibly regular visits to regions where small volumes of wastes are stored for periodic destruction. After the rotary-kiln system passed shakeout tests in New Jersey this spring, EPA contracted with IT Corp. to field-test the unit, probably at the Kin-Buc landfill near Edison, N.J. Polychlorinated-biphenyl-laced oils have been leaching from the site, which is currently undergoing remedial cleanup paid for by Superfund money.

Another mobile incinerator, also intended for wastesite cleanup, has been designed by Pyro-Magnetics Corp. (Whitman, Mass.), and recently got good marks in an evaluation by Battelle Columbus Laboratories (*Chem. Eng.*, June 28, p. 17). The unit operates at 4,600°F (vs. the EPA system's 2,200°F), relying on oxygen-enriched combustion air for its higher temperature. The firm expects EPA approval of the unit for commercial operation.

WET OXIDATION—Another long-standing chemical process that is getting a second look in the post-RCRA era is wet oxidation, now used extensively in municipal wastewater treatments. Ostensibly, the process may have the same advantage for treating hazardous wastes as high-temperature incinera-tion in that it can handle a large variety of combustible wastes. Its proponents believe it has the added benefits of not requiring extensive particulate, or other, pollution controls, and of not needing supplementary fuel to carry out combustion.

At IT Corp., a catalyzed wet-oxidation process has been on the shelf for several years, but has not yet seen a fullscale demonstration. "We've done all the necessary laboratory work," says IT's Exner, "now it is a marketing problem, to try to find the right wastes to treat with it." The IT process, according to published papers, involves the use of an unspecified bromine-nitrate catalyst system, at conditions below the 400°F and 1,000+ psi usually found in wet-oxidation systems. (Even so, all wet-oxidation processes run far below the 2,000-3,000°F usually occurring in incinerators.)

Another wet-oxidation process, originally adapted solely to municipal waste-sludge treatment, is now being successfully applied to hazardous-waste streams. This is the wet-air oxidation system of Zimpro, Inc. (Rothschild, Wis.), which was one of several waste-destruction processes selected as "innovative" last year by the California Office of Appropriate Technology (COAT). (The assessment by COAT is part of a wider program going on in that state to select certain wastes that would no longer be permitted to be landfilled, and to aid in the siting of commercial-scale processes to handle them through other means.)

The Zimpro system is simple in concept—the waste stream (containing 10% or less organics) is heated to 450-530°F, air is bubbled through, and the organics break down or oxidize to less-harmful substances. Sufficient heat is generated to sustain the reaction as well as to provide steam for plant uses. One drawback, though, is incomplete oxidation of some wastes. "It is difficult to achieve 'four nines' destruction," concedes Allen Wilhelmi, regional sales manager for Zimpro. "But in many cases, it [Zimpro's route] affords the customer the opportunity of sending his waste stream to a municipal wastewater-treatment plant [for more-complete biological treatment], rather than paying the higher cost of having the waste hauled away." He adds that Zimpro's system is being used successfully by one pesticides manufacturer to remove over 99% of contained organics from a waste stream, and is also about to be installed at a landfill in southern California, Casmalia Disposal Co., near Santa Barbara.

NEW OXIDATION PROCESSES—Overseas, both Japan and West Germany are pursuing the development of new wet-oxidation processes. In Japan, the government's National Research Institute for Pollution and Resources (Tsukuba City) is engaged in a four-year (fiscal 1981-84) program to develop a process using ozone. Ten to 15 minutes of treatment under pressures of 60-120 atm has resulted in reduction of chemical oxygen demand (COD) by 90-95% in dye-plant wastewater. Further tests on pulpmill wastes are planned.

And in West Germany, Schelde Chemie GmbH, a joint venture between Bayer and Ciba-Geigy, has started up an oxidation process at its new $350-million dyestuffs plant in Brunsbüttel. The 200,000-m.t./yr wastewater-treatment plant, based on Ciba-Geigy technology, decomposes organic waste acids by subjecting them to 280°C and 120-bar pressure while mixing-in oxygen. "The alternative would have been incineration," says a Schelde Chemie manager. "Investment for wet oxidation is about 20% higher, but this is more than offset by energy savings." Once fullscale operation is under way, the wastewater facility at the dyestuffs plant will be self-sustaining in energy.

Back in the U.S., much attention is being focused on a new wet-oxidation process being developed by Modar, Inc. (Natick, Mass.). Called "supercritical water reforming," the technique involves treating wastes (or wastewater with a minimum 5% concentration of organics) in water heated to above 700°F and 218 atmospheres. Under these conditions, water reverses its normal physical relationship with other substances: organics become nearly completely miscible, while salts precipitate out. (The physicochemical reasons for this have to do with the absence of hydrogen bonding with water molecules in the supercritical state.) The addition of air or oxygen and the high temperature combine to decompose organics very quickly—on the order of one minute. Robert Dunlap, executive vice-president of the firm, says Modar is now building a

1,500-gal/d (organics plus process water) demonstration unit that fits on a 45-ft truck bed (see photo). Field tests will be carried out beginning early next year.

COMPREHENSIVE TREATMENT—Although permitting has progressed somewhat since last spring (*Chem. Eng.*, Mar. 22, p. 49), waste-treatment firms are finding their plans for the siting of treatment centers stymied, or greatly delayed. IT Corp. received a partial permit for siting a facility near Burnside, La., in late July, but, according to Schofield, "It will probably be three more years before the site is ready." He points out that "the problem is not in obtaining permits; many facilities are killed in the cradle by the public's fear for their health and safety, even when a permit is issued." Proposed plants in Massachusetts, New York, Texas and other states also are being held up because of public opposition.

The goal of some waste operators

has been to see the European model of hazardous-waste control adopted in the U.S. One good example of European practice is the Kommunekemi plant of Chemcontrol A/S (Copenhagen, Denmark) near Nyborg, which on a large site has incinerators, chemical- and biological-treatment facilities, storage areas and landfills. All types of wastes can be brought directly to the site, or transferred there from storage points within a certain region.

A visit to the Kommunekemi plant itself has been scheduled into a tour for U.S. waste managers later this month. Chemcontrol will undoubtedly be showing off the latest additions to the plant (see photo, p. 57), including a new rotary kiln that can handle halogenated compounds. With the new incinerator, total capacity of the facility, which has been in operation since the early 1970s, is 60,000 m.t./yr.

But even as the European plants grow, U.S. observers foresee a somewhat different tack being taken in this

Modar Inc's mobile supercritical-water reforming unit will go on the road early next year to perform onsite waste-treatment demonstration tests

country, once public acceptance of waste facilities is established. Robert Pojasek of Weston, which has done waste-management surveys for the state and provincial governments of Minnesota and Ontario, says, "The publicly-owned sites [in Europe] are somewhat inflexible in how they handle wastes. Their prices are very high, and there is a high degree of dependence on incineration. They don't operate in the real world."

This opinion is amplified by Richard Moon of SCA. "Sites that rely on an incinerator tend either to be oversized to take the heat surges of all kinds of wastes, and are therefore very expensive to build and operate, or the incinerators are undersized, and can be easily damaged if operators are not careful," he says. "The Danish philosophy of putting everything into a rotary kiln is not economical to us." He asserts that waste-management firms must first "back-integrate" from the wastes present to the treatment techniques needed, then build a suitable facility. He envisions, smaller, specialized facilities, with a pool of customers within a 150-mile radius.

"The key to this concept is waste segregation," he says. Wastes must be first classified before being sent to a facility. Then, they can be neutralized, oils separated, solids chemically treated to remove toxics, and leftover wastewaters treated by wet oxidation and biological means. Only final residues would be landfilled. Not all such treatment centers would have landfills or incinerators—partly for site-specific reasons, and partly for economics. "A $100-million site is doomed to economic failure," he feels.

INORGANIC TREATMENT—Most incineration methods, essentially, are meant for flammable, organic wastes rather than for inorganics. Among inorganic materials, many neutralization or separation techniques have been practiced for years, especially when high-value metals could be recovered. Still, new techniques are being put forth for treatment of this category of waste as well.

For example, Inco Metals Co. (Toronto) has just announced a new method of treating cyanide-bearing waste streams, from ore-refining or electroplating operations (*Chem. Eng.*, Aug. 9, p. 19). Its method uses a sulfur dioxide-in-air mixture to oxidize metal-cyanide complexes plus free

New construction proceeds at Chemcontrol's Kommunekemi plant in Denmark

cyanide to harmless cyanates and metal hydroxide sludge. It has been successfully lab-tested on zinc and cadmium plating rinsewaters, coke-oven and blast-furnace effluents, and flotation slurries from metal refining, and is currently being installed at a British Columbia gold refiner.

A similar method, using ferrous sulfate and sodium metabisulfite, was announced to be under consideration for use at Homestake Mining Co.'s Lead, S.D., gold mine. (*Chem. Eng.*, June 14, p. 17). Roy Neville, president of Engineering & Technical Consultants, Inc. (Redwood City, Calif.) and developer of the process, says it is under active consideration at several other firms as well.

Both E&TC's and Inco's processes remove certain metal-cyanide complexes, especially ferrocyanide, that the developers claim cannot be precipitated by chlorination processes that are presently available.

For at least one metal processor,

however, methods like these two have the drawback of producing hard-to-dispose-of metal sludges. Moreover, the metal can be recovered more readily by extracting it directly out of the electroplating solution, before it becomes part of a sludge. "We're throwing dollars out the door when we don't reclaim metals," says Thomas Ameen, environmental engineer at Gould Foil Div. (Cleveland, Ohio). At his company (which produces copper foil sheets for electronic products), he is trying to commercialize an electrowinning process that can recover copper from solution prior to cyanide cleanup. "The main difficulty of this electrolytic recovery is that the copper tends to collect as a powder around the cathode, which lowers electrolytic efficiency," he observes. An improved process that avoids this problem is in the developmental stage, and Gould expects to demonstrate it as commercially viable later this year.

Nicholas Basta

Pyrolysis process converts waste polymers to fuel oils

This noncatalytic thermal cracking technique is currently being used to recover almost 94% of the available fuel value from a waste plastic stream from U.S.S. Chemicals' polypropylene plant. Research shows that other plastic wastes can be similarly treated.

Jeet Bhatia and *Robert A. Rossi, Procedyne Corp.*

☐ In July, Procedyne Corp. (New Brunswick, N.J.), U.S.S. Chemicals (Pittsburgh, Pa.) and the U.S. Dept. of Energy formally unveiled the first successfully operating commercial process for converting atactic polypropylene (the noncrystalline polymer byproduct from propylene manufacture) to liquid fuels and other organic chemicals. A 17-million-lb/yr atactic conversion plant has been operating since March 1982 at U.S.S. Chemicals' La Porte, Tex., polypropylene facility (*Chem. Eng.*, July 26, 1982, p. 10). Worldwide patents have already been applied for by Procedyne, developer of the new technique.

The heart of the process is a tubular, plug-flow pyrolysis reactor that is immersed in a gas-heated fluidized bed of sand. At the U.S.S. Chemicals facility, molten atactic polymer at 400°F flows through the reactor, where thermal cracking to gaseous and liquid components takes place. Unlike in other fluidized-bed pyrolysis and gasification processes, no char byproduct is produced, and the only effluents are the hydrocarbon products and clean combustion stack-gas. About 94% of the available fuel value in the feed polymer is converted into usable energy, with 90% being liquid and 4% gas. The remaining 6%, also gas, is recycled to heat the pyrolysis reactor.

Procedyne started bench-scale process developments in 1975 and, after initial success, approached DOE for partial funding of the program. A cost-sharing contract was signed in 1977 to develop a 500-lb/h conversion facility that would convince U.S. polypropylene producers of the method's commercial viability. Following the pilot plant's success, U.S.S. Chemicals contracted with Procedyne to build a conversion plant at La Porte in 1979. DOE provided the funding for the startup and operating analysis of the plant.

Polypropylene constitutes a significant portion of U.S. polymer production (about 4 billion lb in 1980). About 7% of this amount is atactic resin, which is not biodegradable and has traditionally been dumped, burned, or sold for use in such products as hot-melt adhesives and carpet backing. Thus, when this process was first conceived, the goal was very specific—convert the atactic material into a liquid fuel, for in-plant boilers.

INITIAL PROBLEMS—U.S.S. Chemicals' commercial-scale demonstration/development plant was started up during early 1980, and operated for a few months, successfully generating 30,000 gal of fuel oil. But two problems arose that forced curtailment of production: (a) carbon formed on the inner surface of the tubular reactor, and (b) some difficulties were encountered in cold weather when handling the high-viscosity No. 6-type fuel oil produced.

The first problem was solved by developing a proprietary decoking technique, in which instead of just one reactor coil being installed within the

Process saves company the expense of disposing of polypropylene waste

Originally published October 4, 1982

fluidized bed, extra duplicate coils are preset to use when fouling occurs. The fouled coil then undergoes decoking while the cracking operation continues. The cold-weather handling problem was solved by redesign of the equipment.

After these design changes were made and additional tests conducted, the system was restarted last March and has been operating since.

THREE BASIC STEPS—The entire process comprises three basic steps: feed preparation, pyrolysis reaction, and product separation. The waste plastic is first granulated and fed to an agitated tank in which melting takes place at about 400°F, depending on the polymer. In the case of the atactic operation at U.S.S. Chemicals, the molten plastic comes from the process directly. With high-viscosity plastics, recycling and mixing of some of the liquid product assists in overcoming pumping problems.

The melted plastic is pumped into the reaction coil, which is immersed in the heated fluidized-bed furnace. It is here that the plastic is heated to approximately 800-950°F at 50-250 psi. Residence time is less than 30 min. The specific choice of operating temperature and pressure depends upon the plastic being processed.

By having pyrolysis take place in a closed reactor and not in the fluidized bed, the cracking products are kept separate from the fluidizing gases, and product separation can take place more economically.

The reaction products are pressure-reduced and fed to a flash distillation tank. Here, gases and more-volatile liquids are separated from less-volatile ones, similar to a No. 6 fuel oil.

The lighter fraction proceeds to a condenser, where separation of the gaseous component from the liquids takes place. The liquid is cooled and sent to storage. Typically, 10% or less of the feed becomes gaseous fuel. The heavy-oil fraction from the flash distillation is also cooled and stored.

RESEARCH WORK—Conversion of atactic polypropylene to useful chemicals is just a small part of the potential to convert plastic wastes to usable products. Procedyne and DOE cooperated on another project in 1981 to investigate conversion of crystalline polypropylene and polyethylene waste and scrap (from resin producers, fabricators and converters) into fuel oils.

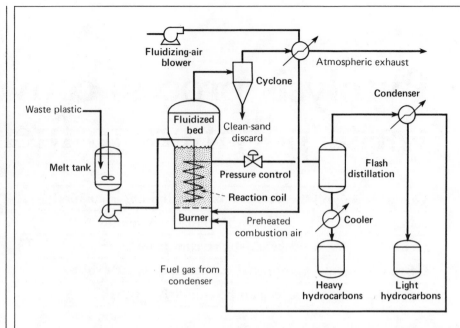

The pyrolysis reactor is made up of three tubular coils (2-in. dia.). Two are used during manufacture, while the third is decoked by oxidation.

These are not insignificant waste streams: about 850 million lb/yr of polyolefin waste is generated.

The research work confirmed that this scrap could also be converted to No. 6 and No. 2 type fuel oils. More importantly, under controlled conditions of temperature, pressure and residence time, these wastes could be converted into gasoline additives, lighter fluid, spot remover, solvents, and petrochemical feedstocks. The initial results yielded hydrocarbons with carbon numbers between C_4 and C_{29}.

It is felt that this development impacts favorably on the new process' operating economics, since such products have higher market value than fuel oil products.

Current R&D effort is being directed toward optimizing pyrolysis conditions to obtain specific products from polypropylene and polyethylene scrap. Future investigations will encompass polyvinyl chloride, polystyrene, and polyethylene terephthalate (PET) bottles, all of these readily identifiable as waste plastic sources.

ECONOMICS—The table above provides an estimate of the economics of building a conversion plant for atactic polypropylene, which itself is assumed to have no economic value. The pyrolysis products are No. 2 and No. 6 liquid fuels.

Leonard J. Kaplan, Editor

Atactic-polypropylene conversion-plant economics

Basis: 25 million lb/yr of atactic polymer (800 h/yr of operation)

Products	
No. 6 fuel oil	6.9 million lb/yr
No. 2 fuel oil	15.1 million lb/yr
Gaseous fuels (net)	1.0 million lb/yr
Capital investment (est.)	$3.06 million
Operating costs	
Utilities	
Electricity, 100 hp	$33,600
Cooling water, 60 gpm	3,000
Direct labor (1/3 of a person per shift)	26,000
Maintenance, overhead and G&A	286,050
Other	147,000
Total	496,250
Payback	2.4 yr

The Authors

Jeet Bhatia, *manager of process technology development at Procedyne Corp. (P.O. Box 1286, New Brunswick, NJ 08903), holds both B.S. and M.S. degrees in chemical engineering from the University of Baroda, India, as well as an M.B.A. from Rutgers University. His professional experience includes work in fluid-bed processes for high-temperature gas/solid reactions, calcination, and solvent removal from polymers and foodstuffs.*

Robert A. Rossi *is process sales manager at Procedyne Corp. He obtained a B.S.Ch.E. from the New Jersey Institute of Technology. With a background that includes work in process development, unit operations optimization and thermal-processing systems design, he is responsible for marketing and sales of the company's fluid-bed process systems. He is a member of AIChE, ACS, and TAPPI.*

Section XI
Wood, Pulp and Paper

Anthraquinone pulping of wood

The use of a catalyst in alkaline wood-pulping processes

can reduce energy requirements, increase yields,

and minimize air pollution problems.

Trevor I. Tenn, Canadian Industries Ltd.

☐ The kraft process, an alkaline method for pulping wood, accounts for about 70% of all chemical pulp produced in North America. This process has a number of advantages in that it produces quality pulp, is energy self-sufficient and can be used with nearly all wood species. But the technology has a number of disadvantages: pulp yields are low, sulfurous gases are produced as a byproduct, and the kraft process has a high capital cost/ton of production.

Anthraquinone can have a catalytic effect on the delignification reactions that are a critical part of alkaline pulping. Canadian Industries Ltd., CIL, (Montreal, Que.) has patented (in March 1977) and now licenses a technique based on this phenomenon.

The CIL process, which can be retrofitted to an existing plant, offers a number of advantages: anthraquinone pulping yields are higher than equivalent kraft yields (see table); delignification reaction times are shorter, or reaction temperatures are lower, so energy savings and higher pulp production are possible with existing equipment; alkali-to-wood ratios are lower; and with anthraquinone pulping it is possible to reduce the amount of sulfides required and thereby limit the generation of sulfurous gases that can create an air pollution problem.

A full-scale mill trial of anthraquinone pulping was conducted in 1977 at Interstate Paper's Corp.'s Riceboro, Ga., pulp mill. The trial indicates that 0.05% anthraquinone increases the pulp yield of southern softwood by 2-3%, decreases the H-factor (a measure of the combined effect of cooking time and temperature) by 25-35%, and decreases alkali consumption by 5%. The pulp produced was equal in quality to the regular kraft pulp and has been run on paper machines without adverse effects.

One Swedish mill has adopted the new technique this year; three Latin American firms have been using it since 1978; and a firm in Japan and two in Australia also are using it.

In North America, most pulp mills are awaiting the U.S. Food and Drug Administration's approval of anthraquinone technology for the production of food-container-grade pulps—in the event that the end-use of the mills' products might be a container of that type. The FDA approval is expected in January 1980.

FASTER REACTIONS—In regular kraft pulping, wood chips are introduced to a caustic pulping liquor containing sodium hydroxide and sodium hydrosulfide. The caustic attacks the lignin in the wood, breaking down cross-linked lignin macromolecules into lower-weight units that are soluble in the liquor. This removal of the lignin allows cellulosic fibers to separate from the wood matrix.

However, the delignification reactions are not completely selective, and cellulose as well as lignin is attacked. The cellulosic fibers' polysaccharide components suffer degradation. Cellulose chains are repetitively split, causing a loss in fiber strength. The kraft process also suffers yield losses from what is called a "peeling reaction" in which the terminal groups of the cellulosic chains are split off, leaving a reducing, aldehydic end-group.

Sulfide in the cooking liquor helps to accelerate the delignification reactions and to minimize the exposure of the polysaccharides to the degradation reactions.

However, with the CIL technique, the catalytic effect of anthraquinone is so great that sulfide levels in the pulping liquor can be lowered—in some cases, it may be possible to eliminate sulfides completely. In general, alkali requirements can be reduced (see table). Also, anthraquinone protects against the peeling reaction.

Anthraquinone oxidizes aldehydic end-groups, increasing the stability of the cellulose molecules and thereby limiting yield losses through the peeling reaction. It is believed that the oxidation forms 9,10 dihydroxy anthracene, which cleaves lignin molecules to form anthraquinone and a lignin stabilized to side-reactions. And, since anthraquinone is regenerated, only a small quantity is needed.

The equipment required for anthraquinone pulping is minimal: a storage tank and metering pump capable of handling a mildly abrasive slurry, and possibly flow measuring and totaling devices to monitor dosage rates (see figure).

Before the digesters, powdered anthraquinone (about 50 wt% as a slurry in water) is added either to wood chips or to the pulping liquor. This promotes uniform mixing of wood, alkali and anthraquinone. The catalyzed delignification reactions take place in the digester. And after the reaction, the pulp follows the normal kraft-process steps of washing, screening and bleaching.

MORE CAPACITY—An economic analysis of anthraquinone pulping may not show a significant cost benefit if wood costs are low. A cost/benefit analysis must be run for individual situations, taking into account environmental benefits as well as the process-

Originally published December 3, 1979

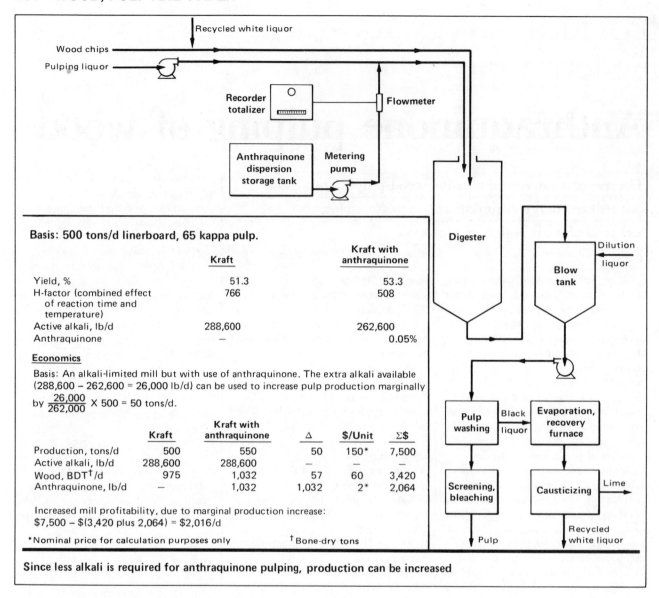

Recycled white liquor

Wood chips

Pulping liquor

Recorder totalizer

Flowmeter

Anthraquinone dispersion storage tank

Metering pump

Digester

Dilution liquor

Blow tank

Pulp washing — Black liquor → Evaporation, recovery furnace

Screening, bleaching

Causticizing — Lime

Pulp

Recycled white liquor

Basis: 500 tons/d linerboard, 65 kappa pulp.

	Kraft	Kraft with anthraquinone
Yield, %	51.3	53.3
H-factor (combined effect of reaction time and temperature)	766	508
Active alkali, lb/d	288,600	262,600
Anthraquinone	—	0.05%

Economics

Basis: An alkali-limited mill but with use of anthraquinone. The extra alkali available (288,600 − 262,600 = 26,000 lb/d) can be used to increase pulp production marginally by $\frac{26,000}{262,000} \times 500 = 50$ tons/d.

	Kraft	Kraft with anthraquinone	Δ	$/Unit	Σ$
Production, tons/d	500	550	50	150*	7,500
Active alkali, lb/d	288,600	288,600	—	—	—
Wood, BDT†/d	975	1,032	57	60	3,420
Anthraquinone, lb/d	—	1,032	1,032	2*	2,064

Increased mill profitability, due to marginal production increase:
$7,500 − $(3,420 plus 2,064) = $2,016/d

*Nominal price for calculation purposes only   †Bone-dry tons

Since less alkali is required for anthraquinone pulping, production can be increased

ing bottlenecks that may be alleviated.

In a mill, digestion capacity can limit total production. Anthraquinone pulping will decrease the H-factor, so that pulp production can be increased in proportion to the shorter cooking schedule. Here the technique may be credited with a yield gain, decreased alkali consumption, and income associated with increased production.

If the recovery furnace is being operated at maximum capacity, pulp production is limited because organics generated during the kraft process cannot be incinerated. With anthraquinone pulping, fewer organics (about 4% less) are produced per ton of pulp, and yields are higher, so that the furnace capacity does not have to be increased, even though more pulp is produced.

If the pulp mill is limited by its ability to produce enough alkali (NaOH and Na₂S), pulp production can still be increased, since with the new technology more wood can be digested with the same amount of pulping liquor.

Also, it is possible to reduce sulfide levels with the technology, so that kraft-odor-abatement equipment can be omitted or downsized. Where air pollution equipment is in place, the adoption of anthraquinone pulping may allow the facility to meet more-stringent regulations.

An economic analysis of the CIL technology is given in the table. The estimate shown is based on an alkali-limited mill that, with the use of anthraquinone, will produce an additional 50 tons/d of pulp. The increased production will improve mill profitability by $2,016/d, based on a marginal pulp selling-price of $150/ton (less associated handling costs).

Over the past 50 years, anthraquinone has been used without adverse effect on workers in the dyestuffs industry as a dye intermediate. In pulp mill applications, this organic chemical is offered as a stable dispersion of powder in water (with minor amounts of FDA-approved dispersants and preservatives) in order to permit metering control and to avoid the hazard of dust explosions associated with the dry powder.

Reginald Berry, Editor

The Author

Trevor I. Tenn is Technical Service Representative, Pulp and Paper Group, Canadian Industries Ltd. He is a Registered Engineer in Quebec and holds a B.Eng. degree in chemical engineering from McGill University.

Wood: An ancient fuel provides energy for modern times

Wood contains no sulfur, can be processed like coal, is renewable, and is becoming increasingly economical. Small wonder that industry is more frequently turning to it as a fuel.

☐ After millennia of use, wood, it seems, is being rediscovered. Industries and utilities that never before even *considered* wood energy are beginning to plan major wood-fueled installations; meanwhile, long-time users of wood as a supplemental fuel, such as pulp and paper makers, are relying on it more than ever.

Of course, the interest in wood energy—both for home owners and for business—is rooted in other fossil fuels' rapidly rising prices and sometimes uncertain availability. Depending on plant siting, wood is cheaper in many cases than coal, oil or gas, and at least as abundant.

And the fuel's merits go past price. Like coal, wood can be burned directly, gasified to produce a low-to-medium fuel gas, or liquefied to make a fuel oil. Unlike coal and oil, wood contains virtually no pollution-producing sulfur.

There are, however, some limits to wood use. One is the absence of some kind of collection system to bring chips or residues to potential consumers. Also, the farther a wood-based plant is sited from a good source, the quicker the economics deteriorate because of the transportation costs involved.

RECENT DEVELOPMENTS—Despite these drawbacks, wood burning now provides the equivalent of 140,000 bbl/d of oil in the U.S., and the U.S. Dept. of Energy (DOE) predicts that this could double by the year 2000.

One of the largest wood-fueled installations to date was announced just last month by California Power and Light Corp. (Fresno). Some 700 tons/d of pelletized fuel will be burned in a new $70-million electric generating plant to be built in Madera, Calif. According to CP&L, the 40-50-MW facility will be the world's largest wood-powered electricity plant when it starts up in 1981. Construction is scheduled to begin in July, with Bechtel Power Corp. (San Francisco) expected to receive the engineering and construction contract.

Meanwhile, late last year, Dow Corning Corp. (Midland, Mich.) announced plans for a $30-million cogeneration plant that will burn 180,000 dry tons/yr of wood. The plant, which will start up in 1982, will be located in Midland and is set to provide 22.4 MW of electricity and 275,000 lb/h of 1,250-psi steam for the company's silicone-products facility there. Construction is due to start this month.

Another large installation also was announced last year. In Vermont, the Burlington Electric Dept. unveiled plans to build an $80-million, 50-MW electric generating plant that will feed on 500,000 tons/yr of green wood. When this starts up in 1983, it will be in the same league as the CP&L installation.

And pulp and paper companies are now taking even greater advantage of wood's fuel value. By 1982, for example, Chesapeake Corp. (West Point, Va.) will install a new 350,000-lb/h wood-waste-fueled boiler at its pulp and paper mill there. The unit will cut the site's oil consumption from 460,000 bbl/yr to 115,000 bbl; during normal operations, no oil will be used for steam- or power-generation.

EASING THE WAY—As utilities and industries switch to wood burning, a number of firms are attempting to smooth the transition.

To make it easier to use waste wood in existing coal-fired boilers, for instance, Bio-Solar Research and De-velopment Corp. (Eugene, Ore.) has a patented process called Woodex that converts the material into 9,100-Btu/lb pellets. By itself, and through licenses, the company has put nine pelletizing plants onstream—eight in the past two years. Another ten plants are now under construction. Woodex, in fact, will fuel the new California Power and Light facility.

In the Woodex process, waste is pulverized, then dried in a rotary drum. The dried material is extruded at 30,000 psi, driving off more moisture and producing pellets about ⅜ in. long by ⅛ in. dia.; water content averages 10-14%, vs. up to 60% in the original wood.

Another pelletizing process is said to upgrade the heating value of wood to a level (13,000-14,250 Btu/lb) that makes it economical to ship the fuel across country—a previously unheard-of claim. Invented by Edward Koppelman of Encino, Calif., and developed in pilot-plant work at SRI International (Menlo Park, Calif.), the process reportedly will soon be commercialized; in fact, Koppelman says he'll be announcing a plantsite this month.

Meanwhile, to allow oil-fired boilers to be more readily switched to wood, Forest Fuels Manufacturing (Antrim, N.H.) introduced a wood gasifier last year that is designed to plug into existing furnaces.

INTEREST IN THE FORESTS—Much of this wood-burning activity stems from the skyrocketing prices of other fuels, and the ready availability of wood in most parts of the U.S.

As Table I shows, the annual Btu value of total unused wood residues is a sizable percentage of the usage of other fossil fuels. In the southern states alone, more than 1,600 trillion Btu/yr of wood residues are available—about 60% of the total fossil fuel used in that region in 1974.

"There are about 500 million dry tons/yr* of material that is wasted," says John I. Zerbe, program manager, energy research and development, for the U.S. Dept. of Agriculture's Forest Service. "It decays in the forest to be recycled back into the soil. About half of this material is recoverable at a reasonable price—about $40/ton." At approximately 17 million Btu/dry ton

*This is a more recent estimate than the one appearing in Table I.

Originally published April 21, 1980

Comparison of industrial fossil-fuel use and quantity of unused wood residues   Table I

Region	1974 Fossil-fuel use, 10^{12} Btu/yr					Total unused wood residues	
	Coal	Residual fuel oil	Distillate fuel oil	Natural gas	Total fossil fuel	10^6 DTE*	10^{12} Btu/yr†
Northeast	461.4	430.2	29.2	241.1	1,162	21.4	364.5
North Central	740.3	201.4	62.5	696.4	1,701	19.9	337.5
Southeast	124.8	240.8	17.6	108.2	491	43.6	740.9
South Central	184.5	168.1	14.5	1,795.3	2,162	51.0	866.4
Pacific Northwest	16.5	39.0	1.0	113.0	170	47.0	799.5
Pacific Southwest	2.7	58.8	3.3	250.2	315	15.0	254.4
So. Rocky Mountain	16.5	17.1	1.3	55.2	90	10.8	184.0
No. Rocky Mountain	66.4	26.0	6.6	99.8	199	3.7	63.0
U.S. Totals	1,613.0	1,181.4	135.9	3,359.2	6,290	212.4	3,610

*DTE=Dry tons equivalent.
†Conversion factor=8,500 Btu/dry lb, or 17×10^6 Btu/DTE.   Source: Battelle Columbus Laboratories

The economics of wood as a fuel   Table II

Price comparison (mid-1979 basis)

Fuel	Estimated heat value	Boiler efficiency	Current price	Cost per million Btus, allowing for efficiency
No. 2 Fuel oil	139,000 Btu/gal	82.5%	$0.67/gal*	$5.84
Natural gas	1,000 Btu/ft³	82.5%	$2.36/$10^6$ Btu	$2.86
Wood 50% moisture (Wet basis)	4,300 Btu/lb	66.7%	$12/ton	$2.09
Wood 13% moisture (Wet basis)	7,490 Btu/lb	78.0%	$15/ton	$1.28

*This has gone up to about $1/gal as of April (CE estimate).
Source: Georgia Institute of Technology

of wood, that price converts to roughly $2.35/$10^6$ Btu—comparable with natural gas (see Table II) and cheaper than oil (currently approaching $5/$10^6$ Btu).

It's this bright price picture that is motivating the drive to burn wood. Dow Corning, for instance, expects to cut costs by 30% compared to $2.40/$10^6$ Btu for natural gas. The economics are so attractive, in fact, that the firm finds it will be worthwhile to pay to have wood harvested from privately owned forests, and to buy sawmill residues, and scrap material from commercial clearing and building-demolition operations.

Aside from economics and good supplies, wood has another advantage: it burns fairly cleanly. Because sulfur is not a normal component of wood, sulfur dioxide is not generated. And,

"compared to fossil-fuel-fired units of the same capacity, wood-fired boilers emit less oxides of nitrogen," points out John Milliken, of the U.S. Environmental Protection Agency's Industrial Environmental Research Laboratory in Research Triangle Park, N.C.

NOT FOR EVERYONE—Despite all those pluses, however, wood has enough drawbacks to limit its industrial potential.

"More industrial users have not pursued this fuel because there is no wood chip brokerage," states Elton H. Hall, senior research scientist in the Energy and Environmental Systems Assessment Section of Battelle Columbus Laboratories (Columbus, Ohio). Generally there is no infrastructure for getting wood chips or residues to potential users, he explains.

And transportation is another prob-

lem. "The energy costs for a 20-ton truck would almost preclude going over 20 miles," says Joseph G. Massey, assistant professor at Texas A&M University's Dept. of Forest Science. Indeed, transportation costs can run 5¢/ton-mile or more. And, since wood contains a great deal of water, the transportation costs get even worse on a cost-per-million-Btu basis.

Finally, wood's high moisture content also adversely affects combustion efficiency, with much of the heat produced going to drive off water: An 8,200-Btu/lb oven-dry wood typically rings in at just 4,000 Btu/lb on a wet basis. (As mentioned before, pelletizing processes are emerging to solve problems of high water content.)

OTHER WAYS—Although today's commercial applications simply involve direct burning of wood, gasification, liquefaction and pyrolysis may make inroads in the future.

Because existing oil- or gas-fired boilers and furnaces are not easily converted to a solid fuel, it would be more economical to gasify or liquefy wood than to use it in solid form, many experts believe.

Last summer, Mitre Corp. (McLean, Va.), under contract to DOE, completed an evaluation of wood gasification's potential. At that time, Mitre felt that low-Btu gas produced from wood would be cost-competitive in 1985. Now, says Mitre's Abu Talib, "if the current trend in oil price escalation is considered, the low-Btu gas ... is cost-competitive today."

Wood gasification is not new—some of the technology, in fact, is decades old. But a bevy of newer, more-

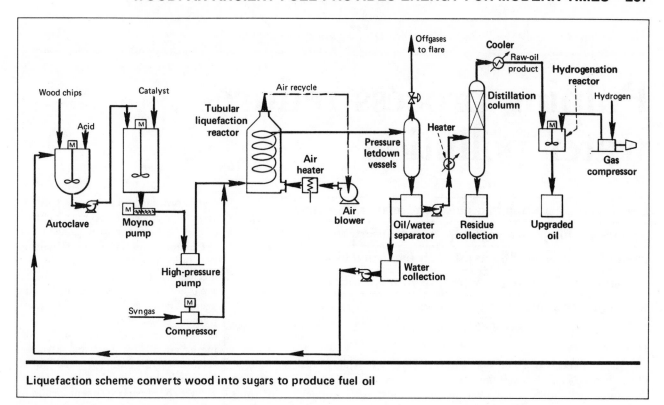

Liquefaction scheme converts wood into sugars to produce fuel oil

efficient processes are being developed that rely on catalysts and/or higher pressures.

Since 1977, for instance, DOE has funded the development of a catalytic gasification route at Battelle Northwest Laboratories. At 550°C and atmospheric pressure, steam is reacted with wood in the presence of a nickel catalyst to produce a gas containing 20-35% methane. This would then be methanated to make a substitute natural gas. The process will be tested with a fluid-bed reactor this year.

Wright Malta Corp. (Ballston Spa, N.Y.), also funded by DOE, plans to build a 4-ton/d unit this year to demonstrate its high-pressure, catalytic process that produces a 450-Btu/ft³ gas. Here, wood is fed to a stationary kiln that uses an auger to move the material. The reaction takes place at 330 psi and 1,100-1,200°F. According to the firm, 95% of the wood is converted to gas that contains 1% CO, 20% H_2, 25% CO_2, 18% CH_4 and 8% ethane and higher hydrocarbons.

Meanwhile, in Hudson Bay, Saskatchewan, a $250,000 wood-gasification demonstration plant has been operating since January 1979. Jointly funded by the province and the Canadian federal government, the gasifier has a feedrate of 290 kg/h of bonedry chipped wood and generates 4 million Btu/h of 150-Btu/ft³ gas. At the moment, planning calls for the testing of a unit that will drive a 65-kW generator.

PRODUCING OIL—Wood can also be liquefied to produce a Bunker C type of oil that can be used directly in furnaces.

DOE now spends about $25 million/yr on its wood programs, with much of the money going to a woodliquefaction project in Albany, Ore. Last August, its first barrel of oil was produced by a process developed by the University of California's Lawrence Berkeley Laboratory.

In the process, an aqueous slurry containing 25% wood is acidified and reacted at 180°F and 150 psi to form an emulsion of sugars (hexoses and pentoses). The stream then is reduced with a synthesis gas (a 50-50 CO, H_2 mixture) at 3,000 psi and 700°F.

From 100 lb of wood, 30-35 lb of 15,700-Btu/lb oil are obtained. The product, roughly half phenolics, has a specific gravity of about 1.1 and resembles No. 6 fuel oil. The researchers estimate that oil can be produced at $6/10⁶ Btu.

In Canada, researchers at the University of Toronto have hydrogenated wood in the presence of a Raneynickel catalyst (*Chem. Eng.*, Feb. 11, p. 55). Shredded wood is slurried in water and reacted with hydrogen at 1,500 psi and 340°C. One kilogram of wood yields approximately 400 g of oil having a viscosity of 5,000-8,000 mPa and a heating value of 35-38 MJ/kg.

Pyrolysis of wood is also catching industry's eye. Enerco Inc. (Langhorne, Pa.), for one, has sold two pyrolysis units—one to the Tennessee Valley Authority, the other to Forest Energy Co. (Beech Glen, Pa.). Both units have been designed to turn 3 tons/h of wood fiber into 40-50 gal of oil, 1 ton of charcoal and 7-8 million Btu of gas.

Other pyrolysis processes are coming on the market, too. Some, like the Thermex reactor of Alberta Industrial Development Ltd. (Edmonton, Alta.), and the Tech Air process (now being licensed by American Can Co., Greenwich, Conn.), introduce air into the reactor to partially oxidize the wood.

Reginald I. Berry

Pulping process reduces water pollution

Developed by a Canadian newsprint maker, this new technology for producing pulp is said to: improve yields from 60 to at least 90%, assist with papermachine runnability, and bring a papermill into compliance with Quebec's pollution guidelines

Bill Schabas, McGraw-Hill World News, Montreal

☐ Q.N.S. Paper Co. (Baie Comeau, Que.) has developed a substitute for the traditional sulfite pulp used in making newsprint. And the company is constructing a 225-metric-ton/d plant to produce the new product (*Chem. Eng.,* Sept. 6, 1982, p. 17). Called the Opco process, the technology takes a mechanically prepared pulp, which has been made by refining wood chips under heat and pressure, and treats it with a weak solution of sodium sulfite. The result, according to the firm, is a high-yield pulp* composed of fibers that exhibit good resistance to the stress of running through a modern, high-speed papermachine.

Q.N.S. Paper reports that because a greater amount of the original wood is recovered as pulp (90-92%) than in other sulfite methods, pollution is reduced. The company expects that the $32-million facility, to be completed in 1984, will produce Opco pulp with an average BOD$_5$ (biochemical oxygen demand) of about 100 lb/ton. But because the new pulp is mixed with virtually effluent-free stone groundwood pulp in a proportion of about 1:2 before being used to make paper, the final BOD$_5$ is expected to be about 55 lb/ton of newsprint—well below the government limit of about 120 lb/ton.

Since the early 1900s, newsprint in Canada has been made by mixing 60-70% of groundwood pulp with chemical pulps, generally low-yield sulfite with lower percentages of more-expensive kraft pulps (also called sulfate pulps). And because groundwood pulp requires primarily electrical energy, plentiful in Quebec, for grinding-machine operation, it is relatively inexpensive. Unfortunately, the grinding of the wood weakens the fibers,

Pilot plant in Finland made pulp for first papermachine trials

* A pulp containing a higher percentage of the original wood than is obtained from traditional sulfite-pulping processes.

and in order to produce a newsprint of acceptable strength, a certain quantity of chemical pulp must be added.

Chemical pulping does not damage the pulp fibers to the degree that mechanical processing does; they are separated by the dissolving of binding agents known as lignins. But chemical pulping costs more, not only because of the chemicals used, but because more than 50% of the original wood is removed with the processing liquid. Although the liquids can be treated to recover the chemical content, including lignins, eastern Canadian newsprint manufacturers have been reluctant to invest in high-cost recovery systems, and usually the pulping waste liquors are simply dumped into waterways downstream, causing severe pollution problems.

CONTROLLING POLLUTION—Canadian mills have tried to solve environmental pollution by coming up with substitutes for sulfite pulps. The goal has been to produce a kind of "super-groundwood" pulp that offers greater strength than the traditional groundwood but provides similar high yields. The different techniques currently being employed for making these pulps use machines called disc refiners, originally developed in Scandinavia during the 1930s. They mechanically separate cellulose fibers by forcing the wood chips through high-speed revolving discs.

Several of these processes are:

1. High-yield sulfite (HYS)—A traditional sulfite cooking process is followed by disc refining at atmospheric pressure. Yields range from 60 to 80%, much better than the approximately 50% of the regular sulfite method, but not nearly as good as the near-100% of groundwood. The pollution problems of traditional methods are not alleviated, and the resulting

Originally published November 1, 1982

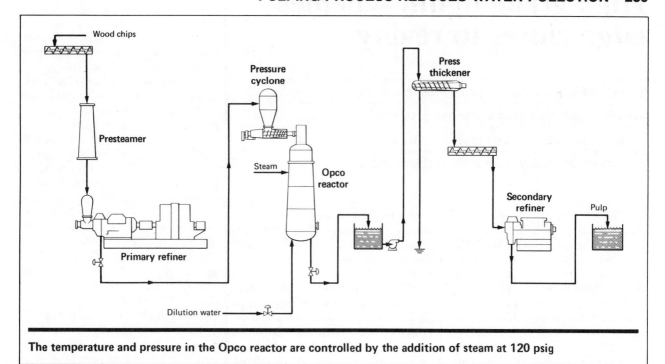

The temperature and pressure in the Opco reactor are controlled by the addition of steam at 120 psig

pulp runs poorly on papermachines—as Q.N.S. discovered when it switched to using an HYS pulp in 1970.

2. Thermomechanical pulp (TMP)—The disc refining of wood chips takes place at elevated temperature and pressure; usually there is a second refining step that occurs at atmospheric pressure. The drawback is higher power consumption.

3. Chemi-thermomechanical pulp (CTMP)—Here a mild sulfite treatment of wood precedes TMP refining. There are several routes, e.g., refining at atmospheric pressure, and using varying degrees of sulfite chemical treatment.

DEVELOPMENTAL WORK—The Q.N.S. mill converted to using a high-yield sulfite pulp for blending in 1970. Though its newsprint maintained acceptable standards for printing, problems were encountered in actually making the paper sheet. As a general rule, increases in yield are paid for with decline in papermachine runnability. The profit from a 30% increase in yield can be lost with a 3% drop in papermaking.

Company researchers concluded that newsprint in its "wet-web" form—that is, when the sheet is still wet and barely able to support its own weight—depends for its runnability on "stretch." This stretch, they hypothesized, was not really a kind of flexibility, but rather a property of the pulp

that is gradually used up as it responds to stresses of the papermachine. When the stretch is used up, the sheet breaks and there is a costly shutdown.

The Opco process is said to maximize pulp stretch. After limited testing on experimental papermachines, the company decided to gamble on an expensive mill-run trial. From a survey of North American and European newsprint mills, the only satisfactory combination of facilities was found at the United Paper Mills in Kaipola, Finland. At this mill, newsprint was made with a pulp comprised of 82% groundwood, 9% low-yield sulfite and 9% semibleached kraft pulps. A test was set up in 1979 in which Opco pulp replaced the standard on the plant's fourdrinier papermachine. The machine appeared to be unaware of the switch in feeds, and continued to produce newsprint without major difficulty for over seven hours.

JOINT VENTURE—Q.N.S. Paper Co., Hymac Ltd. of Montreal, and United Paper Mills formed a joint venture to develop the process, and a pilot plant was constructed adjacent to the Kaipola mill. Tests were conducted to determine the different modes of operation of the Opco process.

TREATMENT AFTER REFINING—The innovation claimed in the Opco process is the use of chemical treatment after the first or second stage of refining. It is said that if the treatment step

comes between the first and second stages, electrical consumption can be reduced for the entire operation by about 20-25% because the pulp requires less work in the second stage after being softened by the chemical processing.

The process begins with the pre-steaming of wood chips (see flowsheet) to make the primary refining step easier. After leaving the primary disc refiner (at about 40 psig), the pulp enters a pressure cyclone where some of its steam vapor is removed before it passes to the reactor. Q.N.S. says the reactor must be large enough to allow a retention time of 30-60 min., and unlike in lower-yield processes, perfect plug flow is not required.

Normally, between 5-10% sodium sulfite (by weight of wood) is used, and the reactor is kept at 160°C at 75-80 psig. There must be sufficient chemical to leave enough at the end of the reaction to keep the pH from dropping below 3, in order not to cause a change in fiber structure. The pulp leaves the reactor to have additional water removed, in the press thickener, before being conveyed to the secondary refiner, which may be either pressurized or at atmospheric pressure. The completed Opco pulp is then blended with both sulfite and ground-wood pulps before being sent to the papermaking machines.

Leonard J. Kaplan, Editor

High-grade lignin schemes edge closer to reality

Several biomass-refining processes that yield a premium-quality lignin are moving toward commercialization. The result may be a new source of aromatics, specialty chemicals and polymers.

☐ The decades-old dream of making chemicals from lignin—the agent that binds wood constituents (cellulose, hemicellulose) into a material highly resistant to breakdown—has always been defeated by the reality of uncertain economics and difficult processing. But there's more hope for success now, because U.S. developers of biomass-refining processes that recover mainly cellulose (for use as animal feed or conversion to ethanol)—well aware that their woody raw materials contain 17-30% lignin, and that this ingredient could hold the key to profitability of the entire project—are focusing their efforts on its recovery in a form easily convertible into marketable chemicals.

To be sure, the availability of a "new" lignin—lower in molecular weight, and therefore less disrupted than the intractable product from kraft pulpmills—will have to wait until those biomass refining processes (*Chem. Eng.*, Jan. 26, 1981, p. 51) reach the commercial stage. And this is not likely to happen soon: Most of the techniques have not gone much beyond the demonstration stage. Still, there is optimism among the would-be biomass refiners; an improved economy, they say, could bring several experimental technologies to commercial maturity within 5-6 years.

PRESENT VS. FUTURE USE—Right now, most of the 22 million tons/yr of lignin generated by the U.S. pulp industry is used as a low-cost boiler fuel, although some of it is converted into asphalt stabilizers, emulsion stabilizers, dyestuff dispersants, cement processing additives and other chemical products. But biomass refiners with an eye on the future now realize that they'll have to convert most of their lignin to chemicals. The reason, says chemical engineer Henry R. Bungay of Rensselaer Polytechnic Institute (Troy, N.Y.), is that "you can't afford to throw away 25% of your feedstock."

The lignin that comes out of conventional kraft and sulfite pulping operations—and some of the wood-to-ethanol systems—is not, however, easy to convert into chemicals. Kraft lignins, for example, are recovered from the black liquor of pulpmills as high-molecular-weight polymers that have been structurally damaged by the pulping process. And lignins obtained from sulfite pulpmills contain sulfonate groups that must be removed before further treatment.

By contrast, some of the newest biomass-refining schemes yield high-quality lignin with a molecular weight anywhere from 200 to 1,000, compared with 1,000 to 50,000 for conventional lignin, and little sulfonation. The low molecular weight makes the material easier to dissolve in solvents and incorporate into chemical products; and the low sulfonation eases conversion into derivatives. Moreover, the high-quality lignin is marked by little structural damage, facilitating conversion into phenols, cresols and other valuable commodities.

Developers of the biomass-refining systems that produce quality lignin expect it to find its way into several new markets. The most immediate outlet is likely to be as a replacement for increasingly expensive phenol in adhesives formulations. In fact, one Canadian company, Forintek Canada Corp. (Ottawa), has found that it can substitute more than 80% of the phenol in its adhesives with the new type of lignin without any loss of adhesive properties.

Another use for the high-grade lignin is as a substitute for polyols in polyurethane formulations. Indeed, many researchers have high hopes for this application. "I could see something like a new Bakelite emerging from this work," says Bungay.

Virtually every wood conversion process, whether aimed at ethanol or chemicals, uses some form of pretreatment. But opinions vary about which method is best. "Many people feel that steam pretreatment is ideal" for disruption of the lignocellulose linkages in wood, particularly when cellulose or sugars for cattle feed are the desired products, says Edward S. Lipinsky of Battelle Columbus Laboratories (Columbus, Ohio). On the other hand, he adds, solvent pretreatments "appear to do a good job of getting 'clean' products." In the long run, says Lipinsky, no single process or family of processes will prevail.

STEAM TECHNIQUES—Perhaps the most advanced lignin-oriented process that uses steam pretreatment is the one developed by Canada's Iotech Corp. (*Chem. Eng.*, Jan. 26, 1981, p. 51). The technique of the Ottawa-based firm involves exposure of wood chips in a gun reactor to high-pressure steam (240-300°C at 500-1,000 psi) for periods from 5 s to 5 min, followed by a sudden release of the steam. The resulting "explosion" depolymerizes the bonds between lignin and cellulose, yielding a material resembling peat moss.

This mixture of cellulose, hemicellulose and lignin is highly amenable to further treatment, such as enzymatic hydrolysis of the cellulose to glucose for animal feed. The lignin fraction in particular can be recovered at the end of the process by extraction with dilute alkali.

Iotech, which was formed in 1975, ran its process in a 1,000-L pilot plant during 1981 at a Gulf Oil facility in Pittsburgh, Kan. Since then, says Iotech vice-president Peter Smith, the company has been looking for a market and trying to optimize the biotechnology aspect (finding a better microorganism for fermenting sugars to ethanol) of the process. Iotech is now attempting to get the backing of New York state for a feasibility study on a demonstration-scale plant.

While Iotech declines to elaborate on the status of these negotiations, Larry Hudson, a project manager at the New York State Energy Research

Originally published December 27, 1982

Lignin hydrocracking picks up steam

While it will be easy to transform the high-grade lignin from future biomass-refining systems into phenols, cresols, benzene and other valuable products, chemical producers face a major challenge in converting low-grade lignin from pulpmills into salable commodities. One technology for doing this, the Lignol process, developed by Hydrocarbon Research, Inc. (Lawrenceville, N.J.), recently marked a milestone by completing a successful demonstration run in a pilot plant (*Chem. Eng.*, Oct. 4, p. 19).

The Lignol process, under development since the late 1960s, is a continuous hydrocracking operation that starts with lignin derived from the black liquor of kraft pulpmills. Among the key products are phenol and cresols, as well as benzene and alkyl benzenes. The cresols can be distilled and sold as such or they can be hydrodealkylated to make more phenol. And the hydrocarbon gases that the process produces are sufficient to generate the hydrogen requirements.

While most experts consider the Lignol process technologically attractive, HRI has had trouble selling it. The reason is that for a number of years, phenol, one of its main products, was depressed in price. But with phenol now selling at an all-time high of about 36¢/lb, HRI believes it's time to take the process off the shelf.

According to HRI's manager of new processes, Derk Huibers, the cresols produced by the Lignol process "are even more attractive than phenol from a commercial point of view." The reason, he says, is that a lot of cresols are derived from coal tar, which has been in somewhat short supply. The shortage has developed because the traditional coal-tar source—coke ovens in steel mills—has been underutilized as a result of the slump in the steel industry. Of course, he adds, "in principle, phenols are a much bigger market than cresols."

Right now, HRI is trying to sell its process to conventional pulpmills. In its pitch to potential customers, HRI is pointing out that part of the lignin of the concentrated black liquor could be extracted and fed into its hydrocracking reactor. The black liquor remaining after the extraction would enter the mill's normal recovery furnace in order to reclaim inorganic pulping chemicals and a portion of the steam for the pulping operations.

While HRI believes that pulpers using the Lignol system could add to their profits by selling the aromatic products, the company says that the process could also boost pulpmill capacity. The reason, according to HRI, is that the Lignol technique would permit the recovery furnace, which is often the bottleneck of the whole operation, to use a more concentrated black-liquor feed, thereby raising efficiency.

Despite such pluses, Huibers concedes that HRI has a tough selling job ahead. Although he reports that some pulp producers are interested, he says that a natural conservatism in the industry is causing hesitation about using the technique. But he nonetheless remains optimistic about the longterm outlook. "It will be a major undertaking to convince the first firm, but after that others may follow."

many respects. The distinguishing feature of the Xerox technique is the very short amount of time—60 s—during which the wood is exposed to high-pressure steam (at 200-238°C and 500-700 psi).

The lignin that emerges is low enough in molecular weight to be soluble in such simple agents as methanol, ethanol, acetone and dimethyl sulfoxide. And the cellulose is readily accessible to enzymatic hydrolysis.

The Iotech and Xerox processes, which are batch operations, have yet to go commercial. But a continuous steam-explosion wood treatment system, developed by Stake Technology, Ltd. (Ottawa), has been on the market for several years (*Chem. Eng.*, May 5, 1980, p. 57). The treatment increases the digestibility of cellulose by farm animals.

Industry sources believe that the Stake process will soon be challenged by the up-and-coming steam explosion technologies, though it is hard to predict which will win out in the end. As far as lignin is concerned, says one wood-conversion researcher, "both the Iotech and Stake systems have a bright future because they provide easily accessible lignin."

SOLVENT PROGRESS—Even as the steam explosion processes move closer to large-scale reality, alternative techniques that employ solvent pretreatment are also advancing. One company in particular, Biological Energy Corp. (BEC), says it's ready to take its solvent-based process—developed originally as a collaboration between General Electric Corp. and former University of Pennsylvania chemist Kendall Pye—to the market.

The BEC process (*Chem. Eng.*, Jan. 21, 1981, p. 55) involves cooking wood chips in an acidic or alkaline ethanol/water mixture, which dissolves the lignin and hydrolyzes the hemicellulose to soluble sugars. The undissolved cellulose fibers can be used as a high-grade pulp. Meanwhile, a solvent-recovery distillation causes precipitation of the lignin, which is suitable for specialty chemical applications. The syrup remaining after distillation is rich in pentoses and can be used for cattle feed.

Originally, BEC had the ambitious goal of converting biomass into ethanol fuels as well as chemicals and animal feed. But the company has

and Development Authority, says talks are now going on between his agency and Motor Energy, Inc.—a subsidiary of Missouri Terminal Oil Co. (St. Louis), which has a U.S. license for the Iotech process. The aim of the discussions, says Hudson, is to build a "several million dollars" demonstration plant capable of processing 150 tons/d of dry wood. "We're

hoping to see the beginnings of a plant within two or three years," he says.

Another steam-explosion process with good yields of commercially exploitable lignin is under study by Robert H. Marchessault and Shardi Malhotra at the Xerox Research Centre of Canada (Mississauga, Ont.). Their process resembles Iotech's in

recently decided to deemphasize the fuel aspect of its project.

BEC's assessment of its progress is in marked contrast to reports from some outsiders. For example, while vice-president Mark DeNino asserts that in the last few months "we've taken a quantum leap forward," and that successful tests have been made in a kilogram-scale pilot plant, a non-BEC source says that the company has found no immediate commercial prospect for its process, and that GE has given BEC a deadline of early 1983 to prove commercial feasibility before it pulls out of the venture.

PHENOL SOLVENT—Another solvent-pretreatment process, under study in Switzerland by Battelle-Geneva Laboratories, is now completing initial laboratory tests. The Battelle process resembles BEC's in many respects, but instead of an ethanol/water mixture, it uses a phenol/water solvent in the pretreatment step.

The wood chips are cooked at moderate temperature (100°C) and atmospheric pressure. The right proportions of phenol and water are critical, Lipinsky says, otherwise "you get a sticky mess."

The acidic solvent causes lignin to dissolve and the hemicellulose to hydrolyze to pentose sugars; but the temperature is low enough to preclude formation of furfural (an undesirable byproduct that can react with phenol in the solvent).

The aqueous phase of the mixture separates into two fractions after cooling. The phenol-containing portion is rich in the lignin; the remaining fraction contains the sugars from hemicellulose hydrolysis. The solid residue contains mostly cellulose, which can be sold as a high-grade pulp.

Lignin separated out of the mixture can be hydrocracked to such valuable aromatics as phenol (3% yield from feed), benzene and cresols. In any commercial system, says Battelle, some of the phenol could be recycled to the reactor. On the other hand, Battelle is weighing the possibility of selling the lignin for specialty chemical applications.

One major problem facing the Battelle process is how to remove all the phenol from the sugar-containing portion of the reaction mixture in order to make it suitable for animal or

Worker fills a hopper with wood chips prior to steam explosion

human consumption. (Sale of the sugars will be necessary to make the process economically viable.) The Geneva group is now trying a variety of solvent-recovery techniques.

Gordon M. Graff

Section XII
Chemical Engineering's Kirkpatrick Awards

1979 KIRKPATRICK CHEMICAL ENGINEERING ACHIEVEMENT AWARDS

For outstanding group effort in new chemical engineering
technology commercialized during the past two years

Winner of the Award

UNION CARBIDE CORP.

Winners of Honor Awards

OAK RIDGE NATIONAL LABORATORY
FLUOR ENGINEERS AND CONSTRUCTORS, INC.
UNION CARBIDE CORP., LINDE DIV.
CHEMISCHE WERKE HÜLS AG

New route to low-density polyethylene

Coupling innovative reactor design with special catalyst compositions, Union Carbide Corp. engineers have opened up a simpler, shorter path to the versatile resin.

☐ Many, perhaps most, future low-density polyethylene plants will look a lot different from present ones, if Union Carbide's Unipol process lives up to the acclaim it has received. Much of the equipment that has been standard in the older processes (tubular or autoclave reactors, extra-thick-wall piping, heavy-duty machinery, monomer strippers, extruders, pelletizers, etc.) has been left out of the low-pressure, gas-phase, fluid-bed reactor process that Carbide has dubbed Unipol.

This new look has been achieved mainly via a drastic reduction in operating pressure. Whereas the established processes operate at pressures ranging from 30,000 to 50,000 psi, polymerization in the Unipol reactor takes place at between 100 and 300 psi. This sharply lower pressure has resulted in a startling compression in plant size. Now, a Unipol facility occupying much less than 1,000 ft² can produce as much polyethylene as an older plant taking up more than an acre of space, about a 90% contraction.

Furthermore, the 100°C temperature inside the Unipol fluid-bed reactor falls significantly below the approximately 300°C temperature at which the high-pressure processes operate.

The flowsheet highlights the radical departure of the Unipol design from the tubular-reactor process.

Originally published December 3, 1979

Big breakthrough on costs

Lowering the pressure and temperature of the low-density process has done much more than simply change its appearance and size. Of weightier impact have been giant reductions in capital and operating costs.

Eliminating some equipment, reducing the number of some, and replacing heavy-duty types with conventional ones has reportedly chopped the capital cost of a Unipol plant a whopping 50% below that of a high-pressure facility of similar capacity.

With the fluid-bed reactor, the polymer (variously described as resembling white coffee grounds or coarsely ground sugar) is removed dry from the gas stream, so only residual monomers need be separated from it. Consequently, the strippers for removing monomer from the molten polymer can be dispensed with.

As an example of equipment being reduced in quantity and size—the conventional process requires a huge 11,000-hp compressor that is fed by smaller 2,000-hp compressors, compared with only a single 2,000-hp recirculating pump in the Unipol process.

By such changes, Carbide engineers have lopped 75% off the energy consumption and about 50% off the operating costs of making low-density polyethylene.

Still other benefits follow quite naturally from the enormous drop in operating pressure. For one thing, the more-ordinary low-pressure equipment suitable for the Unipol process can be ordered "off the shelf," therefore at less cost and with shorter lead-time. And easier maintenance and enhanced safety are obvious fallouts. An important advantage is the extremely low level of hydrocarbon emissions. This makes the process readily acceptable environmentally, which means that environmental impact permits for plant construction will be obtained fairly easily.

New catalysts make it all possible

Key to the development of the Unipol process is a new family of catalysts that trigger the polymerization reaction at the lower pressure. Without these catalysts, the giant step in reducing the operating pressure would not have been possible. The company has not disclosed the composition of, or otherwise identified, the catalysts.

It is primarily by means of catalyst composition that resin properties are controlled in the Unipol process. In the high-pressure processes, resin properties are manipulated chiefly by varying the rate of polymer throughput in the reactor and by the means taken to initiate the reaction. Thus, certain product properties can only be achieved in the older processes at the expense of production efficiency.

Comonomer is added along with the ethylene entering the fluid-bed reactor, to lower the density of the polymer. This practice has been standard for older low-pressure processes, even though lowering resin density much below 0.925 g/cm^3 by adding comonomer has been difficult.

Unipol reportedly can make polymers as low as 0.918 to 0.920 density that meet all polyethylene grades, including injection molding, film, wire and cable, blow molding, pipe, and roto-molding. Exceptional stress-cracking resistance is claimed for the injection-molding resin, and 25% higher bursting strength for the pipe grade.

One reported chief disadvantage of the gas-phase reaction is higher comonomer (butene-1, or more costly, higher-boiling 1-hexene or 1-pentene) consumption, which could range from 4% all the way up to 10% (*Chementator,* Nov. 19, p. 79).

Many attempts have been made in the past to produce low-density polyethylene via low-pressure processes with either hydrocarbon diluent or gas-phase technologies, but none had had broad commercial success. A major problem with slurry processes was swelling of the polymer in the slurry medium, which severely limited resin densities and production rates. In solution processes, molecular-weight ranges were narrowly curtailed. Also, both types of processes required a sizable investment for equipment to separate residual solvent from the polymer and, usually, to remove oligomer and wax. With the Unipol process, Carbide's engineers and scientists have overcome these process and product limitations.

By combining special catalyst compositions of high activity and selectivity with innovative process design, Carbide engineers and scientists have made possible operation at high rates of polymerization to produce gran-

Twin reactors will add 300-million-lb/yr capacity to output from existing revamped plant (shown opposite page).

ular resins of controlled particle-size distribution over a wide range of densities, molecular weights and molecular-weight distributions.

Process description

In the Unipol process, ethylene and an alpha-olefin comonomer are copolymerized in the presence of a chain transfer agent to produce polymers having desired melt indices, densities and molecular-weight distributions. Monomers are fed continuously into the fluid-bed reactor. Before entering the reactor, the monomers are subjected to varying degrees of pretreatment (depending on their sources) to protect the catalyst against impurities that could poison it. The catalyst is added separately.

The fluid bed in the reactor is made up of granular polyethylene polymer—product of the polymerization reaction. Circulated up through the bed, the gas stream—made up of ethylene and comonomer—passes out of the reactor through an enlarged top section designed to disengage most of the fine particles. It then goes to a cycle compressor and through an external cooler before returning to the reactor.

Dry free-flowing solid product is removed intermittently from the continuously growing fluid bed through a discharge system, in such a way as to keep the volume of the bed approximately constant.

Although most of the unreacted monomer is recovered and recycled, some residual hydrocarbons are purged from the granular product so that it can be safely conveyed in air. As an optional final process step, one or more conventional additives (e.g., antiblocking, antislipping, antioxidizing, ultraviolet-light-stabilizing) may be joined to the granular product before it is stored or shipped.

The overall, combined conversion rate of ethylene and comonomer is approximately 97%. The average polymer residence time is 3 to 5 h, during which the particles grow to an average size of about 1,000 microns.

Reaction pressure is controlled normally at about 21 kg/cm^2; reaction temperature, from about 75 to 100°C. Density is regulated by copolymerizing with appropriate concentrations of the alpha-olefin comonomers, which varies the introduction of short-chain branches.

Molecular weight is influenced by the reaction temperature and the concentration of chain-transfer agent in the circulating gas. Molecular-weight distribution is manipulated primarily by catalyst type and composition and, to a much lesser extent, by reactor operating conditions.

How the development progressed

Union Carbide's search for a new low-density polyethylene process began simultaneously with its quest for a low-pressure process for high-density polyethylene. Success in the former proved far more elusive.

Exploratory research on the low-density process got underway with a bench-scale reactor at the corporation's technical research center at South Charleston, West Virginia, in the 1960s.

Previous work had shown that polymerizations in a slurry of hydrocarbon diluents limited polymer density to a minimum of about 0.925 to 0.935 g/cm^3. Excessive swelling of the polymer particle at practical operating temperatures sets this bottom boundary.

Although pilot-plant work in solution polymerization did not indicate a floor on density, only lower molecular weights (i.e., melt index above 1) could practically be achieved. Solution polymerization presents the additional drawbacks of high investment costs and difficulty in reducing the residual dissolved hydrocarbons to the low level now required.

Efforts first concentrated on resolving the problems associated with high- and medium-density polymers during the scaling up of a fluid-bed, pilot-plant reactor in 1965 and the initial operation of an 8-ft-dia. prototype fluid-bed reactor at Seadrift, Tex., in 1968. At that time, the polymer structures obtained with the catalysts then developed were not suitable for the plastic's major end-use—as film.

Thereafter, an aggressive catalyst research program was mounted at the company's laboratories at Bound Brook, N.J., aimed at producing improved catalysts capable of generating the required polymer structures. Periodic testing of new catalysts formulated in the pilot-plant fluid-bed reactor, and further modifications

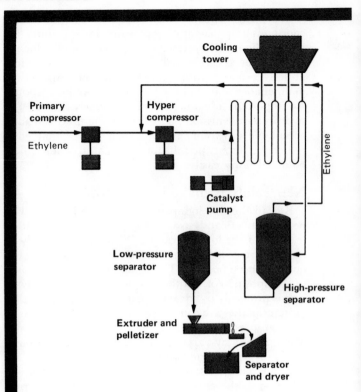

High-pressure steps to low-density polyethylene

Until the development of the Unipol process, low-density polyethylene was being produced in essentially the same way as when introduced nearly 40 years ago.

In the established processes, ethylene is catalytically polymerized in either tubular or stirred autoclave reactors. Only the first type of reactor is depicted in the flowchart.

Ethylene, the basic raw material, enters the process (along with comonomers for modifying polymer properties) via the primary compressor, which performs the first step in building up the pressure of the reactant stream. Next, a large compressor completes the job—boosting the pressure to as high as 50,000 lb/in^2.

Catalyst is injected, by a separate system of pumps, into this highly pressured feedstream just before it goes into the reactor. The tubular reactor consists of as much as a mile of extra-thick-walled piping, which twists back and forth in a precisely designed configuration. For safety, the reactor must be contained within heavy walls of reinforced concrete.

The polymerization reaction that takes place in this mile of pipe generates such an enormous quantity of heat that it must be dissipated in huge cooling towers.

Molten polymer, along with unconverted reactants, leaves the reactor to undergo cleaning up. First, unreacted monomers are drawn off from the molten polymer in a high-pressure separator. These monomers are then cycled to the booster (hyper) compressor. Remaining monomers are removed in the low-pressure separator.

Afterward, the cleaned-up molten polymer is fed into an extruder, out of which it leaves as strands that are immediately chopped into pellets. Cold water rapidly cools the pellets. The pellets are separated from the water and dried by a stream of air.

In the conventional processes, variations in product properties are controlled by the type and concentration of catalyst, reaction temperature, throughput rate in the reactor, the chain terminator (the comonomer), the method the catalyst and monomer are introduced, and the configuration of the reactor.

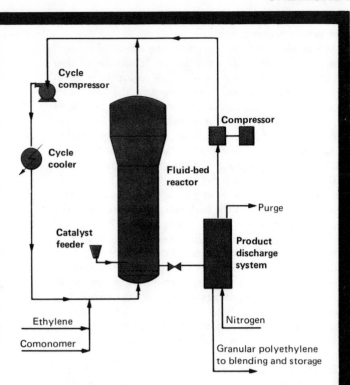

Low-pressure shortcut to low-density polyethylene

Low-pressure shortcut to low-density polyethylene

Other than process basics, there is little that is similar between conventional high-pressure tubular-reactor systems and Union Carbide's revolutionary new low-pressure gas-phase reactor system. The departure in plant size and shape is dramatic. However, the really important differences are in what happens inside the reactor.

In the Unipol process, gaseous ethylene and comonomer are fed continuously into the fluidized-bed reactor. Catalyst is added separately.

Instead of the extremely high pressures required with the conventional low-density polyethylene processes, the pressure inside the Unipol reactor need only be about 100 to 300 lb/in². Instead of the approximately 300°C temperature at which the older processes are operated, the temperature inside the fluid-bed reactor is controlled at about 100°C.

Circulated by a small compressor (almost small enough to be a pump), the gas stream passes through a cooler, which removes the exothermic heat of reaction from the fluidized bed of polymer. Cooled monomer is then recycled to the reactor.

Polyethylene is removed directly from the reactor—a major difference from conventional processes—through a gas-lock chamber. Only a small amount of residual monomer accompanies the granular polymer into this chamber, and is easily and safely purged. Catalyst productivity is so high that the removal of catalyst residue is not necessary. Because low-molecular-weight fractions are produced in such small quantity, these also need not be removed.

Upon leaving the product discharge system, the polymer granules are ready for shipment to customers—with or without conventional additives but without any further processing. No longer formed into pellets, the polymer product resembles laundry detergent.

The proprietary catalyst plays the key role in the new process. Polyethylene resin properties are controlled primarily by catalyst composition. Reactor operating conditions remain the same, regardless of what product is being made.

in the process, led to an initial trial of low-density polyethylene in the prototype reactor during 1975.

After continued large-scale experimentation and market testing, the first embodiments of the Unipol low-pressure process attained full commercial status in late 1977 and early 1978, with the production and sale of many million pounds of the new and improved low-density resin.

Obstacles that were overcome

Although low-pressure processes using diluents have successfully produced high-density polyethylene for several decades, major obstacles have blocked the application or adaptation of these processes to the economical manufacturing of a broad range of commercially acceptable low-density polyethylenes. Certain of the problems inherent in solution and slurry processes can be avoided by resorting to a fluid-bed technique. However, a fluid-bed process also imposes particular limitations.

To bring the fluid-bed gas-phase low-density polyethylene process to full commercial status, Union Carbide engineers had to solve such major problems as preventing the agglomeration of the polymer bed, controlling polymer particle size and shape under a broad range of operating conditions, and developing highly productive catalysts that efficiently copolymerize ethylene with other olefins to yield a range of low-density products having the desired molecular weight and other properties.

Polymerization in a bed of fluidized polymer granules is more difficult with lower-melting-point low-density polyethylene than with high-density polyethylene, because the difference between the melting point and polymerization temperatures is much narrower.

Agglomeration will occur if the temperature at any point in the fluid bed exceeds the melting point of the polymer in the bed. Therefore, the reaction heat must be uniformly generated and removed during all phases of operation, such as startup, equilibrium conditions, and normal and emergency shutdowns. Additionally, uniform bed composition must be achieved and maintained with respect to both time and space.

Carbide engineers found that the catalyst types and equipment design that had been developed for high-density polymers caused low-density particles to agglomerate. And this, in turn, generated excessive hot-spots and lumps, and caused granular bulk density to be much too low.

Engineers turn to modeling

It became apparent that prior engineering solutions for achieving macroscopic temperature uniformity were inadequate. Thus, it became necessary to look into heat generation and removal on a microscopic (single particle) scale. By modeling the heat transfer from an individual particle to the gas stream, engineers solved the problem of achieving and maintaining uniformity of heat generation on the microscopic scale. They then selected catalysts on the basis of uniformity of activity, and thereby avoided the higher-than-melting-point surface temperatures that caused polymer particles to agglomerate.

Catalyst is added separately to the Unipol reactor from feed hoppers by means of special pumps.

Also via modeling, together with designed experimentation, the engineers determined the reactor conditions that would assure uniform heat removal at all times and under all circumstances, and established safe conditions of reaction rate, temperature, gas velocity and polymer composition.

For instance, they found it advantageous to prevent the reaction surges that followed occasional spikes of catalyst poisonings. A guard system that removes impurities from the monomers effectively smoothes out the reaction surges. (These disturbances are particularly likely to happen when ethylene is received via pipeline from several different suppliers or storage domes.)

Close particle-shape control is necessary to achieve acceptable heat removal, and heat transfer between the fluidized particles and the fluidizing gas. It is also essential so as to obtain a granular product of the desired bulk density.

Because the polymer product's size distribution and particle shape are related to properties of the catalyst, control of the individual activity and productivity of catalyst particles is important. By this means, the size distribution and shape of the polymer in the fluid bed can be tailored for good operation: i.e., low minimum fluidizing velocity, and the absence of fine particles that could be carried overhead into the recycle line, and of oversized particles that could settle out of the bed to cause plugging and other discharging problems. Of course, a product without overly fine and coarse particles is also important to customers, because they must convey and store it.

Importance of composition

In addition to particle shape and size distribution, still another property of the catalyst is critical in the Unipol process. Catalyst composition must be optimized for high copolymerization rates and for minimum generation of low-molecular-weight fractions (oligomers). This optimization was accomplished by means of designed experiments, from the results of

which computer models were also developed for the necessary reaction conditions.

As a result of extensive study of the effects of catalyst composition and reactor conditions, it was found that the required polymer structure with regard to molecular-weight distribution, branching and the low-molecular-weight fraction could be provided via control of catalyst composition, in combination with the factors already discussed for regulating the heat generation and fluidization (or heat removal) characteristics of the catalyst and consequently of the polymer particles.

To achieve acceptable heat transfer between the resin particles and the fluidizing gas, as well as to maintain proper fluidization of the bed, ways to control both the size and shape of the growing polymer particles had to be found. Such control had to be possible at a wide range of production rates and under the varying reaction conditions necessary for producing a full scope of low-density product grades. Having these means of control, Carbide engineers were able to manipulate particle-size distribution, and thus simultaneously achieve proper fluidization and the optimum packing geometry necessary for getting a product of an acceptably high bulk density.

To attain densities below 0.926 g/cm^3, low-density polyethylenes now on the market contain fairly high concentrations of olefin comonomers. Therefore, it was apparent that a catalyst capable of really superior copolymerization efficiencies had to be formulated.

A particularly critical aspect of catalyst development was the problem of achieving adequate control of chain-branching distribution. If this could not be regulated adequately, it would not be possible to provide the wide range of properties needed in low-density polyethylene to satisfy all of its commercial applications.

In the end, Carbide's scientists solved the problem of optimizing catalyst compositions to obtain desired molecular-weight-distributions, and thereby achieved the product properties desired and avoided the low-molecular-weight fractions not desired. The oligomers cannot be tolerated in the process because of their surface stickiness, or in the final product because they can be extracted at levels unacceptable in food packaging.

Further, this selectivity for high polymers was attained while the catalysts were also optimized for control of particle size and shape. As a result, a free-flowing, high-bulk-density granular resin can be produced that does not need further processing into pellets to make it suitable for manufacturing into products.

Also, the efficiency with which residual monomers are purged from the resin overcomes some of the problems created by the high solubility of comonomers in lower-density polyethylene. Measurements of hydrocarbon solubility and desorption kinetics led to a model for predicting rates of desorption in the purging system. This model was verified in the pilot and scaled-up plants. As a result, the comonomer dissolved in the resin could be lowered to the levels required for emission control and safety considerations.

Reaching commercial status

In November 1977, Union Carbide announced that "embodiments" of its low-pressure process had attained

Computer control can be readily applied to the Unipol low-pressure process because of the system's simplicity.

commercial status, with the production and sale of millions of pounds of its new low-density polyethylene. Introduced commercially (in substantial volume) during this time were commodity products for film, injection and blow molding, and specialty high-performance products for wire and cable insulation, pipe, and rotational molding.

The fully commercial status of the Unipol process was further confirmed in April 1979, with the announcement of the first licensing agreement, this with Exxon Chemical Co. Exxon will build a 340-million-lb/yr plant at Mont Belvieu, Tex., based on the process. The plant is scheduled for startup sometime in late 1981.

Shortly after Union Carbide's announcement of the commercial realization of its new process, Exxon had reiterated plans to build a plant based on improved tubular-reactor technology. Exxon's reversal appears to attest to the attractions of the Unipol process.

African Explosives and Chemicals Inc., South Africa, has become a second licensee (*Chementator,* Oct. 8, p. 49).

Union Carbide is also proceeding with its own plans to add nearly 1 billion lb/yr of low-density polyethylene capacity in the U.S. by 1982. The first phase of this expansion program—the construction of a 300-million-lb/yr plant—is underway at the company's Seadrift petrochemical complex. Construction is scheduled to be completed in the first quarter of 1980.

Until this plant comes onstream, Carbide will continue to operate the revamped high-density unit at Seadrift that produced the initial quantities of the new low-density product for commercial introduction.

Supplementing the Seadrift plant, a 500-million lb/yr plant is to be built at Taft, La., with construction to start early next year at a new site adjacent to the company's existing petrochemical complex. Named the Star Plant, this facility is expected to come onstream in 1982. Provisions have been made to allow for a doubling of capacity. Quick expansions at the two plants can add another 150 million lb/yr of capacity.

Virtually every major supplier of low-density polyethylene has considered taking a license for the process from Union Carbide. Serious negotiations are underway with a number of producers worldwide, and the company expects that it will grant additional licenses for the process before long.

Ever since the supply of oil tightened, producers of the low-density resin have been predicting that its cost would climb so sharply as to undermine its traditional low-price attraction.

Unipol's coming should allay these fears. Investment prices required for the new process are expected to be 6¢ to 12¢/lb lower, depending on grade, than for conventional high-pressure processes. This advantage could hold the price of the low-density resin to 15-25% below that previously forecasted for 1980, enabling the resin to retain its reputation as an economically superior raw material and maintain its strong growth pattern.

Recovering uranium from phosphoric acid

Oak Ridge National Laboratory has developed an economic process that retrieves uranium while removing this radioactive contaminant from fertilizers.

☐ Wet-process phosphoric acid contains a significant amount of uranium. Indeed, this uranium totals more than 1,500 tons/yr in current U.S. acid output—and is worth about $135 million. And projections put the uranium level at 8,000 tons/yr in the year 2000.

Since the phosphoric acid is a major raw material for fertilizers, uranium finds its way into those products and is effectively lost as a resource, while adding to the amount of radioactive material that can contaminate the food chain.

So, resource-conservation and environmental considerations both make recovery of the uranium from phosphoric acid desirable. Making such recovery economic is the achievement of Oak Ridge National Laboratory (Oak Ridge, Tenn.).

Thanks to Oak Ridge's development of the DEPA-TOPO extraction process, a growing proportion of the uranium is being retrieved. Freeport Minerals Corp. (New York City) and Wyoming Mineral Corp. (Denver, Colo.) commercialized the process during the last quarter of 1978—Freeport bringing online a 425,000-lb (U_3O_8)/yr plant at Uncle Sam, La., and Wyoming starting up a 690,000-lb/yr unit at Bartow, Fla. International Minerals & Chemical Corp. (Libertyville, Ill.) is now shaking down a 750,000-lb/yr facility at New Wales, Fla., and CF Industries, Inc. (Chicago) is installing a total of 1,260,000-lb/yr capacity at Bonnie, Fla., and Plant City, Fla.

When all these units are running, more than half of the uranium in U.S. wet-process phosphoric acid will be recovered.

And the DEPA-TOPO extraction process is moving ahead elsewhere. Pilot plants have been operated in France, Spain and Taiwan.

Originally published December 3, 1979

Process development

A solvent-extraction process for recovery of uranium from wet-process acid was commercialized briefly in the early 1950s. Unfortunately, the route suffered from a number of drawbacks. The instability of the process extractant was one of the most important ones—about one third of the extractant decomposed each cycle, necessitating that fresh extractant be added continuously. Also, a costly iron-reduction step was needed to prevent saturation of the solvent with ferric iron. And the uranium product was of low quality and difficult to upgrade to a usable product. Because of these drawbacks, the process could not compete economically with recovery of uranium from Western U.S. deposits.

That is how the situation remained until Oak Ridge developed the DEPA-TOPO process, which boasts costs that are competitive with those for winning uranium from Western ores.

Oak Ridge had started studies on recovering uranium from wet-process acid in 1967. It drew upon decades of its own work on recovery of metals by solvent extraction—during which time it had developed a wide variety of extraction reagents and processes for winning uranium from solutions.

However, extending this work to phosphoric acid posed problems. The extractants used successfully for recovering uranium from Western-ore sulfate liquors did not work well for wet-process acid because of the strong uranium-complexing power of phosphate in solution. So, many extractants were tested, and Oak Ridge eventually selected the combination of di (2-ethylhexyl) phosphoric acid (DEPA) and trioctylphosphine oxide (TOPO) as the most suitable.

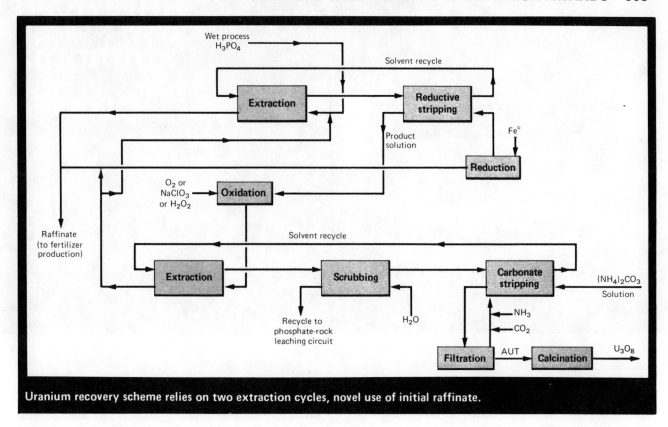

Uranium recovery scheme relies on two extraction cycles, novel use of initial raffinate.

The choice of DEPA-TOPO as extractant raised another difficulty: Because this solvent has such strong uranium-extraction power, stripping the uranium conventionally via acid solutions is hard. And retrieval of the uranium from these solutions would involve added complexities.

Prior work had shown that ammonium carbonate solution readily strips uranium. However, Oak Ridge discovered that use of the carbonate involved high ammonia consumption, and that the system suffered from poor phase separation, difficult iron filtration, and problems in reaching quantitative uranium recovery from the stripping solution.

A breakthrough

The dilemma of how to remove the uranium from the DEPA-TOPO solvent was resolved when Oak Ridge developed a novel reductive-stripping method based on the use of the wet-process acid itself.

In this method, a small amount of raffinate from the solvent-extraction system is "borrowed," and the iron that it contains is reduced to the ferrous state with iron metal. Contact of this solution with the extract then results in the reduction of the uranium to the tetravalent state, whereupon the uranium transfers quantitatively to the aqueous phase. With appropriate control of the organic-to-aqueous flow ratios and of the ferrous iron concentration, the uranium can be concentrated in two to three stages by a factor of 50 to 100, compared with its level in the phosphoric acid.

Use of the wet-process acid is inexpensive, since the acid is not consumed and is returned to the raffinate (and subsequently can be used for fertilizer manufacture). Also, the acid contains about 0.15-M iron, and

only a relatively small amount of iron metal reductant need be added to produce the desired ferrous iron concentration in the solution. Further, the solution contains about 1-M fluoride, which catalyzes uranium reduction.

The flowscheme

As shown in the diagram above, the process starts by passing wet-process phosphoric acid to a four-stage extractor, where at $40°C$ to $45°C$ the acid contacts 0.5-M DEPA/0.125-M TOPO in kerosene. The uranium-laden extract then goes through two to three stages of reductive stripping at $50°C$. It is stripped by raffinate of a specific ferrous iron concentration, which has been achieved by the prior addition of metallic iron.

From this step, the DEPA-TOPO solvent recycles to the extractor. Meanwhile, product solution is oxidized with either sodium chlorate, oxygen or hydrogen peroxide, and is sent through a second, low-volume DEPA-TOPO extraction cycle at $25°C$ to $45°C$. Here, 0.3-M DEPA/0.075-M TOPO is used, and the solution is kept relatively highly loaded with uranium. This high loading cuts down on the extraction of contaminants, particularly iron, and lowers the amount of ammonia required in a subsequent stripping step.

The extract then undergoes two to three stages of scrubbing with water at $25°C$ to $45°C$ to remove impurities. Next, the uranium is removed from the solution at $35°C$ to $45°C$ by two stages of stripping with ammonium carbonate. A relatively pure ammonium uranyl tricarbonate (AUT) precipitates. It is filtered out of solution, and calcined to yield high-grade U_3O_8, an attractive feed for producing uranium hexafluoride.

Overall uranium recovery from phosphoric acid should run 90% or better.

Improved ammonia process

Removing CO_2 with a physical absorbent eliminates solvent regeneration by heating. Via innovative design, the heat saved drives the refrigeration unit.

☐ The advantages of a new process development are not always obvious from the beginning. It often takes years of experimentation to match an innovation with the process the researchers had in mind.

Such was the case with Fluor Engineers and Constructors, Inc.'s improved ammonia process. Back in the 1950s, Fluor researchers undertook a major program to find an organic physical solvent for carbon dioxide that could be regenerated merely by depressurizing. Such a solvent, they believed, would have a major impact on the economics of synthesis gas generation for hydrogen and ammonia plants. Propylene carbonate was chosen as most suitable for such an application.

Benefits of the improved process

Now, more than 20 years later, events have proved the researchers were right. Two 1,200-ton/d ammonia plants that started up in 1978 have shown that Fluor's process:

• Uses 5–10% less energy per ton of ammonia than the conventional process, which regenerates chemical solvents by heating.

• Lowers the capital cost of the plant. (Fluor declines to be specific on this because of the various bases that may be used for cost estimates.)

• Offers an advantage when a urea plant is included in the complex, in that roughly half of the carbon dioxide removed from the process gas is available at a pressure of around 100–200 psi.

• Permits reliable and essentially trouble-free plant operation because propylene carbonate is noncorrosive, even though carbon steel is used in the plant.

It took time

However, the time that elapsed between the conception of the recipe and the proof of the pudding was

Originally published December 3, 1979

taken up with evolutionary developments and necessary modification of the process. For example, the energy savings are achieved largely because physical solvents, unlike chemical solvents, do not need heat for regeneration. The catch is that in conventional plants much of the heat used for regeneration is waste heat from the process gas stream. Fluor solved the problem of what to do with the waste heat it had "saved" by using it to drive an ammonia-absorption refrigeration unit that condenses out the ammonia product gas at the end of the process. This eliminated the separate closed-cycle ammonia-compression refrigeration system used in conventional plants.

Fluor process vs. standard process

Fluor's ammonia process, like the conventional system, starts out with two-stage steam reforming of natural gas. First, the natural gas is reformed at about 450 psi to produce a synthesis gas that contains hydrogen and carbon monoxide. In the second step, a controlled amount of air is added to provide the approximately 3 to 1 hydrogen-to-nitrogen ratio required for ammonia. At the same time, more hydrogen is created, along with undesirable carbon dioxide. This is the point where the Fluor process begins to differ.

The standard technique for removal of CO_2 is chemical absorption. Potassium carbonate is a popular absorbent because it operates at around 250°F and, compared with other chemical solvents, requires less cooling of the synthesis gas, whose post-reforming temperature is around 400–500°F. (Amine chemical absorbents were widely used at one time, but are less common now because they need a temperature of around 100°F.)

The potassium bicarbonate resulting from the CO_2 removal step is regenerated in a stripper tower, where steam drives off the CO_2. Heat for the stripper reboiler

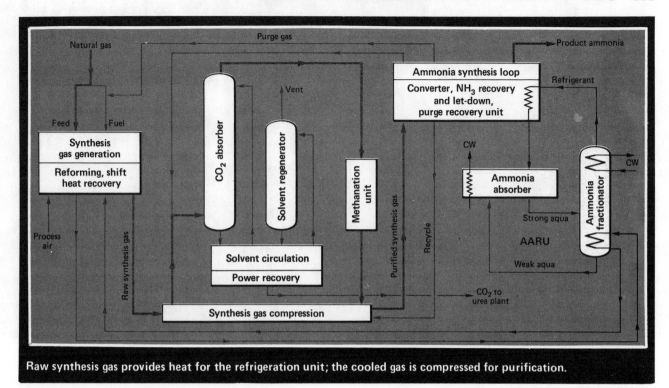

Raw synthesis gas provides heat for the refrigeration unit; the cooled gas is compressed for purification.

is partially supplied by the synthesis gas stream prior to potassium carbonate treatment. Following CO_2 absorption, the gas goes through a methanator, where remaining traces of CO and CO_2 react with some of the hydrogen to form methane. The gas is then compressed to around 2,000–3,000 psi in a multistage centrifugal compressor, heated to 700–800°F, and reacted with an iron catalyst to form ammonia, which is condensed out by water and refrigeration.

In Fluor's process, CO_2 is removed by propylene carbonate, as noted earlier. Unlike the chemical solvent, the physical solvent works better under pressure. Fluor says the pressure used is somewhere between the 300 psi of the steam shift conversion and the 2,000–3,000 psi of ammonia synthesis, but declines to be more specific. The company points out that physical solvents have little or no advantage over chemical absorbents at CO_2 partial pressures of less than 5 bar, but that their absorption capacity increases rapidly as pressure increases.

Stripping of the CO_2 from the physical solvent is much easier than is the case with a chemical solvent. Most of the CO_2 is released simply by a quick, three-stage depressurization, with the rest then being removed by an air stripper. If the CO_2 is needed for a urea plant, it may be removed at 100–200 psi, which saves much of the energy that would otherwise be required to compress the gas for the urea process.

Innovative engineering

Fluor's biggest challenge in ammonia plant design was to find an economical use for the heat saved by switching from a chemical to a physical solvent. This was solved by designing the plant so that the heat could drive an ammonia-absorption refrigeration system to condense out the process ammonia.

The combination of the physical solvent with ammonia-absorption refrigeration saves energy and reduces the capital cost of the plant. A Fluor spokesman notes that absorption refrigeration is not common in process plants, but feels it is now more attractive than before and could find a place in other process applications.

An apparent drawback in the process is that the entire gas stream (including the unwanted CO_2) must be pressurized prior to CO_2 removal. However, the company's earlier experience with hydrogen processing had already indicated this would not be a significant factor in the process economics.

Fluor designed a propylene carbonate CO_2 removal system for a Chevron U.S.A. commercial hydrogen plant based on steam/methane reforming in Pascagoula, Miss. In this case, a centrifugal compressor was justified as a replacement for the reciprocal compressor normally used in hydrogen plants. The switch would have been uneconomical for hydrogen alone, but for the hydrogen/CO_2 mixture the extra power required for centrifugal compression was outweighed by the fact that reciprocals need more maintenance and one or two standby units must be included in the plant. This experience made it easier for Fluor to justify compression for ammonia plants, in which centrifugal compressors are standard equipment.

The two ammonia plants that use Fluor's technology are located on the island of Trinidad and at Woodward, Okla. Both use natural gas as feedstock. W. R. Grace & Co., a principal owner of both plants, contributed to the plant design.

It has been shown that propylene carbonate may be used for years without deteriorating, says Fluor, and only a very small amount is vaporized and lost during CO_2 removal. Other advantages, according to the company, are that the solvent is noncorrosive and nontoxic.

Large quantities of high-purity hydrogen

A pressure-swing adsorption process developed by Union Carbide will help meet the need for plentiful, cheap hydrogen.

☐ During the past decade, demand has been increasing for large quantities of hydrogen to make chemicals, fertilizer and, particularly, high-octane gasoline. In 1975, Union Carbide Corp.'s Linde Division (New York City) set out to find a cheaper route to high-purity hydrogen than the conventional wet-process purification technology. The result was the Polybed PSA (pressure-swing adsorption) process.

The method went commercial in August 1977, when Wintershall AG opened up a 41-million-scfd unit at its Lingen, West Germany, oil refinery. The hydrogen produced is 99.999% pure, compared with 97–98% for conventional systems. The major advantage of Union Carbide's process is economics—while such a facility costs about as much to build as a conventional plant, operating expenses are less: Polybed PSA is cheaper by about 7–10% on a production-cost basis. Here, a comparison is made on the basis of a 50-million-scfd conventional plant that produces a 97% product, and employs high- and low-temperature shift conversion, carbon-dioxide washing, and methanation.

Improved recovery

In developing the process, Union Carbide engineers started with conventional PSA technology, which while producing a high-purity product, was limited by a low capacity and a low recovery rate. Capacity was improved from about 13 million to about 50 million scfd, and recovery was increased from 70–75% to 80–85%. Recovery at the Wintershall plant is 86%.

Other advantages over traditional PSA and hydrogen processing include:

■ *Turndown flexibility*—The system can turn down from 100 to 0%. High-efficiency operation is maintained down to 25% of capacity.

■ *Efficient heat recovery*—Heat recovery is enhanced

since losses to cooling water are minimized. Conventional wet-process plants require cooling before carbon dioxide removal and before methanation.

■ *Increased reliability*—The process requires no external heat input, uses a long-life adsorbent, involves fewer processing units and eliminates problems associated with temperamental catalysts and wet-scrubbing.

■ *Design flexibility*—The design of the reforming furnace is independent of product-purity considerations. The process is a single-step operation, in which all constituents except hydrogen are removed at once. Changes in the makeup of the mixture of the reformer effluent have only a small effect on the design of the PSA plant, thus allowing for this independence.

■ *Quick startup*—Pure hydrogen can be withdrawn from the unit within a couple of hours after the introduction of the feed gas.

■ *Clean operation*—The process has only two outgoing streams—product and tail gas. The tail gas is used as a fuel for the steam reformer, rendering the process clean and nonpolluting.

Also, the high-purity hydrogen saves energy when it is used later on in other processes. If necessary, purity can be greater than 99.999%.

Process development

The Polybed PSA process, shown in the diagram, is a refinement of the basic PSA process. The basic process is a cyclical one in which a gaseous product is separated from a mixture at moderate to high pressure, and the adsorbent is regenerated at low pressure.

With conventional PSA systems, once the adsorbent bed becomes saturated with impurities, it is depressurized in a flow direction countercurrent to the feed. This is followed by a purge—also countercurrent—to complete the removal of impurities. Then, the unit is pres-

Originally published December 3, 1979

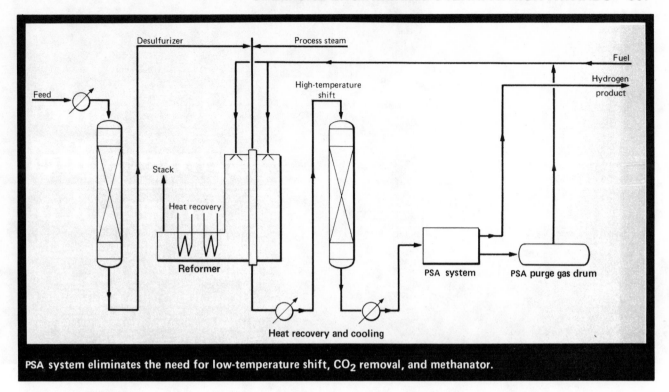

PSA system eliminates the need for low-temperature shift, CO$_2$ removal, and methanator.

surized with product gas and is put back into operation. Conventional PSA units employ from two to four beds, one of which is on adsorption while the others are regenerating.

Such a cycle is inefficient for two reasons: (1) the blowdown from adsorption pressure to low waste-pressure results in large losses of the product gas that is stored in the bed at high pressure; and (2) some product gas is cycled to waste, since this gas is used as a purge.

This basic cycle was improved to overcome such problems. The innovation consisted of withdrawing product-quality gas from an adsorber that has completed its adsorption step, by depressurizing the bed cocurrently with the feed to some intermediate pressure. By control of the pressure and hence the mass-transfer fronts of the impurities, this lower-pressure gas is kept at a high level of purity and it can therefore be used to perform some of the pressurization and all of the purging of other adsorbers. Thus, product gas that was previously lost in these functions is recovered. Also, since blowdown starts at lower pressure, losses of product stored within the bed are reduced.

At least seven adsorbers (in a staggered sequence) are used in the improved system. The Wintershall plant uses ten adsorbers, three of which are on adsorption at any one time, while the other seven are regenerating. The two sections of five adsorbers can be decoupled to permit operation of a single train of five beds.

Tail-gas variation

Variations in tail-gas flow and composition had to be overcome to integrate the Polybed system with the steam reformer. Because of the cyclic nature of the process, tail gas rejected by the Polybed unit during blowdown and purging has substantial flow and composition variations. Such fluctuations would have made

it impossible to use the tail gas as a fuel. The use of a buffer/mixing tank has evened out such fluctuations, keeping variation of heating value to within ±1%.

Key engineering steps

In developing the process, Union Carbide engineers:

■ Performed an analysis of laboratory data and the data base for conventional PSA systems.

■ Developed and refined a model to reflect the transient behavior of the PSA process.

■ Extended the model to systems with as many as 12 adsorbers, with multiple intermediate pressure-levels and flowrates.

■ Designed a mechanical component package to accommodate the expanded process. The mechanical system has a high degree of reliability and needs minimum operator attention.

■ Designed a system to provide accurate control of the integrated process. A reliable control system is important due to the large number of stages and the use of several intermediate pressures in each stage.

Other applications

Another hydrogen plant using the process was brought onstream last year at Ashtabula, Ohio, by Union Carbide's Linde Division. The system produces 9 million scfd of 99.9999%-pure hydrogen. Several other hydrogen plants that integrate steam reforming with the Polybed process are in various stages of design and construction. The route is not restricted to steam-reforming hydrogen plants. Other applications include:

■ Refinery hydrogen-upgrading—as a substitute for cryogenic processing.

■ Ammonia loop-vent systems—hydrogen or a hydrogen/nitrogen mixture can be recovered, while inerts are rejected.

More punch for gasoline

Catalytic etherification yields methyl *tert*-butyl ether (MTBE), a substitute for toluene and other aromatics in gasoline.

☐ If necessity were ever the mother of invention, then certainly the oil crunch fostered the new business in MTBE—methyl *tert*-butyl ether—a high-octane blending component. Starting from near zero when the 1970s began, MTBE capacity, installed or planned, has grown to roughly 1 million metric tons/yr. One of the earliest proponents, in terms of both new technology and new capacity, was present honor-award winner Chemische Werke Hüls AG (Marl, West Germany).

Hüls took a well-known chemical synthesis—the acid-catalyzed reaction of methanol and isobutylene, in the presence of other normal butenes, to form MTBE—and worked it up in a way that gets around difficulties with the methanol/MTBE azeotrope, enabling economic production of the ether. As an added bonus, what is left behind is a "raffinate" of mixed normal butenes, which can be easily distilled to its pure components to serve as feedstocks for methyl ethyl ketone, C_4-oxoalcohols and maleic anhydride. Were isobutylene to remain in this stream, the separation of butene-1 and isobutylene (boiling points $-6.3°C$ and $-6.6°C$, respectively) by conventional distillation would be virtually impossible.

Not just a fuel

Indeed it was this sort of modified distillation to remove isobutylene before distillation of the remaining components that led to the first MTBE plant at Anic S.p.A.'s Ravenna, Italy, plant in the early 1970s. Snamprogetti developed the process employed there. The unit was built with high-octane fuel in mind right from the start. Plant capacity was 100,000 metric tons/yr. Leftover C_4s consisting primarily of butene-1 and butene-2 were at that time burned as fuel gas. Snam-

progetti conceived the technique as an alternative to improved reforming and catcracking processes that were making quick gains at that time.

Because of its high octane—about 110 [(RON + MON)/2]—MTBE offered a way out for refiners strapped by government-enforced phasedowns of lead anti-knock, a problem that was worsened by competition in the chemical-feedstock market for high-octane aromatics. Furthermore, MTBE has good vapor-pressure and other physical properties when used in gasoline component.

At about the same time, Hüls had the complementary objective of producing high-quality butene-1 and butene-2 feedstocks. It sought to do this by removing isobutylene from mixed C_4s via MTBE synthesis. So Hüls developed its own process based on similar chemistry and built a 50,000-metric-ton/yr plant feeding on mixed butylenes. Subsequently, it built the first plant feeding on mixed C_4. As a result, Hüls' second Marl plant, commissioned in May 1978, is able to produce MTBE in three grades, covering all possible markets from fuel to chemical feedstocks to solvents for highly sophisticated chemical reactions.

The Reaction

One of the greatest difficulties faced by Hüls in the design of its MTBE plant was the wide variability in mixed-C_4 feedstock, which can contain isobutylene in concentrations as low as 15% if obtained from refinery catalytic crackers, or as high as 50% from ethylene steam crackers. The consequences for gas-handling equipment in the plant are significant, amounting to capital costs as much as double for a plant feeding on cat cracker gas (*Chem. Eng.*, May 21, pp. 91–93).

Originally published December 3, 1979

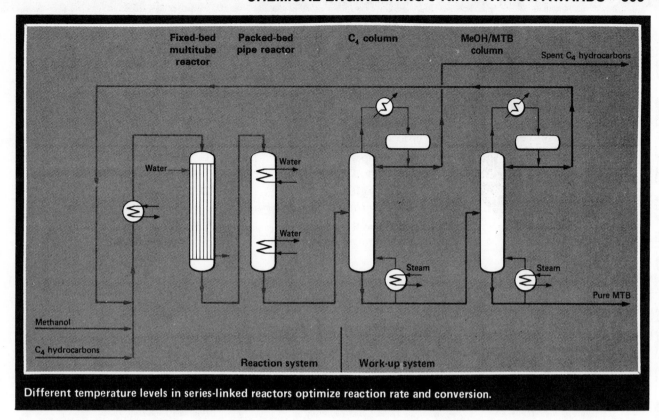

Different temperature levels in series-linked reactors optimize reaction rate and conversion.

Cost considerations aside, Hüls had enough difficulty with the reaction chemistry and subsequent workup of reaction products.

One problem had to do with the proper excess of methanol in the reactor. The kinetics favor an excess, though the larger the excess, the larger the reactor has to be. And workup is all the more complicated, particularly because methanol also forms an azeotrope with the MTBE.

Another major difficulty had to do with reaction temperature. Equilibrium conversion is favored by decreasing the temperature; however, reaction time then becomes longer. Hüls solved this problem through the proper choice of catalyst and the use of a dual-reactor split-temperature approach.

The catalyst, a strongly acidic, macroporous polystyrene-divinylbenzene ion-exchange resin—the same type as used in competing MTBE processes as well—provides the best combination of activity and selectivity. Most importantly, its useful life is practically indefinite, and side-reactions are barely significant under 70°C.

The exothermic etherification reaction requires a different design in the first reactor—the hotter—than in the second. The initial high temperature is provided in order to speed the reaction rate, while the second, lower temperature satisfies requirements for higher equilibrium conversion.

Hüls gave a great deal of consideration to the temperature profile across the fixed catalyst bed, and chose a packed, multiple-tube reactor cooled by water in the initial stage. The second stage contains a single water-cooled bed. The temperatures range from 40 to 70°C between the beds, and pressures vary from 5 to 8 bars. Isobutylene conversion comes to 99.9%.

Product workup

The first step following the reactors is easy—distill away spent C_4 hydrocarbons. The small amount of methanol that combines azeotropically with this mixture is extracted by water, redistilled, and recycled.

The larger problem is separating methanol from product MTBE. This azeotrope contains 14% alcohol.

Separation problems of course affect upstream reactor design, since an excess of methanol improves the reaction kinetics. However, too much of the alcohol calls for prohibitively large separation equipment.

Hüls got around this dilemma by taking advantage of the pressure dependence of the azeotrope's methanol/MTBE ratio. At 20 bars, methanol content rises to 42%, which translates to smaller separation equipment. Hüls' three process variations differ mainly in the type of workup, hence the degree of MTBE dryness.

Byproducts' future

Right now, MTBE's future seems assured by its usefulness in gasoline. However, no prognostication would be complete without considering the consequences for byproduct normal butenes.

In this case, a fortunate coincidence of technology points to an equally bright outlook for the byproducts, particularly butene-1. This year's award-winning process, the Unipol polyethylene route, requires much larger amounts of comonomer butene-1 than previous low-density polyethylene technologies. Butene-2 is easily isomerized to butene-1, so the byproduct stream of mixed normal butenes should find easy markets. Already, a couple of large butene-1 plants have been announced (*Chem. Eng.*, Oct. 22, p. 79).

Board of Judges

James Wei, Chairman
Massachusetts Institute of Technology

Lloyd Berg
Montana State University

C.D. Holland
Texas A&M University

J.R. Couper
University of Arkansas

C. Judson King
University of California, Berkeley

L. C. Eagleton
The Pennsylvania State University

D.T. Wasan
Illinois Institute of Technology

J.K. Ferrell
North Carolina State University

J.W. Westwater
University of Illinois

Committee of Award

D. K. Anderson
Michigan State University

George Austin
Washington State University

J. T. Banchero
University of Notre Dame

D. H. Barker
Brigham Young University

K. B. Bischoff
University of Delaware

J. R. Bowen
University of Wisconsin

Howard Brenner
University of Rochester

R. W. Coughlin
University of Connecticut

S. H. Davis
Rice University

P. F. Dickson
Colorado School of Mines

Nicholas Dinos
Ohio University

L. T. Fan
Kansas State University

T. Fort, Jr.
Carnegie-Mellon University

A. L. Fricke
University of Maine

J. D. Goddard
University of Southern California

D. B. Greenburg
University of Cincinnati

H. L. Greene
University of Akron

R. B. Grieves
University of Kentucky

Deran Hanesian
New Jersey Institute of Technology

David Hansen
Rensselaer Polytechnic Institute

W. J. Hatcher
University of Alabama

G. M. Hoerner
Lafayette College

G. W. Preckshot
University of Missouri

D. W. Hubbard
Michigan Technological University

J. L. Hudson
University of Virginia

H. F. Johnson
The University of Tennessee

K. H. Keller
University of Minnesota

L. B. Koppel
Purdue University

L. E. Lahti
The University of Toledo

M. Larson
Iowa State University of Science and Technology

H. Lee
University of Florida

C. F. Long
Dartmouth College

H. A. McGee
Virginia Polytechnic Institute and State University

T. B. Metcalfe
University of Southwestern Louisiana

R. L. Motard
Washington University

A. L. Myers
University of Pennsylvania

J. O. Osburn
University of Iowa

Thomas Owens
University of North Dakota

Elton Park, Jr.
University of Mississippi

J. T. Patton
New Mexico State University

D. R. Paul
The University of Texas

C. A. Plank
University of Louisville Speed Scientific School

G. W. Polhein
Georgia Institute of Technology

R. L. Sandvig
South Dakota School of Mines and Technology

W. R. Schowalter
Princeton University

R. A. Servais
University of Dayton

R. E. Slonaker
Bucknell University

J. C. Smith
Cornell University

S. S. Sofer
University of Oklahoma

W. F. Stevens
Northwestern University

D. L. Stinson
University of Wyoming

M. R. Strunk
The University of Missouri

H. H. Szmant
University of Detroit

A. Ralph Thompson
University of Rhode Island

R. E. Thompson
The University of Tulsa

W. D. Threadgill
Vanderbilt University

R. A. Troupe
Northeastern University

K. A. Van Worner
Tufts University

John W. Walkinshaw
University of Lowell

Thomas J. Ward
Clarkson College of Technology

R. Weaver
Tulane University

R. E. White
Villanova University

C. E. Wicks
Oregon State University

J. L. Zakin
Ohio State University

1981 KIRKPATRICK CHEMICAL ENGINEERING ACHIEVEMENT AWARDS

For outstanding group effort in developing significant
chemical engineering technology commercialized during the past two years

Winner of the Award

MONSANTO CO.

Winners of Honor Awards

AIR PRODUCTS AND CHEMICALS, INC.

CELANESE CHEMICAL CO.

KERR-McGEE REFINING CORP.

**SOCIÉTÉ NATIONALE ELF AQUITAINE/
SOCIÉTÉ DE RECHERCHES TECHNIQUES ET INDUSTRIELLES**

UNION CARBIDE CORP.

Monsanto Co.

Unique membrane system spurs gas separations

☐ Monsanto Co. has won the 1981 Kirkpatrick Chemical Engineering Achievement Award. The St. Louis firm garnered the honor for its efforts in introducing an ingenious hollow-fiber system that, for the first time, allows the practical use of permeable membranes for large-scale gas separations.

The company's new Prism separator, in fact, represents a simple, low-cost option for performing such separations. It particularly suits the splitting off of such fast-permeating gases as hydrogen and/or carbon dioxide, and promises to have a major impact in diverse areas, ranging from ammonia and petrochemical processes to tertiary oil-recovery operations.

A number of accomplishments distinguish the development. First of all, Monsanto invented a membrane whose structure flies in the face of convention. Then, too, the firm had to devise a viable means of employing the membrane; this required the design of a special vessel, and coming up with a suitable sealant for the membrane tubes, among other challenges.

The results attest to the company's success: Prism separators today are handling industrial hydrogen-removal jobs on a virtually maintenance-free basis, with minimal, if any, energy consumption. And they are providing other operating advantages as well.

The units can provide hydrogen of 86% to 96% or even 99% purity (depending upon feed-gas composition and the available hydrogen differential partial pressure) from feeds containing over 30% hydrogen.

Originally published November 30, 1981

Yet, for all their capabilities, the separators are disconcertingly simple. Basically they consist of hollow fibers arranged in a bundle within a vertically-oriented tubular pressure vessel.

Rapid development

This seeming simplicity camouflages the massive amount of work required to bring the system to commercialization. Monsanto reckons that it invested almost half a million manhours on the project.

The membrane technology that is the key to the system was discovered at the Corporate Research Department in mid-1974 by Dr. Jay M. S. Henis and Mary K. Tripodi. By October of that year, bench-scale testing was taking place on plant slip-streams.

The two researchers did more than come up with a revolutionary membrane. Early on, they provided the stimulus for its commercialization.

Hénis and Tripodi were among the first people at Monsanto to recognize that supplies of hydrogen would become increasingly limited and costly. And they also realized that conventional hydrogen-recovery techniques—cryogenic systems, pressure-swing adsorption, and absorption in liquids—had limitations that would blunt their appeal even in this situation. All this added up, they reasoned, to an attractive and well-focused potential market for the new membrane.

Monsanto agreed, and in 1975 marshaled a major development program. The Corporate Research De-

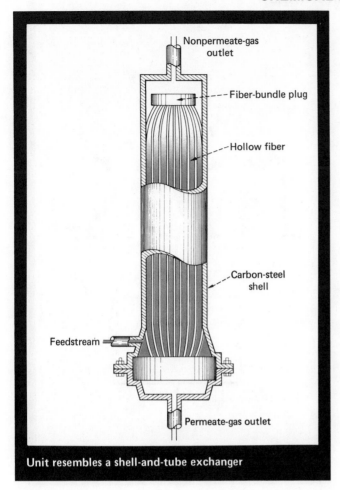

Nonpermeate-gas outlet

Fiber-bundle plug

Hollow fiber

Carbon-steel shell

Feedstream

Permeate-gas outlet

Unit resembles a shell-and-tube exchanger

The Award

The Kirkpatrick Chemical Engineering Achievement Award was initiated in 1933, and is bestowed biennially to recognize outstanding group effort in developing and commercializing important chemical-engineering technology.

The award is named after the late, Editorial Director of CHEMICAL ENGINEERING, Sidney D. Kirkpatrick, and is sponsored by the magazine. However, the selection of the winners is done independently.

Finalists are chosen by ballot of the chairmen of chemical engineering departments at U.S. colleges and universities. The winner then is picked by a Board of Judges drawn from among these chairmen. A complete list of these distinguished educators appears at the end of this report.

This year marks the 26th presentation of the Kirkpatrick Award—won by Monsanto Co. Recent winners have included:

1979—Union Carbide Corp.
1977—Davy Powergas Ltd./
      Johnson Matthey and Co./
      Union Carbide Corp.
1975—Amoco Oil Co.

Notable work by an individual is honored by the Personal Achievement in Chemical Engineering Award, which is also presented biennially, alternating with the Kirkpatrick Award.

partment headed the effort (with Eli Perry in charge), but a host of other groups within the company also contributed their skills.

Since it was breaking new technical ground, the company decided to demonstrate the technology on a fullscale basis in its own plants before offering systems commercially. The first fullscale unit was installed at its Texas City, Tex., petrochemicals complex in March 1977, only two and a half years after the first field test. This represented a scaleup, in terms of number of fibers, on the order of 10^5 to 10^6. (That system is still running, with no noticeable change in performance.)

Other internal installations soon followed and, in all, five different fullscale demonstrations took place before the company brought the technology to the market in November 1979. Only five years had elapsed between invention and commercial introduction.

(Now, more than twenty systems are in operation or in the works.)

Contrasts

This timing appears even more remarkable considering that the company was pioneering an entirely new type of membrane.

To appreciate this, consider the status of conventional composite membranes (which consist of a very porous but inert substrate covered by a polymer surface layer that separates components): First demonstrated in the laboratory in the late 1940s, these membranes did

not achieve any major industrial role until the 1960s. Then, the introduction of monolithic asymmetric membranes (ones whose densities are not uniform) provided high-enough flux rates for units to attain a commercial niche in desalting.

However, flux rates still were too low for economic separations of many liquids and most gas mixtures. Besides, the method of making the membranes (solution spinning in water) invariably produced pores in the surface, rendering gas separations impossible.

To reach the more-than-an-order-of-magnitude improvement in flux rates required for economic separations of gases, conventional membranes would have to resort to a much thinner surface layer—one even more susceptible to pore formation. "Densifying" the surface layer to remove its pores was possible, but would add to its thickness, and cut the flux rate.

Indeed, providing a thin-enough polymer while plugging its pores to allow gas separations seemed, to many researchers, to pose irreconcilable requirements.

Special hollow fibers

Yet, Monsanto has been able to achieve this feat with its membrane—by taking the novel approach of using the separating material, usually a polymer such as polysulfone, as the substrate instead of as the coating. This substrate is chosen on the basis of its separation characteristics and its environmental resistance.

A hollow fiber of it is spun such that it has the thin-

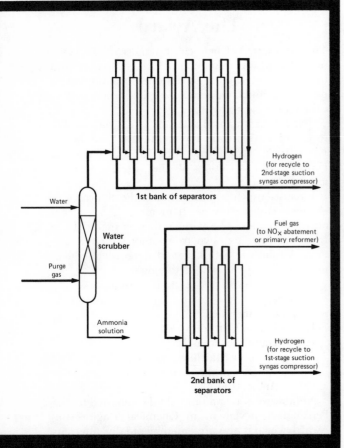

Hydrogen recovery from ammonia-plant purge gas

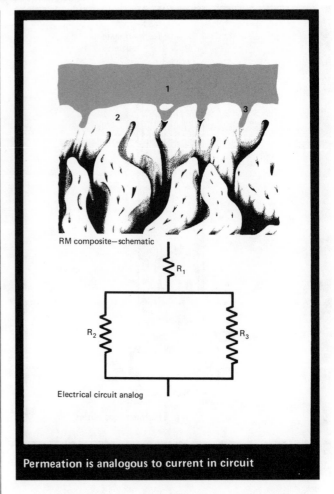

RM composite—schematic

Electrical circuit analog

Permeation is analogous to current in circuit

nest-possible separating layer (typically 500Å to 1,000Å), although its overall wall thickness may reach 300μ. This membrane, like any such asymmetric one, is porous. However, instead of densifying the separating layer, a coating (generally of a poor separator) is applied. This coating may be 10 to 20 times thicker than the separating layer of the substrate, and plugs the pores. It also serves to protect the substrate from damage due to abrasion and normal handling.

Because of the coating, the separating layer can be made as thin as it is without worry about pore problems. This increases available flux rates to the extent that Prism separators are 10^3 to 10^4 times faster per unit area than the only other system proposed for commercial gas separations, notes Monsanto. It also makes possible selection from a much wider variety of substrate polymers, since economically viable membranes can be produced from polymers with much lower intrinsic permeabilities than would otherwise be permissible.

The composite membranes also are more uniform in terms of rate and separation factor than equivalent simple asymmetric fibers produced by similar spinning methods.

Membrane mechanism

Gas permeation through the new membrane is akin to the flow of electricity through a series/parallel arrangement of resistors. In fact, Monsanto calls its membrane an RM (for resistance model) composite.

Gas first flows through the coating layer—labeled (1)

on the illustration, and the equivalent of R_1 on the circuit diagram. Then it encounters the separating-polymer surface. This consists of separating polymer (2) and pores, which have been filled to a certain depth with the coating (3). So, instead of having the bulk of the gas pass through open channels, as would be the case if the pores were not filled, the flow splits in inverse proportion to the resistance to permeability offered by the substrate polymer and the filled pores.

The coating generally is much more permeable than the substrate but occupies a very much smaller cross-sectional area (the filled-in pores). So, the resistance to permeation actually is significantly greater for the pores, and most all flow occurs through the substrate.

Thus, the RM composite provides the unique characteristic of having the substrate, not the coating, control separation. Monsanto has received patents both for the general principle and for specific composites.

Producing fibers

Developing the concept of RM composites was only the starting point to successful commercial application. The company had to surmount a host of other important difficulties.

First of all, the concept had to be translated into commercially viable composites. This not only required the identification of appropriate substrates but also the development of coating formulations compatible with them. Procedures also had to be devised for improving adhesion of the coatings to the substrates.

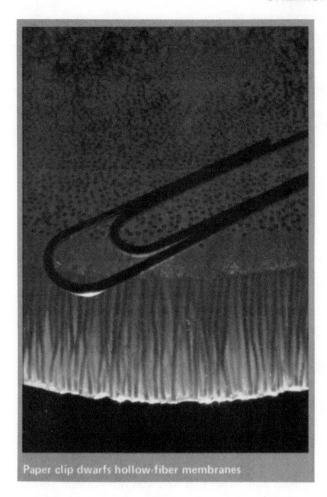

Paper clip dwarfs hollow-fiber membranes

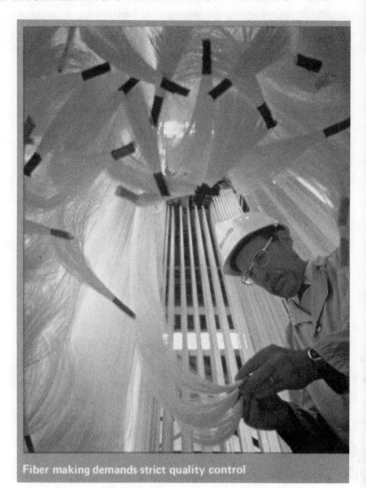

Fiber making demands strict quality control

Then the hollow fibers had to be produced on a large scale, with tightly controlled porosity and other morphological properties. So, critical parameters and variables had to be identified and specifications set. At the same time, prototype equipment had to be developed for fiber spinning, drying, coating, and assembly into bundles, as well as for making of seals and modules. Equipment for handling the fibers individually without breakage or damage had to be devised, too, along with valid test methods for final products.

Module scaleup

Monsanto could draw upon little in the technical literature to help scale up the fiber modules. Extensive work had been done, of course, on desalting; however, this proved to be of only minor value because of the great differences in viscosities, recoveries, purities, diffusion rates and safety requirements between gas and liquid separations.

Successful scaleup depended to a considerable extent on minimizing the pressure drop on both the shell and bore sides of the hollow fibers during operation. This is important because any increase in pressure drop results in reduced driving forces and recoveries.

The company minimizes pressure drop by feeding gas to the shell side of the fibers, which themselves have large bore-diameters.

The shell is far more open than the bore of the fiber. So, the feed undergoes only a minor pressure drop. However, assuring a uniform distribution and flow of feed through the shell is more difficult. Stagnant areas in the shell have to be eliminated; this is accomplished by proprietary design and fiber-shaping procedures.

Meanwhile, the bore only has to handle the much smaller flow of permeate gas. This lower flow, coupled with the larger bore-diameter, results in very low bore pressure-drops as well.

Scaleup posed other challenges. For instance, gas had to be prevented from bypassing the fibers, not only initially but even after years of operation, when fiber movement, packing or settling may have occurred.

The fibers also had to be sealed effectively. Such a seal had to: resist impurities in the gases; penetrate the fiber bundle when applied, and adhere well to the fibers to prevent leakage; be safe; have a low enough exotherm so that the fibers would not be affected during the curing of the sealant; and demand minimum fabrication equipment, to keep costs low. Monsanto had to develop special sealant formulations with low viscosities, controlled exotherms, and very high strength and environmental resistance when cured.

Advantages

The units are mechanically simple, without any moving parts. This reduces the possibility of malfunction and the need for supervision by operators.

The company calls the separators virtually maintenance-free, and adds that the systems should operate for about five years without significant decline in their performance.

Modules now operating at ammonia plant

most instances. This contrasts with cryogenic separations and many absorption processes, which require much more stringent pretreatment.

Unlike most other gas-separation techniques, which are limited in the combinations of recovery and composition that they can achieve, the new units allow these parameters to be varied over a wide range, by adjustment of the feed flowrate and/or the bore backpressure.

Prism separators also boast an inherent stability toward upsets that sets the units apart from rivals.

Typically, if cryogenic or adsorption systems undergo changes in feed composition, or process upsets that cause loss of capacity, product output also will change. Prism separators do not respond this way, thanks to the nature of the hollow fibers and the system design. Instead, once a given set of operating conditions has been established, large fluctuations in feed composition or flowrate will result in only minor fluctuations in permeate-gas composition.

Similarly, loss of fiber capacity, through upset or failure, prompts only minor losses in permeate-gas recovery and even smaller changes in permeate composition. For instance, if 25% of the fiber in a module of a typical hydrogen purge-recovery unit becomes inoperative, hydrogen recovery might drop 5%, and purity 1%.

This stability makes it possible to remove individual modules, or even banks of them, from service without significantly disrupting output.

Another major operating plus of the units is their ease of startup. An ammonia-purge unit, for instance, typically reaches equilibrium in 2 h or less. This compares, notes Monsanto, with 24 to 48 h or more for cryogenic systems and adsorption and absorption installations.

Potential applications

The Prism unit's advantages have already enabled it to make inroads in handling hydrogen recovery or removal in ammonia and petrochemical plants and refineries. However, the firm foresees much greater impact:

The U.S. alone uses an estimated 10 billion std. ft^3/d of hydrogen now. And 5 to 15% of this winds up in purge streams or is burned as fuel, for want of an economic means of recovering the hydrogen. Prism separator systems could provide as much as 500 billion std. ft^3/yr from these purge streams.

And this does not factor in the possibilities related to the increasing use of hydrogen as a reactant, as in synfuels processing, where Prism separators can serve to adjust gas composition. The systems also may allow the use of greater hydrogen recycle for some reactions, enabling higher per-pass conversions.

Carbon dioxide recovery offers another significant prospect for utilizing the technology. Consider tertiary oil-recovery operations. Some 15 to 30 billion bbl of potentially recoverable oil lie in fields appropriate for tertiary recovery using carbon dioxide flooding. Such flooding yields the oil and also a methane-contaminated gas rich in carbon dioxide, but requires the injection of 5,000 to 12,000 std. ft^3 of carbon dioxide per barrel of oil. Removing this carbon dioxide with Prism separators would allow it to be reinjected, while upgrading the methane for other uses.

Mark D. Rosenzweig

Because of their compactness, Prism separators can be retrofitted in many existing facilities.

Installing the shop-fabricated, skid-mounted units is straightforward; only a concrete support pad and the appropriate tie-ins need be provided. And capacity can be expanded at any time by hooking up added modules in parallel.

Importantly, the Prism separators generally impose no added energy burden on a plant, except for any limited demands for operation of control instrumentation.

This is because the differential pressure between the shell and the bore of the fibers provides the driving force for separation, and feed gases to the unit are almost always available at an adequate pressure so that extra compression usually is not required. Prism separators can operate with feed pressures ranging from 150 to 2,000 psig, and differential pressures of 50 to 1,500 psig—at temperatures from 0° to 55°C.

Even when compression is required for the permeate gas, notes Monsanto, the incremental capital and energy costs frequently are far less than those associated with cooling in cryogenic separations or stripping in adsorption processes.

Flexible operation

The Prism separators typically tolerate most contaminants at levels up to several hundred ppm in the process stream, and even much higher levels. So, the necessity for pretreatment is reduced, if not eliminated, in

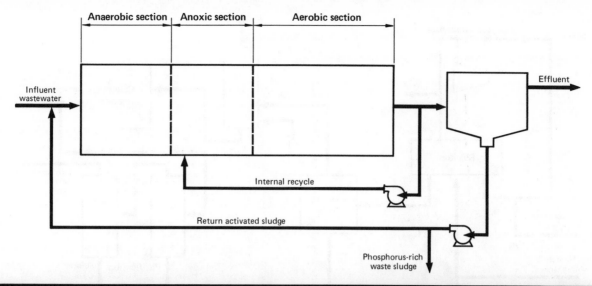

Anaerobic section | Anoxic section | Aerobic section

Influent wastewater → Effluent

Internal recycle

Return activated sludge

Phosphorus-rich waste sludge

Air Products and Chemicals, Inc.

Wastewater treatment scores by getting bugs in

□ The ingenious use of bacteria that store polyphosphate has led to a completely biological system for removing phosphates, nitrates and biochemical oxygen demand (BOD) from wastewater.

Developed by Air Products and Chemicals, Inc. (Allentown, Pa.), the A/O process has proven itself since 1979 at a municipal treatment plant at Largo, Fla., and now has been selected for several other installations.

The route outwardly resembles a conventional activated-sludge system, except that an anaerobic section has been added at the front end.

Wastewater and recycled sludge flow into this anaerobic zone. There, the polyphosphate-containing bacteria (because they can hydrolyze their polyphosphate to provide the energy to assimilate BOD) hold a definite edge over other microorganisms. Indeed, most of the BOD winds up being taken by these bacteria.

If denitrification is required, the stream and some liquor exiting from the downstream aerobic zone pass to an "anoxic" section (one void of free oxygen). BOD sorbed in the anaerobic zone reduces nitrate and nitrite, with the extent determined by the amount of liquor recycle.

Phosphate removal occurs in the aerobic section. The polyphosphate-storing bacteria metabolize BOD to resynthesize polyphosphate and to form new microorganisms as well.

Liquor goes to a conventional gravity clarifier, which yields clear water and a sludge.

Originally published November 30, 1981

Advantages

Energy efficiency—The A/O system operates with minimal mixing energy at low levels of dissolved oxygen. The recycle-sludge pumping requirement amounts to only about one-quarter that of a conventional activated-sludge process. Sludge dewatering also takes less energy because A/O sludge is denser than conventional ones, and dewaters readily to a drier cake.

For denitrification, the route offers additional aeration-power savings, since reduction is not aerobic.

The Largo plant boasts a net energy saving of 50%, compared with conventional units, notes Air Products.

Completely biological process—Unlike conventional methods, nutrient and BOD removal are accomplished without resorting to chemicals and generating a chemical sludge.

Fast reaction—Processing typically requires from $1\frac{1}{2}$ to 3 h residence time in the reactor, about half that of conventional BOD-removal systems. Residence time in the anaerobic section is short, inhibiting the growth of filamentous bacteria that are hard to remove in the clarifier.

Fertilizer-quality sludge—The sludge can be dried to produce a higher-value fertilizer than that from a conventional process. A/O-derived product typically contains 6 to 8% fixed nitrogen, and 10 to 20% phosphorus (expressed as P_2O_5). The Largo facility values its A/O sludge at $110/ton, versus $80 for conventional.

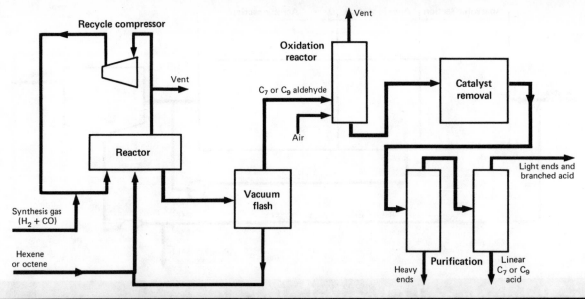

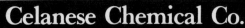

Celanese Chemical Co.

Fatty-acids process sets supplies straight

☐ Polyol esters are attracting increasing interest as high-performance synthetic lubricants, and more. Now, the esters' prospects are likely getting a further greasing, thanks to the successful commercialization of a new process to make some key raw materials—linear short-chain fatty acids.

The new route provides a safe, low-energy and high-efficiency synthesis of highly linear heptanoic (C_7) and nonanoic (C_9) acids, notes its developer, Celanese Chemical Co. (Dallas, Tex.).

The first fullscale plant to use the process started up in early 1980 at the company's complex at Bay City, Tex. Rated at 40 million lb of acid/yr, that facility can produce either heptanoic or nonanoic acid. It can simply and speedily switch between the two products, with minimal cross-contamination problems. Acid quality consistently exceeds specifications; straight- to branched-chain ratio typically surpasses 40:1.

Commercialization of the fatty-acids route capped a number of years of intensive research, and owes much to Celanese's extensive knowledge and experience in rhodium oxo and oxidation technology. (The company received a patent license from Union Carbide Corp. in the rhodium oxidation area, but the process employs Celanese-developed technology exclusively.)

The process

The route starts with an α-olefin (either 1-hexene or 1-octene) and synthesis gas. These, along with recycle

gas and catalyst solution, go to an oxo reactor operated at relatively mild conditions (150 psig and 200–300°F). The reactor's design allows catalyst concentration and recycle-gas composition to be varied as required to achieve complete olefin conversion.

Vacuum flashing of the reactor products then separates the aldehydes from the catalyst and heavy ends, which are recycled.

The aldehydes pass to another reactor, where they are oxidized with air in a Celanese-developed mixed-transition-metal-catalyst system at a pressure of under 100 psig. Oxygen conversion is deliberately kept high to minimize any flammability hazards. And special instrumentation is installed to limit risks during startup and upsets.

Conversion of the aldehydes is almost total. Thus, no separation and recycle are necessary, and their attendant losses and costs are avoided.

Product coming from the reactor does contain several ppm of catalyst, and so passes to a unique multi-step catalyst-removal operation devised by the company. This involves solids handling and some chemical consumption, but imposes no energy demands save a minimal one for pumping.

Catalyst-free crude acid goes to a two-column purification train. In the first column, heavy ends are removed, and fed to the plant's boilers. Meanwhile, overhead moves to the second column, where light ends and branched acid are taken off.

Originally published November 30, 1981

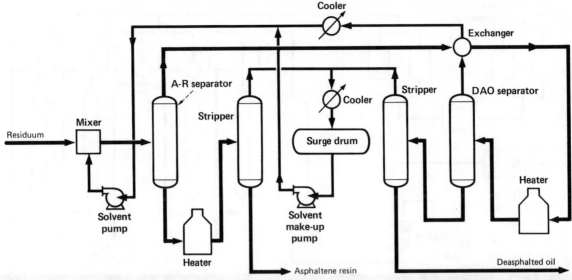

Kerr-McGee Refining Corp.

Novel solvent recovery enhances residuum upgrading

☐ Solvent extraction often gets the nod from refiners for removing undesirable heavy constituents from residuum. However, conventional solvent recovery by evaporation poses significant utility demands. Now, an extraction process avoids this evaporation penalty, while providing recovered oil at high yields and also of high quality.

Dubbed ROSE, for *R*esiduum *O*il *S*upercritical *E*xtraction, by its inventor, Kerr-McGee Refining Corp. (Oklahoma City, Okla.), the route first went commercial at a Pennsylvania refinery in August 1979. Today, two more units are running, and two others are in the works.

The technique typically offers a utility saving of about 50%, compared with conventional extraction with solvent recovery by evaporation, and an investment advantage of around 20%.

Existing extraction installations can easily be converted to the process—this was done for the Pennsylvania unit.

More than that, notes the company, no other solvent-extraction process boasts as much flexibility in choice of solvent, in ability to handle as many different types of feedstocks, or in potential recovery of higher-quality products. Furthermore, the ROSE route enables an economical and effective rejection of metals and carbon from heavy residual feedstocks, which then can be processed in catalytic operations that cannot tolerate these contaminants.

Originally published November 30, 1981

The key

Development of the process stemmed from Kerr-McGee's appreciation that when the temperature and pressure of a solvent/deasphalted-oil fraction are raised into the supercritical region of the solvent, the oil becomes virtually insoluble. This phenomenon of reverse solubility, coupled with the density difference between the oil and the solvent in the supercritical region, makes it possible to recover almost all of the solvent via a simple phase separation.

In practice, residuum enters a mixer, along with a solvent, e.g., pentane. The resultant mixture passes to a vessel in which a heavy asphaltene fraction is withdrawn as bottoms. This then goes to a heater, followed by a stripper, where solvent is recovered for recycle.

Meanwhile, the solvent/deasphalted-oil stream is heated to the supercritical level and then sent to a settling vessel. Substantially-oil-free solvent exits from the top, while the oil bottoms flows to a stripper for removal of traces of solvent.

(If an intermediate resin product is desired, a separator and stripper go after the asphaltene separator.)

Overall, about 90% of the solvent is recovered through phase separation. This eliminates the need to provide heat of vaporization, and also energy for compression of vaporized solvent. In addition, a major portion of the heat of the supercritical solvent is recovered by exchange within the process.

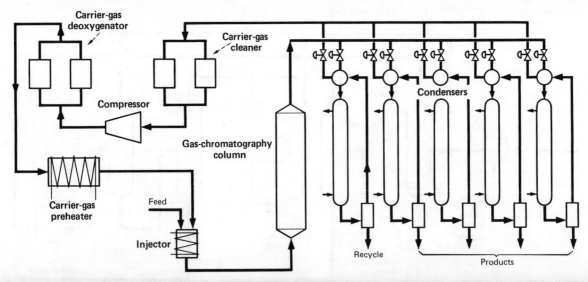

Société Nationale Elf Aquitaine/
Société de Recherches Techniques et Industrielles
Gas chromatography tackles production jobs

☐ Add gas chromatography to the list of industrial unit operations. The technique, long an analytical tool, is now carving a niche in the fullscale production of chemicals.

This development results from the joint efforts of Société Nationale Elf Aquitaine and Société de Recherches Techniques et Industrielles (both of Paris). The first commercial installation of their technology, a 100-metric-ton/yr plant for the purification of flavors and fragrances, went onstream at Jacksonville, Fla., in October 1979. Now, standard units with capacities from 3–5 m.t./yr to 200–300 m.t./yr are available.

Production-scale gas chromatography separates close-boiling products, such as structural isomers; permits selective removal of impurities (even if at low concentrations) from a complex feedstock in one operation; and even fractionates azeotropes and thermally unstable feeds. Indeed, the technique may enable volume purifications heretofore impossible or uneconomic.

Operation is completely automatically controlled, and provides purities of 99.9+% and yields of better than 95%.

The system

Gas chromatography takes advantage of the different travel times of components through a specially packed column. So, precise timing is essential—the production-scale process achieves this via a simple and reliable time-based programmer.

Originally published November 30, 1981

Liquid feed passes at appropriate intervals through a valve controlled by the programmer into an injector. There, it is mixed with carrier gas, usually compressed hydrogen. Heat is then added to vaporize the feed.

The vaporized mixture enters the gas-chromatography column. This column is homogeneously packed with alumina or diatomaceous earth coated with a liquid phase (a variety of polar and nonpolar liquids can be used).

The column boasts more than 1,000 theoretical plates per meter and, because of its homogeneous packing, allows plug flow to be consistently maintained.

In the column, each component moves at a different velocity, according to its affinity for the liquid phase. Thus, each fraction leaves the column at a different time. Because of the precise control of carrier-gas flowrate and other process variables, such as temperature and pressure, these times are reproducible within 0.1% over a period of several days.

Residence times in the column typically are on the order of a minute or two, so components do not thermally degrade.

As each component leaves the column, the programmer, by proper valve sequencing, routes the fraction to an individual condensing and collecting train.

A portion of output from the end of each sequence is recycled to the feed surge tank. Carrier gas is also recycled, after it is cleaned, dried, compressed and passed through a catalytic oxygen-removal step.

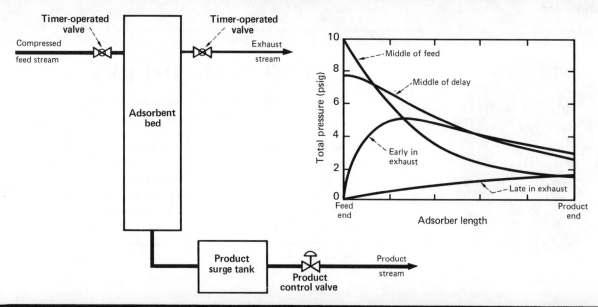

Union Carbide Corp.

Simple gas adsorber features self-regeneration

☐ An innovative single-bed adsorption process provides a continuous output of product, even as the system regenerates itself. The process handles a wider variety of separations than conventional pressure-swing adsorption (PSA), uses only one-fourth to one-fifteenth the adsorbent, and requires less investment.

Called Pressure-Swing Parametric Pumping, the system has been used since January 1979 by its developer, Union Carbide Corp. (Danbury, Conn.), to remove inerts from a vent gas at its Texas City, Tex., ethylene dichloride plant. That same year, the firm's Linde Div. introduced a special version of the unit that provides oxygen for home medical use. The company now is eyeing the method for separating hydrogen from, e.g., synthesis gas, and for removing chlorocarbons.

The process can work with a variety of adsorbents, and can use feed pressures from 10 to 500 psig. Pressure cycles are very short—around 20 s or less, versus several minutes for PSA. So, the technique can handle gases having high heats of adsorption (such as ethylene) that are hard to separate by conventional adsorption because of bed-temperature-rise problems.

Additionally, by achieving continuous operation with only a single bed, the system avoids the increased complexity and cost of conventional multibed units.

Distinctive cycle

The process relies on a unique pressure-versus-distance-and-time profile in the adsorption bed; this contrasts with PSA, where pressure remains relatively uniform within the bed at any given time. Providing appropriate pressure pulses develops gradients in the bed during a cycle—supplying the driving force for continuous product output and also for production of an internal purge gas for bed regeneration.

The adsorbent bed runs for 1 to 4 or more ft, and contains relatively small particles, 40 to 80 mesh for oxygen separation versus the $\frac{1}{16}$-in. pellets typically used in PSA. These small particles make possible the substantial pressure gradients. However, what with the system's high gas velocities and rapidly reversing flows, Union Carbide had to develop a special adsorbent-holddown device. It prevents attrition and minimizes the gas above the bed, while not restricting gas flow.

Feed gas enters the column in a pulse that lasts for 1 s or less; the exhaust valve is kept closed. So, a large pressure gradient develops. Then, the feed valve also is shut for a "delay" period, which ranges from $\frac{1}{2}$ to 3 s. This permits the pressure wave to penetrate further into the bed. Next, the exhaust valve opens for 5 to 20 s. Initially, since pressure has reached a maximum part way down the column at this point, gas flows two ways—one stream, which purges more-adsorbed compounds and regenerates the bed, exits as exhaust, while the product stream also continues to leave. As this exhaust period goes on, the pressure maximum declines and moves toward the product end until, ultimately, all parts of the bed are regenerated. Then the cycle repeats.

Originally published November 30, 1981

Board of Judges

INDEX

waste-sludge treatment in the chemical process industries, 251–265
Solvent de-ashing of liquefied coal, 3–5
Steam reforming for hydrogen production, 44–46
Sulfur dioxide (SO_2) removal, 191–204
  dry scrubbing for removal of SO_2 from coal-fired utilities, 193–194
  flue-gas desulfurization for production of gypsum, 191–192
  fluid-bed combustion (FBC) units, 195–197
  H_2S for reduction of SO_2 in desulfurizing flue gas, 200–201
  new systems for cleaning Claus plant tail gas, 202–204
  SO_2 absorption using a sodium citrate solution, 198–199
Sulfuric acid, contaminated, reconcentration and purification of, 240–241
Supercritical fluids for chemical process industry application 104–106
Synthetic chemicals from coal, 6–9
Synthetic natural gas (SNG) from coal, 6–9
  gasification and methanation in production of, 26–28
Synthetic natural gas (SNG) plant, gas oil as feedstock for, 135–136

Tar sands and shale oil processes, 173–188
  higher oil recovery with new tar-sands extraction technique, 175–176
  improved processes for making jet fuels from shale oil, 182–184
  shale-beneficiation techniques, 185–188
  shale oil commercialization, 177–181
Tar-sands extraction technique, higher oil recovery with, 175–176
Thermochemical decomposition for hydrogen production, 44–46
Titanium dioxide, improved process for, 57–58

Uranium recovery from phosphoric acid, 302–303
Used oil rerefining, 155–162
  new system for recovering lubricants from used oil, 161–162
  waste oil rerefining, 155–160

Waste disposal, solid, 249–280
  hazardous-waste treatment methods, 275–278
  municipal-solid-waste recovery by shredding, 271–272
  petrochemicals recovered by recycling PET beverage bottles, 273–274
  pyrolysis process used to convert waste polymers to fuel oils, 279–280
  sludge drying, 268–270
  waste disposal system produces extra energy, 266–267
  waste-sludge treatment in the chemical process industries, 251–265

Waste gas treatment/recovery processes, 189–216
  cost-saving process for recovery of CO_2 from power-plant fluegas, 215–216
  newly developed cleanup techniques for hot gases, 212–214
  NO_x removal, 205–211
    catalytic burning reduces emission of NO_x, 209–211
    new systems undergoing trials for successful control of NO_x, 205–208
  SO_2 removal, 191–204
    dry scrubbing for removal of SO_2 from coal-fired utilities, 193–194
    flue-gas desulfurization for production of gypsum, 191–192
    fluid-bed combustion (FBC) units, 195–197
    H_2S for reduction of SO_2 in desulfurizing flue gas, 200–201
    new systems for cleaning Claus plant tail gas, 202–204
    SO_2 absorption using a sodium citrate solution, 198–199
Waste oil rerefining, 155–160
Wastewater treatment/recovery processes, 218–247
  activated carbon for removal of pesticides from wastewater, 226–227
  anaerobic wastewater treatment, 242–247
  bacterial cultures for biological treatment of industrial wastes, 219–220
  biological and mechanical methods compete for wastewater cleanup, 223–225
  biological phosphorus removal, 221–222
  new process uses bacteria-containing polyphosphate, 317
  new processes destroy polychlorinated biphenyls, 231–233
  rapid oxidation for the destruction of organics in wastewater, 234–235
  reconcentration and purification of contaminated sulfuric acid, 240–241
  waste treatment with hydrogen peroxide, 236–239
  water hyacinths an economical choice for lower-cost waste treatment, 228–230
Water hyacinths for lower-cost waste treatment, 228–230
Water pollution, new pulping process reduces, 288–289
Wood, pulp and paper, 281–292
  anthraquinone used in alkaline wood-pulping processes, 283–284
  high-grade lignin processes are moving toward commercialization, 290–292
  industry turning more frequently to use of wood as a fuel, 285–287
  new pulping process reduces water pollution, 288–289

Zinc
  new leaching and smelting methods for, 82–84
  simplified process for, 70–72